FORTSCHRITTE DER CHEMIE ORGANISCHER NATURSTOFFE

PROGRESS IN THE CHEMISTRY OF ORGANIC NATURAL PRODUCTS

PROGRÈS DANS LA CHIMIE DES SUBSTANCES ORGANIQUES NATURELLES

HERAUSGEGEBEN VON   EDITED BY   RÉDIGÉ PAR

L. ZECHMEISTER

CALIFORNIA INSTITUTE OF TECHNOLOGY, PASADENA

ACHTER BAND   EIGHTH VOLUME   HUITIÈME VOLUME

VERFASSER   AUTHORS   AUTEURS

Y. ASAHINA · R. B. COREY · A. FREY-WYSSLING · F. GALINOVSKY
G. W. KENNER · L. F. LELOIR · K. MÜHLETHALER · M. PAILER
C. R. RICKETTS · M. ROHDEWALD · H. SCHINZ · M. STACEY
L. ZECHMEISTER

MIT 47 ABBILDUNGEN   WITH 47 ILLUSTRATIONS   AVEC 47 ILLUSTRATIONS

WIEN · SPRINGER-VERLAG · 1951

ISBN-13:978-3-7091-7174-5   e-ISBN-13:978-3-7091-7172-1
DOI: 10.1007/978-3-7091-7172-1

Softcover reprint of the hardcover 1st edition 1951

Titel Nr. 8214

Inhaltsverzeichnis.
Contents. — Table des matières.

Lupinen-Alkaloide und verwandte Verbindungen. Von F. GALI-

Brechwurzel-Alkaloide. Von M. PAILER, II. Chemisches Universitäts-

X-Ray Diffraction Studies of Crystalline Amino Acids and Peptides.

Some Aspects of Enzyme Chromatography. By L. ZECHMEISTER,

The Fine Structure of Cellulose.

By A. Frey-Wyssling and K. Mühlethaler, Zürich.

With 10 Figures and 3 Plates.

Contents.

In volume V of this series E. Pacsu (46) has reviewed the cellulose problem starting from the concept of the cateniform $1,4\text{-}\beta$-polyglucosan molecule and discussing the interrelationship of these chains in crystallized micellae. In the present article we will follow the opposite approach, by proceeding from the microscopically visible raw material of cellulose, the plant cell wall, to the smallest visible elements of its submicroscopic framework which consists of crystallized cellulose.

I. Plant Cell Walls.

1. Identification of Cellulose.

a) Preparation of Cell Walls.

In plant cell walls the cellulose framework is incrusted with variable quantities of two or more of the following substances: pectins, hemicelluloses, lignin (6a, 6b), cutins and ashes (mostly SiO_2 and $CaCO_3$). These "impurities" may amount to over 50% in the wood of broad-leaved trees. Cell walls of relatively pure cellulose are rare (cotton hairs, bast fibers of linen and ramie). In plant histological practice the cell walls are treated with hypochlorite (eau de Javelle) in order to destroy the adherent cytoplasm; thus, the cell walls are bleached and some of the incrusted lignin is removed. Histological sections pretreated in this manner display more conspicuously than untreated cell walls the microchemical reactions which will be discussed below.

For the quantitative determination of the cellulose content in non-lignified cell walls the incrusted substances are removed by boiling with 2% sulfuric acid and then with 2% alkali (54). This method yields excellent preparations for electron microscopic studies (39) (Fig. A–G, p. 11, 15).

b) Reaction of Cellulose with Iodine.

Cellulose yields a violet color with iodine if it has been duly swollen. The swelling agents necessary to obtain this reaction vary for different types of cell walls and, accordingly, the histologists distinguish several varieties of cellulose (14). There are even membranes which can be stained directly with iodine (in potassium iodide) just as well as is starch; consequently, their wall substance is called "amyloid" (61). For other cell walls, such as those of collenchyma tissues, the presence of 25% hydrochloric acid is necessary to produce a coloration with iodine; Ziegenspeck considers such walls to consist of "collose" (14). Whether or not "amyloid" and "collose" belong to the family of the true celluloses and represent different degrees of polymerisation, is questionable. They are only mentioned here in order to indicate how deceptive the iodine reaction of cell walls may be.

For such walls which are known to yield a typical X-ray diagram of cellulose a swelling with zinc chloride is necessary before a positive iodine reaction can appear. The standard reagent for cellulose in histology is, therefore, a mixture of zinc chloride, potassium iodide, and iodine, termed "chlorozinciodide". Since concentrated zinc chloride is capable of producing complete dissolution of cellulose by destroying its crystalline structure, the iodine probably enters in between the individual chains of the locally widened cellulose lattice. The resulting iodine coloration exhibits in the polarizing microscope a conspicuous dichroism, black colorless (7), which proves an oriented iodine adsorption by the chain lattice.

The cell walls of the marine algae *Valonia* and *Chaetomorpha (Siphonoclades)* which are well known for their beautiful X-ray diagrams of cellulose (2, 45), give no reaction with chlorozinciodide, unless they have been previously swollen in cuprammonium. This means that the cellulose of these algae is crystallized in such a manner that zinc chloride is unable to attack its chain lattice.

The most powerful reagent to swell the cellulose crystal lattice is 67% sulfuric acid. On addition of this acid, all types of cellulosic cell walls which have been previously soaked with iodine (in potassium iodide) assume a transient deep-blue color before the cellulose is hydrolyzed.

Lignified, cutinized or heavily mineralized cell walls give yellow colorations with chlorozinciodide, in spite of their cellulosic framework. Evidently, the incrusting substances prevent the necessary swelling of the cellulose lattice, and thus they hinder the penetration of iodine between the cellulose chains. When longitudinal sections of conifer wood are treated with chlorozinciodide, they are first stained yellow; however, when left in the reagent for some time, the violet cellulose reaction appears. Probably, the attack on crystallized cellulose is delayed by the surrounding lignin; and thus the adsorption of iodine is hindered until swelling of the cellulose has taken place.

The yellow coloration of lignin produced by iodine does not show dichroism from which it would seem that the orientation of lignin is less pronounced than that of cellulose. Since the cellulose-iodine coloration produces the dichroism, black/colorless, the position "colorless" in the polarizing microscope can be used to demonstrate the presence of lignin in "cellulosic" cell walls which then show the dichroism, black/yellow. In the ordinary microscope the yellow color of small amounts of lignin s obscured by the strong light absorption of the cellulose-iodine complex.

c) The Behavior of Cellulose toward Direct Dyes and Basic Dyes.

The direct dyestuffs of the class of the diazo benzidines which are able to stain cellulose without any previous treatment, are employed extensively in plant histology. Congo red and benzoazurin are mostly used; since their solutions are of colloidal nature, their power to penetrate increases if some of the incrusting cell wall substances have been destroyed by hypochlorite. There occurs an oriented adsorption of these dyes on cellulose as shown by marked dichroism in the polarizing microscope. It is characteristic for this orientation that the long axis of the benzidine molecule coincides with the direction of the cellulose chain (56). Due to the strong adsorption, the diffusion of congo red within a cellulose fibre is only possible perpendicularly but not parallel to the cellulose strands

[Fig. 1; anisotropic diffusion (*15*)]. The individual congo red molecules fit into the surface of the cellulose chain lattice, the cross-sectional space needed for a cellulose chain being $8.35 \times 3.95 \sin 84°$ Å², and that of a congo red molecule, 8.5×3.8 Å².

Basic dyestuffs are not adsorbed by native cellulose; however, in-crusted pectins are strongly stained by these dyes for which phenomenon the uronic carboxyl groups must be held responsible. Therefore, in plant histological practice methylene blue and ruthenium red are used to differentiate pectic substances from cellulose. However, this microchemical method becomes unreliable in the presence of oxy-celluloses which contain an appreciable number of carboxyl groups. Thus, cell walls which have been heavily bleached or macerated with strong oxidants may react with basic dyes even in the absence of pectins.

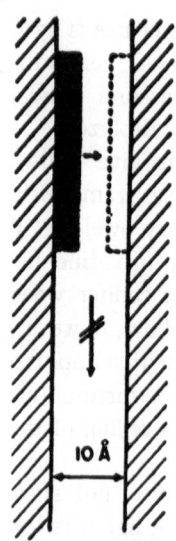

Fig. 1. Congo Red (black) Adsorbed by the Surface of Crystalline Cellulose (hatched). Diffusion is only possible perpendicular but not parallel to the surface (*15*).

d) Dissolution of Cellulose.

Since cuprammonium (SCHWEIZER's reagent) dissolves cellulose, leaving pectins, lignins, cutins and chitin behind, it has been used frequently in order to identify cellulose in cell walls. However, a clear cut dissolution occurs only if the incrusting substances have been previously removed. Otherwise only a more or less pronounced swelling is observable. The presence of insoluble substances prevents the solvated cellulose chains from dispersing into a molecular solution. Thus, relatively small differences in the composition of cell walls may cause marked changes in the solubility. For example, the secondary cell wall of cotton hair is readily soluble in cuprammonium, while its surface. the primary wall is insoluble. This causes the remarkable, repeatedly described and discussed beaded swelling of native cellulose fibers (*22, 53*).

A still better solvent for cellulose than cuprammonium is cupro-di-(ethylenediamine)-hydroxide (*23*).

e) X-ray Analysis.

The most reliable characteristic for the identification of cellulose is its X-ray pattern. Even very young cell walls contain sufficient crystallized cellulose to yield a cellulose pattern, especially when the incrusting materials have been previously removed (*50, 27*). The X-ray method is admittedly not very sensitive since less than 10% crystallized cellulose cannot be detected in unextracted cell walls (*8*). It is probable that the extent of the crystalline structure of cellulose increases by removal of accompanying substances.

The problem whether or not the cell walls of fungi contain cellulose in addition to chitin has been settled by X-ray analysis (8). The result is that a group of primitive fungi *(Oomycetes)* contains a cellulosic framework, while the cell walls of other fungi consist of crystallized chitin. A mixture of these two wall substances has never been found. Although a cell wall may contain several incrusting substances, its framework is evidently built up only by a single compound, either cellulose or chitin.

Cellulose is known to crystallize in four different modifications of the chain-lattice, termed celluloses I, II, III, and IV [(26), there p. 15 ff.]. Of these "native" cellulose I and regenerated cellulose II are the most important representatives. They possess, respectively, the following elementary cells: I, $8.35 : 1.03 : 7.9$ Å, $\beta = 84°$; and II, $8.14 : 10.3 : 9.14$ Å, $\beta = 62°$ (35). The lattice of I has a density of 1.593, and that of II, 1.583. It is assumed that native cellulose I is the stable modification, whereas cellulose II constitutes an intermediate product when alkali cellulose is reconverted into cellulose [(35), there p. 231]. At physiological temperatures the conversion rate, II → I, is said to be too low to yield native cellulose, whereas this transformation can be observed at 100° C., especially when pressure is applied. Since the equilibrium, II $\rightleftarrows$ I, is not effected at room temperature, it is difficult to decide which of the two celluloses is the stable modification under physiological conditions.

This question is of some interest, because living cytoplasm is capable of synthesizing not only "native" cellulose I, but, in exceptional cases, also cellulose II. Whereas in higher plants only I is known to occur, there are some marine algae (e. g. *Halicystis*) whose cell walls consist of cellulose II (51). According to the theory mentioned, this cellulose II ought to have alkali cellulose as a precursor; this possibility cannot be excluded, since its biosynthesis occurs in sea water. On the other hand, Dr NICOLAI (44) detected cellulose II in the fresh water alga *Hydrodictyon* in which alkali cellulose could not possibly be an intermediate product of the cellulose biosynthesis. Since one of the two celluloses I or II must represent the metastable modification, it is evident that the living cytoplasm is capable of synthesizing a cellulose lattice which is richer in energy than the most stable modification.

f) The Double Refraction Test.

Of all cell wall substances only cellulose, chitin, incrusted waxes, and certain special hemicelluloses (in seeds) show intrinsic double refraction, while incrusted pectins, hemicelluloses, lignins, and minerals are isotropic. Consequently, when chitin and waxes can be excluded, the double refraction test may be used as a diagnostic method for the presence of cellulose. Thus, a certain cellulose content of the growing primary walls had been predicted on the basis of their birefringence (*11*), against the views of Hess and co-workers (*28*). At present, the occurrence of cellulose in growing cell walls has been substantiated by X-ray analysis and electron microscopy (Figs. *E* to *G*, p. 15).

The intrinsic double refraction of crystallized cellulose is as high as 0.07 which is almost eight times that of gypsum or quartz (0.009) (*21*).

In contrast to the polarizing microscope which is an excellent instrument for the study of the distribution and the orientation of cellulose in cell walls (*11*), the ultraviolet microscope and the fluorescence microscope are of no help in such investigations, since cellulose does not absorb ultraviolet light, nor does it show any fluorescence; both these features are characteristic of lignin.

2. Ontogeny of Cell Walls.

a) Primary Cell Walls.

Young plant cell walls are three-layered; they consist of an isotropic middle lamella covered on both sides by a thin, faintly birefringent film which is called primary cell wall (*31*). The middle lamella is rich in pectins and is considered as the joining cement between neighboring cells. The primary wall contains some cellulose which is hidden, however, by a considerable amount of accompanying pectins, hemi-celluloses and waxes (*29*), and thus, it does not show either of the mentioned microchemical reactions for cellulose [with iodine or cuprammonium (*55*)], unless a pretreatment for removing non-cellulosic substances has been applied. However, the presence of cellulose in untreated walls can be demonstrated by its double refraction and X-ray pattern. The primary

Table 1. Composition of Coleoptiles with Primary Cell Walls.

	WIRTH (*57*)			NAKA-MURA and HESS (*43*)	THIMANN and BONNER (*54*)
	Maize coleoptile			Maize coleoptile	Oat coleoptile
	9 mm long %	32 mm long %	55 mm long %		
Ether soluble substances...	1.2	10.4	7.8	2.43	—
Water soluble substances ..	44.1	39.5	46.0	61.34	—
Hemicelluloses	9.0	14.5	11.0	—	11.0
Cellulose	8.3	13.2	13.0	10.12	11.9
Pectin	2.2	4.0	4.7	—	2.3
Protein.................	22.17	15.2	13.1	—	3.1
Ash....................	6.9	4.4	3.6	—	—
Total dry matter	100.0	100.0	100.0	100.0	100.0

wall contains less cellulose than the sum of other wall substances (hemi-celluloses, pectins, and waxes). If the ether-soluble fraction is considered to consist mostly of waxes (29), then the cellulose content in the primary wall amounts to about 35% (Table 1). Since the cellulose is not destroyed but preserves its original shape when the wall is completely extracted, it must form a *coherent* frame. This framework was postulated to consist of submicroscopic cellulose strands (11) in a dispersed arrangement (*dispersed texture*, Fig. 2). The presence of such a texture has been sub-stantiated by means of the electron microscope.

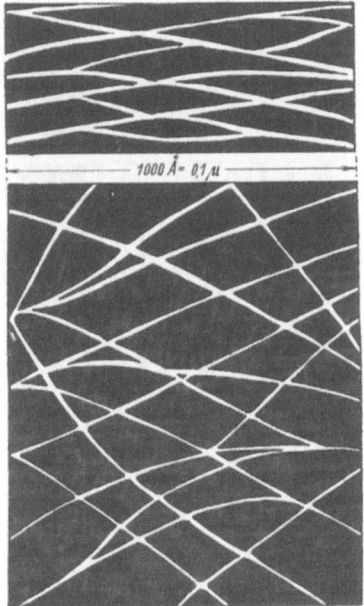

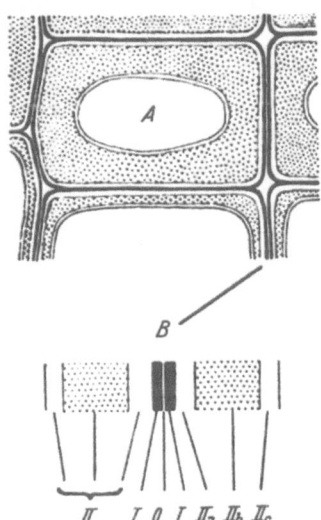

Fig. 2. Texture of the Primary Cell Wall, pro-posed in 1936 (11). Above: cross-section; below: front view.

Fig. 3. Microscopic Lamination of the Plant Cell Wall (31). *o* middle lamella; *I* primary wall; *II* secondary wall, three-layered: *IIa, IIb, IIc.*

The primary wall shows the remarkable faculty of growing in area (16), which phenomenon has always been an enigma to physiologists. It is a puzzling problem, indeed, how a film with a definite solid structure is able to multiply its surface without being liquefied. Not only plant cell walls but also plasmatic surfaces grow in this particular manner, viz. by adding new material to the existing substance. This does not occur at the borders but inside the original surface area. Hence, early cytologists called this phenomenon a *growth by intussusception*, in contrast to the growth by apposition of crystals, starch grains etc. The mechanism of the intussusception has never been clearly explained; only recently has the use of the electron microscope opened up the possibilities for the clarification of some of these hidden processes which take place in the cell wall (Fig. *G*, p. 15).

b) Secondary Cell Walls.

Even before the growth in area of the primary cell wall is completed, the so-called secondary wall is deposited against it (*59*). This causes an

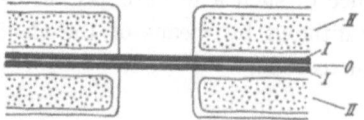

Fig. 4. Longitudinal Section across a Pit; *s* pit membrane (*cf.* Fig. *E*, p. 15).

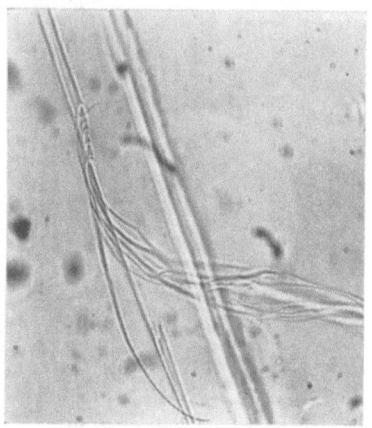

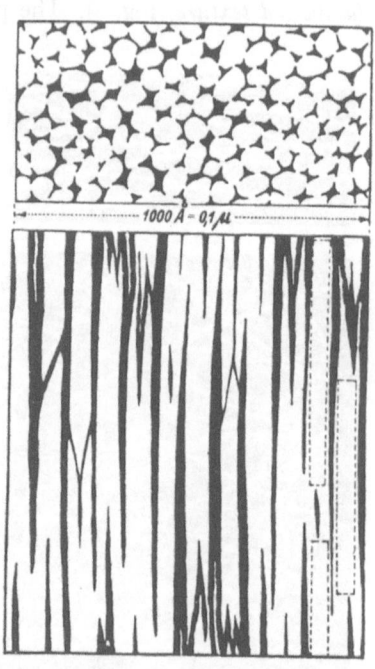

Fig. 5. Mechanical Fibrillization of a Tracheid (Fir Wood) (*11*).

Fig. 6. Texture of the Secondary Cell Wall of a Fiber, proposed in 1936 (*11*). Above: cross-section; below: longitudinal section with indication of regions with ideally crystallized cellulose (the longitudinal scale is considerably shortened in comparison to the transversal scale).

appreciable thickening of the wall and stops its enlargement. The secondary wall grows predominantly by apposition. It may become very thick and is characterized by some conspicuous morphological features. In general, it is visibly laminated, often three-layered as in tracheids and some fibers (*3*) or even multi-laminated as in cotton hairs where the individual layers represent daily growth rings (*5, 30*). The three-layered type appears in Fig. 3, p. 7.

Besides the layers, the secondary wall often shows some perforations, called pits. Between neighboring cells these are communicating pores which are interrupted, however, by the primary wall (Fig. 4). This septum is called the pit membrane. There is some discussion in classical treatises, why the pit holes of adjacent cells do correspond, although they are apparently separated by the primary wall; and whether this pit membrane is of a homogeneous nature or is rather perforated by submicroscopic pores, called plasmodesmata. These questions have been settled by means of the electron microscope (Fig. *E*, p. 15).

Investigations with the polarizing microscope and X-ray studies show that in elongated cells (fibers and tracheids, thus, in the raw material of pulp and cellulose manufacturing), there is a *parallel texture* of elongated cellulose elements (*11*). Consequently, the secondary wall of those cells can easily be split into fibrils (Fig. 5). They have diameters ranging from a few microns to invisibility in the ordinary microscope. In cotton they correspond to the thickness of a single layer, which amounts to about 0.4 μ (*5*).

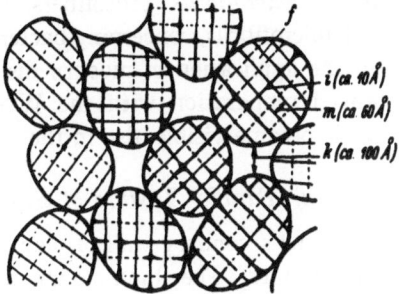

Sometimes the fibrils are arranged parallel to the fiber axis (fiber texture); more often, however, they follow the pitch of a spiral (spiral texture). In the first instance X-ray diffraction yields a so-called fiber diagram with distinct interference

Fig. 7. Section across *Microfibrils* (*f*), proposed in 1937 (*12*). Individual microfibrils of 200—300 Å diameter, composed of crystallized strands *m* which are separated by intermicellar splits *i*; *k* = interfibrillar capillary.

spots, whereas spiral textures produce sickle diagrams due to a circular expansion of the spots whose angle at the center corresponds to the spiral angle.

The fibrils visible in the ordinary microscope display all the properties of cellulose present in intact secondary cell walls, such as X-ray diffraction, double refraction, dichroism with iodine and direct dyes, etc. It must be concluded from these observations that the fibrils mentioned possess both a crystalline structure and a porous texture with spaces available for colloidal dyes.

The crystallinity is due to crystalline rods, the diameter of which, as established by X-ray analysis (*25*), equals about 50 Å. The length of these rods cannot be measured (*12*); it must be very considerable when compared with their diameter; hence we have termed them *"micellar strands"* (*16*). The porous texture has been ascribed to intermicellar spaces as indicated in Fig. 6. Since it is possible to produce silver and gold crystals of more than 100 Å diameter in these pores, they must represent rather coarse capillary spaces. From this fact it was postulated that the secondary wall has a submicroscopic, heterocapillary texture and that the micellar strands are gathered to form *microfibrils* which are invisible in the ordinary microscope (*12*). The microfibrils are considered to contain intermicellar spaces accessible to water, zinc chloride or iodine, whilst the coarser interfibrillar spaces would allow the circulation and deposition of colloidal substances such as direct dyes, pectin, and lignin (Fig. 7, p. 9).

II. Microfibrils.

1. Occurrence of Microfibrils.

The electron microscope allows the direct observation of the submicroscopic texture of cellulose materials. A basic result of such studies was the discovery that natural cellulose appears to exist always in the form of *microfibrils* (protofibers, elementary fibers). The first picture of such microfibrils, concerning bacterial cellulose, was published in 1946 (*17*). By the application of modern disintegration methods (blendor) and improved preparation with the Wyckoff shadow cast technique, microfibrils became visible also in the cell walls (*20*).

a) Parallel Texture of Secondary Walls in Natural Fibers.

In the secondary wall of plant fibers (*37*) the microfibrils run parallel (Fig. *A*, p. 11). They have a diameter of 200–300 Å. Microfibrils of the same diameter have also been found in the secondary wall of some marine algae (*47*). In *Valonia* the microfibrils of each layer of the secondary wall lie parallel, but the orientation in adjacent layers is crossed by about 78° (*2*). This permits a differentiation of the individual layers in the electron microscope. In cotton hairs the lamination is not visible in the electron microscope because the microfibrils of adjacent layers run parallel (Fig. *A*).

The fibrils observable in ordinary microscopes are not individualized in the electron microscope. This means that they are artifacts due to disintegration; consequently, their diameter will be very dependent on the methods used. In contrast, the microfibrils possess true individuality, and they must be considered as fundamental elements in the structure of the cell wall.

In Table 2, p. 12 the morphological elements of a cotton hair are summarized; of these the microscopical layers and fibrils are not essential. The layers can be suppressed by cultivating cotton in a green house under constant conditions [permanent light, constant temperature and

---

Electron Micrographs Showing Microfibrils.

Fig. *A*. Secondary Cell Wall of Cotton Hair with Parallel Texture (*20*).

Fig. *B*. Rayon with Heterogeneous Texture (*40*).

Fig. *C*. Primary Cell Wall with Interwoven Texture, from a Corn Root (*20*).

Fig. *D*. Microfibrils of a Flax Cell Wall Disintegrated by Means of Ultrasonic Waves. Arrows indicate fringes at the surface of the microfibrils (unpublished).

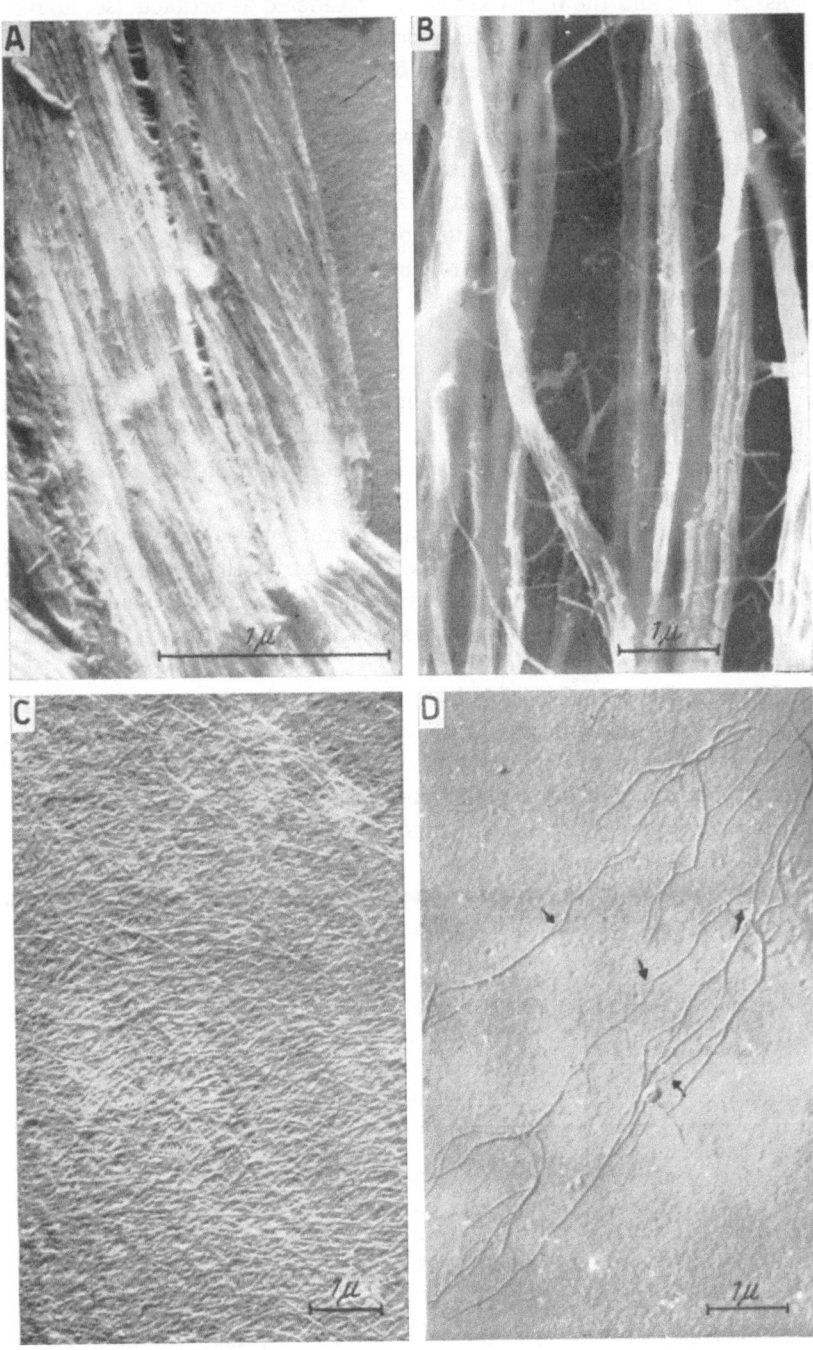

moisture (*1*)]; and the fibrils should be considered as artifacts (see above). Thus, none of the microscopical morphological features are as important structural elements as are the microfibrils which can be observed in the electron microscope.

From a molecular point-of-view the microfibrils are very coarse colloidal strands since they contain at least 2000 chain molecules in the cross-section as indicated in Table 2.

The most conspicuous feature of the microfibrils is their uniform thickness. Although ideally parallelized, they do not coalesce laterally but appear completely individualized, provided the wall is duly macerated. Their length cannot be defined; visible terminations of microfibrils represent mostly broken ends.

The orientation of the parallel microfibrils may change for a short distance and then resume the original direction as demonstrated in Fig. *A*. Such distortions in the cell wall are observed in the ordinary microscope and are known as slip planes in native fibers (*9, 9a*). They never occur in rayon. They are formed when fibers are crushed. Notwithstanding such local structural distortions, the tensile strength of the fiber is not markedly decreased, because the microfibrils are merely bent but not broken. Only a regular parallel texture is capable of producing such deviations across the whole fiber.

In rayon the strands visible in the electron microscope are not ideally parallelized and, furthermore, they have various diameters [Fig. *B*, p. 11 (*40*)]. Therefore, rayon does not show the beautiful and regular internal texture of grown plant fibers. It is interesting that even silk threads which are spun in a manner similar to rayon do not show a regular microfibrillar texture (*24*).

There seem to exist two types of fiber textures; one originating from spinning processes with fibrillar elements of varying thickness, and the

Table 2. Size of Structural Elements and Approximate Number of Cellulose Chain Molecules in the Cross-section of a Cotton Hair with 20 μ Diameter.

	Area of cross-section		Number of chain molecules
Cotton hair	$10^2 \pi \mu^2 =$	$314 \mu^2$	750 000 000*
Layer (medium)	$10 \times 0.4 \pi \mu^2 =$	$12.6 \mu^2$	30 000 000*
Fibril................	$0.4 \times 0.4 \mu^2 =$	$0.16 \mu^2$	500 000
Microfibril	$\mathbf{250 \times 250}$ Å$^2 =$	$\mathbf{62500}$ Å^2	**2000**
Micellar strand	50×60 Å$^2 =$	3000 Å^2	100
Chain molecule	$8.35 \times 3.95 \times \sin 84°$ Å$^2 =$	32.8 Å^2	1

* Due allowance is made to the inclination of the parallel texture of 60° to the cross-sectional plane (spiral angle = 30°) and to the submicroscopic capillarity of the cell wall.

other represented by grown fibers which display textures of uniform filaments.

b) Woven Texture of Primary Walls.

Primary cell walls contain the same microfibrils as secondary walls (20, 39); however, they have no parallel orientation but show a typical dispersion (Fig. C, p. 11) as was established previously by the indirect method of the polarizing microscope. Sometimes they even cross at an angle of approximately 90° (20). Generally speaking, their diameter is the same as that of the microfibrils located in secondary walls. This is an unexpected coincidence since it had been assumed earlier that the low cellulose content (about 35%) in primary walls (Table 1) was due to considerably thinner strands than those which build up the bulky secondary walls (cf. Figs. 2 and 6, p. 7—8) containing 50 to 90% cellulose. (Exceptionally, thinner strands have been found in pollen tubes.)

Seemingly, the plant cell prefers to build up a certain grade of microfibrils by means of which it is able to construct such different textures as the soft, highly elastic, growing primary walls on one hand, and the very strong, bulky, hard secondary walls on the other. With reference to Fig. C it should be taken into consideration that about 65% of the cell wall substances have been removed during the preparation; and, furthermore, that the electron microscope has a considerable focal depth so that the cellulose texture in the living cell is actually of much looser nature than is demonstrated by Fig. C (cf. Fig. 2, p. 11).

The microfibrils in the primary walls are not simply laid down one atop the other but they are mutually interwoven, thus forming a true fabric. We may note that textile manufacturing is by no means a human invention but it has been practiced by nature, in the submicroscopic realm, ever since plant cell walls have been in existence. The woven texture of primary walls is the reason why they cannot be split into finer lamellae, in contrast to the laminated secondary walls. The discovery of this texture poses two interesting problems: (a) how is such a fabric built up by the cytoplasm, and (b) in what manner is such a texture able to grow in area by intussusception?

There are two possible answers to (a), viz. either the microfibrils are subject to tip growth and then their crossed interlaced fabric is the result of a "warp" grown through by a "weft" or else the whole fabric is built up simultaneously inside the cytoplasm which is capable of guiding the cellulose crystallisation in the complicated manner just described. This question touches the problem of morphogenesis which is as yet an unsolved biological enigma. Since no tip growth of the microfibrils has yet been observed, the second possibility mentioned seems to be the more probable one at the present time.

Concerning the intriguing question of the growth in area, we have found some promising indications. It can be observed that the texture of the primary wall is loosened; then new cellulose strands are produced, and thus the weakened fabric is "mended" (Fig. G, p. 15). A tip growth of cells could not proceed in this manner, since the existing area extends by the formation of a new cell wall. The rather scanty evidence that we have obtained of such cell tips shows a texture which is less clearly fibrillated than that of older primary walls. Fig. H shows such a picture which was taken from the very tip of a growing sporangiophore of the fungus *Phycomyces*. It is true that this fungus contains chitin microfibrils instead of a cellulosic cell wall; however, the textures of cellulose and chitin are essentially identical (*19*).

The texture of the growing primary wall is not uniform throughout its whole surface. On the contrary, some areas of a porous fabric can also be observed which develop into pit membranes as seen in Fig. E, p. 15. This beautiful picture demonstrates how microfibrils of the primary wall run into the pit field and help to form pores of about 700 Å diameter. Some others are used to strengthen the border of the pit area; this circular arrangement of cellulose fibrils yields conspicuous pictures in the polarizing microscope. Electron micrographs like Fig. E have settled the question whether the pit membranes are homogeneous septa or sieves with perforating plasmodesmata.

Some of the prepared pit areas formed are closed later by the addition of so many microfibrils that the texture of the normal primary wall is restored (Fig. F, p. 15).

c) Cellulose Slimes.

In a humid environment the walls of epidermal cells are covered frequently with slimes or they even swell up to a slimy mass. The latter phenomenon is especially characteristic for certain seeds (linseed, cress, quince etc.). Plant histologists differentiate between cellulosic slimes, pectic slimes, and callose slimes, depending on staining reactions with iodide, basic dyes or so-called callose reagents (*34*). Whereas cellulose and pectin have been chemically clarified, it has not yet been possible to define that type of carbohydrate to which "callose" belongs.

The electron microscope has furnished a similar picture of all kinds of plant slimes (*41*), in which it seems that cellulose microfibrils are always

Electron Micrographs Showing Microfibrils.

Fig. E. Pit Membrane of a Primary Cell Wall from a Corn Root (*39*).

Fig. F. Pit Field of a Primary Cell Wall from a Corn Root (unpublished).

Fig. G. Growing Cell Wall from Coleoptile of Corn (*20*).

Fig. H. Texture of Chitin at Tip of Growing Sporangiophore (*19*).

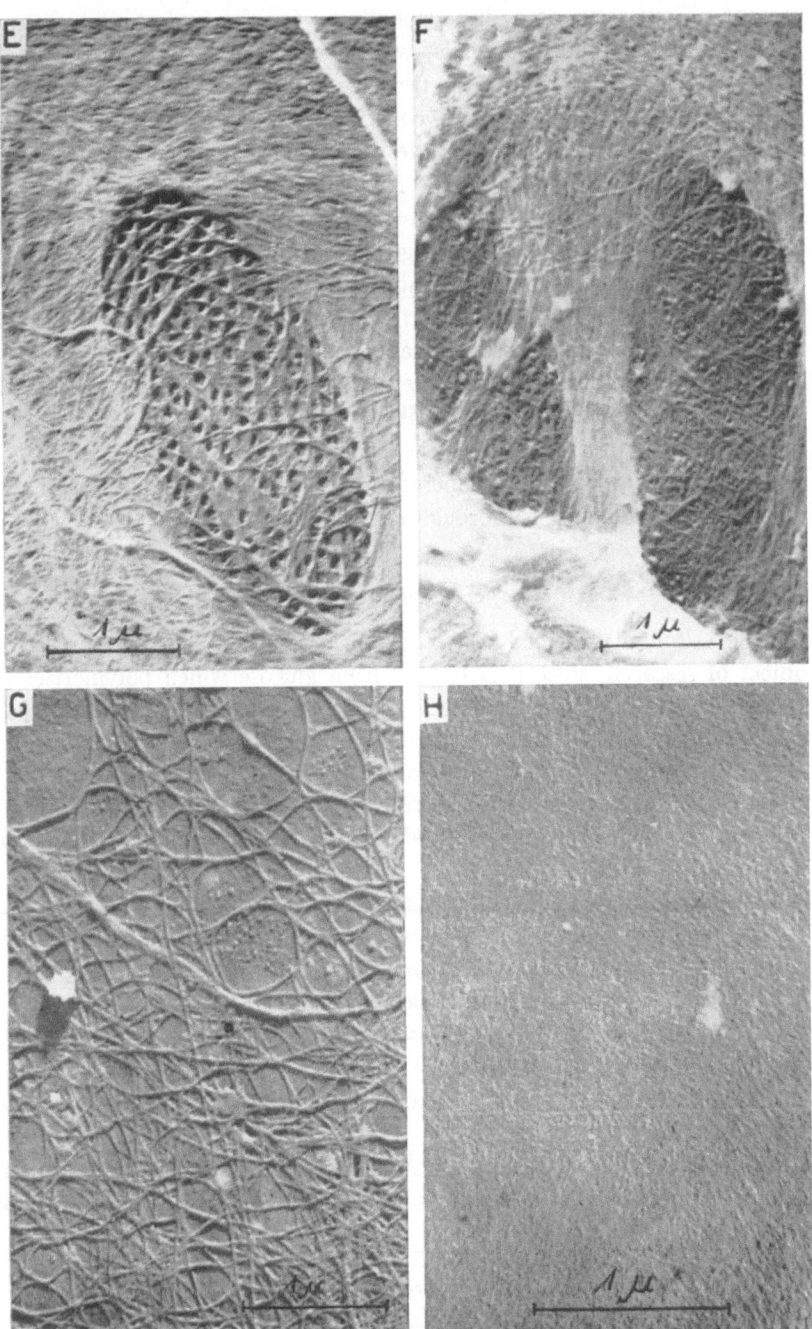

present. They serve as a dispersion frame for the hydrophilic pectins. If the cellulose fraction prevails, we have a "cellulosic slime" (Fig. *J*, p. 17; ex quince seed); and if the pectin fraction prevails, the substance is a "pectic slime", e. g. the slime covering root hairs [Fig. *K* (*18*)]. The cellulosic microfibrils in slimes have the same diameter (200–300 Å) as those occuring in primary walls. This finding is important, because a molecular dispersion of cellulose seemed a priori to be more likely for "cellulosic slimes". The morphological evidence proves that the water binding fraction must be represented by hydrophilic substances other than cellulose, since the microfibrils which are visible in the electron microscope cannot have an appreciable hydration power.

Linseed slime represents the only instance recorded until now, in which a disintegration of cellulosic microfibrils has been observed [Fig. *L* (*41*)], and the question arises whether the inverse process, i. e. the association of visible cellulose particles to microfibrils is also possible. This important question would deserve a thorough experimental study.

d) Bacterial Cellulose.

The cellulose produced by *Bacterium xylinum* is a most interesting object of research, since it is formed from glucose which polymerizes in the environmental medium outside the bacterial cell. This cellulose can easily be collected in form of a gel in large quantities and in a fairly pure state. It was from the investigation of this object that the elementary cell of crystallized cellulose was derived [(*36*), there p. 100]; furthermore, cellulose microfibrils were observed for the first time in this bacterial cellulose (*17*).

When gels of bacterial cellulose are stretched and rolled, preparations of "higher orientation" result, in which not only all of the chain axes run parallel but also the lattice plane (101), and consequently also those of (10$\bar{1}$) and (002) are fairly well parallelized (*36*). It must be concluded from this experiment that there is some lamination (*6*) present in the submicroscopic texture of cellulose I.

These facts made a study of bacterial cellulose in the electron microscope especially timely (*38*). The result is that this cellulose contains the same microfibrils of 200–300 Å diameter as do cell walls and plant slimes. Although a flattened cross-section of these microfibrils was not

---

Electron Micrographs Showing Microfibrils.

Fig. *J*. Quince Slime (*41*).

Fig. *K*. Tip of Corn Root Hair (*18*).

Fig. *L*. Slime of Linseed (*41*).

Fig. *M*. Bacterial Cellulose (*38*).

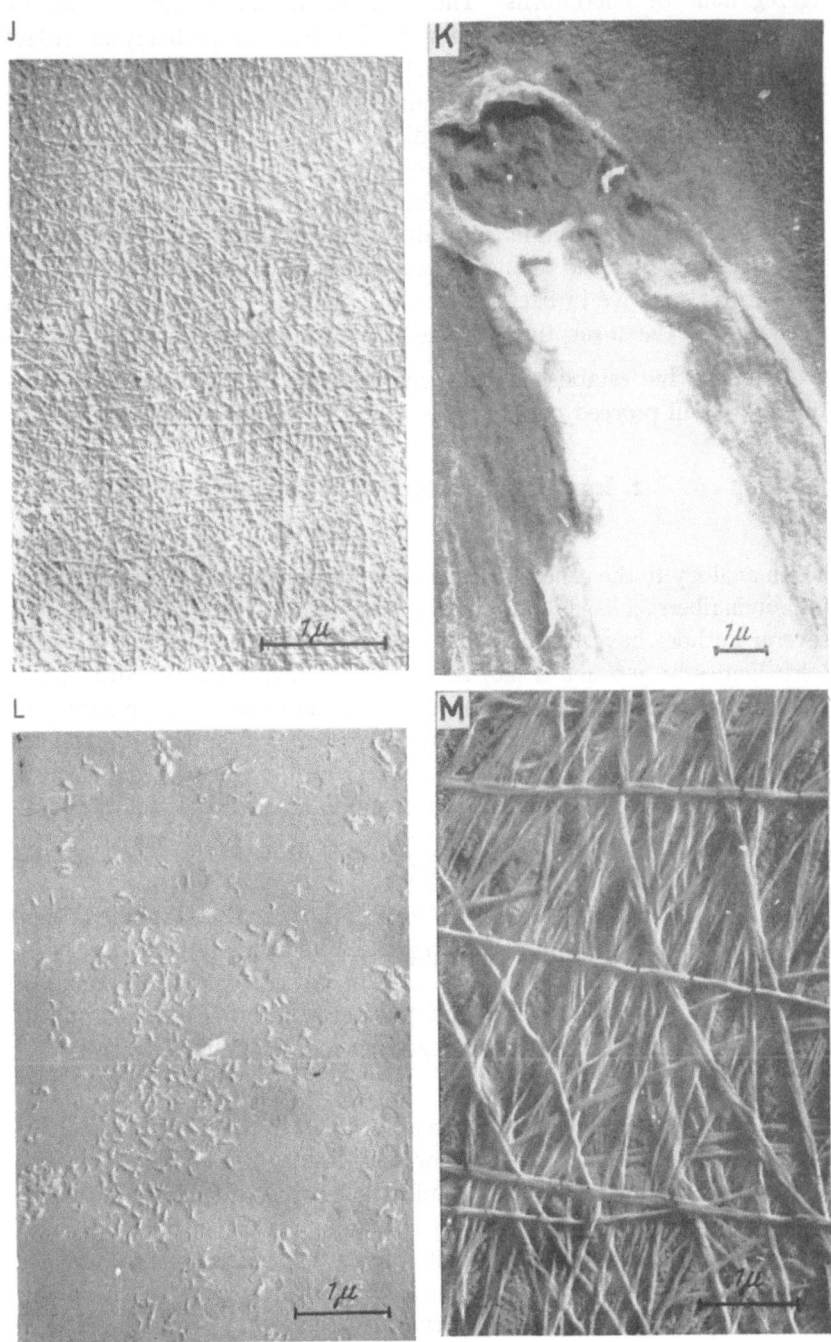

observed, there seems to be some tendency for the formation of lateral aggregations of microfibrils. The resulting fibrillated ribbons display beautiful torsions, as seen in Fig. M, p. 17. This torsion is probably an artifact due to the drying of the cellulose gel in vacuo, since X-ray analysis discloses no such torsional but a laminar arrangement of the crystalline regions in bacterial cellulose as indicated above. We assume that the association of the microfibrils occurs there, where adjacent (10$\bar{1}$)-planes touch each other. It should also be stressed that similar aggregations have never been observed in cell walls or slimes. If the association is due to surface forces of the cellulose chain lattice, it must be concluded that they are only active in bacterial cellulose, whereas in cell walls they are saturated by the adsorption of some other wall substances.

After having established the general occurrence of cellulose microfibrils, we will proceed now to a discussion of their internal structure.

2. Internal Structure of Microfibrils.

a) No Visible Segmentation.

In analogy to the general occurrence of a submicroscopic segmentation in protein fibers (collagen, actomyosin, in smooth muscles, blood fibrin etc.) several authors have claimed that cellulose fibrils are striated as well. Periodicities of 300–400 Å (49) or 150 Å (32) have been recorded in this connection. Such a segmentation may become visible when microfibrils are shadowed with gold which, however, is an inappropriate metal for this purpose, since it recrystallizes under the influence of the electron beam (60). In fully developed microfibrils carefully shadowed with chromium or palladium, we were unable to observe any periodicity in cellulose microfibrils (Figs. A–M, p. 11, 15, 17).

As long as the origin of the microfibrils is not clearly established, there will exist the possibility that they originate from a longitudinal association of individual particles, although the chain lattice structure of cellulose would not favor such a development. However, in full-grown microfibrils of cellulose there is definitely no segmentation visible in electron microscope pictures.

This fact does not constitute a contradiction to the chemical evidence for a periodicity in cellulose as reviewed by Pacsu (46). In those theories it has been assumed that the chemical heterogeneities along the chain molecule (e. g. the lateral bonds) lie in planes perpendicular to the chain axes. However, such an irregularity across the microfibril is amicroscopic and cannot be revealed by the electron microscope whose resolving power is about 50 Å. Any visible periodicity is based on alternating denser and less dense segments of submicroscopic dimensions. This is substantiated in many protein fibers but obviously not in fully developed microfibrils of cellulose, chitin (*19*) or natural silk (*24*).

b) Lamination.

The lattice of cellulose is laminated since all the flat glucose rings lie parallel in planes (002) whose distance apart is 3.95 × sin 84° Å. Therefore, the question arises whether there exists a similar lamination

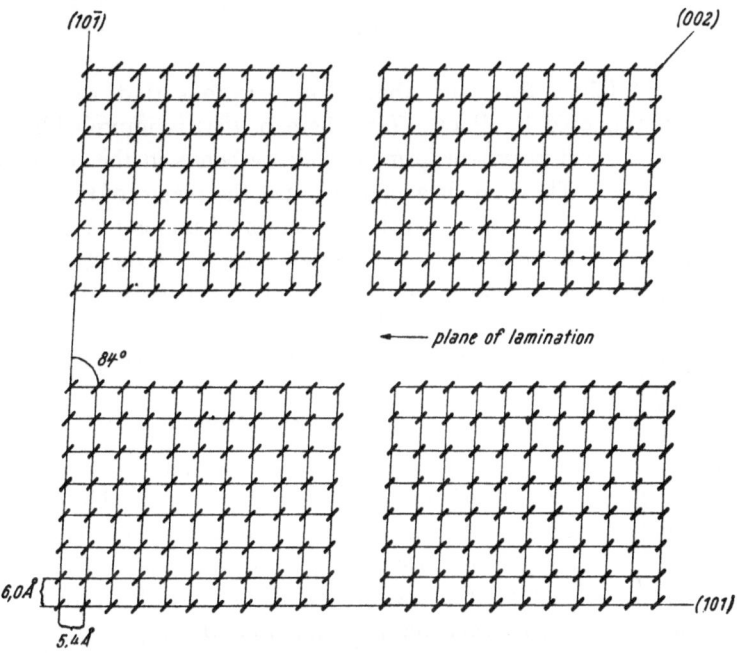

Fig. 8. Lamination of Crystallized Cellulose I along the Plane (101). (NB. In Fig. 89 of (*35*) the two spacings, 5.4 and 6.0, are erroneously interchanged.)

within the microfibrils. The fact that stretched bacterial cellulose and tunicin do yield X-ray diagrams of higher orientation (*36*) proves the existence of lattice planes which are parallelized under tension. This particular plane is, however, not identical with that of the glucose rings (002) but with the plane (101) located diagonally across the unit cell of crystallized cellulose. Such planes (101) have the largest spacing of all possible lattice planes parallel to the chain axis, i. e. 6.0 Å. Furthermore, (101) is the most hydrophilic plane, because the number of its hydroxyl groups per surface unit is higher than that in the planes (002) or (10$\bar{1}$) (Table 3). This may be the reason why in the cell wall of *Valonia* (*52*) and *Chaetomorpha* (*45*) the plane (101), lies parallel to the surface of the protoplasm from which it has originated.

2*

The observations reviewed call for a lamination of the microfibrils, if there is any, as indicated in Fig. 8, p. 19.

Table 3. Hydrophilic Character of the Lattice Planes in Crystallized Cellulose I.

Lattice plane	Area containing 1 OH group	Hydroxyl groups per 10 Å^2
(002)	14.3 Å^2	0.70
(10$\bar{1}$)	10.3 Å^2	0.97
(101)	9.3 Å^2	1.08

It would follow from the above considerations that the cellulose lamination proposed by Pacsu [(46) there p. 164] is improbable, since it postulates primary valence bonds in the plane (002). These bonds would link just the planes of the observed lamination, which are planes of minor resistance and, therefore, cannot be tied together strongly. If there are connecting valence bonds between neighboring cellulose chains, they will more likely bridge the spacing of 3.9 Å which exists between the (002) planes.

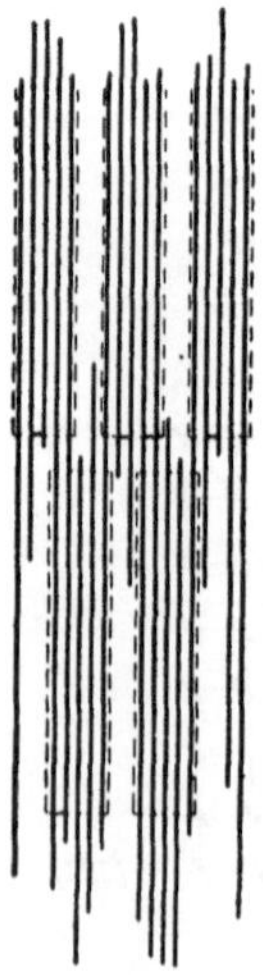

c) Crystallinity.

It is a well established fact that cellulose is partly crystallized and partly paracrystalline. The first suggestion of such a fine structure, made as early as 1935, is indicated in Fig. 9 (cf. also Fig. 6, p. 8). Fig. 9 shows some crystallized regions which correspond to the Nägeli micellae (42) and which are connected to a coherent structure by insufficiently ordered chains

Fig. 9. First Suggestion (1935) of Crystalline and Paracrystalline Regions in the Chain Lattice of Cellulose (10). (The longitudinal scale is considerably shortened in comparison to the transversal scale.)

representing paracrystalline cellulose. Only in 1937 did Kratky and Mark (33) publish their diagram which in some historical reviews is considered the starting point of the "modern outlook" [(26), there p. 29)].

The size of the crystallized regions (25) and the properties of "amorphous" cellulose have been established by X-ray analysis (26). Truly amorphous cellulose, i. e. molecular chains without any relation to neighboring molecules, must be rare. Probably, a part of the paracrystalline cellulose scatters the X-ray beam in a way similar to that of amorphous cellulose. The crystallized cellulose is shaped in rods of about 50–60 Å diameter (25) and undetermined length (12) which have been termed "micellar strands". It is probable that these rods are somewhat flattened as indicated in Fig. 8, and as has been definitely established for cellulose II (6). However, the X-ray method cannot give evidence of a slightly anisodiametric cross-section because the measurement of particle sizes is rather inexact and subject to several sources of error caused by accompanying substances and amorphous portions of the compound. This is why it has been doubted that the measured diameter of 50 to 60 Å corresponds to reality. However, since not only crude, native wood cellulose, bast fibers, bacterial cellulose, and tunicin but also fully purified cotton or ramie cellulose have yielded this same value, it is permissible to accept it as a good approximation (Table 2, p. 12). On the other hand, the astonishingly uniform diameter of micellar strands in so many different preparations of native cellulose has been interpreted as a proof against the reliability of the X-ray method. As a matter of fact, it is difficult to understand why cellulose stops crystallizing when strands of about 250 Å² to 400 Å² cross-section have been reached, and why considerably larger crystallized regions have never been found.

At the present time this uniformity seems less enigmatic, because the microfibrils described in this review behave in a similar manner. Most observed microfibrils of native cellulose possess the same diameter of 200–300 Å, except in membranes with tip growth (Fig. *H*, p. 15), where the microfibrils may not yet be fully grown. It is evidently difficult to find a physical reason for the occurrence of uniform microfibrils; and we must assume that the living cytoplasm intervenes actively when cellulose crystallizes and aggregates to form the fibrillar elements of the cell wall.

What is then the true interrelation of the uniform micellar strands of 50 Å diameter revealed by the X-ray method and the microfibrils of 250 Å diameter as observed in the electron microscope? Since highly purified cellulose from which the totality of interfibrillar substances has been removed, still contains amorphous portions as revealed by diffuse X-ray scattering, the paracrystalline cellulose must be located inside the microfibrils; hence, the latter must consist of crystallized cellulose (e. g. 70%) and of a certain amount of imperfectly oriented cellulose.

There are two possibilities for the distribution of crystallized and paracrystalline cellulose: either (a) parallel micellar strands run along the whole length of the microfibril separated by paracrystalline cellulose, or (b) relatively short crystalline regions are connected by disordered paracrystalline cellulose chains (Fig. 6, p. 8).

In the first case the microfibrils ought to be cleavable. It seemed on the basis of preliminary experiments with ultrasonic waves that the fibrillization obtained with cellulose fibers diminishes continuously to reach invisibility in the electron microscope (58). A closer study showed, however, that the ultrasonic cleavage stops when the disintegration to microfibrils is complete. No further subdivision of the microfibrils occurs but the surface of the microfibrils show small fringes (Fig. *D*, p. 11). This is an indication that there are no continuous micellar strands present but that they are somehow interconnected to form a coherent texture. Consequently, the texture as represented in Fig. 6, which had been thought to be that of a microscopically homogeneous layer of the secondary wall, is probably that of individual microfibrils.

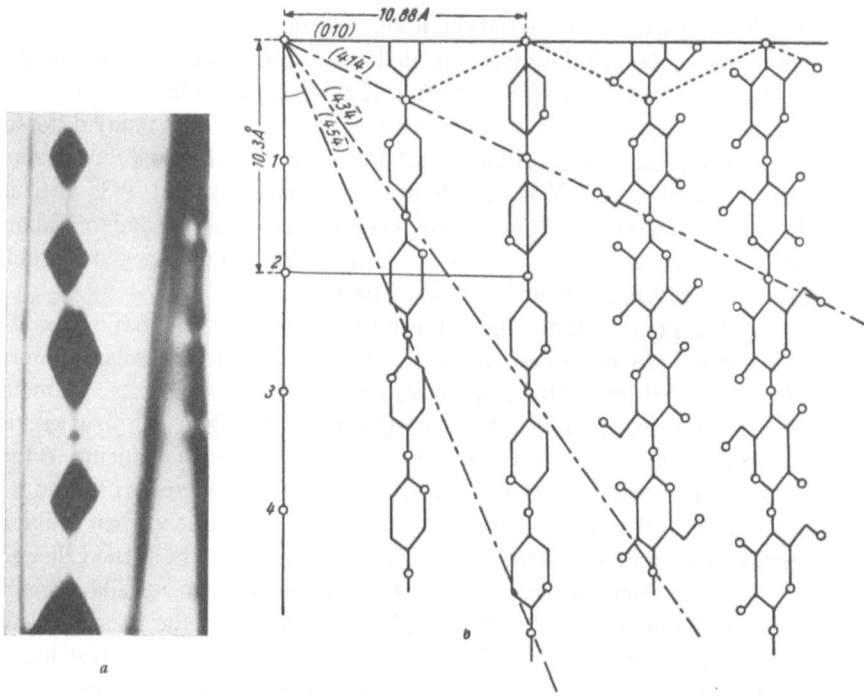

Fig. 10. (a) Wood Fibers Corroded by Wood-Destroying Fungi (4). (b) Section of the Chain Lattice of Cellulose I along the Diagonal Plane (101) (13); – · – · – · – , hydrolysis planes.

Recently, Rånby and Ribi (48) have found crystalline cellulose particles (50 to 100 × 600 Å) in hydrolyzed cellulose, which can be dispersed to stable cellulose sols. They must correspond to the crystallized cellulose in the microfibrils, since the paracrystalline cellulose fraction is probably more readily subject to hydrolysis than crystallized regions. We do not consider these crystallites as an important structural element of the cellulose fine structure, since all of the valuable technical properties of cellulose, such as insolubility, stainability, tensile strength etc. are due to the fact that those "micellae" have no individuality in intact cellulose. They can only be freed from the coherent paracrystalline cellulose by chemical disintegration.

In the absence of any definite evidence of a submicroscopic segmentation of the microfibrils, it seems improbable that crystallized and paracrystalline cellulose would periodically alternate along the microfibril. Present knowledge is rather in favor of irregularly alternating crystallized and paracrystalline regions.

According to the observed length of the colloidal bodies, the crystalline strands measure at least 600 Å, their paracrystalline continuation having been destroyed by hydrolysis. In Fig. *D* individual fringes, which were split off laterally from the microfibrils, with a length of 1000–2000 Å, are shown. Their diameter corresponds to that of the crystalline micellae described by RÅNBY and RIBI (*48*). Since mechanical disintegration yields fringes which are longer than the bodies obtained by hydrolysis, the micellar strands must contain transverse planes along which the hydrolytic cleavage proceeds. Whether these planes are identical with the "chemical" periods of 330 Å (2^6 glucose units $= 64 \times 5.15$ Å) as proposed by PACSU [(*46*), there p. 164], remains open to discussion.

The enzymatic hydrolysis of cellulose fibers by fungi does not proceed across the fiber but along oblique planes (*4*), the inclination of which can be derived from the cellulose chain lattice (Fig. 10). The resulting microscopic corrosion figures on such fibers prove that the crystal lattice of adjacent micellar strands, and even of neighboring microfibrils, must have the same orientation of their *a* and *c* axes (*13*).

d) Intermicellar Spaces.

Since the micellar strands are not individualized and must have the same orientation of their crystal lattice throughout the whole cross-section of a microfibril, it is difficult to understand why they do not coalesce completely to a homogeneously crystallized structure. Possibly, this heterogeneity is due to the water of hydration from the chain molecules which is liberated during the crystallization of the cellulose. Since the cross-section of a microfibril includes about 2000 chain molecules, it is hard to believe that all the liberated water molecules would escape to the surface of the microfibril. Therefore, a certain amount of water is included in the crystallizing portion of the microfibril and gathers there to form splits such as shown in Fig. 6, p. 8. These are the so-called intermicellar spaces. Their diameter amounts only to a fraction of the transverse dimensions of the lattice unit cell; however, they must be accessible to those chemicals which are able to react with cellulose, for example, water, alkali, iodine etc.

When an included water layer reaches a depth which surpasses the attraction distance of neighboring bundles of cellulose chains, the crystallization process comes to an end. The amount of water just mentioned originates from the crystallization of a definite number of cellulose chains. This may be the reason why the crystallized strands of all kinds of cellulose I have the same characteristic diameter of about 50 Å.

No such explanation can be found for the constancy of the diameter of microfibrils (200–300 Å); there the living protoplasm which synthesizes cellulose must limit the size of the cross-section and, by saturating the lattice forces on the surface of peripheric micellar strands, it prevents any further aggregation of the formed microfibrils. Therefore, the inter-microfibrillar cohesion is less than the intermicellar cohesion, and it is clear why a mechanical cleavage of cell walls into microfibrils is possible. Another consequence must be the existence of inter-microfibrillar capillaries of colloidal dimensions. They are accessible to colloidal direct dyes which are adsorbed on the surface of the microfibrils. In the living plant they are responsible for the holopermeability of the cell walls. Furthermore, they represent the spaces into which incrusting substances such as lignin, suberin or cutin are deposited. In heavily lignified cell walls the cellulose texture is completely enclosed by a system of a coherent cement similar to the iron framework in reinforced concrete.

III. Conclusions.

The elements of the fine structure of native cellulose are *microfibrils*. They are individualized and their diameter is astonishingly uniform (200–300 Å). Their internal structure consists of parallelized micellar strands with a diameter of about 50 Å. These crystalline strands are not separated but linked together by paracrystalline cellulose, forming a coherent structure. It has not been possible to fibrilize this structure appreciably by mechanical means but chemical cleavage by hydrolysis of the paracrystalline cellulose has been found possible. As a result, crystalline cellulose particles of 50–100 Å diameter appear.

The orientation of the microfibrils in cell walls is responsible for their physical properties such as swelling, strength, cleavability etc. The microfibrils of the primary cell walls are dispersed and interwoven; in the secondary walls of fibers and tracheids they are parallelized.

As a consequence of this texture there occur in plant fibers two types of spaces available for incrusting substances or topochemical reactions: (a) interfibrillar capillaries which are accessible to colloidal particles (e. g. direct dyes) and in which incrusting substances (e. g. lignin) can be deposited; and (b) intermicellar spaces formed by paracrystalline cel ose which are accessible only to small molecules showing a specific affinity for cellulose, such as water, alkali, and iodine molecules.

References.

1. ANDERSON, D. B. and TH. KERR: Growth and Structure of Cotton Fiber. Ind. Engng. Chem. **30**, 48 (1938).
2. ASTBURY, W. T., T. C. MARWICK and J. D. BERNAL: X-Ray Analysis of the Structure of the Wall of *Valonia ventricosa*. Proc. Roy. Soc. (London), Ser. B **109**, 443 (1932).

3. BAILEY, I. W. and TH. KERR: The Visible Structure of the Secondary Wall and its Significance in Physical and Chemical Investigations of Tracheary Cells and Fibers. J. Arnold Arboretum 16, 273 (1935).

4. BAILEY, I. W. and M. R. VESTAL: The Significance of Certain Wood Destroying Fungi in the M. Study of the Enzymatic Hydrolysis of Cellulose. J. Arnold Arboretum 18, 196 (1937).

5. BALLS, W. L.: The Existence of Daily Growth Rings in the Cell Wall of Cotton. Proc. Roy. Soc. (London), Ser. B 90, 542 (1919).

6. BAULE, B., O. KRATKY u. R. TREER: Einführung der Blättchenmicelle in die Theorie der Deformationsvorgänge. Z. physik. Chem., Abt. B 50, 255 (1941).

6a. BRAUNS, F. E.: Lignin. Fortschr. Chem. organ. Naturstoffe 5, 175 (1948).

6b. FREUDENBERG, K. u. W. DÜRR: Konstitution und Morphologie des Lignins. KLEINS Handbuch der Pflanzenanalyse, Bd. 3, S. 142. Berlin: Springer-Verlag. 1932.

7. FREY, A.: Das Wesen der Chlorzinkjodreaktion und das Problem des Faserdichroismus. Jahrb. wiss. Bot. 67, 597 (1927).

8. FREY, R.: Chitin und Zellulose in Pilzzellwänden. Diss. Eidgen. Techn. Hochschule, Zürich 1950; Ber. Schweiz. bot. Ges. 60, 199 (1950).

9. FREY-WYSSLING, A.: Über die Verschiebungsfiguren zellulosiger Zellwände. Z. wiss. Mikroskop. 51, 29 (1934).

9a. — Über Verschiebungsfiguren in Asbestfasern. Z. wiss. Mikroskop. 56, 309 (1939).

10. — Die Stoffausscheidung der höheren Pflanzen. Berlin: Springer-Verlag. 1935.

11. — Der Aufbau der pflanzlichen Zellwände. Protoplasma 25, 261 (1936).

12. — Über die röntgenometrische Vermessung der submikroskopischen Räume in Gerüstsubstanzen. Protoplasma 27, 372 (1937).

13. — Submikroskopische Struktur und Mazerationsbilder nativer Zellulosefasern. Papierfabrikant 36, 212 (1938).

14. — Über den Zellulosenachweis mit Jod. Verh. Schweiz. Naturf. Ges. 1939, 67.

15. — Anisotropic Diffusion of Direct Dyes in Fibers. J. Polymer Sci. 2, 314 (1947).

16. — The Growth in Surface of Plant Cell Walls. Growth Symp. 12, 151 (1948).

17. FREY-WYSSLING, A. and K. MÜHLETHALER: Submicroscopic Structure of Cellulose Gels. J. Polymer Sci. 1, 172 (1946).

18. — — Über den Feinbau der Zellwand von Wurzelhaaren. Mikroskopie (Wien) 4, 257 (1949).

19. — — Der submikroskopische Feinbau von Chitinzellwänden. Vierteljschr. naturforsch. Ges. Zürich 95, 45 (1950).

20. FREY-WYSSLING, A., K. MÜHLETHALER u. R. W. G. WYCKOFF: Mikrofibrillenbau der pflanzlichen Zellwände. Experientia 4, 475 (1948).

21. FREY-WYSSLING, A. u. K. WUHRMANN: Die Abhängigkeit des Lichtbrechungsvermögens kristallisierter Zellulose von der Temperatur. Helv. chim. Acta 22, 981 (1939).

GRIFFIOEN, K.: Über Quellungsbilder verschiedener Faserarten und deren Bedeutung für die Faserstruktur. Planta 24, 584 (1935).

23. HALLER, R.: Über Verseifungsvorgänge in Acetatkunstseide. Helv. chim. Acta 24, 149 (1941).

24. HEGETSCHWEILER, R.: Über den Feinbau des Seidenfibroins. Diss. Eidgen. Techn. Hochschule, Zürich 1950; Makromol. Chem. 4, 156 (1950).

25. HENGSTENBERG, J. u. H. MARK: Über Form und Größe der Micelle von Zellulose und Kautschuk. Z. Kristallogr. 69, 271 (1928).

26. HERMANS, P. H.: Physics and Chemistry of Cellulose Fibers. New York-Amsterdam: Elsevier Publ. Co. 1949.

27. HESS, K., H. KIESSIG, W. WERGIN u. W. ENGEL: Zur Kenntnis der Bildung von Zellulose in der Zellwand. Ber. dtsch. chem. Ges. **72**, 642 (1939).

28. HESS, K., C. TROGUS u. W. WERGIN: Untersuchungen über die Bildung der pflanzlichen Zellwand. Planta **25**, 419 (1936).

29. HESS, K., W. WERGIN u. H. KIESSIG: Zur Frage des Aufbaus der Primärsubstanz der Baumwollhaare. Planta **33**, 151 (1942).

30. KERR, TH.: The Structure of the Growth Rings in the Secondary Wall of the Cotton Hair. Protoplasma **27**, 229 (1937).

31. KERR, TH. and I. W. BAILEY: The Cambium and its Derivative Tissues. X. Structure, Optical Properties and Chemical Composition of the so-called Middle Lamella. J. Arnold Arboretum **15**, 327 (1934).

32. KINSINGER, W. G. and CH. W. HOCK: Electron Microscopical Studies of Natural Cellulose Fibers. Ind. Engng. Chem. **40**, 1711 (1948).

33. KRATKY, O. u. H. MARK: Zur Frage der individuellen Cellulosemicellen. Z. physik. Chem., Abt. B **36**, 129 (1937).

34. MANGIN, L.: Sur un essai de classification des mucilages. Bull. Soc. bot. France **41**, XL (1894).

35. MEYER, K. H.: Hochpolymere Chemie, Bd. II. Leipzig: Akad. Verlagsges. 1940.

36. MEYER, K. H. u. H. MARK: Der Aufbau der hochpolymeren organischen Naturstoffe. Leipzig: Akad. Verlagsges. 1930.

37. MÜHLETHALER, K.: Electron Micrographs of Plant Fibers. Biochim. et Biophys. Acta **3**, 15 (1949).

38. — The Structure of Bacterial Cellulose. Biochim. et Biophys. Acta **3**, 527 (1949).

39. — Electron Microscopy of Developing Plant Cell Walls. Biochim. et Biophys. Acta **5**, 1 (1950).

40. — An Electron Microscope Study of the Structure of Viscose Silk. Experientia **6**, 226 (1950).

41. — The Structure of Plant Slimes. Exptl. Cell Res. **1**, 341 (1950).

42. NÄGELI, C.: Die Micellartheorie. Ostwald's Klassiker No. 227. Leipzig: Akad. Verlagsges. 1928.

43. NAKAMURA, Y. u. K. HESS: Zur Kenntnis der chemischen Zusammensetzung der Maiskoleoptile. Ber. dtsch. chem. Ges. **71**, 143 (1938).

44. NICOLAI, E. (in Leeds): Oral communication.

45. NICOLAI, E. u. A. FREY-WYSSLING: Über den Feinbau der Zellwand von *Chaetomorpha*. Protoplasma **30**, 401 (1938).

46. PACSU, E.: Recent Development in the Structural Problem of Cellulose. Fortschr. Chem. organ. Naturstoffe **5**, 128 (1948).

47. PRESTON, R. D., E. NICOLAI, R. REED and A. MILLARD: An Electron Microscope Study of Cellulose in the Wall of *Valonia ventricosa*. Nature (London) **162**, 665 (1948).

48. RÅNBY, B. G. u. ED. RIBI: Über den Feinbau der Zellulose. Experientia **6**, 12 (1950).

49. ROSEVEARE, W. E., R. C. WALLER and J. N. NELSON: Structure and Properties of Regenerated Cellulose. Text. Res. J. **18**, 114 (1948).

50. SISSON, W. A.: Identification of Crystalline Cellulose in Young Cotton Fibers by X-Ray Diffraction Analysis. Contr. Boyce Thompson Inst. **8**, 397 (1937).

51. — Some X-Ray Observations Regarding the Membrane Structure of *Halicystis*. Contr. Boyce Thompson Inst. **12**, 31 (1941).

52. SPONSLER, O. L.: Orientation of Cellulose Space Lattice in the Cell Wall. Additional X-Ray Data from *Valonia* Cell Wall. Protoplasma **12**, 241 (1931).

53. STEENBERG, B.: Beating Process Studied by Fiber Swelling. Svenska Träforskningsinst. Medd. (Stockholm) **20** (1947).

54. THIMANN, V. and J. BONNER: The Mechanism of the Action Growth Substance of Plants. Proc. Roy. Soc. (London), Ser. B 113, 126 (1933).

55. TUPPER-CAREY, R. M. and J. H. PRIESTLEY: The Composition of the Cell Wall at the Apical Meristem of Stem and Root. Proc. Roy. Soc. (London), Ser. B 195, 109 (1923).

56. WÄLCHLI, O.: Die Einlagerung von Kongorot in Zellulose. Diss. Eidgen. Techn. Hochschule, Zürich 1945; Schweiz. Arch. angew. Wiss. Techn. 11, 129 (1945).

57. WIRTH, P.: Membranwachstum während der Zellstreckung. Diss.' Eidgen. Techn. Hochschule, Zürich 1946; Ber. Schweiz. bot. Ges. 56, 175 (1946).

58. WUHRMANN, K., A. HEUBERGER u. K. MÜHLETHALER: Elektronenmikroskopische Untersuchungen an Zellulosefasern nach Behandlung mit Ultraschall. Experientia 2, 105 (1946).

59. WUHRMANN-MEYER, K. u. M. WUHRMANN-MEYER: Über Bau und Entwicklung der Zellwände in der Avena-Koleoptile. Jahrb. wiss. Bot. 87, 642 (1939).

60. WYCKOFF, R. W. G.: Electron Microscopy. New York-London: Interscience Publ. 1949.

61. ZIEGENSPECK, H.: Amyloid in jugendlichen Pflanzenorganen als vermutliches Zwischenprodukt bei der Bildung von Wandkohlenhydraten. Ber. dtsch. bot. Ges. 37, 273 (1919).

(Received, July 3, 1950.)

Bacterial Dextrans.

By M. STACEY and C. R. RICKETTS, Birmingham.

Contents.

I. Introduction.

Dextran is a collective name for a series of polyglucoses having usually a high dextrorotation of the order $[\alpha]_D + 180°$. The dextran may be branched or straight chain and possess a high proportion of α-D-glucopyranose units linked through the $1:6$ positions. They are quite remarkable colloids which can be made on a large scale and at low cost from sugar, by fermentation with *Leuconostoc* organisms. They have acquired a new importance since they can be converted into a suitable substitute for blood plasma. Brief accounts have been written on bacterial dextrans by HASSID and DOUDOROFF (22) as well as by EVANS, HAWKINS and HIBBERT (14).

II. Historical.

Sugar refiners have observed the occasional formation of large and troublesome masses of ropiness or slime in their sugar factories for more than a century. Pasteur (43) in 1861 recognised that the slimy fermentation of carbohydrate was due to microbial action. Joubert (36) demonstrated that the slime could be propagated in beet sugar juice only, while Béchamp (4) realised that sucrose and not glucose or invert sugar was responsible (1881). Schiebler (47a) identified the mucilaginous material from infected sugar beet as a polyanhydride of glucose comparable with starch or dextrin. The high dextro-rotation led to the name "dextran". Daumichen (12) prepared the triacetate and tribenzoate establishing the presence of three hydroxyl groups per glucose unit (1892).

At the same time bacteriologists were beginning to classify the *Leuconostoc* species. Cienkowski (9) recognised the essentially biological nature of slime formation in beet sugar juice, and von Treghen (59) gave the first adequate description of one of the causative organisms, naming it *Leuconostoc mesenteroides*. Liesenberg and Zopf (38) obtained pure cultures showing capsules consisting of Schiebler's dextran (1899).

III. The genus Leuconostoc.

1. Classification.

Hucker and Pederson (27) have reviewed this group of organisms and have carried out a biochemical survey of some 80 strains isolated from fermenting sugar solutions, vegetables and milk etc. All were found to ferment glucose, forming *L*-lactic acid, carbon dioxide and volatile products among which ethyl alcohol and acetic acid were identified. Fructose was converted to mannitol and sucrose to dextrans or levans. The fermentations of pentoses and sucrose were found to be reliable characteristics of three species within the genus:

a) *Leuconostoc mesenteroides* which ferments pentoses (arabinose or xylose) and produces a slime of dextran when grown on sucrose.

b) *L. dextranicus* (or *dextranicum*) which does not ferment pentoses but produces some slime on a sucrose medium.

c) *L. citrovorus* which ferments neither pentose nor sucrose.

It is now well established that the dextrans of *L. mesenteroides* and *L. dextranicus* differ in structure and may be regarded as characteristic of the species.

2. Growth Factors.

Hucker and Pederson mention the value of yeast or tomato extract in prompting the growth of the *Leuconostoc* organisms. Stacey and Youd (53) obtained a successful growth in the presence of maple syrup. Using a medium in which all constituents except casein hydrolysate were chemically defined, Gaines and Stahly (18, 19) showed that some vitamins of the B group were essential to growth. Thiamin, calcium pantothenate and nicotinic acid were indispensable to growth while pyridoxin exerted a stimulatory effect. Although growth occurred in the absence of biotin, the avidin inactivation technique demonstrated the biotin requirement of the organism. CARLSON and WHITESIDE-CARLSON (7) consider that to the extent that disaccharide is utilised in dextran formation L. mesenteroides is free of a biotin requirement.

3. Dextran Formation in Leuconostoc Cultures.

TARR and HIBBERT (58) obtained dextran from L. mesenteroides grown on a medium containing sucrose, peptone and phosphate. Growth was slow, a maximum yield being obtained after ten days. The yield increased with increasing sugar concentration. CARRUTHERS and COOPER (8) found that mixed cultures isolated from sugar factory slime rapidly formed dextran. They attributed this to growth factors present in raw beet sugar and molasses. According to JEANES, WILHAM, MIERS and SCHIELTZ (34, 35) in unaerated and unbuffered cultures the maximum yield coincides with the maximum viscosity. From an initial value of 7 the p_H became 4,6 at the maximum yield and then fell to 3,7 when the yield was significantly lower. One of their strains of L. mesenteroides produced a water-soluble dextran and some levan; the other produced only a water-insoluble dextran.

4. Enzymic Synthesis of Dextran.

HEHRE (23) has obtained sterile preparations of the exocellular dextran-synthesising enzyme from Leuconostoc cultures. With these he has explored the kinetics of dextran formation. At sucrose concentrations below 1 per cent the reaction is of the first order; at 5 per cent sucrose the reaction becomes of zero order. The well known inhibitors of glycolysis such as azide, cyanide, fluoride and iodoacetate have no effect on the reaction. Virtually all of the sucrose can be converted to dextran. The course of the reaction resembles the hydrolysis of sucrose by invertase. There is therefore a sharp distinction between the mass action equilibrium of phosphate ions and glucose-1-phosphate associated with starch synthesis, and the complete conversion of sucrose to dextran and fructose. The only common factor is the energy rich bond at carbon atom one of the glucose.

These enzyme preparations caused no hydrolysis, phosphorolysis or levan formation, and dextran production was strictly proportional to the reducing power due to fructose formation. The process of levan formation from sucrose by the levansucrase isolated from Aerobacter levanicum is more complicated (2). In the later stages of the reaction, an unidentified inhibitor comes into play. At suitably low sucrose concentrations levan production accounts for 62% of the total fructose. By symbiotic growth of *L. mesenteroides* and *Saccharomyces cerevisiae*, STACEY (*48*) obtained a sterile enzyme preparation causing extremely rapid formation of a very viscous dextran.

5. Serology of Leuconostoc.

Development in this field originated with the observation of HEHRE, NEILL, SUGG and JAFFE (*25*) that many samples of pure sucrose contained a substance reacting with Type II pneumococcus antiserum. In a further paper (*41*) the observation was extended to *Leuconostoc mesenteroides* antiserum and it was suggested that the material contaminating sucrose was in fact dextran antigen of *L. mesenteroides*. Antisera to pneumococcus Types 12 and 20 were also found to react with dextran. HEHRE (*23*) then prepared sterile enzymes from *Leuconostoc* cultures and showed as already mentioned, that they converted sucrose to a dextran with the known serological and chemical properties. EVANS, HAWKINS and HIBBERT (*14*) have attempted to show that protein-free dextran would react with *Leuconostoc* antiserum. There are difficulties in determining the small quantities of nitrogen in the presence of so much carbohydrate. The lowest nitrogen content they record is 0,08%, and dextran with this minute trace of nitrogen, not necessarily protein nitrogen, reacted with *Leuconostoc* antiserum. Certain strains of streptococci were also shown to produce a substance, believed to be dextran, capable of reacting with both antipneumococcus and antileuconostoc sera.

This work may be compared with that of ZOZAYA (*63*) who reported that *L. mesenteroides* dextran reacted immunologically with antisera of certain members of the *Salmonella* group, of pneumococci and with some strains of streptococci (*viridans*). These were weak cross reactions shown only in the low dilution range, and it was shown that antisera absorbed with the homologous polysaccharides still precipitated with dextran although they no longer gave any homologous specific reaction.

There would appear to be little doubt that relatively crude samples of dextran are antigenic and whether or not the work of some investigators such as FITZGERALD (*16*) is entirely accepted or whether or not the view is confirmed that the presence of prosthetic groups in the dextran molecule causes the cross immunological relationship, there would appear to be a realisation of the growing importance of structural knowledge in this type of polysaccharide. The work on dextran blood plasma substitute makes the position somewhat urgent. BULL and his colleagues (*6*) made use of the fact that antisera to *L. mesenteroides* (Birmingham strain) gave a good precipitin reaction (e. g. at 1/1 million dilution) with degraded dextran in order to test for the presence of dextran in tissues, and were able to decide how long dextran would remain in some tissues. In degraded

dextrans, the point at which the dextran loses its haptenic properties
and the minimum size and shape of dextran haptenes has not yet been
decided, but it is a point of some importance in relationship to the sensitivity of the few individuals who may be reactive to dextran injections.

IV. Derivatives of Dextran.

Numerous esters and ethers of dextran have been prepared by the
well known methods of application to cellulose. Many patents on these
are summarised in a Sugar Research Foundation publication (56a).

Treatment of dextran with acetic anhydride and sodium acetate,
acetyl chloride and pyridine etc., yields the acetate (57). The benzoate
may be obtained by the action of benzoyl chloride and sodium hydroxide etc. (57). The use of chloracetic anhydride as a catalyst enables
dextran to be esterified with long chain fatty acids, e. g., dextran stearate
has been obtained (49).

Special mention is made later of the sulfuric ester of dextran, prepared by treatment of dextran with chlorsulfonic acid and pyridine,
at present under investigation as an alternative to the naturally occurring (20) polysaccharide sulfuric ester, heparin, which functions as a
blood anticoagulant.

The methyl ether of dextran has been investigated with reference
to the structure of dextran. The ethyl (54), ethylenyl, butyl, and benzyl (55)
ethers of dextran have been prepared by treatment of dextran with the
appropriate alkyl halide in the presence of sodium hydroxide. The
carboxymethyl ether, obtained by treatment with chloracetic acid and
sodium hydroxide (57) has unusual properties because of the free carboxyl
group. Carboxymethyl dextran exhibits the electroviscous effect; its
viscosity and precipitation by alcohol are greatly influenced by the
presence of small quantities of electrolytes.

Allyl groups may be introduced with allyl chloride and sodium hydroxide and polymerisation induced (57). Reaction with formaldehyde
yields the methylene ether. Such cross linking reactions may produce
useful modifications in the properties of dextran and its derivatives (57).

Mixed ether-esters have been prepared by partial etherification
followed esterification, for example the butyl ether—benzoyl ester and the
benzyl ether—phthalyl ester (56).

Heparin is present in small quantities in most animal tissues from
which it is extracted for use in the treatment of thrombosis. The sulfuric
esters of several polysaccharides are known to possess some anticoagulant
activity but so far none has been considered suitable for clinical use.
Objections are based mainly upon the formation of an insoluble complex
with fibrinogen and the agglutination of thrombocytes. These undesirable

effects upon vital elements of the blood may be overcome by using sulfuric esters prepared from specially degraded and fractionated dextrans (47).

V. Structure of Dextrans.

The most striking point regarding the structure of the dextrans is the fact that their polyglucose chains possess a high proportion of 1:6-glucosidic linkages, these being mainly of the α type. This important fact was discovered independently in the Birmingham Laboratories and in HIBBERT's Laboratories in Canada. So far no sugar other than glucose has been shown to be a component of any dextran.

FOWLER, BUCKLAND, BRAUNS and HIBBERT (17) using sodium hydroxide and methyl sulfate, methylated a dextran produced by a strain of *L. mesenteroides* and they hydrolysed the methylated product in the usual way with methanolic hydrogen chloride. They identified the methanolysis products as 2:3-di-, 2:3:4-tri-, and 2:3:4:6-tetramethyl methylglucosides and stated that these were present in the proportions of 1:3:1. Accordingly a branched chain structure was proposed for this dextran. Surprisingly however, one of the authors, BRAUNS (5) criticised these results on the grounds that (a) the dextran was incompletely methylated, (b) an inefficient fractionation of the glycosides was carried out and the proportion of the three components was not conclusively established, and (c) there was a considerable loss of material (18,4%) during the fractionation.

With this challenge the senior author HIBBERT, with the assistance of other colleagues (37) re-investigated the whole problem of the structure of the dextran and generally re-affirmed the earlier claims—a view supported later by the independent work of STACEY and SWIFT (52) on another *L. mesenteroides* dextran. HIBBERT and his colleagues in their second paper pointed out the high significance of the need to make sure that polysaccharides were fully methylated before hydrolysis, incomplete methylation giving rise to incorrect ideas of branching. In this work, they completed the HAWORTH type methylation of their dextran by using a modified MUSCAT technique which consisted of suspending the partly methylated dextran in anhydrous anisole followed by treatment with sodium in liquid ammonia. The sodium derivative was then boiled with methyl iodide giving eventually a methylated dextran with a methoxyl content of 45,6%—claimed to be the theoretical amount for a methylated polyglucose.

Cautious methanolysis in sealed glass tubes followed by fractionation of the glycosides again established the 1:3:1 proportions of the di, tri and tetra components and confirmed the branched chain structure postulated.

Stacey and Youd (53) as well as Peat, Schlüchterer and Stacey (44) investigated the dextran from *L. dextranicum* and showed that this possessed a more linear type of structure. The methylation proceeded smoothly and only 0,23% of tetramethyl end group could be isolated. This corresponded to a chain of 550 units whereas osmotic pressure measurements of the molecular weight of the methylated derivative indicated a minimum chain of 200 units.

The significance of the presence of a 2:3-dimethyl methyl glucoside in the distillation products was not clear. The finding that the *L. dextranicum* dextran is essentially a long chain polymer was confirmed by the work of Hibbert and his colleagues who from their strain of *L. dextranicum* dextran obtained a methyl ether, which on hydrolysis yielded no 'tetramethyl' end group at all. The main product was 2:3:4-trimethyl glucose with 10% of unidentified dimethyl glucose. A linear network type of structure was claimed for this dextran.

It is clear that these dextrans should be re-investigated using modern chromatographic techniques.

Further variety in dextran structure was shown by the work of Daker and Stacey (11) on the interesting dextran of H. D. Mayer. This dextran is produced from sucrose by the "Tibi" organism and it was shown by the usual methods to consist of relatively short chains of about 25 glucopyranose units united by α linkages. It was suggested that these were probably aggregated to form larger molecular aggregates in possibly the same way as the basal units in amylopectin are combined. This being so, linkages other than 1:6 will need to be sought. The *L. mesenteroides* (Birmingham strain) dextran investigated by Stacey and Swift (52) was somewhat unusual inasmuch as unlike other dextrans, it could not be obtained in a water-insoluble form.

It was methylated by the sodium hydroxide/methyl sulfate method followed by the sodium/methyl iodide in liquid ammonia technique.

The methylated dextran having a methoxyl content of 45,5 was hydrolysed in the usual way and in this case, the products were separated on a chromatogram.

There were isolated and identified 2:3-di-, 2:3:4-tri-, and 2:3:4:6-tetramethyl glucose in the proportions of 1 : 3 : 1 and thus, in this respect the constitution of the repeating unit of the methylated dextran was identical with that described by Hibbert and his co-workers (37) for the dextran produced by the action of their strain of *L. mesenteroides* on sucrose. The type of structure could be represented by (I) or (II) which are essentially the branched chain formulations of Hibbert and his co-workers (37).

It appeared, therefore that the new organism was *L. mesenteroides*, but it was evidently a different strain, because the dextran it produced had physical properties different from those of the dextran from known strains of *L. mesenteroides*.

It is possible that a repeating unit of the type shown in (I) or (II) may be identical with that of the dextrans produced by various strains of *L. mesenteroides*, although it is likely that the branching occurs at different places. Indeed, such small differences as are shown between (I) and (II) and other variations of the type of structure may account for the physical differences between the polysaccharides formed by the various strains. It is likely also that the number of repeating units in each macromolecule is different. It is an interesting finding that even different strains of an organism give dextrans having different physical properties, and it appears that one may eventually be able to identify an organism and its various strains by determining the finer structure of its extracellular polysaccharides.

Numerous other dextrans are under investigation in the writers laboratories and generally it is being found that they all possess in some measure the branched chain type of structure with 1:6-α-glucosidic linkages predominating. The main evidence on the nature of the α or β linkages comes from studies on the changes in rotation and hydrolysis of polysaccharides. WOLFROM and his colleagues (61) however, have isolated a disaccharide of the gentiobiose type from an *L. mesenteroides* dextran and have designated it 6-α-glucopyranose-*D*-glucopyranoside. The 1:6β type of linkage was shown to exist in the polyglucose "luteose" made from sugar by *P. luteum*

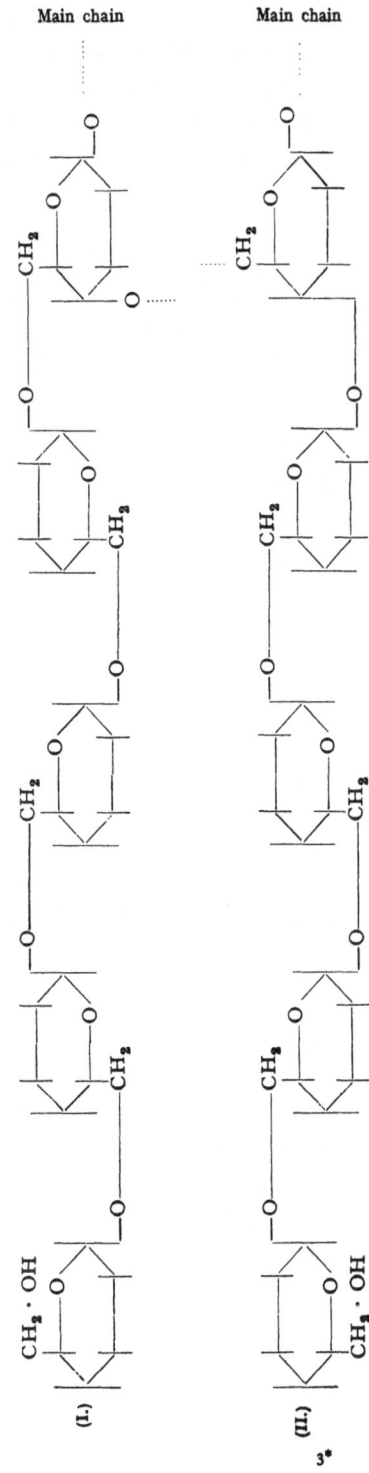

zukal [RAISTRICK and RINTOUL (*46*)]. This luteose is bound up with malonic acid residues forming a mucilaginous complex, termed luteic acid which is apparently possessed of a very high molecular weight. The malonic acid residues are very susceptible to both acid and alkaline hydrolysis.

The type of structure in luteose is shown as follows (III):

(III.)

A good deal remains to be done in regard to the fine structure of dextrans and this is becoming of major importance in the search for an ideal injection fluid as discussed elsewhere in this article (p. 41).

VI. Physico-chemical Studies of Dextran.

1. The Macro Molecule.

From the formula advanced by LEVI, HAWKINS and HIBBERT (*37*), INGELMAN and SIEGBAHN (*32*) estimated the thickness of dextran molecules to be of the order of 50 Å. Using the electron microscope at the Research Institute for Physics in Stockholm which had a resolving power of 30 Å they succeeded in obtaining pictures of dextran showing branched thread-like molecules. Their method was similar to that used to photograph viruses. A drop of 0,002% solution of dextran in distilled water was evaporated on a thin film of zapon lac. Direct electron optical magnification of 7000× and enlargement on fine grain photographic plates 10× was employed. Under these conditions contrast was difficult since the carbon, hydrogen, and oxygen atoms are common to the background and to dextran. Later INGELMAN and HALLING (*31*) using the gold shadowing technique of WILLIAMS and WYCKOFF (*60*) obtained pictures clearly showing the branched molecules of high molecular dextran. Electron-optical magnifications of 7500× followed by final enlargement to 23000× was employed. Distinct pictures could be obtained of partially hydrolysed and fractionated dextran having a molecular weight of 38 million as determined with the ultracentrifuge.

An interesting insight into the fine structure of dextran is the observation of nodes at intervals of about 800 Å or 160 glucose units along the fine molecular threads. One is tempted to speculate on the possibility of correlating this observation with the chain length of the sample as determined by the end group assay. In comparative pictures of dextran and levan taken by INGELMAN and SIEGBAHN (33) the levan molecules do not display the thread-like structure of dextran but are more compact like glycogen in the pictures of HUSEMANN and RUSKA (28). In the picture of levan the variation in particle size can be clearly seen and accords with the polydispersion found with the ultracentrifuge. The way is open for a vivid study of the synthesis and breakdown of these bacterial polysaccharides.

Diffraction of X-rays has revealed orderly association of molecules in suitably prepared samples of several polysaccharides with unbranched chains, amylose and alginic acid for example. As might be expected an orderly arrangement of branched chains is more difficult to achieve but JEANES, SCHIELTZ and WILHAM (34) have obtained X-ray diffraction powder patterns indicative of an ordered molecular arrangement in several preparations of both maize amylopectin and L. mesenteroides dextran. In a series of preparations an increasing degree of orderly association, as shown by X-ray data was accompanied by decreasing solubility in water. Explanation of the remarkable water-insolubility of some dextrans is to be sought along these lines. The analogy with cellulose is evident, and indeed filaments many feet in lenght have been spun from aqueous pastes of L. mesenteroides dextran confirming the fibrous properties to be expected from the molecular arrangement.

2. Partial Hydrolysis.

Dextrans are rather more resistant to hydrolysis by acid than most polysaccharides. Dextrans frequently occur as highly mucilaginous material which when dried is henceforth insoluble in water. Alkali treatment is needed to produce a form henceforth completely water soluble. Any kind of degradation, e. g. with acids or alkalies appeared to remove a prosthetic group which contains nitrogen (53). It may well be that this originates from the dextran-synthesising enzyme. A rapid dissolution in strong hydrochloric acid followed by immediate precipitation with alcohol provides a good method for making a soluble dextran. Dextran can be quantitatively converted to glucose via a series of intermediate poly- and oligo-saccharide stages. A convenient acid is $2\,N$-sulfuric acid at 100°—the hydrolysis being followed by polarimetric readings, (e. g. $[\alpha]_D + 180° \rightarrow 52°$) and by estimation of reducing sugar by standard methods.

A p_H of 2 to 2,5 at 100° C causes partial hydrolysis. When viscosity of the solution is plotted against duration of hydrolysis, the curve falls steeply at first and then flattens out showing that breakdown of the polymer has practically ceased. In this period the reducing power increases by only a few per cent of the value for complete conversion to glucose. Ultracentrifugal examination of the partial hydrolysate showed a wide range of molecular weights (*31*).

Similar changes occur under the influence of adaptive enzymes developed by certain micro-organisms (*29*). Study of the enzymic breakdown of starch and glycogen by the amylases and phosphorylases has, in conjunction with other methods, successfully elucidated the more subtle features in the structure of these polysaccharides.

INGELMAN (*29*) has approached dextran along these lines. He found it possible to train the cellulose degrading bacterium *Cellvibrio fulva* to grow with dextran as the sole source of carbon in the presence of suitable inorganic salts. After several months an enzyme which acted upon dextran could be obtained from the culture. On incubating dextran with the enzyme the viscosity of the solution decreased, the reciprocal of specific viscosity having a linear relation with time in the early stages of breakdown. The sedimentation constant, determined with the ultracentrifuge, also decreased. These changes were accompanied by a slight increase in reducing power, which reached a maximum of 3 per cent. The rate of reaction was closely proportional to the enzyme concentration and the optimum p_H was 5,2. It is thought that the first products of enzyme action were rather large molecules and that endwise degradation of the molecular chains did not occur. INGELMAN emphasises that not all cellulose degrading organisms are able to break down dextran. Extracts of *Aspergillus niger* and *Cytophaga globosa* had no effect on dextran nor indeed does amylase although it breaks the 1:6 links in starch. NORDSTRÖM and HULTIN (*42*) were able to train three moulds, *Penicillium lilacinum* THOM, *P. funiculosum* THOM and *Verticillium coccorum* PETCH to grow on dextran and the appropriate inorganic salts. After 10 days incubation, grinding, and autolysis, enabled the extraction of an enzyme acting on dextran. The decrease in viscosity was accompanied by some increase in reducing power. The optimum p_H in this case was about 6. Random scission of the 1:6 linkages is believed to occur.

The development of such adaptive enzymes by microorganisms and careful study of the products of their action should greatly accelerate progress in this field.

The dextran macromolecule has recently been degraded by ultrasonic waves and this method appears to provide a very convenient means of getting smaller units in a controlled manner (*51, 39*).

3. Fractionation.

The general methods of fractionating polydisperse mixtures may be applied to dextran. Gradual addition of alcohol or acetone to a dextran solution precipitates first the larger molecules and then as more is added progressively small molecules come down. A partial hydrolysate may contain molecules with molecular weights from ten thousand to ten million. Clearly if an appreciable yield of a homogeneous fraction is desired the hydrolysis should be discontinued when the hydrolysate contains a maximum proportion by weight of molecules of the required size.

INGELMAN and HALLING (31) took several partial hydrolysates of suittable differing molecular composition and subjected them to repeated fractionation. Realising the difficulties inherent in attaining equilibrium after the addition of sufficient alcohol to precipitate a fraction they employed temperature as the operative variable. Alcohol was added slowly to a 2 per cent solution of dextran at 20° C until the solution became turbid. The temperature was then raised to 25–30° C until the precipitate redissolved. On slowly decreasing the temperature to 20° C the first fraction slowly separated under conditions which minimise contamination with smaller molecules. The progress of the fractionation was followed by measuring the intrinsic viscosity of each fraction. The fractions were further studied by ultracentrifugal and adsorption analysis as described below. Using a gradually increasing concentration of acetone as a precipitant, at a constant temperature and with efficient stirring two liquid phases form. Equilibrium between the two phases is rapidly established while denser phase remains in the form of droplets. Temperature control during the coalescence of these droplets to form two liquid layers is an important factor in determining the molecular composition of the dextran in each layer.

RICKETTS, LORENZ and MAYCOCK (40), using these principles fractionated dextran with the object of removing those small molecules able to pass through the glomerular membrane of the kidney. In experiments with rabbits they were able to demonstrate the efficacy of their fractionation.

Dextran, water and acetone appear to form a physical system of the two phase/three component type. Such a system may be described by a triangular diagram of the usual type with one curve corresponding to each molecular weight. Temperature may be plotted along the vertical axis so that the system is completely described within a triangular prism. A detailed investigation of this and similar physical systems is of practical importance in the development of dextran plasma substitute (30, 6).

4. Ultracentrifugal Studies.

INGELMAN and HALLING (31) examined their dextran fractions in the ultracentrifuge. Both sedimentation and diffusion constants were concentration dependent on values obtained by extrapolation to zero concentration were used in SVEDBERG's formula on calculate the molecular weights. Calculation of the dB/dx values according to GRALEN showed that the fractions were not quite homogeneous and that a better approximation to their average molecular weight could be obtained by using weight-average values of the sedimentation constant. Table I reproduces some of the data presented in their paper. The most high molecular fractions of slightly hydrolysed dextran have very high molecular weights. Thus fraction F II, $2a$ has a molecular weight of 38 million and contains particles of a size which is of the same order of magnitude as that of many virus species. Solutions of this fraction showed no sign of molecular breakdown when heated, so it seems probable that these enormous particles are not aggregates but real molecules. Dextran molecules of all molecular weights appear to be considerably elongated and hydrated in solution. There is a linear relationship of intrinsic viscosity with molecular weight for relatively homogeneous fractions having molecular weights between about 50 000 and 300 000.

Table I.

Examination of Some Dextran Fractions in the Ultracentrifuge (*31*).

Fraction	Intrinsic Viscosity	Molecular Weight	f/f_0
U IV	0,08	14 000	1,88
U I	0,18	36 000	2,17
F I, V	0,25	84 000	2,49
F I, IV	0,29	141 000	2,72
F I, III	0,38	240 000	3,15
F II, 10a, β	0,74	1 990 000	3,61
F II, 10a, α	0,86	3 050 000	4,25
F II, 2a, ...	1,26	38 000 000	4,03

5. Adsorption Analysis.

INGELMAN and HALLING (*31*) investigated the adsorption of dextran on active carbon. Solutions of their dextran fractions were treated with carbon at 25° C and after filtration, the concentration of the solution was determined. An adsorption isotherm, expressing the amount of dextran adsorbed per gram of carbon as a function of concentration was plotted for each fraction. The isotherm showed that adsorption increased with decreasing molecular weight. The lowest molecular weight investigated was about 6000–8000.

Using a charcoal adsorption column frontal analysis of the eluate showed that for the series glucose, monosaccharide, disaccharide, etc., derived from dextran adsorption increased with increasing molecular weight. Thus adsorption must pass through a maximum in the region of molecular weight 800 to 8000.

VII. Dextran in Blood Transfusion.

Polysaccharide gums have been used over a long period as blood plasma substitutes, the most successful being the gum saline of Bayliss (3).

These acidic gums suffer from the disadvantage that in certain preparations they are powerful haptènes and they may be antigenic. Dextran appeared to offer several advantages inasmuch as it was shown by Zozaya (62) that when free from nitrogen they were non-antigenic, an observation extended by Fitzgerald (16) and by Stacey and Youd (53). This work showed that the nitrogen component could be removed in the form of a flocculent precipitate by both acid and alkaline hydrolysis causing considerable degradation of the molecule without however breaking it down to very small units. Work on this possible use for dextran and other polysaccharides such as luteose and levans was begun by the senior author in Sir Norman Haworth's laboratories in 1937. He considered that the preponderance of 1:6-glycosidic linkages in dextran and luteose structure would prevent the polysaccharides from being attacked by β-amylase of the body and that there would be no limit to the scale on which the dextran could be produced cheaply from sugar. The project was developed quite independently and applied more rapidly from 1943 onwards by Ingelman and Gronwall in the laboratories of Professor Tiselius in Uppsala [see review by Ingelman (30)].

The requirements of a suitable plasma substitute have been set out as follows by Squire, Maycock and their colleagues (6) who made the first large scale trials of British dextran:

Positive Qualities.

(1) Its colloid solutes should be retained in the circulation until their place can be taken by the natural proteins. This implies that (a) the colloids should not pass readily into the tissue fluids nor be rapidly excreted by the kidney (a molecular size of at least 70 000 is therefore usually desirable); and (b) the colloid substance should not be rapidly metabolised or otherwise removed by the tissues.

(2) The solution used for infusion should have an osmotic pressure and viscosity similar to those of plasma.

(3) The composition from batch to batch should be constant, within narrow and definable limits.

(4) The material should be stable during storage and preferably not require special conditions of temperature.

Negative Qualities.

(1) It must not be toxic, either locally or generally.

(2) It must not induce fever.

(3) It must not induce sensitisation.

(4) It must not be stored for long periods in the tissues.

(5) It must not act as a diuretic. Apart from the absence of specific diuretic properties, this implies that it should not contain large amounts of solutes of low molecular weight.

All these requirements could be met by a solution of non-toxic non-protein material of suitable molecular weight which was slowly metabolised by the tissues. Dextran preparations can fulfil most of these requirements. They are administered primarily to restore the volume of fluid circulating. In this connection its efficiency depends upon its colloid osmotic pressure. This in turn is closely connected with the distribution of molecular weight in the polydisperse mixture. Ideally all the molecules should have molecular weights lying between those of albumin and globulin.

By partial hydrolysis with acid dextran has been prepared free from traces of protein and of approximately the correct molecular weight. The purified polysaccharide has given no indication of antigenicity in numerous experiments with rabbits. The many experiments reported in the literature and the large number of patients transfused in Sweden and Great Britain are evidence of its freedom from toxic and immunogenic action. (A few unfavourable results in America on early Swedish dextran were probably due to incomplete removal of protein material).

The desired range of molecular weights may be attained by fractional precipitation with acetone (40). In this way the larger molecules having no appreciable colloid osmotic pressure and those which are too small to remain in the circulation may be removed.

The available evidence indicates that dextran molecules are considerably elongated in shape (31). This is believed to be associated with the tendency of red cells to aggregate in the form of rouleaux in the presence of dextran and indeed all of the numerous colloids suggested as substitutes for plasma. Opinion differs on the importance to be attached to this factor but there can be no doubt that a plasma substitute which did not have this effect would have a substantial advantage over known substitutes. In all other important respects suitably purified and fractionated dextran satisfactorily fulfils the function of plasma in blood transfusion.

Indeed where a relatively small scale plasma transfusion is needed as in prophylaxis for operations, less severe shock and burns, dextran plasma has some advantages over natural plasma itself. A major advantage is its ease of sterilisation and its excellent keeping properties over a period of years. Natural plasma will always be costly to collect, difficult to sterilise and keep, and will always carry a risk of virus infection. The enormous size of the dextran molecule and the variability of the many structures possible by choice of suitable strains of dextran forming organisms, gives hope that from the dextran-type molecule, we may be able to achieve in size and shape, the ideal plasma substitute. The method of degrading the macromolecule would appear to be of profound importance and recently a new major advance has been made. This consists in the harnessing of supersonic sound waves to degrade the macromolecule (*51, 39*). This physical method is under sensitive control even in the large scale and is proving of unusual importance in the manufacturing and processing of the plasma substitute. The significance of this in saving lives particularly in time of war can well be appreciated.

Several subsidiary applications of dextran in the medical field have arisen and no doubt others will occur. High-molecular degraded dextrans which aggregate red cells are being explored by Dr. E. J. COHN (private communication) for the processing of whole blood to give natural plasma and white cells. PICKUP (*45*) added dextran to defibrinated blood causing the red cells to form rouleaux and sediment rapidly enabling a 95% recovery of lymphocytes to be obtained from the supernatant serum. Lymphocytes are required for transfusion in some circumstances so this observation may have therapeutic significance. ELKES (*13*) used dextran to prepare oil in water type of emulsion with glycerides, which might enable fats to be given intravenously. The artificial chylomicrons were 0,5 to 1,0 μ in diameter and disappeared from the circulation after one hour. GRUBB (*21*) suggested the use of dextran as a medium for demonstrating the agglutinating effect of anti-Rh-antibodies in serological testing.

The ease with which these polysaccharide colloids can be made on a large scale from sucrose suggests that further uses for injection purposes will be found for them.

VIII. Conclusion.

It is hoped that this summary has indicated the significance of dextrans and will stimulate others to do research in this interesting field of polysaccharide chemistry.

References.

1. ANDERSON, C. G., W. N. HAWORTH, H. RAISTRICK and M. STACEY: Polysaccharides Synthesised by Micro-organisms. IV. The Molecular Constitution of Luteose. Biochemic. J. **33**, 272 (1939).

2. AVINERIA-SHAPIRO, S. and S. HESTRIN: The Mechanism of Polysaccharide Production from Sucrose. 2. Biochemic. J. **39**, 167 (1945).

3. BAYLISS: Gum Saline for Injection. British Med. Res. Council Rep. No. 25 (1919).

4. BÉCHAMP, A.: Sur la viscose ou substance gommeuse de la fermentation visqueuse: équation de cette fermentation. C. R. hebd. Séances Acad. Sci. **93**, 78 (1881).

5. BRAUNS, F. E.: The Structure of Dextran. Canad. J. Res. **16**, 73 (1938).
6. BULL, J. P., C. R. RICKETTS, J. R. SQUIRE, W. D'A. MAYCOCK, S. J. L. SPOONER, P. L. MOLLISON and J. C. S. PATERSON: Dextran as a Plasma Substitute. Lancet 134 (1949).
7. CARLSON, W. W. and V. WHITESIDE-CARLSON: Biotin-carbohydrate Inter-relationships in the Metabolism of Leuconostoc. Proc. Soc. exp. Biol. Med. **71**, 416 (1949).
8. CARRUTHERS, A. and E. A. COOPER: Enzyme Formation and Polysaccharide Synthesis by Bacteria. II. Biochemic. J. **30**, 1001 (1936).
9. CIENKOWSKI: In: F. LAFAR, Handbuch der technischen Mykologie, Bd. 1 und 2. 2. Aufl. Jena: G. FISCHER. 1904-1914.
10. DAKER, W. D. and M. STACEY: Investigation of a Polysaccharide Produced from Sucrose by *Beta-Bacterium vermiformé* (WARD-MAYER). Biochemic. J. **32**, 1946 (1938).
11. — — Polysaccharides. XXX. The Polysaccharide Produced from Sucrose by *Beta-Bacterium vermiformé* (WARD-MAYER). J. chem. Soc. (London) **132**, 585 (1939).
12. DAUMICHEN, P.: Über Dextran. Chem. Zbl. **61**, 1000 (1890).
13. ELKES, J.: Fats in Human Nutrition. Nature (London) **164**, 1037 (1949).
14. EVANS, T. H., W. L. HAWKINS and H. HIBBERT: Reactions Relating to Carbo-hydrates and Polysaccharides. LXIV. Antigenicity of Dextran Produced by *Leuconostoc mesenteroides*. J. exp. Medicine **74**, 511 (1941).
15. FAIRHEAD, E. C., M. J. HUNTER and H. HIBBERT: The Structure of Dextran Synthesised by *Leuconostoc dextranicus*. Canad. J. Res. **16**, 151 (1938).
16. FITZGERALD, J. G.: On the Nature of Antigens. Trans. Roy. Soc. Canada, Sect. V **27**, 1 (1933).
17. FOWLER, F. L., I. K. BUCKLAND, F. E. BRAUNS and H. HIBBERT: Studies on Reactions Relating to Carbohydrates and Polysaccharides. LIII. Structure of the Dextran Synthesised by the Action of *Leuconostoc mesenteroides* on Sucrose. Canad. J. Res. **15**, 486 (1937).
18. GAINES, S. and G. L. STAHLY: The Use of *L. mesenteroides* as an Assay Agent for Several Members of the Vitamin B Group. J. Bacteriol. **45**, 35 (1943).
19. — — Growth Requirements of *L. mesenteroides* and its Use as an Assay Agent. J. Bacteriol. **46**, 441 (1943).
20. GRONWALL, A., B. INGELMAN and H. MOSIMANN: Sulphuric Ester with Heparin Activity. Upsala Läkareören. Förh. **50**, 397 (1945).
21. GRUBB, R.: Dextran as a Medium for the Demonstration of Incomplete Anti-Rh-agglutinins. J. Clin. Pathology **2**, 223 (1949).
22. HASSID, W. Z. and M. DOUDOROFF: Enzymatically Synthesised Polysaccharides and Disaccharides. Fortschr. Chem. organ. Naturstoffe **5**, 101 (1948).
23. HEHRE, E. J.: Studies on the Enzymic Synthesis of Dextran from Sucrose. J. biol. Chemistry **163**, 221 (1946).
24. HEHRE, E. J. and D. M. HAMILTON: Bacterial Conversion of Dextrin into a Polysaccharide with the Serological Properties of Dextran. Proc. Soc. exp. Biol. Med. **71**, 336 (1949).
25. HEHRE, E. J., J. M. NEILL, J. Y. SUGG and E. JAFFE: Serological Studies on Sugar. 1. Reactions Between Solutions of Reagent Sucrose and Type II Anti-pneumococcus Serum. J. exp. Medicine **70**, 427 (1939).
26. HEHRE, E. J. and J. Y. SUGG: Serological Reactivity of Dextran Plasma Substitute. Federat. Proc. (Amer. Soc. exp. Biol.) **9**, 383 (1950).
27. HUCKER, G. J. and C. S. PEDERSON: Studies on Coccae XVI Leuconostoc. New York State agric. Exp. Stat., techn. Bull. **167**, 3 (1930).

28. Husemann, E. u. H. Ruska: Versuche zur Sichtbarmachung von Glykogen-molekülen. J. prakt. Chem. **156**, 1 (1940).

29. Ingelman, B.: Enzymatic Breakdown of Dextran. Acta chem. scand. **2**, 803 (1948).

30. — Investigations on Dextran. Upsala Läkarefören. Förh. **54**, 107 (1949).

31. Ingelman, B. and M. S. Halling: Some Physico-chemical Experiments on Fractions of Dextran. Ark. Kemi **1**, 61 (1949).

32. Ingelman, B. and K. Siegbahn: An Electron-Microscopic Study of Dextran Molecules. Ark. Kem., Mineral. Geol. **18 B**, No. 1 (1944).

33. — — Dextran and Levan Molecules Studied with the Electron Microscope. Nature (London) **154**, 237 (1944).

34. Jeanes, A., N. C. Schieltz and C. A. Wilham: Molecular Association in Dextran and in Branched Amylaceous Carbohydrates. J. biol. Chemistry **176**, 617 (1948).

35. Jeanes, A., C. A. Wilham and J. C. Miers: Preparation and Characterization of Dextran from *Leuconostoc mesenteroides*. J. biol. Chemistry **176**, 603 (1948).

36. Joubert: In: F. Lafar, Handbuch der technischen Mykologie, Bd. 1 und 2. 2. Aufl. Jena: G. Fischer. 1904-1914.

37. Levi, I., W. L. Hawkins and H. Hibbert: Studies on Reactions Relating to Carbohydrates and Polysaccharides. LXVI. Structure of the Dextran Synthesized by the Action of *Leuconostoc mesenteroides* on Sucrose. J. Amer. chem. Soc. **64**, 1959 (1942).

38. Liesenberg, C. u. W. Zopf: Über den sogenannten Froschlaichpilz *(Leuconostoc)* des europäischen Rübenzuckers. Centr. prakt. Parasit. **12**, 659 (1892).

39. Lockwood, A. R., F. G. Pautard and A. James: Studies on the Breakdown Product of Dextran Formed by Ultrasonic Vibration. Research **4**, 46 (1951).

39 a. Lockwood, A. R. and G. Swift: Improvements in the Manufacture of Dextran. Brit. Pat. 583378.

40. Lorenz, L., W. d'A. Maycock and C. R. Ricketts: Molecular Composition of Dextran Solutions for Intravenous Use. Nature (London) **165**, 770 (1950).

41. Neill, J. M., J. Y. Sugg, E. J. Hehre and E. Jaffe: Serological Studies on Sugar. II. Reactions of the Antiserums Type 2 Pneumococcus and of *Leuconostoc mesenteroides* with Cane and Beet Sugars and with Cane Juice. Amer. J. Hyg. **34** B, 65 (1941).

42. Nordström, L. and E. Hultin: Dextranase, a New Enzyme from Mold. Svensk kem. Tidskr. **60**, 283 (1948).

43. Pasteur, L.: Sur la fermentation visqueuse et la fermentation butyrique. Bull. Soc. chim. France **1861**, 30.

44. Peat, S., E. Schlüchterer and M. Stacey: Polysaccharides. XXIX. Constitution of the Dextran Produced from Sucrose by *Leuconostoc dextranicum* *(Betacoccus arabinosaceous haemolyticus)*. J. chem. Soc. (London) **1939**, 581.

45. Pickup, D. M.: Extraction of Lymphocytes from Rabbits' Blood. Nature (London) **164**, 959 (1949).

46. Raistrick, H. and W. Rintoul: Studies on the Biochemistry of Microorganisms. Philos. Trans. Roy. Soc. London, Ser. B **220**, 1 (1931).

47. Ricketts, C. R. and K. Walton: (in preparation for press).

47 a. Schiebler, In: F. Lafar, Handbuch der technischen Mykologie. Bd. 1 und 2. 2. Aufl. Jena: G. Fischer. 1904-1914.

48. Stacey, M.: Enzymatic Production of Bacterial Polysaccharides. Nature (London) **149**, 639 (1942).

49. — Unpublished.

50. — Aspects of Immunochemistry. Quart. Rev. chem. Soc. **1**, 179 (1947).

51. — Degradation of Dextran by Ultrasonic Waves. Research **4**, 48 (1951).

52. Stacey, M. and G. Swift: Structure of the Dextran Synthesised from Sucrose by a New Strain of *Betacoccus arabinosaceous*. J. chem. Soc. (London) **1948**, 1555.

53. Stacey, M. and F. R. Youd: A Note on the Dextran Produced from Sucrose by *Betacoccus arabinosaceous haemolyticus*. Biochemic. J. **32**, 1943 (1938).

54. Stahly, G. L. and W. W. Carlson: U. S. Pat. 2344179 (1944).

55. — — U. S. Pat. 2203702; 2203703; 2203704 and 2203705 (1940).

56. — — U. S. Pat. 2229941 (1941).

56 a. Sugar Research Foundation Publications, New York.

57. Swift, G.: Thesis, Univ. Birmingham. 1943. Private communication.

58. Tarr, H. L. A. and H. Hibbert: Studies on Reactions Relating to Carbohydrates and Polysaccharides. XXXVII. The Formation of Dextran by *Leuconostoc mesenteroides*. Canad. J. Res. **5**, 414 (1931).

59. Treghen, von: In: F. Lafar, Handbuch der technischen Mykologie, Bd. 1 und 2. 2. Aufl. Jena: G. Fischer. 1904-1914.

60. Williams, R. C. and R. W. G. Wyckoff: Electron Shadow-micrography of Virus Particles. Proc. Soc. exp. Biol. Med. **58**, 265 (1945).

61. Wolfrom, M. L., L. W. Georges and I. L. Miller: Crystalline Derivatives of Isomaltose. J. Amer. chem. Soc. **71**, 125 (1949).

62. Zozaya, J.: Carbohydrates Adsorbed on Colloids. J. exp. Medicine **55**, 325 (1932).

63. — Immunological Reactions Between Dextran Polysaccharides and Some Bacterial Antisera. J. exp. Medicine **55**, 353 (1932).

(Received, December 28, 1950.)

Sugar Phosphates.

By L. F. LELOIR, Buenos Aires.

With 3 Figures.

Contents.

Introduction.

"Since the original work of Harden and Young on the function of phosphates in alcoholic fermentation, the idea that the formation of phosphoric esters may be an essential stage not only in the biochemical degradation of hexoses but also in the condensation of these sugars to the polysaccharides has taken firm hold of biochemical imagination."

Since Robison and Morgan (*161*) wrote the above paragraph in 1928, biochemical imagination has been busily engaged on the subject and many new facts have been discovered. According to some sceptics, the importance of phosphoric esters in metabolism is due to the ease with which phosphate can be estimated; nevertheless, interest in the subject continually increases. Several new natural esters of carbohydrates have been discovered and many others have been prepared synthetically.

Although this review should be dedicated to natural products, the synthetically obtained esters have also been included because their chemistry is illuminating for the others, and also because it cannot be predicted which of the purely synthetic substances will be found in nature as research proceeds.

Previous reviews of our field have been written by ROBISON and MACFARLANE (159) as well as by COURTOIS (25). For Nucleotides cf. p. 96.

I. Preparation of Sugar Phosphates.

Isolation from Natural Sources.

This work was initiated by HARDEN who discovered hexose-diphosphate in collaboration with YOUNG, and a monophosphate with ROBISON. Several of the monophosphates were later identified by ROBISON.

The usual procedure consisted in allowing air or acetone dried yeast or yeast juice to ferment sugar in the presence of phosphate. Under these conditions phosphate became esterified according to the HARDEN and YOUNG equation:

$$2 \text{ hexose} + 2 \text{ phosphate} \rightarrow \text{hexose-diphosphate} + 2 \text{ ethanol} + 2 \text{ CO}_2.$$

According to the conditions hexose-monophosphate also accumulated in amounts ranging from 20 to 50% or more. In experiments with yeast juice less than half of the ester mixture was hexose-diphosphate. When air or acetone dried yeast was used, the diphosphate yield represented 80—90% of the total esterified phosphate. According to ROBISON and MORGAN (162), the production of the hexose-monophosphate is at the maximum when inorganic phosphate is in excess and a high rate of fermentation is maintained. When the fermentation is continued at low inorganic phosphate concentration, the amount of trehalose monophosphate increases and that of hexose-monophosphate decreases. Finally, after very prolonged fermentation, the trehalose phosphate disappears while hexose-diphosphate accumulates. Separation in two fractions was effected by precipitating the diphosphate with neutral lead acetate or barium ions, followed by precipitation of the monophosphate from the supernatant liquid with basic lead acetate or barium ions in 50% or stronger ethanol [see ROBISON and MORGAN (162)]. The separation of the different monophosphates was then carried out by fractional crystallization of the brucine salts [ROBISON and KING (158)].

Besides the many esters isolated by Robison and his co-workers, Levene et al. were able to prepare 3- and 5-ribose-phosphates by hydrolysis of nucleotides. Later, glucose-1-phosphate was isolated by Cori et al. as a product of the phosphorolysis of glycogen. Galactose-1-phosphate was found by Kosterlitz in the liver of rabbits which had been fed galactose. Ribose-1-phosphate was obtained by phosphorolysis of nucleosides by Kalckar, and glucose-diphosphate was isolated from crude fructose-diphosphate preparations and found to possess the properties of a coenzyme.

Preparation of Synthetic Esters.

The synthesis of carbohydrate phosphoric esters was first effected by phosphorylating free (unprotected) sugars, and the products thus obtained were probably mixtures of several isomers containing the phosphate groups in various positions. In this manner Neuberg and Pollak (*130, 131*) obtained saccharose- and glucose-phosphates. Some success has been attained by the use of partially protected sugars. Thus Levene and Raymond (*98—99*) were able to obtain glucose-6-phosphate from 1,2-isopropylidene-glucose, in which the hydroxyl groups 3, 4, and 6 are free. However, as it occurs with several other substituents, the phosphate group was preferentially introduced into the 6-position.

Glucose-6-phosphate was later prepared using a glucose which was fully acetylated except in position 6. The structure of glucose-6-phosphate was thus proved; however, the possibility of acyl migration has repeatedly induced workers to seek additional proofs of the structure.

Phosphorylation of the Hemiacetalic Hydroxyl. The methods which have been described consist in reacting the acetobromo sugar with silver phosphate or silver benzyl- or phenylphosphate. Depending on the experimental conditions either the α- or the β-derivatives are formed.

The α-form of glucose-1-phosphate was successfully synthesized by Cori, Colowick and Cori (*24*) by reacting trisilver phosphate with acetobromo-glucose in benzene solution. The product, tri-(tetracetyl-glucose-1-)phosphate was then submitted to acid hydrolysis in methanol until about 20% of the organic phosphate became inorganic. The mixture was then made slightly alkaline with barium hydroxide, thus completing the deacetylation and precipitating the glucose-1-phosphate in form of its barium salt. This method has been described in detail by Krahl and Cori (*78*).

Somewhat later Zervas (*189*) made use of silver dibenzyl phosphate and acetobromo-glucose, and obtained a different compound. Wolfrom et al. (*185, 186*) compared the esters of Cori and Zervas and proved that they represent the α- and β-forms of glucose-1-phosphate, respectively. During the acid hydrolysis of the Cori ester, the solution became more levorotatory while the ZERVAS ester solution became more dextrorotatory. The observed final value was the same for both compounds and corresponded to that of free glucose.

Similarly, REITHEL (*152*) found that acetobromo-galactose yields the α-form with trisilver phosphate and the β-form with silver dibenzylphosphate or monosilver phosphate.

POSTERNAK (*145*) described another method for the preparation of the α-1-phosphates which consists of treating acetobromo-aldoses with silver diphenylphosphate. The phenyl groups are then removed by catalytic hydrogenation and the acetyl groups by alkali. The yield is reported to be about five times higher than with the trisilver phosphate procedure.

Since acetobromo-glucose is considered to be the α-anomer, it appears that substitution of the bromine atom by phosphate involves an inversion to form the β-phosphate; and that with most reagents two competing reactions take place: one with inversion and one without it. As in the synthesis of glycosides [ZEMPLÉN (*188*)], the relative yields of the two processes can be modified but not fully controled. A thorough study of the influence of such factors as the temperature, solvent, etc., which favor one or the other reaction type would be desirable. It may be mentioned here that in the case of acetylated glycosides it is possible to convert the β-into the α-form by a treatment with titanium tetrachloride [PACSU (*135—137*)].

The success of the CORI procedure for the preparation of α-glucose-1-phosphate may be not wholly due to the phosphorylation process itself but in part to the subsequent acid hydrolysis which leads to the accumulation of the more stable α-compound.

Phosphorylation of Alcoholic Hydroxyls. In the past the most widely used reagent was phosphoryl chloride (phosphorus oxychloride), first proposed by NEUBERG and POLLAK (*130*), then by E. FISCHER (*41*). LEVENE and his co-workers (*86—108*) prepared several carbohydrate phosphates by reacting phosphoryl chloride with the suitably protected sugar derivatives in pyridine, at low temperature.

More recently, other reagents have come into use, such as diphenyl and dibenzyl chlorophosphonate, which have the advantage of reacting with only a single hydroxyl group. The protecting phenyl or benzyl groups can be easily removed by hydrogenation.

Diphenyl chlorophosphonate, which was introduced by BRIGL and MÜLLER (*11*), is prepared from phenol and phosphorylchloride, and is used mostly in pyridine solution. A slight modification of the preparative procedure has been described recently by BAER (*6*).

Dibenzyl chlorophosphonate, which has been used extensively by TODD and co-workers (*3, 4*), is obtained by the action of chlorine or sulfuryl chloride [Atherton, Howard and Todd (2)] on dibenzyl hydrogen phosphite ($C_6H_5 \cdot CH_2O)_2P \cdot OH$) which in turn is prepared from phosphorus trichloride, benzyl alcohol and dimethyl aniline [Atherton, Openshaw and Todd (3)]. The use of this reagent is not limited to the preparation of simple phosphoric esters. Thus, the synthesis of pyrophosphate derivatives has been achieved according to the following equations:

$$ROH + (C_6H_5CH_2O)_2POCl \longrightarrow RO \cdot PO(OCH_2C_6H_5)_2 + HCl \qquad \text{(I.)}$$

$$RO \cdot PO(OCH_2C_6H_5)_2 \longrightarrow RO \cdot PO(OCH_2C_6H_5) \cdot OH + HO \cdot CH_2C_6H_5 \qquad \text{(II.)}$$

$$RO \cdot PO(O \cdot CH_2C_6H_5) \cdot OAg + (C_6H_5 \cdot CH_2 \cdot O)_2POCl \longrightarrow$$
$$\longrightarrow RO \cdot PO(OCH_2C_6H_5)-O-PO(OCH_2C_6H_5)_2 + AgCl \qquad \text{(III.)}$$

The mono-debenzylation (II) was carried out using a tertiary base such as 4-methyl morpholine [Baddiley, Clark, Michalski and Todd (4)]. Using this procedure it has also been possible to prepare adenosine triphosphate [Baddiley, Michelson and Todd (5)].

Catechol chloro-phosphonate has also been recommended as a phosphorylating agent which is easy to prepare (from catechol and phosphoryl chloride), and from which the catechol can be removed by treatment with water [Reich (151)].

Phosphorylation of Sugar Anhydrides. This method was first used by Bailly (7) who succeeded in phosphorylating 3-hydroxypropylene oxide (glycide), $CH_2OH \cdot \overset{\lceil O \rceil}{CH \cdot CH_2}$, by treatment with sodium phosphate in aqueous solution. Ethylene oxide (3) and propylene oxide (3, 82) have been likewise esterified to ethylene glycol phosphate and β-propanediol phosphate, respectively:

$$\overset{\lceil O \rceil}{CH_2 \cdot CH_2} + K_2HPO_4 \longrightarrow CH_2 \cdot OH \cdot CH_2OPO_3K_2 \qquad \text{(IV.)}$$

$$\overset{\lceil O \rceil}{CH_2 \cdot CH \cdot CH_3} + K_2HPO_4 \longrightarrow CH_2OH \cdot CHOPO_3K_2 \cdot CH_3 \qquad \text{(V.)}$$

Similarly Lampson and Lardy (81) treated 1,2-isopropylidene-5,6-anhydro-glucofuranose (VI) in water with K_2HPO_4 and, after hydrolytic elimination of the acetone group with HBr, obtained glucose-6-phosphate (VIII). While the yield was lower than with other methods, the authors recommend this procedure in the special case when it is desired to introduce labelled phosphate, because it avoids the preparation of special phosphorylating agents.

Phosphorylations Followed by Group Migration. Levene and Raymond (105) tried to prepare xylose-3-phosphate by the phosphorylation of 1,2-isopropylidene-5-benzoyl-xylose, which contains a free hydroxyl in the 3-position. After the reaction with phosphoryl chloride and hydrolysis of the protecting groups, the product obtained was not xylose-3-phosphate but xylose-5-phosphate. Phosphorylation of 1,2-isopropylidene-5-benzoylcarboxy-xylose also led to xylose-5-phosphate.

HC
  \
   O
     C(CH$_3$)$_2$
HC—O   O
  |
HOCH
  |
HC
  |
HC
  \
   O
H$_2$C
          (VI.)
1,2-Isopropylidene-5,6-anhydro-
        glucofuranose.

$\xrightarrow{\text{HPO}_4^-}$

HC
  \
   O
     C(CH$_3$)$_2$
HC—O   O
  |
HOCH
  |
HC
  |
HCOH
  |
H$_2$COPO$_3^=$
        (VII.)

$\xrightarrow{\text{HBr}}$

HOHC
  |
HCOH
  |
HOCH   O
  |
HCOH
  |
HC
  |
H$_2$COPO$_3^=$
        (VIII.)
Glucose-6-phosphate.

It appears that during the phosphorylation of ribosides similar changes may take place; thus, starting with sugar derivatives believed to have only the position 2 free, the resulting products have been the 5-phosphates [BROWN, HAYNES and TODD (*13*)].

II. Determination of the Structure of Sugar Phosphates.

The problems arising in the clarification of the structure of phosphate esters include the identification of the carbohydrate component, the location of the phosphate group, the determination of the ring type, and the assignment of the α- or β-configuration.

The Identification of the Carbohydrate Component is particularly easy for those esters which are acid-labile. For those which are more resistant, such as aldose-6-phosphates, the sugar component may be partly altered during hydrolysis, although in general no great difficulties have been encountered. The classical methods of sugar identification have now been supplemented and partly replaced by paper chromatography with the great advantage of requiring only micro amounts.

Changes in the configuration during the simple acid hydrolysis of phosphate esters have not been observed so far. This point will be discussed in more detail in connection with the hydrolysis of these esters (p. 59).

The Position of the Phosphate Group in aldose-1-phosphates is recognized by the lack of their reducing power, the stability in alkali, and the easy acid hydrolysis with the appearance of reducing power.

In contrast, the 1-ketose esters are reducing compounds; they are destroyed by alkali and hydrolyzed by acids somewhat more slowly than the aldose-1-phosphates; furthermore, they lose the phosphate group on osazone formation.

The only known aldose-2-phosphate is that of glucose. It is reducing but less so than the free sugar; it is destroyed by alkali, and is less acid labile than glucose-1-phosphate.

The properties of glucose-3-phosphate are probably similar. During osazone formation, the phosphate group is hydrolyzed and the process is accompanied by the formation of a 3,6-anhydro ring and inversion. Therefore, the osazone obtained does not correspond to that of the original sugar.

For the determination of the structure of the 4-, 5-, and 6-phosphates, some methods based on the ring types of the sugars or of the corresponding aldonic acids have been used. A 4-ester is only compatible with a pyranose ring, and a 5-ester with a furanose ring, while a 6-ester may contain either of these two ring types.

A method which has been used in order to detect the presence of a furanose or pyranose ring consists of preparing the corresponding methyl glycosides by treatment with methanol and hydrogen chloride. The furanosides can then be differentiated from the pyranosides by the greater acid sensitivity of the former type. Such a method has been applied to glucose-6-phosphate [KING, McLAUGHLIN and MORGAN (69)]. Both a furanoside and a pyranoside were formed, thus proving that the positions 4 and 5 were free.

Another method which is based on the study of the lactones of the corresponding aldonic acids has also been useful. As well known, aldonic acids can be easily prepared from sugars. The free acids form lactones, and the rate of the latter conversion can be easily followed polarimetrically. Usually a γ-lactone is obtained containing a five-membered ring, or a δ-lactone with a 6-membered ring. The initial speed of the δ-lactone formation is mostly about 8 times higher than that of the γ-lactone [LEVENE and SIMMS (107)]. Some useful information on the phosphoric esters has been obtained by this means.

At the time when the structures were not yet clarified, LEVENE and RAYMOND (98, 99) prepared the aldonic acids of some synthetic and naturally occurring glucose-6-phosphates, and examined the rate of lactone formation. The curves were identical for the two samples and showed also that two lactones were formed in each case. LEVENE and RAYMOND concluded that the positions 4 and 5 must be free. Therefore, the phosphate group could not occupy either of these two positions. Moreover.

it was observed that the lactone formation curve for the aldonic acid
of glucose-3-phosphate was quite different from the corresponding 6-phos-
phate curves.

Periodate Oxidation. A procedure which usually gives clearcut results
consists in the oxidation with periodate which has been successfully
applied in the structural determination of many carbohydrates and also
of some phosphoric esters.

Compounds having two hydroxyl groups attached to adjacent carbon
atoms are oxidized as follows:

$$R_1 \cdot CHOH \cdot CHOH \cdot R_2 + HIO_4 \longrightarrow R_1CHO + R_2CHO + H_2O + HIO_3 \quad \text{(IX.)}$$

If more than two hydroxyls are adjacent, the middle ones yield formic
acid:

$$R_1 \cdot CHOH \cdot CHOH \cdot CHOH \cdot R_2 + 2 HIO_4 \longrightarrow R_1CHO + R_2 \cdot CHO +$$
$$+ HCOOH + H_2O + 2 HIO_3 \quad \text{(X.)}$$

When one of the hydroxyls is contained in a primary alcohol group,
formaldehyde appears:

$$R \cdot CHOH \cdot CH_2OH + HIO_4 \longrightarrow R \cdot CHO + HCHO + H_2O + HIO_3 \quad \text{(XI).}$$

Substances containing carbonyl or amino groups instead of hydroxyls,
are also oxidized by periodate; for more details cf. COURTOIS (*27*), JACK-
SON (*61*), PIGMAN and GOEPP (*142*).

In the periodate oxidation of glucose-1-phosphate 2 moles of periodate
were used and 1 mol. of formic acid appeared but no formaldehyde as
would be expected [WOLFROM and PLETCHER (*185*)]:

Glucose-1-phosphate.

(XII.)

Fructose-1,6-diphosphate used 3 moles of periodate, giving 2 moles
of formic acid, 1 mol. of phosphoglycolic aldehyde, and 1 mol. of phosphoglycolic acid [COURTOIS (*26*)]:

$$
\begin{array}{c}
\mathrm{CH_2OPO_3H_2} \\
| \\
\mathrm{CO} \\
| \\
\mathrm{HO\cdot CH} \\
\vdots \\
\mathrm{HCOH} \\
| \\
\mathrm{HCOH} \\
| \\
\mathrm{CH_2OPO_3H_2} \\
\text{Fructose-1,6-diphosphate.}
\end{array}
\;+\; 3\,\mathrm{HIO_4} \;\longrightarrow\;
\begin{array}{c}
\mathrm{CH_2OPO_3H_2} \\
| \\
\mathrm{COOH} \\
| \\
\mathrm{CHO} \\
| \\
\mathrm{CH_2OPO_3H_2}
\end{array}
\;+\; 2\,\mathrm{HCOOH} + 3\,\mathrm{HIO_3}
$$

(XIII.)

The periodate oxidation of fructose-6-phosphate [Courtois and Ramet (*29*, *147*)] occurs as shown in the following equation:

$$
\begin{array}{c}
\mathrm{CH_2OH} \\
| \\
\mathrm{CO} \\
| \\
\mathrm{(CHOH)_3} \\
| \\
\mathrm{CH_2OPO_3H_2} \\
\text{Fructose-6-phosphate.}
\end{array}
\;+\; 3\,\mathrm{IO_4H} \;\longrightarrow\;
\begin{array}{c}
\mathrm{CH_2OH} \\
| \\
\mathrm{COOH} \\
| \\
\mathrm{CHO} \\
| \\
\mathrm{CH_2OPO_3H_2}
\end{array}
\;+\; 2\,\mathrm{HCOOH} + 3\,\mathrm{IO_3H}
$$

(XIV.)

Since formaldehyde can originate only from primary alcoholic groups, it follows that the method can be used to detect substituents in this position. Thus, periodate oxidation of glucose-6-phosphate or ribose-5-phosphate yields no formaldehyde while that of ribose-3-phosphate does [Euler, Karrer and Becker (*39*)].

The Structure of Sugar Phosphates in Solution. Solutions of reducing sugars are considered to be equilibrium mixtures of the open chain form with α- and β-furanose and pyranose forms. When the formation of the pyranose ring is not impeded by a substituent in the position 5, this is the predominant form in solution.

The pyranose ring may assume two chair and six boat forms as shown in Figure 1. Several lines of evidence [Reeves (*150*)] suggest that boat forms would be unstable, and that most glycosides in solution exist in the C 1 chair form. The conformation of α- and β-methylglucosides is shown in Figure 2. The other chair form (1 C of Reeves) is also presented in the Figure in order to demonstrate how the substituents which are in equatorial position in C 1 become polar in 1 C. The equatorial substituents [lying position of Hassel and Ottar (*53*)] are essentially in the plane of the ring, while the polar substituents (erected position of Hassel and Ottar) project perpendicularly to the plane of the ring.

No such studies have yet been carried outh with sugar phosphates but as shown in Table 1 (p. 58) the molecular rotation of aldose-1-phosphates is very similar to that of the corresponding glycosides. This may indicate that their configuration is the same.

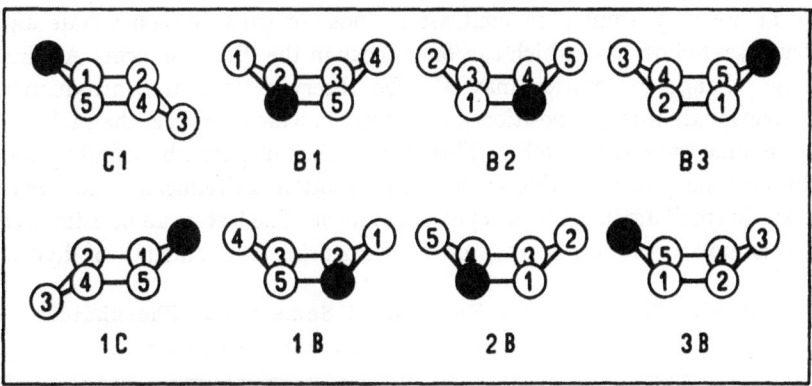

Fig. 1. The conformation of the eight pyranose strainless rings. Dark circles represent oxygen atoms.
[REEVES (150).]

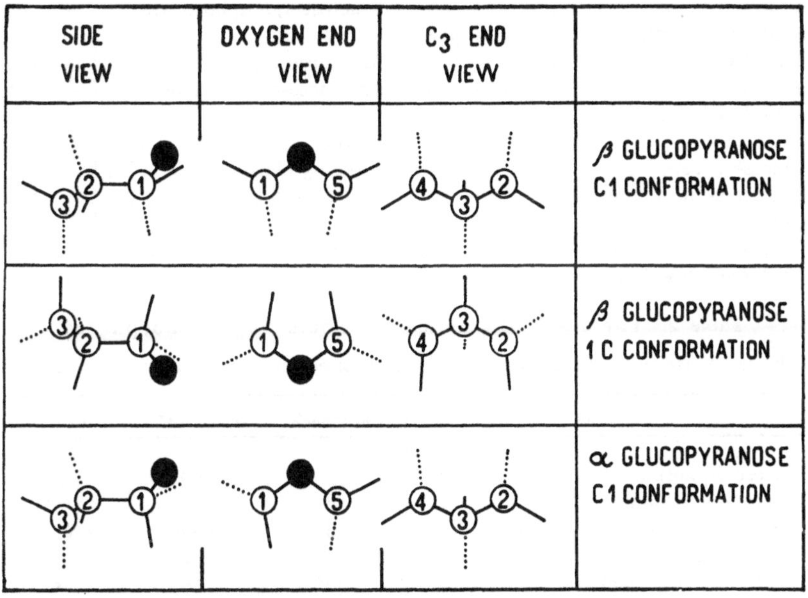

SIDE VIEW	OXYGEN END VIEW	C₃ END VIEW	
			β GLUCOPYRANOSE C1 CONFORMATION
			β GLUCOPYRANOSE 1 C CONFORMATION
			α GLUCOPYRANOSE C1 CONFORMATION

Fig. 2. Conformation of the ring in glucopyranosides. Oxygen atoms in black. Carbon atoms are represented as empty circles with the corresponding number. CO and $C_{(5)}$—$C_{(6)}$ valencies: full lines; CH valencies: dotted lines.

    With reference to the other glucose phosphates, Table 1 shows that their molecular rotations lie all around the value for the equilibrium mixture of α- and β-glucose. The observation that the value for glucose-5-phosphate, which contains a furanose ring, is similar to the equilibrium value mentioned is rather surprising, since a change in the ring type would be expected to cause a large shift in rotation. For instance, the values for free fructose and fructose-1-phosphate which may assume the pyranose form are very similar, in contrast to those of fructose-6-phosphate and fructose-diphosphate which can only occur in the furanose form. A fairly large change in rotation may also be observed upon the introduction of phosphate into the position 5 of arabinose which compels the molecule to assume the furanose form. This change is smaller for ribose and xylose. Besides the predominating cyclic form, solutions of reducing sugars may contain small amounts of the open chain form. The latter can be estimated either polarographically (*14*) or by measuring the amount of hydro-

Table 1. The Molecular Rotation of Some Sugar Phosphates.

(The figures denote $[\alpha]_D \times$ molecular weight $\times$ 0,1.)

α-Glucose-1-phosphate	+ 2960	Fructose	— 1600
α-Methyl glucoside	+ 3040	Fructose-1-phosphate	— 1540
α-Mannose-1-phosphate	+ 1400	Fructose-6-phosphate	+ 142
α-Methyl mannoside........	+ 1530	Fructose-1,6-diphosphate ...	+ 195
α-Galactose-1-phosphate	+ 3640		
α-Methyl galactoside	+ 3800	Arabinose	— 1570
α-Xylose-1-phosphate	+ 2400	Arabinose-5-phosphate	— 690
α-Methyl xyloside	+ 2540		
		Ribose	— 300
Glucose	+ 950	Ribose-3-phosphate	— 265
Glucose-2-phosphate	+ 505	Ribose-5-phosphate	+ 182
Glucose-3-phosphate	+ 1040		
Glucose-5-phosphate	+ 590	Xylose	+ 280
Glucose-6-phosphate	+ 790	Xylose-5-phosphate........	+ 88

Table 2. The Reducing Power of Some Sugar Phosphates.

(The values are given in per cent of those valid for equimolecular amounts of glucose.)

Compound	Ferricyanide*	Hypoiodite**	Reference
Glucose-6-phosphate	78	100	(*157*)
Fructose-6-phosphate	78	4	(*157*)
Mannose-6-phosphate	78	60—70	(*156, 62*)
Glucose-1-phosphate	0	0	—
Fructose-1,6-diphosphate	40***	7	(*157*)
Glucose-1,6-diphosphate..............	0	0	—
Trehalose-monophosphate	0	0	(*157*)
Fructose-1-phosphate	57	2	(*176*)

* Hagedorn and Jensen procedure with the addition of 0,5 ml. of 0,5 N-NaOH (*162*).

** Method as described by Macleod and Robison (*112*).

*** With the method of Ost (*129*) it gives 48% of the value obtained for fructose.

cyanic acid which combines rapidly with the sugar (*109*). The amount of this form is of the order of 0,1 per cent. Determinations by means of the hydrocyanic acid method were carried out by STEPANOV and STEPANENKO with solutions of glucose-6-phosphate (*168*) and fructose mono- and diphosphate (*169*). It was found that the phosphoryl group greatly increases the amount of hydrocyanic acid which combines rapidly with these compounds.

By generalization of what is known on the behavior of pyranosides in solution [REEVES (*150*); HASSEL and OTTAR (*53*)] it may be expected that of all the possible forms (boats and chairs) phosphoric esters will exist predominantly in a pyranose chair form with the hydroxyl and phosphate groups extending equatorially out of the ring.

III. Hydrolysis of Phosphoric Esters.

Measurements of the rate of hydrolysis of phosphoric esters have been carried out for analyzing mixtures, as a test of homogeneity and as a means of identification. LOHMANN (*110*) initiated this work by measuring the rate of hydrolysis in 1 N-acid at 100° of the NEUBERG ester which is obtained by partial hydrolysis of fructose-diphosphate, and of the EMBDEN ester which is a mixture of fructose- and glucose-6-phosphate. He found that neither hydrolysis followed a first order equation. From the curves obtained LOHMANN could calculate that the NEUBERG ester contained 70% and the EMBDEN ester 30% of a more acid-labile component which was the fructose phosphate. He also found that the saccharose-phosphate prepared by NEUBERG and POLLAK (*131*) did not behave as a homogeneous compound. Useful as this test of purity is, it may fail in some instances. Thus, PATWARDHAN (*140*) reports that phosphogluconic acid started to hydrolyze with $K = 0,27 \times 10^{-3*}$ in normal acid at 100° while towards the end of the hydrolytic process the constant changed to $K = 0,15 \times 10^{-3}$ Since, in acid solution, phosphogluconic acid forms a lactone, he concluded that the final rate was that of the lactone. The lactones were therefore prepared and submitted to hydrolysis. The K value were, $0,12 \times 10^{-3}$ for both the γ- and δ-lactone.

Changes in the constant during hydrolysis were also observed by FARRAR (*40*) using β-hexahydrobenzylglucoside-2-phosphate which started to hydrolyse with $K = 0,76 \times 10^{-3}$ in 0,1 N acid at 100°. After 30 minutes this value increased to $1,59 \times 10^{-3}$. From this it was inferred that the glucoside is more slowly hydrolyzed than the glucose-2-phosphate to which it gives rise.

---

* See the footnote following Table 3 for the units of the hydrolysis constant (p. 61).

Acid Hydrolysis of Aldose-1-phosphates. The phosphate group, when located at a hemiacetal hydroxyl is like the glycosides in being easily hydrolyzed in acid but resistant in alkaline medium. Also like the glycosides, the β-anomers are usually more acid-labile than the α-forms. For instance, in $1\,N$ acid at $33°$ the hydrolysis constants are: 5×10^{-3} for the α-form of glucose-1-phosphate, and 15×10^{-3} for the corresponding β-form. For glucose-1,6-diphosphate the K values at $30°$ in $1\,N$ acid are: $0,78 \times 10^{-3}$ for the α- and $3,15 \times 10^{-3}$ for the β-form [Posternak (144)].

The introduction of another phosphate group into the position 6 of glucose stabilizes the phosphate in position 1. Thus, glucose-1-phosphate is hydrolyzed about four times more rapidly than glucose-1,6-diphosphate, in $0,25\,N$ acid at $25°$ [Cardini et al. (18)].

A study of the mechanism of hydrolysis of aldose-1-phosphates was carried out by using labelled oxygen [Cohn (20)]. It was found that when glucose-1-phosphate was hydrolyzed in acid and H_2O^{18}, no labelled oxygen appeared in the inorganic phosphate. Evidently, the hydrolysis proceeded through a cleavage of the C—O bond. Similar results were obtained on studying the phosphorylase reactions, while in phosphatase catalyzed conversion the cleavage occurred at the O—P bond.

Acid Hydrolysis of the Phosphates at the Alcoholic Groups. Removal of the phosphate group from a primary alcohol group with acid requires a fairly drastic treatment which usually leads to a partial destruction of the sugar. Thus, in order to obtain 80% of free mannose starting from its 6-phosphate, Robison (156) kept the ester in $0,1\,N$-acid at $100°$ for 93 hours. The K values for the different esters appear in Table 3.

It is rather surprising that for mannose- or glucose-6-phosphate the rate of hydrolysis is about the same for the ester at its own acidity as in $1\,N$-acid solution. This is not valid, however for fructose-6-phosphate, where the hydrolysis constant increases from $0,4 \times 10^{-3}$ to $4,36 \times 10^{-3}$ upon the addition of acid to attain normality. Robison (156) suggested that the acid-resistant form contains the pyranose ring and that this form is more abundant in N-acid solution. The fructose ester, which cannot exist in pyranose form, is thus hydrolyzed more easily. Robison and King (158) further pointed out that the phospho-osazone common to all these esters and which cannot exist in ring form, is hydrolyzed even more rapidly than is fructose-6-phosphate. Robison further suggested that the changes in the specific rotation of glucose-6-phosphate with p_H might be due to a change in the amount of the pyranose form present (158).

The available data do not allow general conclusions concerning the dependence of the rate of hydrolysis on the position of the phosphate group. For ribose-phosphates the 3-ester has been found to be hydrolyzed five to nine times faster than the 5-isomer.

Table 3. Observed Hydrolysis Constants of Some Sugar Phosphates.

Compound	Normality of acid	Temperature	$K \times 10^3$ *	Reference
Ribose-1-phosphate	0,5	25	1200	(66)
Ribose-3-phosphate	0,01	100	~ 1,7	(108)
,, ...	0,25	100	~ 4,5	(1)
Ribose-5-phosphate	0,01	100	~ 0,3	(108)
,, ...	0,25	100	~ 0,5	(1)
Xylose-1-phosphate	0,1	36	6,21	(116)
Xylose-5-phosphate	1	100	3–4	(104)
α-D-Glucose-1-phosphate	0,1	36	4,36	(116)
,, ...	0,25	37	1,30	(24)
,, ...	1	33	5,0	(186)
β-D-Glucose-1-phosphate.........	1	33	15,0	(186)
D-Glucose-2-phosphate	0,1	100	2,18	(40)
2-Phosphogluconic acid	1	100	0,3	(147, 31)
Glucose-6-phosphate............	0,1	100	0,13	(158)
,, ...	OA**	100	0,167	(158)
,, ...	1	100	0,23	(156)
,, ...	OA**	100	0,28	(156)
6-Phosphogluconic acid	1	100	0,26–0,15	(140)
α-Glucose-1,6-diphosphate	0,25	37	a) 0,31	(18)
,, ...	1	30	a) 0,78	(144)
β-Glucose-1,6-diphosphate........	1	30	a) 3,15	(144)
Fructose-1-phosphate............	OA**	100	0,62	(176)
,, ...	1	100	70	(176)
,, ...	0,1	100	9	(176)
Fructose-6-phosphate............	OA**	100	0,4–0,5	(156)
,, ...	1	100	4,36	(156)
Fructose-1,6-diphosphate.........	1	100	a) 52 / b) 4,2	(113)
α-Galactose-1-phosphate	0,25	25	0,90	(75)
,, ...	0,25	37	5,9	(75)
β-Galactose-1-phosphate	0,25	37	5,6	(152)
Mannose-6-phosphate............	0,1	100	0,13	(156)
,, ...	1	100	0,29	(156)
,, ...	OA**	100	0,28	(156)
6-Phosphomannonic acid	1	100	0,199–0,131	(140)
6-Phosphomannonic acid γ-lactone	1	100	0,12	(140)
L-Sorbose-6-phosphate...........	1	100	4,8	(115)
Maltose-1-phosphate.............	0,1	36	3,21	(116)
Ketoheptose-monophosphate	1	100	4	(159)

* The constants are calculated with the formula $K = \dfrac{1}{t} \log_{10} \dfrac{a}{a-x}$, or more usually, $K = \dfrac{1}{t_2 - t_1} \log_{10} \dfrac{a-x_1}{a-x_2}$. The time is in minutes, and a is the initial concentration of the substance. The time for 50% hydrolysis, $t_{1/2} = \dfrac{0,30}{K}$, the time for 98% hydrolysis, $t = \dfrac{1,7}{K}$.

** OA means that the free acid was heated at its own acidity.

Table 4. The Specific Rotation of Some Sugar Phosphates.

(W = water.)

Compound	Salt	Solvent	Light	$[\alpha]$	Reference
D-Ribose-3-phosphate......	Na$_2$	W	D	− 9,7	(90)
,, ...	Na$_2$	0,5 sat. boric acid	—	+ 38	(90)
D-Ribose-5-phosphate......	free acid	W	D	+ 16,5	(108)
,, ...	Ba	W	D	+ 5	(122)
D-Arabinose-5-phosphate ...	Ba	W	D	− 18,8	(86)
,, ...	Brucine	50% pyridine	D	− 48,6	(86)
D-Xylose-1-phosphate......	Ba	W	D	+ 65	(116)
,, ...	K$_2$	W	D	+ 76	(116)
D-Xylose-5-phosphate......	Na$_2$	W	D	+ 3,2	(104)
,, ...	Na$_2$	0,5 sat. borax	D	+ 4	(104)
α-D-Glucose-1-phosphate ...	free acid	W	D	+ 120	(24)
,, ...	K$_2$	W	D	+ 78,5	(185, 50)
,, ...	K$_2$	W	5461	+ 90	(185)
,, ...	Ba	W	D	+ 75	(24)
,, ...	Brucine	W	D	+ 0,5	(185)
β-D-Glucose-1-phosphate ...	Brucine	W	D	− 20	(185)
α-L-Glucose-1-phosphate ...	Ba	W	D	− 73,2	(146)
,, ...	K$_2$	W	D	− 78,2	(146)
D-Glucose-2-phosphate.....	K$_2$	W	D	+ 15	(40)
,, ...	K$_2$	0,1 N sulfuric acid	D	+ 35	(40)
Glucose-3-phosphate	free acid	W	5461	+ 39	(64)
,, ...	free acid	W	D	+ 39,5	(98)
,, ...	Brucine	50% pyridine	D	− 14,5	(99)
,, ...	Ba	W	D	+ 26,5	(98)
,, ...	Ba	W	5461	+ 27	(64)
Glucose-4-phosphate	Brucine	Pyridine	D	− 45,3	(148)
,, ...	Brucine	20% ethanol	D	− 9,8	—
Glucose-5-phosphate	Ba	—	D	+ 15	(65)
Glucose-6-phosphate	free acid	W	D	+ 35,1	(158)
,, ...	free acid	W	5461	+ 41,4	(158)
,, ...	Ba	W	D	+ 18	(158)
,, ...	Ba	W	5461	+ 21,2	(158)
,, ...	K$_2$	W	D	+ 21,2	(83)
Phosphogluconic acid	Ba	W	5461	− 1,5	(158)
,, ,, ...	free acid	W	5461	+ 0,2	(158)
Phosphogluconic lactone ...	—	W	5461	+ 21	(158)
α-Glucose-1,6-diphosphate ..	free acid	W	D	+ 83	(144)
β-Glucose-1,6-diphosphate ..	free acid	W	D	− 19	(144)
Fructose-1-phosphate	free acid	W	5461	− 64,2	(176)
,, ...	Ba	W	5461	− 39	(176)
,, ...	Brucine	W	5461	− 52,1	(176)
Fructose-6-phosphate	Ba	W	D	+ 3,6	(129)
Fructose-1,6-diphosphate ...	free acid	W	D	+ 4,1	(129)
α-Galactose-1-phosphate....	free acid	W	D	+ 148,5	(75)
,, ...	K$_2$	W	D	+ 108	(75)

Compound	Salt	Solvent	Light	[α]	Reference
α-Galactose-1-phosphate....	Ba	W	5461	+ 113	(75)
,, ...	Ba	W	D	+ 92	(75)
β-Galactose-1-phosphate....	Ba	W	D	+ 31,3	(152)
Mannose-1-phosphate	free acid	W	D	+ 58	(21)
,, ...	Ba	W	D	+ 36	—
Mannose-6-phosphate	free acid	W	5461	+ 15,1	(156)
,, ...	Ba	W	5461	+ 3,5	—
Phosphomannonic γ-lactone	—	W	5461	+ 54,1	(140)
Phosphomannonic δ-lactone	—	W	5461	+ 60,6	(140)
D-Tagatose-6-phosphate	Ba	W	D	+ 5,6	(177)
L-Sorbose-1-phosphate	K₂	W	D	— 16,5	(115)
L-Sorbose-6-phosphate	Ba	W	D	— 12,0	(115)
Trehalose-monophosphate ..	free acid	W	5461	+ 185	(161)
,, ..	Ba	W	5461	+ 132	(161)
,, ..	Brucine	W	5461	+ 31	(161)
Ketoheptose-monophosphate	Ba	W	5461	+ 8	(160)

Glucose-2-phosphate is much more acid -labile than the 6-phosphate, and FARRAR (40) suggests that its hydrolysis may occur after a migration of the phosphate group from position 2 to 1.

Alkaline Hydrolysis. Data referring to this point are somewhat scanty. Aldose-1-phosphates are very resistant to alkali, while the reducing esters are rapidly decomposed. Fifty per cent of glucose-2-phosphate is hydrolyzed in 97 minutes in 0,1 N-alkali at 100° [FARRAR (40)]. Glucose-6-phosphate is hydrolyzed to the extent of 60% in 0,2 N-alkali, at 100°, in 3 minutes. Under the same conditions fructose-diphosphate is quantitatively decomposed [KIESSLING (68)]. Observation on the hydrolysis of some substituted sugar phosphates were reported by FARRAR (40).

Inversion During Hydrolysis. A case of inversion during the hydrolysis of the phosphate group was observed by LEVENE, RAYMOND and WALTI (106). When fructose-3-phosphate was treated with phenylhydrazine and acetic acid, the phosphate was split off and the osazone formed was not glucosazone but the 3,6-anhydro-allosazone, i. e. the configuration at $C_{(3)}$ had become inverted.

Glucose-3-phosphate also yields 3,6-anhydro-allosazone on treatment with phenylhydrazine [RAYMOND and LEVENE (149)]; however, the hydrolysis of glucose-3-phosphate with phosphatase and subsequent osazone formation gave glucosazone (98) and not allosazone as would have been expected if inversion had taken place.

The alkaline hydrolysis of a glucose-3-phosphate with protected reducing group was studied by PERCIVAL and PERCIVAL (141) and no inversion or anhydride formation was observed . Similar results were obtained with a glucose-6-phosphate derivative.

The hydrolysis of carboxylic esters has never been found to produce inversion. This is understandable because the bond which breaks is the one marked A. Inversion would be exspected only in the case that bond B were involved.

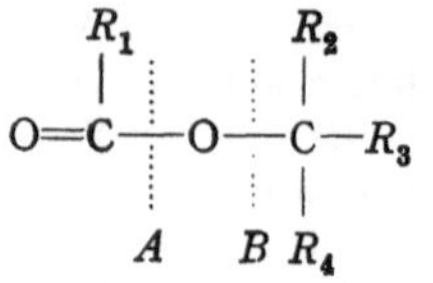

$$O=C \overset{R_1}{\underset{A}{|}} O \overset{}{\underset{B}{|}} \overset{R_2}{\underset{R_4}{\overset{|}{C}}} R_3$$

The hydrolysis of trimethyl phosphate has been studied by Blumenthal and Herbert (10) using water enriched with heavy oxygen as tracer. Their results showed that during hydrolysis in alkaline solution the $P—OCH_3$ bond was broken but that in acid or neutral solution the $O—CH_3$ bond was split preferentially.

IV. The Acid Strength of Sugar Phosphates.

Some pertinent data appear below in Table 5.

The esters are stronger acids than free phosphoric acid, thus both pK'_1 and pK'_2 have smaller values. The factors which may produce this effect were discussed by Kumler and Eiler (79).

Wolfrom and Pletcher (185) have pointed out that the third dissociation constant of phosphoric acid ($pK'_3 = 11{,}74$) (12) is of the same

Table 5. Dissociation Constants of Some Sugar Phosphates.*

Compound	pK_1'	pK_2'	Reference
Phosphoric acid...............	1,95–2,00	6,83–6,93	(179, 12, 120, 24)
Xylose-1-phosphate............	1,25	6,15	(116)
Glucose-1-phosphate	1,10	6,13	(24)
Glucose-3-phosphate	0,84	5,67	(117)
Glucose-6-phosphate	0,94	6,11	(117)
Fructose-6-phosphate	0,97	6,11	(117)
Fructose-1,6-diphosphate	1,48	6,29**	(120)
Galactose-1-phosphate	1,00	6,17	(76)
Maltose-1-phosphate	1,52	5,89	(116)

* The values are those calculated with the formula: $K_1' = \dfrac{H^+ [B + H^+]}{C - [B + H^+]}$
and $pK_2' = pH - \log \dfrac{B}{C-B}$ [Van Slyke (179), Hastings and Van Slyke (54)], where H^+ represents the hydrogen ion concentration, B the amount of alkali added to the acid and C the concentration of the acid.

** The pK values for fructose-diphosphate are average values of the constants of the two phosphate groups. Meyerhof and Lohmann (117) calculated the values of each of the secondary dissociation constants as $pK_{2a}' = 6{,}1$, and $pK_{2b}' = 6{,}5$.

order of magnitude as that of polyhydric alcohols. Therefore, the aldose-1-phosphates may be considered to resemble an acetal or glycoside structure more than the ester of a strong acid.

V. Methods of Estimation and Identification.

In the successful separation of ester mixtures carried out by ROBISON and co-workers (p. 49), each fraction was analyzed for phosphate and for reducing power toward ferricyanide and hypoiodite. The ketose content (SELIWANOFF reaction) and the specific rotation were also measured. In this manner a fair estimate of the composition of the mixture was made.

In the estimation of reducing power with ferricyanide ROBISON obtained more reproducible results by adding 0,5 ml. of 0,5 N-NaOH to the usual HAGEDORN and JENSEN ferricyanide mixture. For the hypoiodite titration the use of carbonate is recommended [MACLEOD and ROBISON (112)]. The values obtained with these methods appear in Table 2 (p. 58). The figures obtained using the BERTRAND and the WILLSTÄTTER-SCHUDEL methods are given by MEYERHOF and LOHMANN (117).

The difference in rate of hydrolysis of the various esters provides in some instances a method for their estimation in rather simple mixtures. Thus, the aldose-1-phosphates which are hydrolyzed in normal acid solution at 100°, within a few minutes, can be easily differentiated from the 6-phosphates which are stable under these conditions.

In contrast, by a treatment with 0,2 N-alkali at 100° for three minutes glucose-1-phosphate remains unaffected while glucose-6-phosphate liberates 60% of its phosphate groups, and fructose-diphosphate even 100% [KIESSLING (68)].

DESJOBERT (34) proposes as an additional test for the detection of glucose-1-phosphate in mixtures the lack of hydrolysis after a 30 minutes heating at p_H 5.

The values for the hydrolysis constants of the different esters appear in Table 3 (p. 61).

Some other methods depend on specific color reactions, such as the reaction with resorcinol [ROE (163)], which permits the estimation of ketose esters or the pentose reaction with orcinol. In connection with the latter it may be mentioned (1) that the rate of the color development permits a differentiation between ribose-3- and -5-phosphate.

A method which has been proposed for the estimation of fructose-diphosphate is based on the determination of the phosphate groups liberated during osazone formation [DEUTICKE and HOLLMANN (35)]. Fructose-1-phosphate should be estimated in the same way, and also triose phosphates. However, the latter can be separately determined as "alkali-labile phosphate". Glucose-3-phosphate also loses the phosphate during osazone formation.

The most specific methods consists, however, in the application of suitable enzymes. A procedure for the estimation of fructose diphosphate, making use of aldolase, was described by Meyerhof and Wilson (*121*); and a method for glucose-6-phosphate was reported by Haas (*48*). Even very small amounts of glucose-diphosphate may be estimated due to its property of acting as the coenzyme of phospho-glucomutase (*18*). Many other enzymatic methods may be developed in the future by making use of the purified enzyme preparations now available.

Paper chromatography provides a valuable micro-method for the identification of the sugar component of phosphate esters [Partridge (*139*); Jermyn and Isherwood (*63*); Hirst et al. (*59*)]. The method has also been applied to ester mixtures [Hanes and Isherwood (*51*)]; Cohen and McNair Scott (*19*)]. A good separation of the di- from the monophosphates has been achieved but monophosphate mixtures do not seem to have been well resolved so far. However, it is likely that better separations will be obtained eventually. The addition of boric acid to the solvents is stated to retard the movement of ribose-phosphate as compared with arabinose-phosphate [Cohen and McNair Scott (*19*)]. This effect was attributed to the combination of boric acid with the *cis* hydroxyls of ribose.

Separations of different phosphate esters have also been effected by means of countercurrent distribution. The solubility in the organic phase was increased by the addition of long chain fatty amines [Plaut, Kuby and Lardy (*143*)].

VI. The Coenzymatic Action of Some Esters.

Besides the well known role of phosphoric esters as intermediates in carbohydrate metabolism, recent work has shown that in some cases they may display a coenzymatic action. The identification of glucose-1,6-diphosphate as the coenzyme of phosphoglucomutase disclosed the mechanism of action of this enzyme and was the first known case of a reaction type which may be fairly common in metabolism. Phosphoglucomutase catalyzes the following conversion:

$$\text{Glucose-1-phosphate} \rightleftharpoons \text{glucose-6-phosphate}$$

If glucose-diphosphate is absent, the reaction does not take place; and it was suggested by Leloir et al. (*84*) that the process catalyzed by phosphoglucomutase was not a direct intramolecular transference of phosphate but an intermolecular reaction between glucose-1-phosphate and the diphosphate. As shown in Figure 3, the phosphate group in position 1 of the diphosphate is transferred to position 6 of glucose-1-

phosphate. The reaction products are, glucose-6-phosphate and a new molecule of glucose diphosphate.

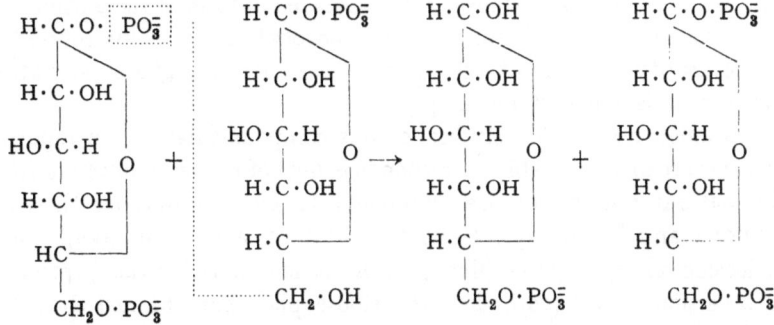

Fig. 3. Interaction between glucose-1,6-diphosphate and glucose-1-phosphate as catalyzed by phosphogluco-mutase. [LELOIR et al. (84).]

Confirmation of this mechanism was afforded by experiments of SUTHERLAND et al. (172, 173) who used glucose-1-phosphate labelled with both radioactive phosphate and carbon. When this substrate was mixed with the enzyme and glucose-diphosphate, the labelled elements appeared in the glucose-diphosphate molecules.

A similar mechanism has been found to be applicable to the enzymatic interconversion of 2- and 3-phosphoglyceric acid, in which 2,3-diphospho-glyceric acid acts as the coenzyme [SUTHERLAND, POSTERNAK and CORI (174)].

Another transformation which may be similar is that of ribose-1-phosphate into ribose-5-phosphate, and here it may be predicted that the coenzyme should be ribose-1,5-diphosphate. Analogous changes probably take place also in the desoxyribose series.

For fructose, the 1- and 6-phosphates are known as well as the diphosphate but no enzyme has yet been described which would catalyze their interconversion.

Many other isomerizations will probably be found to follow a mechanism similar to that caused by phosphoglucomutase, and glucose-diphosphate may thus constitute the first member of an extended family of new coenzymes.

VII. Pentose Phosphates.

Ribose Phosphates.

Ribose-1-phosphate. KALCKAR (66) discovered this ester which is formed by enzymatic action on some nucleosides. It was prepared, using a partially purified nucleoside phosphorylase, according to the following reaction:

Inosine + phosphate $\rightleftharpoons$ ribose-1-phosphate + hypoxanthine

Since the reaction is reversible, hypoxanthine was removed from the system with xanthine-oxidase, which converts it into xanthine and uric acid. The ribose-phosphate was then separated in form of the barium salt. The substance is so acid-labile that its phosphate group is hydrolyzed off at the acidity of some of the methods used for phosphate estimation; however, it is fairly stable at p_H 4.

Although no detailed studies of this compound have been reported, its structure most probably corresponds to β-ribofuranose phosphate (XV). The natural nucleosides are considered to be β-furanosides [Davoll, Lythgoe and Todd (33); Howard, Lythgoe and Todd (60)], and if nucleoside phosphorylase (like polysaccharide phosphorylase) produces no inversion [Cohn (20)], then the ribose phosphate formed should be a β-furanoside.

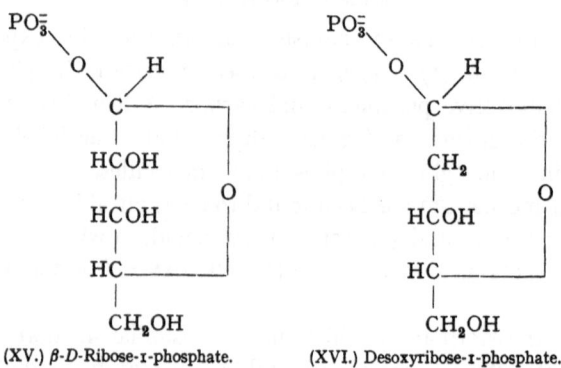

(XV.) β-D-Ribose-1-phosphate.   (XVI.) Desoxyribose-1-phosphate.

A synthetic ribopyranose-1-phosphate has been tested with negative results as a possible substrate for the enzyme [Kalckar (67)].

Desoxyribose-1-phosphate was detected by a procedure similar to that used for ribose-1-phosphate [Friedkin, Kalckar and Hoff-Jörgensen (45); Friedkin and Kalckar (44)]. A preparation of liver nucleoside phosphorylase catalyzed the following reaction:

Guanine-desoxyriboside + phosphate $\rightleftarrows$ desoxyribose-1-phosphate +
+ guanine.

This conversion is reversible and was driven to the right side of the equation by removing the guanine with guanase which converts it to (insoluble) xanthine. The ester (XVI) was obtained as a crystalline cyclo-hexylamine salt [Friedkin (43)] and was found to be even more unstable than ribose-1-phosphate in acid solution. At p_H 4 and room temperature approximately 50% of the ester is split in about 15 minutes. The ester is hydrolyzed by the acid used in all the current methods for phosphate estimation. It estimates therefore as „inorganic phosphate". The only way of differentiating this type of ester phosphate from inorganic phosphate is to precipitate the latter with magnesia mixture. The ester remains then in the supernatant and can be subsequently determined.

Ribose-3-phosphate. LEVENE and JORPES (*94*) found that in adenosine-3'-phosphate the acid hydrolysis of the base takes place at about the same rate as that of the phosphate. Thus, it seemed impossible to prepare ribose-3-phosphate from adenylic acid.

Later, LEVENE et al. (*87, 88, 89*) observed that if a solution of xanthylic acid was incubated at p_H 1,9 at 50°, hydrolytic cleavage occurred with the formation of ribose-3-phosphate.

The xanthylic acid was prepared by treatment of guanylic acid with nitrous acid. After this deamination the sugar-base linkage becomes more labile. For the preparation of ribose-3-phosphate the hydrolysis was allowed to proceed during 3 or 4 days. The base which precipitated was filtered off and the separation of nitrogen containing substances was completed by the addition of mercuric sulfate. After removing the excess mercuric ions, adding barium and concentrating, barium ribose-3-phosphate was precipitated with one volume of alcohol.

The preparation of ribose-3-phosphate can be also carried out by starting with adenosine-3'-phosphate [LEVENE and HARRIS (*90*)], by de-amination to the corresponding inosinic acid followed by hydrolysis of the free acid at 95—97° for one hour.

The corresponding ribonic acid (XVIII) was prepared by oxidation with bromine and was found to be hydrolyzed about twice as rapidly as 5-phosphoribonic acid. Moreover, the curves obtained by following polarimetrically the methyl glycoside formation indicated that both a furanoside and a pyranoside were formed. Since position 4 had to be free for the furanoside and position 5 for the pyranoside formation [cf. (XVII)], it was concluded that the phosphate group must occupy the position 2 or 3. The problem was finally solved in an elegant manner by reducing to ribitol-3-phosphate (XIX) with hydrogen and a platinum catalyst. The compound formed was found to be optically inactive as would be expected for ribitol-3-phosphate, according to LEVENE and HARRIS (*89, 90*)].

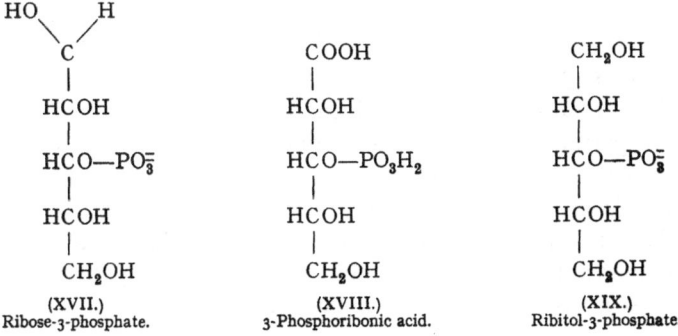

(XVII.)
Ribose-3-phosphate.

(XVIII.)
3-Phosphoribonic acid.

(XIX.)
Ribitol-3-phosphate.

Ribose-5-phosphate (XX) was first obtained by LEVENE and JACOBS (*92, 93*). Inosinic acid was hydrolyzed in 0,1 *N*-HCl at 100° for 1 hour, the hypoxanthine was precipitated with mercuric sulfate and barium ribose-5-phosphate was obtained as a crystalline salt containing five molecules of water.

By oxidation with bromine or nitric acid a phosphoribonic acid was prepared. If position 5 had been free, nitric acid should have produced a trihydroxyglutaric acid; therefore, the observation just mentioned indicated that the phosphate group occupied the position 5. As to the ring type of the original compound, it had to be a furanose derivative, since the presence of a pyranose ring would not be compatible with a 5-substituted pentose.

The rate of lactone formation was also studied [LEVENE and MORI (*95*)]. Based on previous studies of LEVENE and SIMMS (*107*), it was predicted that 4-phosphoribonic acids should give the corresponding δ-lactone which is formed rapidly, while 5-phosphoribonic acid should yield a γ-lactone whose formation is slow. The 2- or 3-phosphoribonic acids would yield both lactones mentioned. Experiments showed that the lactonisation of the phosphoribonic acid originating from inosinic acid is a slow process. Hence the product must be a γ-lactone, and the phosphate group occupied the position 5.

Reduction of the phosphoribose to phosphoribitol gave an optically active compound, thus excluding position 3 for the phosphate group [LEVENE, HARRIS and STILLER (*91*)]. Moreover, the preparation of the methyl glycoside yields only a furanoside.

D-Ribose-5-phosphate has been synthesized by phosphorylating 2,3-isopropylidene-methyl-*D*-ribofuranoside with phosphoryl chloride [LEVENE and STILLER (*108*)] or with dibenzylchlorophosphonate [MICHELSON and TODD (*122*)]. It was found to be vigorously metabolized by some animal and yeast extracts [DICKENS (*36, 37*); DICKENS and GLOCK (*38*)], while arabinose-5-phosphate and xylose-5-phosphate were only slightly attacked. On the other hand, yeast extracts may convert 6-phosphogluconate into ribose-5-phosphate [COHEN and McNAIR SCOTT (*19*)].

Arabinose Phosphates.

D-Arabinose-5-phosphate (XXI) was prepared by LEVENE and CHRISTMAN (*86*) by treating 1,2-isopropylidene arabofuranose with phosphoryl chloride in pyridine solution. It was obtained as the amorphous barium salt which could be converted into a crystalline brucine salt. The preparation of isopropylidene arabinose was effected through the following intermediates: *D*-arabinose → *D*-arabinose diethyl-mercaptal → 5-tosyl-arabinose-mercaptal → 5-tosyl-arabinose → 1,2-isopropylidene-5-tosyl-arabinose → 1,2-isopropylidene-*D*-arabo-furanose.

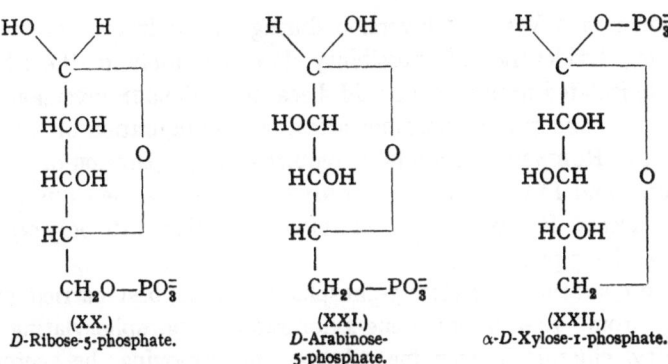

(XX.)
D-Ribose-5-phosphate.

(XXI.)
D-Arabinose-
5-phosphate.

(XXII.)
α-D-Xylose-1-phosphate.

Arabinose-5-phosphate can be fermented by yeast extracts [DICKENS (36, 37); DICKENS and GLOCK (38)]. It is probably formed from 6-phosphogluconate [COHEN and McNAIR SCOTT (19)].

Xylose Phosphates.

D-Xylose-1-phosphate (XXII) is known only as a synthetic product at the present time [MEAGHER and HASSID (116)]; it was prepared from bromoacetylxylose and trisilver phosphate. As expected, its periodate oxidation required 2 moles of oxidant and yielded 1 mol. of formic acid. The specific rotation, hydrolysis constants and dissociation constant are given in the corresponding tables (p. 61—64).

The xylose-1-phosphate molecule is identical with that of glucose-1-phosphate except for the fifth carbon which carries a hydrogen atom instead of a —CH_2OH group. Some experiments were therefore conducted in order to detect a possible polysaccharide formation in the presence of potato phosphorylase; however, no reaction could be observed.

(XXIII.)
D-Xylose-3-phosphoric acid.

(XXIV.)

(XXV.)
D-Xylose-5-phosphoric acid.

D-Xylose-5-phosphate (XXV). The interest in xylose phosphate largely arose from ROBINSON's suggestion (155) that xylose might be a primary constituent of nucleic acids. Ribose would thus be formed as the product of a WALDEN inversion during the hydrolysis of xylose-3-phosphate. Evidently, this possibility does not apply to the ribose-5-phosphate isolated from inosinic acid, because a WALDEN inversion at the primary alcoholic group would not affect the configuration of the sugar.

A test of ROBINSON's hypothesis required the preparation of xylose-3-phosphate; and LEVENE and RAYMOND (*102—105*) tackled this problem without success. Phosphorylation of xylose derivatives with only position 3 free gave the 5-phosphate.

The preparation of xylose-5-phosphate can be best carried out by starting from 1,2-isopropylidene-xylofuranose, phosphorylating with phosphoryl chloride in pyridine solution and removing the acetone by heating in 2 *N*-sulfuric acid at 80° for 2 hours.

Xylose-5-phosphate, on treatment with methyl alcohol containing hydrogen chloride, yielded a curve of glycoside formation similar to that given by furanosides; a 3-phosphoxylose would be expected to give both a furanoside and a pyranoside.

Preparation of xylose-3-phosphate was attempted with phosphoryl chloride and 1,2-isopropylidene-5-benzoyl-xylofuranose. Unexpectedly, the product was xylose-5-phosphate. The same result was obtained when, instead of benzoyl-, the corresponding acetyl- or benzylcarboxy-derivatives were treated.

LEVENE and RAYMOND (*102, 103, 105*) studied carefully the isopropylidene-xylose derivatives mentioned above in order to ascertain the correctness of the postulated structure. The only 3-phospho derivative which they were able to isolate was 1,2-isopropylidene-3-phospho-5-methyl-xylose. This compound is not a satisfactory material for the preparation of the phosphosugar, because there is no way of removing the methyl group without splitting off the phosphate at the same time. Its rate of hydrolysis was found to be many times higher than that of xylose-5-phosphate.

LEVENE and RAYMOND (*105*) attributed the ease of migration of the phosphoryl group in xylose to the close proximity of the hydroxyls at the positions 3 and 5, and they proposed a mechanism similar to that suggested by FISCHER (*42*) for acetyl migrations, i. e. the intermediary formation of a bridge as shown in the formulas (XXIII), (XXIV), and (XXV).

VIII. Hexose Phosphates.

Glucose Phosphates.

α-D-Glucose-1-phosphate (XXVI) (CORI Ester) was first isolated by CORI, COLOWICK and CORI (*24*) as a product of the action of frog muscle enzymes on glycogen and inorganic phosphate. The crude ester also contained a certain amount of glucose-6-phosphate which had to be eliminated by fractional crystallization of the brucine salts. Better results were eventually obtained by carrying out the enzymatic reaction in the absence of magnesium ions which are known to accelerate the conversion of the 1- into the 6-ester.

Rabbit muscle extracts were dialyzed and then incubated with glycogen, phosphate buffer and adenylic acid. The latter served as an activator of the enzyme. Under these conditions, glucose-1-phosphate was practically the only ester formed.

Proteins were precipitated with mercuric chloride and the ester was precipitated in form of the barium salt, which is water soluble and insoluble in 50% ethanol.

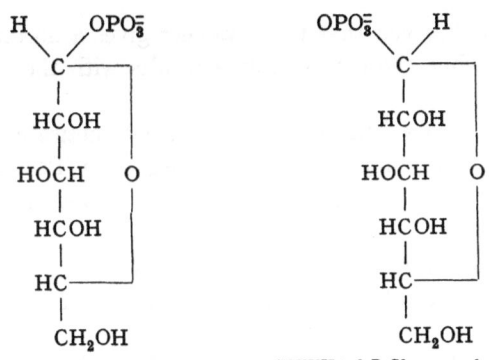

(XXVI.) α-D-Glucose-1-phosphate.   (XXVII.) β-D-Glucose-1-phosphate.

The synthesis of the CORI ester (*24*) was achieved by treatment of acetobromoglucose with silver phosphate in benzene. The tri-(tetracetyl-glucose-1) phosphoric acid obtained was then converted into glucose-1-phosphoric acid by acid methanol; other deacetylation methods were unsuccessful.

The naturally occurring and the synthetic ester samples were found to be identical and both gave glucose upon hydrolysis with acid or phosphatase. In these studies the amorphous barium salt was used. The crystalline dipotassium salt was described by KIESSLING (*68*).

Since the CORI ester is alkali-stable and non-reducing and liberates glucose on acid hydrolysis, there is no doubt that it is glucose-1-phosphate. CORI et al. had suggested that it constituted the α-form and contained a pyranose ring which was confirmed by WOLFROM and PLETCHER (*185*) who studied the periodate oxidation. As expected for a hexopyranose-1-phosphate, two moles of periodate were consumed and one mole of formic acid was formed but no formaldehyde. As to the α-structure, it was indicated by the high dextrorotation, and further proof was obtained by comparison with synthetic β-D-glucose-1-phosphate.

Several procedures for the enzymatic preparation of glucose-1-phosphate have been described; cf. CORI, COLOWICK and CORI (*24*); KIESSLING (*68*); HANES (*49*); BERNFELD, DE TRAZ and GAUTIER (*9*); SUMNER and SOMERS (*171*); WEIBULL (*183*);

for the purification of the ester see McCready and Hassid (*114*), and for the synthesis: Posternak (*145*); Krahl and Cori (*78*). The latter paper contains detailed information.

β-D-Glucose-1-phosphate (XXVII) as mentioned previously was prepared by interaction of acetobromoglucose with silver dibenzyl-phosphate (*186*) or with monosilver phosphate (*152*). Wolfrom et al. followed the changes in the rotatory power during the hydrolysis of β-glucose-1-phosphate and that of the α-form. They observed that while the β-compound became increasingly dextrorotatory, the α- form behaved inversely. Thus, there remains no doubt concerning the configuration of either compound.

It was also observed that the β anomer gives a normal rotatory dispersion curve while the corresponding results with the α-form are anomalous.

α-L-Glucose-1-phosphate was obtained synthetically following the Cori procedure but using *L*- instead of *D*-glucose [Potter et al. (*146*)]. The properties are the same as those of α-*D*-glucose-1-phosphate, except for the rotation.

The specific rotation of the dipotassium salt, $[\alpha]_D = -78,2$, is almost exactly the opposite of that of the corresponding *D*-glucose derivative, $[\alpha]_D = +78,5$.

Oxidation with periodate consumed 2 moles of the oxidant and yielded 1,1 moles of formic acid. Tested with potato phosphorylase or saccharose phosphorylase, the substance was found to remain unchanged.

Glucose-2-phosphate. Acid hydrolysis of the saccharose phosphate, obtained by the action of phosphoryl chloride on unprotected saccharose [Neuberg and Pollak (*130, 131*)] yields a mixture of fructose and glucose esters usually known as Hatano's ester (*55, 30, 28*).

Courtois and Ramet (*31, 147*) fractionated this mixture by means of the crystalline quinine salts and obtained a glucose ester which was studied by periodate oxidation. Their data showed that it was neither glucose-1,5- nor -6-phosphate. The corresponding aldonic acid was prepared by bromine oxidation (*32, 147*). The results of a study indicated that the substance was 2-phosphogluconic acid which could be oxidized by periodate as shown in (XXVIII).

$$
\begin{array}{l}
\text{COOH} \\
| \\
\text{HCO—PO}_3^= \\
| \quad\quad\quad + 3\ \text{IO}_4\text{H} \longrightarrow \\
\text{(CHOH)}_3 \\
| \\
\text{CH}_2\text{OH}
\end{array}
\quad
\begin{array}{l}
\text{COOH} \\
| \\
\text{HCO—PO}_3^= \\
| \\
\text{CHO}
\end{array}
+ 2\ \text{HCOOH} + \text{HCHO} + 3\ \text{IO}_3\text{H} + \text{H}_2\text{O}
$$

2-Phosphogluconic **acid.**

(XXVIII.)

Thus, it was concluded that the main constituent of HATANO's ester is glucose-2-phosphate.

An unequivocal synthesis of the latter was carried out by FARRAR (*40*) by reacting 1,3,4,6-tetraacetyl-β-D-glucose (*52*) with diphenyl chlorophosphonate followed by hydrogenation and deacetylation with potassium methoxide. The resulting substance obtained as the crystalline potassium salt was found to reduce FEHLING's solution decidedly more slowly than did glucose.

Glucose-3-phosphate was obtained synthetically by the interaction of 1,2–5,6-diisopropylidene-glucofuranose and phosphoryl chloride [NODZU (*134*); KOMATZU and NODZU (*72*); LEVENE and RAYMOND (*96, 149, 98*); JOSEPHSON and PROFFE (*64*)]. It has been repeatedly tested for fermentation with negative results.

Since its positions 4 and 5 are free, the compound should be able to give both a furanose and a pyranose ring which was confirmed by the glycoside formation [LEVENE and RAYMOND (*98*)]. Moreover, the aldonic acid prepared from glucose-3-phosphate gave a lactone formation curve which indicated that both a γ- and a δ-lactone were formed.

Removal of the phosphate group with phosphatase gave glucose while the loss of phosphate during osazone formation led to 3,6-anhydro-allosazone. Non-reducing derivatives of glucose-3-phosphate, when treated with alkali, lost phosphate without inversion or anhydride formation [PERCIVAL and PERCIVAL (*141*)]. The preparation of the calcium salt from the dibrucine salt was described by LOUGH and SPENCER (*111*).

Glucose-4-phosphate has been obtained synthetically only by phosphorylation of 1,2,3,6-tetraacetyl-glucose. The latter compound was prepared by HELFERICH et al. (*56, 58*) by acetyl migration of 1,2,3,4-tetra-acetyl-glucose.

RAYMOND (*148*) used phosphoryl chloride in the phosphorylation and prepared the tetracetyl-glucose by the sequence, 4,6-benzylidene-glucose → 1,2,3-triacetyl-4,6-benzylidene-glucose → 1,2,3-triacetyl-glucose → → 1,2,3,6-tetraacetyl-glucose. In view of the mentioned tendency to group migration, the properties of the resulting glucose phosphate were specially studied. It was found that it did not give a crystalline phenylhydrazine salt of the osazone as does glucose-6-phosphate but only a crystalline acid barium salt of the osazone was obtained. The curve of glycoside formation was different from that of glucose-3- or- 6-phosphate and resembled more that of tetramethyl-glucopyranose which does not have a free 4-position.

REITHEL and CLAYCOMB (*153*) reported more satisfactory results when using the HELFERICH procedure for the preparation of 1,2,3,6-tetraacetyl-glucose, and diphenyl-chloro-phosphonate as phosphorylating agent.

Glucose-4-phosphate is not fermented by acetone dried yeast [RAY-MOND (*148*)].

Glucose-5-phosphate (XXIX). In 1929 it was suspected that the mono-phosphate formed in yeast fermentation might be glucofuranose-5-phosphate. This substance was synthesized by JOSEPHSON and PROFFE (*65*) and it was found that the specific rotation differed from that of the natural ester. Moreover, it did not decrease the induction period of fermentation in dried yeast and was not appreciably fermented.

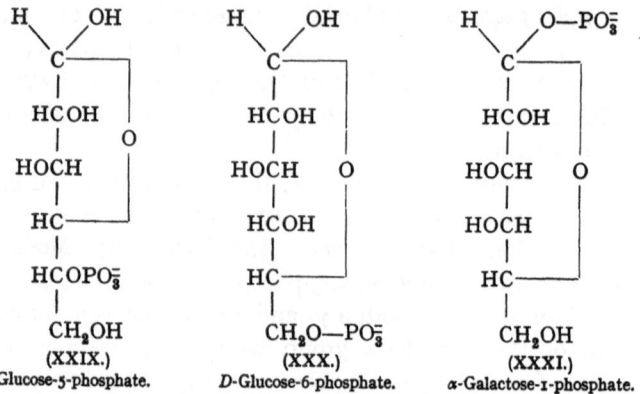

(XXIX.)
Glucose-5-phosphate.

(XXX.)
D-Glucose-6-phosphate.

(XXXI.)
α-Galactose-1-phosphate.

The synthesis of glucose-5-phosphate was carried out as follows: First, 1,2,5,6-diisopropylidene-glucofuranose was acetylated in position 3, then partial hydrolysis with acetic acid gave 1,2-isopropylidene-3-acetyl-glucofuranose. The position 6 was then covered by tritylation. The product, viz. 1,2-isopropylidene-3-acetyl-6-trityl-glucofuranose, was phosphorylated with phosphoryl chloride. After acid hydrolysis of the protecting groups, glucofuranose-5-phosphate was isolated in the form of an amorphous barium salt. From the latter a crystalline strychnine salt could be obtained.

Glucose-6-phosphate (ROBISON Ester) (XXX). Although HARDEN and ROBISON detected the hexose-monophosphate fraction in the products of yeast juice fermentation in 1914, it was not till 1931 that ROBISON and KING (*158*) were able to prepare glucose-6-phosphate in pure form. During that period of years the crude hexose-monophosphate was recognized to be a mixture of an aldose and a fructose component. The assignment of the position to the phosphate group was no easy matter. Positions 1 and 2 were excluded because the osazone contained phosphate. Position 3 was regarded as a possibility for a time until glucose-3-phosphate was synthesized and found to be different from the natural substance [JOSEPHSON and PROFFE (*64*); LEVENE and RAYMOND (*96, 98*)]. The positions 4 and 5 were excluded on the ground that two types of methyl glycosides could be prepared: on acid hydrolysis one of them behaved as a furanoside and the other as a pyranoside [KING, McLAUGHLIN and MORGAN (69)]. Position 6 was uncertain for some time due to erroneously observed differences in the melting point of the osazones of the ROBISON ester and of fructose mono- and diphosphate. LEVENE and RAYMOND (98—100) prepared an ester by the interaction of phosphoryl chloride and 1,2-isopropylidene-glucose in pyridine solution. Although the positions 3, 4, and 6 are free for reaction in this compound, it was known that some substitutions take place preferentially in 6-position. The authors mentioned expected, therefore, to obtain glucose-6-phosphate which was confirmed by the experiment.

This ester was readily fermented by yeast and gave a phospho-hexosazone which was identical with that obtained from the HARDEN and YOUNG or NEUBERG ester. The aldonic acid prepared by bromine oxidation of the ROBISON ester and of the ester from monoacetone glucose, gave identical curves of lactone formation.

The specific rotation of the synthetic substance did not quite agree with that of the naturally occurring ester, but these differences became smaller on purification; and finally, the main objection to the acceptance of position 6 was removed when ROBISON and KING prepared from their purified natural ester an osazone which was identical with that given by the NEUBERG ester.

The procedure for the isolation of glucose-6-phosphate from yeast followed by ROBISON and MORGAN (162) has been mentioned previously. The same procedure, with slight improvements, was also used by WARBURG and CHRISTIAN (180) who obtained a crystalline calcium salt. Of course, the crystalline nature of this salt does not guarantee its purity since many complex salts are formed by phosphoric acid derivatives and divalent cations. Thus, SMYTHE (166), applying WARBURG and CHRISTIAN's method, obtained a calcium salt which after recrystallization gave only 50% of the theoretical aldose quantity when titrated with hypoiodite, and which was in fact an equimolar mixture of glucose-6-phosphate and glycerophosphate.

A procedure for improving the yield in the fermentation as described by SMYTHE (167) consists in allowing hexose diphosphate to accumulate during the fermentation process and then adding a dye such as rosinduline GG. The fermentation then takes place as follows:

$$4 \text{ glucose} + \text{hexose-diphosphate} + 2 \text{ phosphate} = 4 \text{ hexose-monophosphate} + 2 \text{ alcohol} + 2 CO_2.$$

The hexose-monophosphate yield was thus increased about threefold and the product was, as usual, a mixture of glucose-, fructose-, and mannose-monophosphates.

A method of preparation starting from glucose-1-phosphate and using phosphoglucomutase was reported by COLOWICK and SUTHERLAND (22). SWANSON (175) described a similar procedure starting from starch and using phosphorylase and phosphoglucomutase; a good yield was reported. Acid hydrolysis of potato starch affords another method of preparation (145 a).

The synthetic methods can be summarized as follows. The procedure starting from 1,2-isopropylidene-glucose and phosphoryl chloride (LEVENE and RAYMOND) leads to an impure product which has to be purified by fractional crystallization of the brucine salt. Dibenzyl chlorophosphonate as a phosphorylating agent was used by ATHERTON et al. (2). LEVENE and RAYMOND (100) described another method in which use was made of 1,2,3,4-tetraacetyl-glucose. The latter can be obtained by first preparing 6-trityl-glucose, acetylating and then removing the trityl group. LARDY and FISCHER (83) recommend diphenyl chlorophosphonate instead of phosphorylchloride; they obtained a pure product without any special purification.

6-Phosphogluconic acid was prepared by bromine oxidation of glucose-6-phosphate [ROBISON and KING (158), WARBURG and CHRISTIAN (181)] and by enzymatic oxidation [WARBURG, CHRISTIAN and GRIESE (182)]. The preparation is described by ROBISON and MACFARLANE (159). Some enzymes from yeast or animal tissues are able to convert it into 6-phospho-5-keto-gluconic acid [DICKENS (36, 37), DICKENS and GLOCK (38)].

Glucose-1,6-diphosphate. The presence, in fermentation products, of another diphosphate besides fructose-diphosphate has often occupied the mind of workers in the field. Thus, MACLEOD and ROBISON (113) wrote in 1933: "It is remarkable that while at least three hexose-monophosphoric esters, those of glucose, fructose and mannose, are formed during the fermentation of sugar by yeast juice, only one diphosphoric ester, the hexosediphosphate of HARDEN and YOUNG, has been discovered among the fermentation products".

The presence of glucose-1,6-diphosphate was finally detected on the basis of its coenzymatic activity mentioned above, and the substance was isolated from the ester mixture formed by toluene treated baker's yeast incubated with phosphate and sugar [LELOIR et al. (84), CARDINI et al. (18)]. The amount formed is about 0,5% of that of fructose-diphosphate; and the compound accompanies the latter when precipitated with lead or barium ions. The separation was carried out by destroying the fructose ester with alkali which leaves glucose-diphosphate unchanged. Further purification by fractional precipitation with divalent metals was ineffective but better results were obtained by fractional precipitation of the free acid with acetone. Finally a barium salt was isolated which was non-reducing but became reducing on mild acid hydrolysis with simultaneous liberation of one half of the phosphate content. The compound resulting after the elimination of one phosphate group was identified as glucose-6-phosphate on the basis of its behavior towards enzymes, acid and alkali and by osazone formation. The specific rotation of the glucose-diphosphoric acid, $[\alpha]_D =$ $= + 63°$ to $70°$, decreased during acid hydrolysis to values agreeing with those of glucose-monophosphoric acid.

A first synthesis of glucose-diphosphate from 1,6-dibromotriacetyl-glucose and silver phosphate gave a coenzymatically active mixture but was not successful with respect to yields or purity of the product. It is much more satisfactory to start from a fully protected glucose-6-phosphate, to transform it into the 1-bromo derivative and then to introduce the phosphate into the 1-position. The compound required for this purpose is 1,2,3,4-tetraacetyl-6-diphenyl-phosphonoglucose which had been prepared by LARDY and FISCHER (83). A treatment with hydrogen bromide gave the 1-bromo compound which was then phosphorylated with trisilver phosphate or silver diphenylphosphate. After deacetylation and removal of the phenyl groups by catalytic hydrogenation, glucose-1,6-diphosphate was obtained [REPETTO et al. (154), POSTERNAK (144, 145)].

The biosynthesis of glucose-diphosphate in yeast and muscle takes place according to the reaction [PALADINI et al. (137 a)]:

adenosine triphosphate + glucose-1-phosphate →
→ glucose diphosphate + adenosine diphosphate.

In *Eschericia coli* a biosynthesis has been detected in which adenosine phosphate is not involved. There occurs probably a transphosphorylation between two glucose-1-phosphate molecules, yielding glucose-diphosphate and free glucose [LELOIR et al. (85)].

Galactose Phosphates.

α-D-Galactose-1-phosphate (XXXI). KOSTERLITZ (73) isolated from the liver of rabbits fed galactose an alkali-stable, acid-labile and non-reducing galactose ester which was later found to be identical with a synthetic sample [KOSTERLITZ (75)].

Its synthesis was carried out by reacting acetobromogalactose with trisilver phosphate [COLOWICK (21), KOSTERLITZ (74)] or with silver diphenylphosphate [POSTERNAK (145)]. For crystalline brucine and di-potassium (+ 2 H_2O) salts see (74).

β-D-Galactose-1-phosphate is formed in the enzymatic phosphorylation of galactose by adenosine-triphosphate [CAPUTTO, LELOIR and TRUCCO (16), TRUCCO et al. (178), WILKINSON (184)].

Like other 1-phosphates, this ester is stable to alkali but practically complete hydrolysis is attained in 2 min. in 0,09 *N*-acid solution, at 100° (*74*).

It is fermented by galactose-adapted yeast extracts [KOSTERLITZ (*77*)].

The transformation of galactose-1-phosphate into glucose-1-phosphate is effected by a yeast enzyme in the presence of uridine-diphosphate-glucose which acts as coenzyme [CARDINI et al. (*17*), CAPUTTO et al. (*15*)].

β-Galactose-1-phosphate was prepared by REITHEL (*152*) who treated α-acetobromo-*D*-galactose with either silver dibenzyl-phosphate or monosilver phosphate.

Galactose-6-phosphate was synthesized by LEVENE and RAYMOND (*101*) starting from 1,2,3,4-di-isopropylidene-galactose and phosphoryl chloride. It has been tested repeatedly, with negative results, for fermentability with yeast juice (from galactose-adapted yeast) [LEVENE and RAYMOND (*101*); GRANT (*47*); TRUCCO et al. (*178*)].

Mannose Phosphates.

Mannose-1-phosphate was prepared by COLOWICK (*21*) who treated acetobromomannose with trisilver phosphate. Its properties are similar to the other 1-phosphates. It was found to remain unchanged on incubation with animal or yeast enzymes.

Mannose-6-phosphate (XXXII) was first obtained by ROBISON (*156*) by fractional crystallization of the brucine salt of the hexose monophosphate mixture formed by the yeast juice fermentation of sugar. Mannose was identified after hydrolysis in 0,1 *N*-acid at 100° for 93 hours; and since the ester gave the same osazone as fructose-diphosphate or glucose or fructose-6-phosphate, it was decided that the phosphate group occupied the position 6.

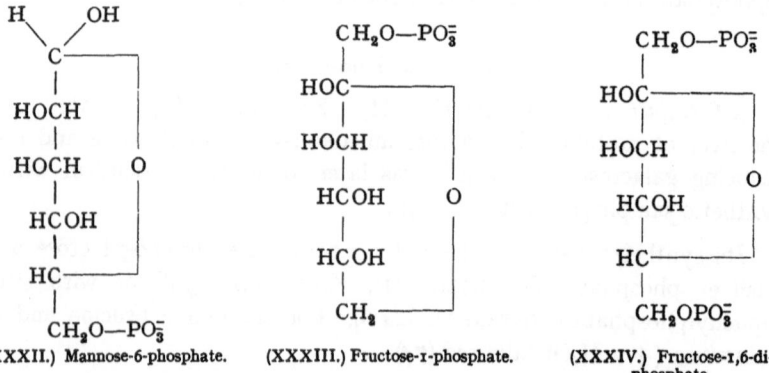

(XXXII.) Mannose-6-phosphate.   (XXXIII.) Fructose-1-phosphate.   (XXXIV.) Fructose-1,6-di-phosphate.

On addition of phenylhydrazine to a solution of the free acid an insoluble phenylhydrazine salt of the hydrazone is formed (m. p. 144°

to 145°). The fructose- and glucose-6-phosphates do not form such insoluble hydrazones.

Studying the fermentation products of different sugars, JEPHCOTT and ROBISON (62) observed that when mannose was fermented by dried brewer's yeast at 38° the monophosphate fraction amounted in some instances to 76% of the total esterified phosphate, and that about 90% of this monophosphate corresponded to the mannose ester. When fructose or glucose were fermented, only 20–30% of monophosphate was formed. When yeast juice or acetone-dried yeast or dried baker's yeast was used, the monophosphate fraction of the mannose fermentation mixture was not as rich in mannose-phosphate as that obtained with dried brewer's yeast [PATWARDHAN (140)].

Furthermore, JEPHCOTT and ROBISON (62) found that the phenyl-hydrazone of mannose-phosphate could be directly obtained from mixtures containing only small amounts of this ester. This fact was used for preparative purposes. The phenylhydrazone could be decomposed with benzaldehyde and from the solution of mannose phosphoric acid a crystalline barium salt could be obtained [PATWARDHAN (140)].

Mannose-6-phosphate is formed by yeast hexokinase in the presence of mannose and adenosine-triphosphate [BERGER et al. (8); KUNITZ and McDONALD (80)], and is converted to fructose-6-phosphate by a specific enzyme [SLEIN (165)].

6-Phosphomannonic acid was prepared by PATWARDHAN (140) from mannose-6-phosphate by oxidation with bromine in the presence of barium hydroxide. The curves of its lactonization indicated that two lactones were formed.

Both lactones were prepared, viz. the γ-lactone by heating the acid at 100° for one hour, and the δ-lactone by evaporating a solution of the acid in vacuo at room temperature. The γ-lactone of 6-phosphomannonic acid gave, $[\alpha]_{5461} = = + 54,1°$; and the δ-lactone, $[\alpha]_{5461} = + 60,6°$. The curves of hydrolysis of each of these lactones were also studied as well as the rate of hydrolysis of their phosphates (see Table 3, p. 61).

Fructose Phosphates.

Fructose-1-phosphate (ROBISON-TANKO Ester) (XXXIII). Acid hydrolysis of fructose-1,6-diphosphate leads to fructose-6-phosphate because the phosphate group attached to the position 1 is split off much faster than that in position 6. Hydrolysis with phosphatase splits off both phosphate groups at the same rate and thus, half of the monophosphate formed is fructose-1-phosphate [McLEOD and ROBISON (113)]. When preparing fructose-1-phosphate TANKO and ROBISON (176) found that it was not convenient to use highly purified phosphatase because the crude enzyme also catalyzes the interconversion, fructose-6-phosphate ⇌ glucose-6-phosphate. The glucose-phosphate could be easily oxidized with bromine to 6-phosphogluconic acid and the latter separated in form of its water insoluble barium salt. There remained thus less fructose-6-phosphate to be separated from the 1-phosphate than if pure phosphatase had been used. The separation of the two fructose-monophosphates was carried out by fractional crystallization of the brucine salts followed by conversion to the barium salt.

Fructose-1-phosphate was later obtained by the condensation of phosphodihydroxy-acetone and D-glyceraldehyde which was catalyzed by aldolase [MEYERHOF, LOHMANN and SCHUSTER (119)].

The fructose-phosphate synthesized by RAYMOND and LEVENE (149) as well as by BRIGL and MÜLLER (11), starting from the so called "β-di-acetone-fructose", which has a free 1-position, is probably identical with the ester formed enzymatically. It was found to decrease the induction period of yeast as do glucose- and fructose-6-phosphate [LEVENE and RAYMOND (96)].

Fructose-1-phosphate was detected in the liver [CORI (23), PANY (138)] and in the intestine during fructose absorption [KJERULF-JENSEN (70, 71)].

The reducing power of fructose-1-phosphate, estimated by the HAGEDORN and JENSEN method is about two-thirds that of glucose-6-phosphate [MCLEOD and ROBISON (113)].

Fructose-3-phosphate is a synthetic product obtained from 1,2,4,5-di-isopropylidene-fructose and phosphoryl chloride (106, 96, 149) or diphenyl-chlorophosphonate (11). It is not fermentable by acetone dried yeast.

Fructose-6-phosphate (NEUBERG Ester) was first prepared (125) by partial hydrolysis of fructose-diphosphate and later isolated by ROBISON (156) from the hexose-monophosphate fraction obtained by yeast juice fermentation.

The phosphate group in position 1 of fructose-1,6-diphosphate is hydrolyzed about 12 times faster than that in 6, in 1 N-acid, at 100°. Thus, it is easy to obtain a good yield of the 6-phosphate. However, at that temperature some destruction of fructose-phosphate also takes place. NEUBERG, LUSTIG and ROTHENBERG (129) have improved the procedure by carrying out the hydrolysis at 35° in N-acid during 5 days. The calcium, barium, magnesium, and zinc salts of this compound are water soluble. It can be precipitated by addition of basic (but not neutral) lead acetate [NEUBERG (125)].

Fructose-6-phosphate is formed when yeast hexokinase is incubated with fructose and adenosine-triphosphate [BERGER et al. (8), KUNITZ and McDONALD (80)].

Fructose-1,6-diphosphate (HARDEN-YOUNG Ester) (XXXIV). The allocation of a phosphate group to position 1 followed from the observation of YOUNG (187) that, when hexose-diphosphate was heated with hydrazine, an osazone was formed with the simultaneous elimination of one phosphate group.

The position of the second phosphate group remained uncertain for some time. Evidently, the 1,6-fructose-diphosphate structure is compatible only with a furanose ring system. Consequently, if the type of the ring could be ascertained, then the location of the second phosphate group was likely to follow.

MORGAN (123) as well as SCHLUBACH and BARTELS (164) prepared the α- and β-methyl-fructoside-diphosphates. The rates of hydrolysis of these compounds were measured by LEVENE and RAYMOND (97) and were found to be very high. Since it was known that furanosides hydrolyze faster than pyranosides, it was concluded that the methyl-fructoside-diphosphates contained a furanoside ring. MORGAN and ROBISON (124) eliminated the phosphate groups from these compounds with phosphatase and thus obtained the corresponding α- and β-methyl fructosides. They concluded from the specific rotation and ease of hydrolysis of the latter that they were fructofuranosides. In addition these authors prepared fully methylated derivatives which, after hydrolysis, yielded tetramethyl-fructofuranose.

It has been found, moreover, that fructose-diphosphate gives the same osazone as does glucose-6-phosphate [LEVENE and RAYMOND (98)]. Thus the structure of 1,6-fructose-diphosphate can be definitely assigned to the HARDEN-YOUNG ester.

The preparation of this ester by means of dried yeast was mentioned above. More recently, NEUBERG and LUSTIG (128) described a procedure in which fresh baker's yeast is cytolyzed with an organic solvent such as toluene and then incubated with sugar and phosphate.

For the isolation from the crude mixtures obtained by fermentation, use has been made of the fact that the calcium or barium salts are less soluble at higher temperatures [NEUBERG and SABETAY (132); cf. (128)]. Another method which avoids the decomposition at higher temperatures consists of a fractionation of the acid barium salt with alcohol [ROBISON and MACFARLANE (159)].

The crude barium salt is dissolved in five parts of water and acid is added to p_H 4; 0,25 vol. of alcohol are then added, the mixture is centrifuged, and the supernatant is diluted with 3 volumes of alcohol. The precipitate of the acid barium salt thus obtained can be re-fractionated or converted into the neutral barium salt. The most effective purification is achieved by recrystallization of the crystalline strychnine salt [NEUBERG, LUSTIG and ROTHENBERG (129)].

The dibarium salt is soluble to an extent of 0,6% at p_H 8,2 at 0° but less at higher temperatures [ROBISON and MACFARLANE (159)]. The barium, calcium, lead, and manganese salts are amorphous; the only crystalline salts described so far are those of strychnine [NEUBERG and DALMER (126)], benzidine [NEUBERG and SCHENER (133)], and brucine [SCHLUBACH and BARTELS (164)]. The preparation of the water soluble acid barium and calcium salts is reported by Neuberg, Lustig and Rothenberg (*129*).

Alkali (0,2 *N*-NaOH) liberates in 3 min. at 100° all the phosphate from fructose-diphosphate [Kiessling (*68*)]. A phosphatase believed to be specific for fructose-diphosphate is reported by Gomori (*46*).

The synthesis of fructose-diphosphate has been attempted by starting from 2,3-isopropylidene-fructofuranose (*11*). Treatment with diphenyl chlorophosphonate gave the tetraphenyl ester of 1,6-diphospho-2,3-isopropylidene fructofuranose; however, one of the phosphate groups was lost during removal of the phenyls.

Tagatose Phosphate.

D-Tagatose-6-phosphate (XXXV) was synthesized by the interaction of 1,2,3,4-diisopropylidene-tagatose and diphenylchlorophosphonate, followed by the removal of the phenyls in a hydrogenation process and of the isopropylidene groups by acid hydrolysis [Totton and Lardy (*177*)].

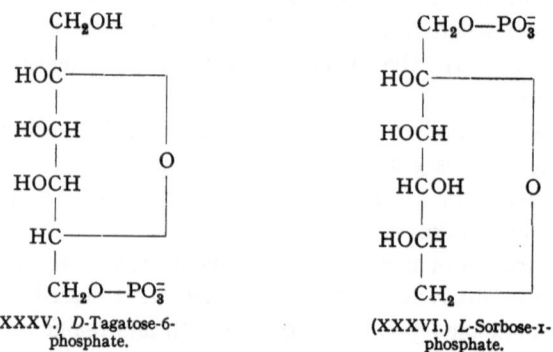

(XXXV.) *D*-Tagatose-6-phosphate.

(XXXVI.) *L*-Sorbose-1-phosphate.

This ester was prepared since it might be an intermediate in the biological interconversion of galactose to glucose; as well known, tagatose is the ketose corresponding to galactose. It was found that tagatose phosphate was phosphorylated by adenosine triphosphate in the presence of brain extract and, furthermore, that triose phosphates were formed.

Sorbose Phosphates.

L-Sorbose-1-phosphate (XXXVI). The configuration of *L*-Sorbose is identical with that of *D*-fructose except for the configuration at carbon atom 5. As mentioned, the enzyme aldolase catalyzes the condensation of *D*-glyceraldehyde with triose phosphate to give fructose-1-phosphate. When *D,L*-glyceraldehyde was used [Meyerhof, Lohmann and Schuster (*119*)], the products were, fructose-1-phosphate and sorbose-1-phosphate. After the action of phosphatase the sugar components were identified as fructose and sorbose respectively. The latter sugar is not fermented by yeast.

Synthetic sorbose-1-phosphate was prepared (*115*) by phosphorylation of 2,3,4,6-diisopropylidene-*L*-sorbose or dibenzylidene-sorbose with diphenylchlorophosphonate; a crystalline potassium salt was obtained.

L-Sorbose-6-phosphate was synthesized by phosphorylation of 2,3-isopropylidene-*L*-sorbose with diphenyl-chlorophosphonate (*115*). The reaction product contained both the sorbose-6-phosphate and 1,6-diphosphate.

IX. Disaccharide Phosphates.

Trehalose-monophosphate. Trehalose is a disaccharide formed by two α-*D*-glucopyranose residues which are linked through their glycosidic hydroxyls. Its monophosphate was discovered by ROBISON and MORGAN (*161*) as a product of dried yeast fermentation of sugar. None could be obtained with yeast juice. This compound was detected by its high dextrorotation and the lack of reducing power. Separation from the hexosemonophosphates was effected by fractional precipitation of the barium salts with 30—70 per cent alcohol, followed by crystallization of the sparingly soluble brucine salt. A crystalline barium salt containing about 5 molecules of water was also described [ROBISON and MORGAN (*162*)]. Hydrolysis for five hours, in 2 *N*-acid at 100°, converted this substance into glucose and a hexose phosphate which showed the properties of glucose-6-phosphate. The linkage between the glucose residues hydrolyzes faster than does the phosphate link. Thus, upon heating at 100° in 0,2 *N*-acid for 15 hours, the reducing power reached half of its final value while only 15% of the phosphate became inorganic.

After removing the phosphate with phosphatase trehalose was obtained in crystalline form.

The conditions of the fermenting mixture which favor the accumulation of trehalose-monophosphate were mentioned above. Some experiments, in which this compound was added to dried yeast, showed that it was fermented slightly faster than saccharose.

HELFERICH, LÖWA, NIPPE and RIEDEL (*57*) prepared a trehalose phosphate by the action of phosphoryl chloride on unprotected trehalose.

Maltose-1-phosphate is identical with α-glucose-1-phosphate except for an α-glucopyranosyl residue joined at position 4. It was obtained synthetically by the interaction of bromoacetylmaltose and trisilver phosphate [MEAGHER and HASSID (*116*)]. The values for the specific rotation, hydrolysis constant and acid strength are given in the tables (p. 61—64). Periodate oxidation consumed three moles of the oxidant and produced one mole of formic acid.

Maltose-1-phosphate was tested with negative results as a possible substrate for potato phosphorylase.

Unidentified disaccharide-monophosphate. NEUBERG and LEIBOWITZ (*127*) incubated fructose-diphosphate with *B. Delbrücki* in the presence of toluene and they isolated from the reaction mixture a barium salt which had the composition of a disaccharide monophosphate. The yield was about 30%. The substance had reducing properties, and the specific rotation was, $[\alpha]_D = + 38°$ for the barium salt, and $+ 55°$ for the free acid.

X. Miscellaneous Phosphate Esters.

Methyltetrose-phosphate was obtained by MEYERHOF, LOHMANN and SCHUSTER (*118*) by the action of aldolase on triose-phosphate and acetaldehyde, in the form of its crystalline silver salt. After treatment with phosphatase, the sugar component could be isolated and was found to be non-fermentable.

Ketoheptose-monophosphate. ROBISON, MACFARLANE and TAZELAAR (*160*) observed that the monophosphate fraction of yeast juice fermentation products gave a green coloration when heated with orcinol and hydrochloric acid. The green substance showed, after extraction with amyl alcohol, a maximum extinction at about 670 mμ and could thus be distinguished from the colored substance produced by pentoses in the absence of ferric ions. It was found that mannoketoheptose gave the same reaction.

Purification of the esters was carried out by treatment with bromine in neutral solution. The acids formed from the aldose esters could then be removed as insoluble barium salts. Upon fractional crystallization of the brucine salts an apparently homogeneous product was obtained, and its composition agreed with that of a ketoheptose monophosphate. Hydrolysis with bone phosphatase yielded a non-fermentable sugar which could not be identified and which gave the orcinol reaction mentioned.

The rate of acid hydrolysis (Table 3, p. 61) was practically identical with that of fructose-6-phosphate so that ROBISON et al. were inclined to believe that the phosphate group did not occupy the position 1.

It was suggested that ketoheptose-monophosphate might be formed by aldolase from dihydroxyacetone phosphate and a tetrose, followed by phosphate group migration.

References.

1. ALBAUM, H. G. and W. W. UMBREIT: Differentiation Between Ribose-3-phosphate and Ribose-5-phosphate by Means of the Orcinol-Pentose Reaction. J. biol. Chemistry **167**, 369 (1947).

2. ATHERTON, F. R., H. T. HOWARD and A. R. TODD: Studies on Phosphorylation. IV. Further Studies on the Use of Dibenzyl Chlorophosphonate and the Examination of Certain Alternative Phosphorylation Methods. J. chem. Soc. (London) **1948**, 1106.

3. ATHERTON, F. R., H. T. OPENSHAW and A. R. TODD: Studies on Phosphorylation. I. Dibenzyl Chlorophosphonate as a Phosphorylating Agent. J. chem. Soc. (London) 1945, 382.

4. BADDILEY, J., V. M. CLARK, J. J. MICHALSKI and A. R. TODD: Studies on Phosphorylation. V. The Reaction of Tertiary Bases with Esters of Phosphorous, Phosphoric and Pyrophosphoric Acids. A New Method of Selective Debenzylation. J. chem. Soc. (London) 1949, 815.

5. BADDILEY, J., A. M. MICHELSON and A. R. TODD: Nucleotides. II. A Synthesis of Adenosine Triphosphate. J. chem. Soc. (London) 1949, 582.

6. BAER, E.: d, l-Glyceraldehyde-3-phosphoric Acid. In: CARTER, Biochemical Preparations, Vol. I, p. 30. New York: Wiley & Sons. 1949.

7. BAILLY, O.: The Phosphoric Esters of Glycerol. Ann. Chim. 6, 133 (1916) [Chem. Abstr. 10, 3069 (1916)].

8. BERGER, L., M. W. SLEIN, S. P. COLOWICK and C. F. CORI: Isolation of Hexokinase from Baker's Yeast. J. gen. Physiol. 29, 379 (1945/46).

9. BERNFELD, P., CL. DE TRAZ et CH. GAUTIER: Recherches sur l'amidon. XXVII. Prescription pour la préparation de glucose-1-phosphate. Helv. chim. Acta 27, 843 (1944).

10. BLUMENTHAL, E. and J. B. M. HERBERT: The Mechanism of the Hydrolysis of Trimethyl Orthophosphate. Trans. Faraday Soc. 41, 611 (1945).

11. BRIGL, P. u. H. MÜLLER: Zur Synthese von Phosphorsäureestern. I. Mitt. Ber. dtsch. chem. Ges. 72, 2121 (1939).

12. BRITTON, H. T. S. and R. A. ROBINSON: The Use of the Glass Electrode in Titrimetric Work and Precipitation Reactions. The Application of the Principle of the Solubility Product to Basic Precipitates. Trans. Faraday Soc. 28, 531 (1932).

13. BROWN, D. M., L. J. HAYNES and A. R. TODD: Synthetic Ribonucleoside-2'-phosphates: a Correction. J. chem. Soc. (London) 1950, 408.

14. CANTOR, S. M. and Q. P. PENISTON: The Reduction of Aldoses at the Dropping Mercury Cathode: Estimation of the Aldehydo Structure in Aqueous Solutions. J. Amer. chem. Soc. 62, 2113 (1940).

15. CAPUTTO, R., L. F. LELOIR, C. E. CARDINI and A. C. PALADINI: Isolation of the Coenzyme of the Galactose Phosphate-glucose Phosphate Transformation. J. biol. Chemistry 184, 333 (1950).

16. CAPUTTO, R., L. F. LELOIR and R. E. TRUCCO: Lactase and Lactose Fermentation in Saccharomyces Fragilis. Enzymologia 12, 350 (1948).

16a. CAPUTTO, R., L. F. LELOIR, R. E. TRUCCO, C. E. CARDINI and A. C. PALADINI: The Enzymatic Transformation of Galactose into Glucose Derivatives. J. biol. Chemistry 179, 497 (1949).

17. CARDINI, C. E., A. C. PALADINI, R. CAPUTTO and L. F. LELOIR: Uridine Diphosphate Glucose: the Coenzyme of the Galactose-glucose Phosphate Isomerization. Nature (London) 165, 191 (1950).

18. CARDINI, C. E., A. C. PALADINI, R. CAPUTTO, L. F. LELOIR and R. E. TRUCCO: The Isolation of the Coenzyme of Phosphoglucomutase. Arch. Biochemistry 22, 87 (1949).

19. COHEN, S. S. and D. B. MCNAIR SCOTT: Formation of Pentose Phosphate from 6-Phosphogluconate. Science (New York) 111, 543 (1950).

20. COHN, M.: Mechanisms of Cleavage of Glucose-1-phosphate. J. biol. Chemistry 180, 771 (1949).

21. COLOWICK, S. P.: Synthetic Mannose-1-phosphoric Acid and Galactose-1-phosphoric Acid. J. biol. Chemistry 124, 557 (1938).

22. COLOWICK, S. P. and E. W. SUTHERLAND: Polysaccharide Synthesis from Glucose by Means of Purified Enzymes. J. biol. Chemistry 144, 423 (1942).

23. CORI, C. F.: Phosphorylation of Glycogen and Glucose. Biol. Symposia 5, 131 (1941).

24. CORI, C. F., S. P. COLOWICK and G. T. CORI: The Isolation and Synthesis of Glucose-1-phosphoric Acid. J. biol. Chemistry 121, 465 (1937).

25. COURTOIS, J.: Les esters phosphoriques des oses et holosides. Bull. Soc. Chim. biol. 23, 133 (1941); Annales Fermentat. 6, 1 (1941).

26. — Action de l'acide périodique sur l'acide hexose-diphosphorique. Bull. Soc. chim. France 9, 136 (1942).

27. — Emplois analytiques de l'acide périodique dans la préparation et l'essai des médicaments. Produits pharm. 2, 5, 65 (1942).

28. — Recherches sur le saccharosephosphate. I. Étude de l'ester phosphorique libéré par hydrolyse oxalique du saccharose-phosphate. Annales Fermentat. 8, 105 (1943).

29. COURTOIS, J. et M. RAMET: Recherches sur le diosephosphate. V. Sa formation par oxydation du fructose-6-phosphate. Bull. Soc. chim. France 11, 539 (1944).

30. — — Les produits d'hydrolyse acide du saccharosephosphate. C. R. hebd. Séances Acad. Sci. 218, 360 (1944).

31. — — Recherches sur le saccharosephosphate. II. Oxidation par l'acide périodique du glucosephosphate obtenu par l'hydrolyse acide du saccharosephosphate. Bull. Soc. Chim. biol. 27, 610 (1945).

32. — — Recherches sur le saccharosephosphate. III. Étude de l'acide phosphogluconique dérivant du glucosephosphate libéré par l'hydrolyse du saccharosephosphate. Bull. Soc. Chim. biol. 27, 614 (1945).

33. DAVOLL, J., B. LYTHGOE and A. R. TODD: Experiments on the Synthesis of Purine Nucleosides. XII. The Configuration at the Glycosidic Centre in Natural and Synthetic Pyrimidine and Purine Nucleosides. J. chem. Soc. (London) 1946, 833.

34. DESJOBERT, M. A.: L'hydrolise chimique du glucose-1-phosphate (Ester de CORI). Bull. Soc. Chim. biol. 32, 19 (1950).

35. DEUTICKE, H. J. u. S. HOLLMANN: Über das Vorkommen von Hexosediphosphorsäure im Skelettmuskel. Hoppe-Seyler's Z. physiol. Chem. 258, 160 (1939).

36. DICKENS, F.: Oxidation of Phosphohexonate and Pentose Phosphoric Acids by Yeast Enzymes. I. Oxidation of Phosphohexonate. II. Oxidation of Pentose Phosphoric Acids. Biochemic. J. 32, 1626 (1938).

37. — Yeast Fermentation of Pentose Phosphoric Acids. Biochemic. J. 32, 1645 (1938).

38. DICKENS, F. and G. E. GLOCK: Direct Oxidation of Glucose-6-phosphate by Animal Tissues. Nature (London) 166, 33 (1950).

39. EULER, H. v., P. KARRER u. B. BECKER: Charakterisierung von Zuckerphosphorsäuren und Konstitution der Pentose-phosphorsäure aus Cozymase. Helv. chim. Acta 19, 1060 (1936).

40. FARRAR, K. R.: Glucose-2-phosphate: its Preparation and Characterization by Hydrolysis Studies. J. chem. Soc. (London) 1949, 3131.

41. FISCHER, E.: Über Phosphorsäureester des Methyl-glucosids und Theophyllinglucosids. Ber. dtsch. chem. Ges. 47, 3193 (1914).

42. — Wanderung von Acyl bei den Glyceriden. Ber. dtsch. chem. Ges. 53, 1621 (1920).

43. FRIEDKIN, M.: Desoxyribose-1-phosphate. II. The Isolation of Crystalline Desoxyribose-1-phosphate. J. biol. Chemistry 184, 449 (1950).

44. FRIEDKIN, M. and H. M. KALCKAR: Desoxyribose-1-phosphate. I. The Phosphorolysis and Resynthesis of Purine Desoxyribose Nucleoside. J. biol. Chemistry 184, 437 (1950).

45. FRIEDKIN, M., H. M. KALCKAR and E. HOFF-JÖRGENSEN: Enzymatic Synthesis of Desoxyribose Nucleoside with Desoxyribose Phosphate Ester. J. biol. Chemistry 178, 527 (1949).

46. GOMORI, G.: Hexosediphosphatase. J. biol. Chemistry 148, 139 (1943).

47. GRANT, G. A.: The Metabolism of Galactose. I. Phosphorylation During Galactose Fermentation and its Relation to the Interconversion of Hexoses. Biochemic. J. 29, 1661 (1935).

48. HAAS, E.: A Colorimetric Determination for Studies Involving Coenzymes. J. biol. Chemistry 155, 333 (1944).

49. HANES, C. S.: The Breakdown and Synthesis of Starch by an Enzyme System from Pea Seeds. Proc. Roy. Soc. (London), Ser. B 128, 421 (1939).

50. — The Reversible Formation of Starch from Glucose-1-phosphate Catalysed by Potato Phosphorylase. Proc. Roy. Soc. (London), Ser. B 129, 174 (1940).

51. HANES, C. S. and F. A. ISHERWOOD: Separation of the Phosphoric Ester on the Filter Paper Chromatogram. Nature (London) 164, 1107 (1949).

52. HARDEGGER, E. u. J. DE PASCUAL: Glucoside und β-1,3,4,6-Tetracetylglucose aus Triacetyl Glucosan-α-(1,2)-β-(1,5). Helv. chim. Acta 31, 281 (1948).

53. HASSEL, O. and B. OTTAR: The Structure of Molecules Containing Cyclohexane or Pyranose Rings. Acta chem. Scand. 1, 929 (1947).

54. HASTINGS, A. B. and D. D. VAN SLYKE: The Determination of the Three Dissociation Constants of Citric Acid. J. biol. Chemistry 53, 269 (1922).

55. HATANO, J.: Über die partielle Hydrolyse der Rohrzucker-phosphorsäure zu d-Fructose und d-Glucose-phosphorsäure. Biochem. Z. 159, 175 (1925).

56. HELFERICH, B. u. W. KLEIN: Zur Acylwanderung. Liebigs Ann. Chem. 455, 173 (1927).

57. HELFERICH, B., A. LÖWA, W. NIPPE u. H. RIEDEL: Über die Einwirkung von Fermenten auf Schwefelsäure- und Phosphorsäure-ester der Zucker und ihrer Derivate. Hoppe-Seyler's Z. physiol. Chem. 128, 141 (1923).

58. HELFERICH, B. u. A. MÜLLER: Über den Methyl-glykosid eines neuen Anhydro-Zuckers; zugleich Beitrag zur Acyl-Wanderung bei partiell acylierter Glucose. Ber. dtsch. chem. Ges. 63, 2142 (1930).

59. HIRST, E. L., L. HOUGH and J. K. N. JONES: Quantitative Analysis of Mixtures of Sugars by the Method of Partition Chromatography. II. The Separation and Determination of Methylated Aldoses. J. chem. Soc. (London) 1949, 928.

60. HOWARD, G. A., B. LYTHGOE and A. R. TODD: A Synthesis of Cytidine. J. chem. Soc. (London) 1947, 1052.

61. JACKSON, E. L.: Periodic Acid Oxidation. Organic Reactions, Vol. 3, p. 341. New York: Wiley & Sons. 1944.

62. JEPHCOTT, C. M. and R. ROBISON: Mannose Monophosphate. II. The Fermentation of Mannose by Dried Yeast. Biochemic. J. 28, 1844 (1934).

63. JERMYN, M. A. and F. A. ISHERWOOD: Improved Separation of Sugars on the Paper Partition Chromatogram. Biochemic. J. 44, 402 (1949).

64. JOSEPHSON, K. u. S. PROFFE: Über Umlagerungsreaktionen in der Kohlenhydratgruppe. Zur Kenntnis synthetischer Hexosephosphorsäureester. Liebigs Ann. Chem. 481, 91 (1930).

65. — — Zur Kenntnis synthetischer Hexosephosphorsäureester. II. Biochem. Z. 258, 147 (1933).

66. KALCKAR, H. M.: The Enzymatic Synthesis of Purine Nucleosides. J. biol. Chemistry 167, 477 (1947).

67. — The Biological Incorporation of Purines and Pyrimidines Into Nucleosides and Nucleic Acid. Biochim. biophys. Acta 4, 232 (1950).

68. KIESSLING, W.: Über die Reindarstellung von Glucose-1-phosphorsäure (CORI-Ester). Biochem. Z. 298, 421 (1938).

69. KING, E. J., R. R. McLAUGHLIN and W. T. J. MORGAN: The Methylation of Hexosemonophosphoric Ester. Biochemic. J. 25, 310 (1931).

70. KJERULF-JENSEN, K.: The Hexosemonophosphoric Acids Formed Within the Intestinal Mucosa During the Absorption of Fructose, Glucose and Galactose. Acta physiol. Scand. 4, 225 (1942).

71. — The Phosphate Esters Formed in the Liver Tissue of Rats and Rabbits During Assimilation of Hexoses and Glycerol. Acta physiol. Scand. 4, 249 (1942).

72. KOMATZU, S. and R. NODZU: Synthesis of the Phosphoric Acid Esters. II. Synthesis of Some Glucose Monophosphoric Esters and their Behaviour Towards Yeast. Mem. Coll. Sci., Kyoto Imp. Univ., Ser. A 7, 377 (1924) [Chem. Abstr. 19, 2811 (1925)].

73. KOSTERLITZ, H. W.: The Presence of a Galactose-phosphate in the Livers of Rabbits Assimilating Galactose. Biochemic. J. 31, 2217 (1937).

74. — Synthetic Galactose-1-phosphoric Acid. Biochemic. J. 33, 1087 (1939).

75. — The Structure of the Galactose-phosphate Present in the Liver During Galactose Assimilation. Biochemic. J. 37, 318 (1943).

76. — The Apparent Dissociation Constants of Galactose-1-phosphoric Acid. Biochemic. J. 37, 321 (1943).

77. — The Fermentation of Galactose and Galactose-1-phosphate. Biochemic. J. 37, 322 (1943).

78. KRAHL, M. E. and C. F. CORI: The α-Glucose-1-phosphates. In: CARTER: Biochemical Preparations, Vol. 1, p. 33. New York: Wiley & Sons. 1949.

79. KUMLER, W. D. and J. J. EILER: The Acid Strength of Mono and Diesters of Phosphoric Acid. The n-Alkyl Esters from Methyl to Butyl, the Esters of Biological Importance, and the Natural Guanidine Phosphoric Acids. J. Amer. chem. Soc. 65, 2355 (1943).

80. KUNITZ, M. and M. R. McDONALD: Crystalline Hexokinase (Heterophosphatase). J. gen. Physiol. 29, 393 (1946).

81. LAMPSON, G. P. and H. A. LARDY: Phosphoric Esters of Biological Importance. II. The Synthesis of Glucose-6-phosphate from 1,2-Isopropylidene-5,6-anhydro-D-glucofuranose. J. biol. Chemistry 181, 693 (1949).

82. — — Phosphoric Esters of Biological Importance. III. The Synthesis of Propanediol Phosphate. J. biol. Chemistry 181, 697 (1949).

83. LARDY, H. A. and H. O. L. FISCHER: Phosphoric Esters of Biological Importance. I. The Synthesis of Glucose-6-phosphate. J. biol. Chemistry 164, 513 (1946).

84. LELOIR, L. F., R. E. TRUCCO, C. E. CARDINI, A. C. PALADINI and R. CAPUTTO: The Coenzyme of Phosphoglucomutase. Arch. Biochemistry 19, 339 (1948).

85. — — — — — The Formation of Glucose Diphosphate by Escherichia coli. Arch. Biochemistry 24, 65 (1949).

86. LEVENE, P. A. and C. C. CHRISTMAN: Synthesis of 5-Phospho-d-Arabinose. J. biol. Chemistry 123, 607 (1938).

87. LEVENE, P. A. and A. DMOCHOWSKI: The Comparative Rates of Hydrolysis of Adenylic, Guanylic and Xanthylic Acids. J. biol. Chemistry 93, 563 (1931).

88. LEVENE, P. A. and S. A. HARRIS: The Ribosephosphoric Acid from Xanthylic Acid. J. biol. Chemistry 95, 755 (1932).

89. — — The Ribosephosphoric Acid from Xanthylic Acid. II. J. biol. Chemistry 98, 9 (1932).

90. — — The Ribosephosphoric Acid from Yeast Adenylic Acid. J. biol. Chemistry 101, 419 (1933).

91. LEVENE, P. A., S. A. HARRIS and E. T. STILLER: d-Ribitol-5-Phosphoric Acid. J. biol. Chemistry 105, 153 (1934).

92. LEVENE, P. A. u. W. A. JACOBS: Über die Inosinsäure. Ber. dtsch. chem. Ges. 41, 2703 (1908).

93. — — Über die Inosinsäure. Ber. dtsch. chem. Ges. 44, 746 (1911).

94. LEVENE, P. A. and E. JORPES: The Rate of Hydrolysis of Ribonucleotides. J. biol. Chemistry 81, 575 (1929).

95. LEVENE, P. A. and T. MORI: On Inosinic Acid. IV. The Structure of the Ribose Phosphoric Acid. J. biol. Chemistry 81, 215 (1929).

96. LEVENE, P. A. and A. L. RAYMOND: Hexosephosphates and Alcoholic Fermentation. J. biol. Chemistry 79, 621 (1928).

97. — — Hexosediphosphate. J biol. Chemistry 80, 633 (1928).

98. — — Glucose-3-Phosphate, Glucose-6-Phosphate and their Bearing on the Structure of ROBISON's Ester. J. biol. Chemistry 89, 479 (1930).

99. — — Hexosemonophosphate (ROBISON). Natural and Synthetic. J. biol. Chemistry 91, 751 (1931).

100. — — Hexosemonophosphates. Synthetic ROBISON Esters. J. biol. Chemistry 92, 757 (1931).

101. — — Hexosemonophosphates. Galactose-6-phosphate. J. biol. Chemistry 92, 765 (1931).

102. — — Derivatives of Monoacetone Xylose. J. biol. Chemistry 102, 317 (1933).

103. — — 3-Methyl Xylose and 5-Methyl Xylose. J. biol. Chemistry 102, 331 (1933).

104. — — Phosphoric Esters of Xylose and of 5-Methyl Monoacetone Xylose. Their Bearing on the Nature of the Pentose of Yeast Nucleic Acid. J. biol. Chemistry 102, 347 (1933).

105. — — Xylose Phosphoric Acids. II. J. biol. Chemistry 107, 75 (1934).

106. LEVENE, P. A., A. L. RAYMOND and A. WALTI: A New Case of WALDEN Inversion in the Hexose Series. J. biol. Chemistry 82, 191 (1929).

107. LEVENE, P. A. and H. S. SIMMS: Lactone Formation from Mono and Dicarboxylic Sugar Acids. J. biol. Chemistry 65, 31 (1925).

108. LEVENE, P. A. and E. T. STILLER: The Synthesis of Ribose-5-phosphoric Acid. J. biol. Chemistry 104, 299 (1934).

109. LIPPICH, F.: Die Reaktion zwischen Zucker und Cyankalium in ihren Beziehungen zum Problem der Zuckermodifikationen in wässeriger Lösung. Biochem. Z. 248, 280 (1932).

110. LOHMANN, K.: Über die Isolierung verschiedener natürlicher Phosphorsäure-Verbindungen und die Frage ihrer Einheitlichkeit. Biochem. Z. 194, 306 (1928).

111. LOUGH, A. S. and V. E. SPENCER: The Preparation of Calcium Glucose-3-phosphate from Dibrucine Glucose-3-phosphate. J. org. Chemistry 3, 541 (1939).

112. MACLEOD, M. and R. ROBISON: The Application of the Iodimetric Method to the Estimation of Small Amounts of Aldoses. Biochemic. J. 23, 517 (1929).

113. — — The Hydrolysis of Hexosediphosphoric Ester by Bone Phosphatase. A New Fructose Monophosphate. Biochemic. J. 27, 286 (1933).

114. McCready, R. M. and W. Z. Hassid: The Preparation and Purification of Glucose-1-phosphate by the Aid of Ion Exchange Adsorbents. J. Amer. chem. Soc. **66**, 560 (1944).

115. Mann, K. M. and H. A. Lardy: Phosphoric Esters of Biological Importance. V. The Synthesis of *L*-Sorbose-1-phosphate and *L*-Sorbose-6-phosphate. J. biol. Chemistry **187**, 339 (1950).

116. Meagher, W. R. and W. Z. Hassid: Synthesis of Maltose-1-phosphate and *D*-Xylose-1-phosphate. J. Amer. chem. Soc. **68**, 2135 (1946).

117. Meyerhof, O. u. K. Lohmann: Über die enzymatische Milchsäurebildung im Muskelextrakt. IV. Die Spaltung der Hexosemonophosphorsäuren. Biochem. Z. **185**, 113 (1927).

118. Meyerhof, O., K. Lohmann und Ph. Schuster: Über die Aldolase, ein Kohlenstoff-verknüpfendes Ferment. I. Aldolkondensation von Dioxyaceton-phosphorsäure mit Acetaldehyd. Biochem. Z. **286**, 301 (1936).

119. — — — Über die Aldolase, ein Kohlenstoff-verknüpfendes Ferment. II. Aldolkondensation von Dioxyacetonphosphorsäure mit Glyceraldehyd. Biochem. Z. **286**, 319 (1936).

120. Meyerhof, O. u. J. Suranyi: Über die Dissoziationskonstanten der Hexose-diphosphorsäure und Glycerinphosphorsäure. Biochem. Z. **178**, 427 (1926).

121. Meyerhof, O. and J. R. Wilson: Studies on Glycolysis of Brain Preparations. IV. Arch. Biochemistry **17**, 153 (1948).

122. Michelson, A. M. and A. R. Todd: Nucleotides. III. Mononucleotides Derived from Adenosine, Guanosine, Cytidine and Uridine. J. chem. Soc. (London) **1949**, 2476.

123. Morgan, W. T. J.: The Chemistry of Hexosediphosphoric Acid. I. Biochemic. J. **21**, 675 (1927).

124. Morgan, W. T. J. and R. Robison: Constitution of Hexose-diphosphoric Acid. II. The Dephosphorylated α- and β-Methylhexosides. Biochemic. J. **22**, 1270 (1928).

125. Neuberg, C.: Überführung der Fructose-diphosphorsäure in Fructose-mono-phosphorsäure. Biochem. Z. **88**, 432 (1918).

126. Neuberg, C. u. O. Dalmer: Kristallisierte Salze einiger physiologisch wichtiger Zucker-Phosphorsäure-Verbindungen. Biochem. Z. **131**, 188 (1922).

127. Neuberg, C. u. J. Leibowitz: Biochemische Darstellung eines Disaccharid-mono-phosphorsäure-esters. Biochem. Z. **193**, 237 (1928).

128. Neuberg, C. and H. Lustig: Preparation of *D*-Fructose-1,6-diphosphate by Means of Baker's Yeast. J. Amer. chem. Soc. **64**, 2722 (1942).

129. Neuberg, C., H. Lustig and M. A. Rothenberg: Fructose-1,6-diphosphoric Acid and Fructose-6-monophosphoric Acid. Arch. Biochemistry **3**, 33 (1944).

130. Neuberg, C. u. H. Pollak: Über Kohlenhydratphosphorsäureester. I. Über Saccharosephosphorsäure. Biochem. Z. **23**, 515 (1910).

131. — — Über Phosphorsäure und Schwefelsäure von Kohlenhydraten. Biochem. Z. **26**, 514 (1910).

132. Neuberg, C. u. S. Sabetay: Über lösliche und unlösliche Salze der Hexose-di-phosphorsäure. Biochem. Z. **161**, 240 (1925).

133. Neuberg, C. u. M. Schener: Verbindungen der Fructose Diphosphorsäure. Biochem. Z. **249**, 478 (1932).

134. Nodzu, R.: The Synthesis of the Phosphoric Acid Esters. III. Synthesis of Some Hexose-monophosphoric Acid Esters and their Behaviour Towards Yeast. J. Biochemistry **6**, 31, 49 (1926) [Chem. Abstr. **21**, 924 (1927)].

135. PACSU, E.: Action of Titanium Tetrachloride on Derivatives of Sugars. II. Preparation of Tetra-acetyl-beta-normal-hexylglucoside and its Transformation to the Alpha Form. J. Amer. chem. Soc. **52**, 2563 (1930).

136. — Action of Titanium Tetrachloride on Derivatives of Sugars. III. Transformation of Tetra-acetyl-beta-cyclohexyl-glucoside to the Alpha Form and the Preparation of Alpha-cyclohexylglucoside. J. Amer. chem. Soc. **52**, 2568 (1930).

137. — Action of Titanium Tetrachloride on Derivatives of Sugars. IV. Transformation of Hepta-acetyl-beta-methyl-cellobioside to the Alpha Form and the Preparation of Alpha-methylcellobioside. J. Amer. chem. Soc. **52**, 2571 (1930).

137a. PALADINI, A. C., R. CAPUTTO, L. F. LELOIR, R. E. TRUCCO and C. E. CARDINI: The Enzymatic Synthesis of Glucose-1,6-diphosphate. Arch. Biochemistry **23**, 55 (1949).

138. PANY, J.: Über die Isolierung von Fructose-1-phosphorsäure aus biologischem Material. Hoppe-Seyler's Z. physiol. Chem. **272**, 273 (1942).

139. PARTRIDGE, S. M.: Filter-paper Partition Chromatography of Sugars. I. General Description and Application to the Qualitative Analysis of Sugars in Apple Juice, Egg White and Foetal Blood of Sheep. Biochemic. J. **42**, 238 (1948).

140. PATWARDHAN, V. R.: Mannose Monophosphate. III. Phosphomannonic Acid and its Lactones. Biochemic. J. **28**, 1854 (1934).

141. PERCIVAL, E. E. and E. G. V. PERCIVAL: Carbohydrate Phosphoric Esters. I. The Alkaline Hydrolysis of α-Methylglucopyranoside-6-phosphate, Methylglucofuranoside-3-phosphates and Iso-propylidene Glucofuranose-3- and -6-phosphates. J. chem. Soc. (London) **1945**, 874.

142. PIGMAN, W. W. and R. M. GOEPP, Jr.: Chemistry of the Carbohydrates. New York: Academic Press. 1948.

143. PLAUT, G. W. E., S. A. KUBY and H. A. LARDY: Systems for the Separation of Phosphoric Esters by Solvent Distribution. J. biol. Chemistry **184**, 243 (1950).

144. POSTERNAK, T.: Synthesis of α- and β-Glucose-1,6-diphosphate. J. biol. Chemistry **180**, 1269 (1949).

145. — Synthesis of α-D-Glucose-1-phosphate and α-D-Galactose-1-phosphate. J. Amer. chem. Soc. **72**, 4824 (1950).

145a. — Sur le phosphore des amidons. Helv. chim. Acta **18**, 1351 (1935).

146. POTTER, A. L., J. C. SOWDEN, W. Z. HASSID and M. DOUDOROFF: α-L-Glycose-1-phosphate. J. Amer. chem. Soc. **70**, 1751 (1948).

147. RAMET, M.: Recherches sur les Esters Phosphoriques des Sucres. Contribution a l'étude de leur Oxydation par l'acide périodique. Thèse, Faculté Pharm., Paris (1945).

148. RAYMOND, A. L.: Hexosemonophosphates. Glucose-4-phosphate. J. biol. Chemistry **113**, 375 (1936).

149. RAYMOND, A. L. and P. A. LEVENE: Synthetic Hexosephosphates and their Phenylhydrazine Derivatives. J. biol. Chemistry **83**, 619 (1929).

150. REEVES, R. E.: The Shape of Pyranoside Rings. J. Amer. chem. Soc. **72**, 1499 (1950).

151. REICH, W. S.: A New Method of Phosphorylation. Nature (London) **157**, 133 (1946).

152. REITHEL, F. J.: β-D-Galactose-1-(barium phosphate). J. Amer. chem. Soc. **67**, 1056 (1945).

153. REITHEL, F. J. and C. K. CLAYCOMB: The Synthesis of Derivatives of Glucose-4-phosphoric Acid. J. Amer. chem. Soc. **71**, 3669 (1949).

154. Repetto, O. M., R. Caputto, C. E. Cardini, L. F. Leloir and A. C. Paladini: La Síntesis del Glucosa Difosfato. Ciencia e Invest. (Buenos Aires) **5**, 175 (1949); An. Asoc. quím. argent. **37**, 187 (1949).

155. Robinson, R. A.: An Aspect of the Biochemistry of the Sugars. Nature (London) **120**, 44 (1927).

156. Robison, R.: Hexosemonophosphoric Esters: Mannose-monophosphate. Biochemic. J. **26**, 2191 (1932).

157. — The Significance of Phosphoric Esters in Metabolism. New York: University Press. 1932.

158. Robison, R. and E. J. King: Hexosemonophosphoric Esters. Biochemic. J. **25**, 323 (1931).

159. Robison, R. u. M. G. Macfarlane: Biologisch wichtige Derivate der Zucker, Zwischen- und Endprodukte beim Abbau: b) Phosphorhaltige. In: Bamann-Myrbäck: Methoden der Fermentforschung, S. 296. Leipzig: G. Thieme. 1941.

160. Robison, R., M. G. Macfarlane and A. Tazelaar: A New Phosphoric Ester Isolated from the Products of Yeast Juice Fermentation. Nature (London) **142**, 114 (1938).

161. Robison, R. and W. T. J. Morgan: Trehalose Monophosphoric Ester Isolated from the Products of Fermentation of Sugars with Dried Yeast. Biochemic. J. **22**, 1277 (1928).

162. — — The Phosphoric Esters of Alcoholic Fermentation of Sugars with Dried Yeast. Biochemic. J. **24**, 119 (1930).

163. Roe, J. H.: A Colorimetric Method for the Determination of Fructose in Blood and Urine. J. biol. Chemistry **107**, 15 (1934).

164. Schlubach, H. H. u. H. E. Bartels: Über das β-Methyl-fructofuranosid. Liebigs Ann. Chem. **541**, 76 (1939).

165. Slein, M. W.: Phosphomannose Isomerase. J. biol. Chemistry **186**, 753 (1950).

166. Smythe, C. V.: Phosphoric Acid Esters from Yeast Extract. The Isolation of a Crystalline Calcium Salt Consisting of an Equimolar Mixture of Glucose-monophosphate and Glycerophosphate. J. biol. Chemistry **117**, 135 (1937).

167. — An Improved Method of Preparing Hexosemonophosphate from Yeast Extract. J. biol. Chemistry **118**, 619 (1937).

168. Stepanov, A. V. and B. N. Stepanenko: Active Form of Simple Sugars. IV. Reaction Capacity of Glucose-6-phosphate (Comparative Study of the Addition of Hydrocyanic Acid to Glucose-6-phosphate and to Glucose). Biokhimiya **2**, 917 (1937) [Chem. Abstr. **32**, 2513 (1938)].

169. — — The Active Form of Simple Sugars. VI. The Reactivity of Fructose-1-phosphate. Biokhimiya **5**, 198 (1940) [Chem. Abstr. **35**, 1385 (1941)].

170. — — Active Form of Simple Sugars. VII. Reactivity of Fructose-1,6-di-phosphate. Biokhimiya **5**, 567 (1940) [Chem. Abstr. **35**, 4742 (1941)].

171. Sumner, J. B. and F. G. Somers: Preparation of Glucose-1-phosphate. Arch. Biochemistry **4**, 11 (1944).

172. Sutherland, E. W., T. Posternak and C. F. Cori: The Mechanism of Action of Phosphoglucomutase and Phosphoglyceric Acid Mutase. J. biol. Chemistry **179**, 501 (1949).

173. Sutherland, E. W., M. Cohn, T. Posternak and C. F. Cori: The Mechanism of the Phosphoglucomutase Reaction. J. biol. Chemistry **180**, 1285 (1949).

174. Sutherland, E. W., T. Posternak and C. F. Cori: Mechanism of the Phosphoglyceric Mutase Reaction. J. biol. Chemistry **181**, 153 (1949).

175. Swanson, M. A.: Phosphatases of Liver. J. biol. Chemistry **184**, 647 (1950).

176. TANKO, B. and R. ROBISON: The Hydrolysis of Hexosediphosphoric Ester by Bone Phosphatase. II. a) The Participation of Phosphohexokinase. b) The Isolation of Pure Fructose-1-phosphate. Biochemic. J. **29**, 961 (1935).

177. TOTTON, E. L. and H. A. LARDY: Phosphoric Esters of Biological Importance. IV. The Synthesis and Biological Activity of *D*-Tagatose-6-phosphate. J. biol. Chemistry **181**, 701 (1949).

178. TRUCCO, R. E., R. CAPUTTO, L. F. LELOIR and N. MITTELMAN: Galactokinase. Arch. Biochemistry **18**, 137 (1948).

179. VAN SLYKE, D. D.: On the Measurement of Buffer Values and on the Relationship of Buffer Value to the Dissociation Constant of the Buffer and the Concentration and Reaction of the Buffer Solution. J. biol. Chemistry **52**, 525 (1922).

180. WARBURG, O. u. W. CHRISTIAN: Über ein neues Oxydationsferment und sein Absorptionsspektrum. Biochem. Z. **254**, 438 (1932).

181. — — Abbau von ROBISON-Ester durch Triphospho-pyridin-nucleotid. Biochem. Z. **292**, 287 (1932).

182. WARBURG, O., W. CHRISTIAN u. A. GRIESE: Wasserstoffübertragendes Co-Ferment, seine Zusammensetzung und Wirkungsweise. Biochem. Z. **282**, 206 (1935).

183. WEIBULL, C.: The Purification of Potato Phosphorylase and of Glucose-1-phosphate (CORI Ester). Ark. Kem., Mineral. Geol., B **21**, No. 2 (1945).

184. WILKINSON, J. F.: The Pathway of the Adaptive Fermentation of Galactose by Yeast. Biochemic. J. **44**, 460 (1949).

185. WOLFROM, M. L. and D. E. PLETCHER: The Structure of the CORI Ester. J. Amer. chem. Soc. **63**, 1050 (1941).

186. WOLFROM, M. L., C. S. SMITH, D. E. PLETCHER and A. E. BROWN: The β-Form of the CORI Ester (*d*-Glucopyranose-1-phosphate). J. Amer. chem. Soc. **64**, 23 (1942).

187. YOUNG, W. J.: Über die Zusammensetzung der durch Hefepreßsaft gebildeten Hexosephosphorsäure. Biochem. Z. **32**, 178 (1911).

188. ZEMPLÉN, G.: Neuere Richtungen der Glycosid-Synthese. Fortschr. Chem. organ. Naturstoffe **1**, 1 (1938).

189. ZERVAS, L.: A New Phosphorylation Method. 1-Glucosylphosphate. Naturwiss. **27**, 317 (1939) [Chem. Abstr. **33**, 7279 (1939)].

(Received, December 22, 1950.)

The Chemistry of Nucleotides.

By **G. W. KENNER**, Cambridge.

Contents.

Introduction.

In agreement with SCHLENK (*178*) *"nucleotide"* is here taken to mean a phosphoric ester of an N-glycoside of a heterocyclic base. Although they fall outside the scope of this definition, phosphorylated derivatives of riboflavin are added in recognition of their close relation to the main body of nucleotides.

A complete review of even the organic chemistry of nucleotides and their derivatives would now form a substantial monograph. The smaller and less characteristic fragments of the nucleotides, namely the hetero-cyclic bases, the component sugars and the derived sugar phosphates, are therefore excluded from this article.* Their chemistry is of longer standing and is covered in standard works. On the other hand knowledge of the *"nucleosides"*, the N-glycosides of heterocyclic bases, has expanded very considerably in recent years and an account of the relevant work is therefore included. With regard to the nucleotides themselves attention is focussed on work published in the last few years leading to improved characterisation, more exact structural knowledge and increased availability. Only such enzymatic experiments as bear directly on structural problems are discussed and therefore much fascinating work by KALCKAR, FRIEDKIN et al. and by KORNBERG has been omitted.

It would be unthinkable to present a review on nucleotides without mentioning that the foundations of the subject were laid by P. A. LEVENE. His monograph with BASS (*136*) was published in 1931 and two reviews by LYTHGOE (*150, 151*) cover the intervening period until 1944. Since then a considerable portion of the advances in the field have come from the work of A. R. TODD and his collaborators (cf. *185*). It is a pleasure for the author to record that he has benefited from the advice and criticisms of Professor TODD in preparing this manuscript. Although it is not claimed that the present article is complete, as other reviews of a slightly different character have appeared in the meantime (e. g. *66, 178, 184*), it is intended to provide a continuation of LYTHGOE's reviews up to the end of 1950.

The plan of the article will be clear from the table of contents. Outside a conventional classification of nucleotide work a section is devoted to a discussion of methods for handling nucleotides and a sub-section to methods for preparing nucleotides from nucleosides. Broadly speaking the main body reviews work which has more or less completely solved problems in the nucleotide field, whilst immediate future progress will depend on the application of the methods described in these two sections.

The structural formulae of most of the common nucleosides and nucleotides are to be found in the sections dealing with work on their constitution. The *method*

---

* For Sugar Phosphates cf. p. 47 of the present volume.

of numbering the pyrimidine and purine nucleoside ring systems will be apparent from the following diagrams:

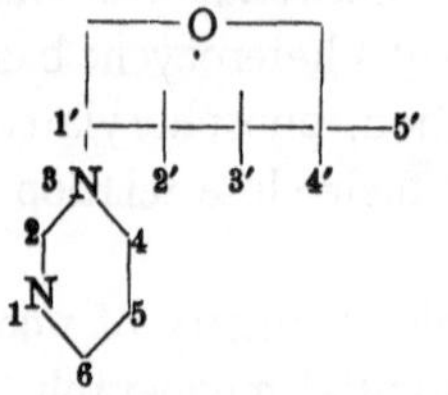

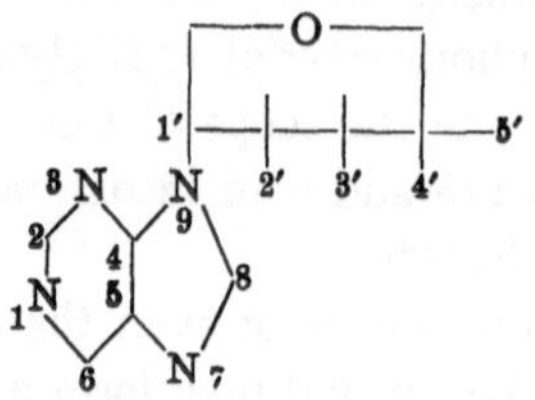

Customary *abbreviations* for the names of the more complex nucleotides are:

ADP for adenosine-5′-diphosphate (LXXIII, p. 134).

ATP for adenosine-5′-triphosphate (LXXVI, p. 134).

DPN for diphosphopyridine nucleotide (coenzyme 1) (L, p. 124).

FAD for flavin adenine dinucleotide.

FMN for flavin mononucleotide (LXX, p. 132).

UDPG from uridine diphosphate glucose (coenzyme of "galactowaldenase") (LII, p. 124).

TPN for triphosphopyridine nucleotide (coenzyme 2) (LI, p. 124).

I. Methods.

The paucity of reliable chemical information about the polynucleotides is due to the fact that they are combinations of three very different chemical classes, the heterocyclic bases, the sugars, and the phosphoric acids. Only in recent years have methods come to the fore which are capable of handling these intractable substances and has considerable progress been made with the simpler nucleotides. No apology is therefore needed for devoting a special section to methods of wide application.

Three important methods for the separation of nucleosides and nucleotides have been introduced. In order of decreasing theoretical simplicity they are counter-current distribution, ion exchange, and partition chromatography. The first may be used preparatively and gives evidence of the homogeneity of substances. The second is capable of very high resolving powers, but in preparative work involves the handling of larger quantities of solutions. The third is essentially a method of rapid micro-analysis.

The purified substances may be identified and estimated through their behaviour towards ultraviolet, infrared or X-radiation. Periodate oxidation gives valuable information about the ring-structure and configuration of the sugars and the location of attached phosphate groups.

1. Countercurrent Distribution.

One of the most useful methods of purifying complex substances is that of countercurrent distribution developed by Craig (*62—64, 194*). In this, one solvent phase is allowed to flow discontinuously over a second with which it is in equilibrium, equilibrium of solutes between the phases being ensured at each step. The attainment of high resolving power through many steps is tedious and requires complex apparatus, but the fundamental simplicity of the technique permits an easy and exact correlation with theory and thus a test of the homogeneity of substances separated by it. For efficient operation the partition coefficient of the substance between the two solvent phases should be about unity. This condition is fairly easy to satisfy in the case of the nucleosides, which may be extracted from aqueous solution even by chloroform-ethanol mixtures (*176*).

TINKER and BROWN (*183*) have demonstrated the applicability of the method to nucleosides using the system *n*-butanol/M-phosphate buffer p_H 6,5. The method might well find greater application in the separation of nucleotides, which are much less easy to purify and characterise by other means, for instance crystallisation and melting point, and which occur in more complex mixtures of closely related compounds. Their phosphoric ester groups however make the nucleotides more reluctant to dissolve in the organic phase and so to give a suitable partition coefficient. In a comprehensive study of the partition of riboflavin between different solvent pairs GREENE and BLACK (*88*) found that high partition coefficients with respect to an aqueous phase could be obtained with quinoline and particularly with phenols. The only recorded counter-current distribution of a nucleotide is that of diphospho-pyridine nucleotide (L, p. 124) (DPN) in phenol/water, to which had been added small amounts of ether or chloroform to adjust the partition coefficient (*100*). Despite the notorious tendency of phenolic systems to emulsify this provides a convenient process, readily adaptable to large scale work, for the preparation of pure DPN in 70–80% yields from crude concentrates. It is clear that the simultaneously polyacidic and polybasic nature of the nucleotides will make their partition coefficients sensitive to variations in p_H. Therefore it is desirable to use a buffered aqueous phase (*63*), although the recovery of the products is now more complicated than with simple phenol/water systems. PLAUT, KUBY and LARDY (*168*) have described solvent systems, consisting of an alcohol (e. g. *n*-octanol), a buffer and a long chain amine (e. g. octade-cylamine), which would appear to be very suitable for the separation of nucleotides. Reports of their application to actual separations will be awaited with interest.

2. Ion Exchange.

The ion exchange resins (*67, 80, 125, 186*) are commonly divided into two groups, cation and anion exchangers, depending on whether they contain acidic functions and therefore tend to retain or exchange cations or whether they contain basic functions and thus retain different anions to varying extents. Cation exchange power may be derived from the presence of sulphonic acid groups in the resin (as in Dowex-50, Amberlite IR-105) or from carboxylic acid groups (Amberlite IRC-50, Zeokarb 216). Anion exchange resins correspondingly may contain quaternary ammonium groups (Dowex-1,2, Amberlite IRA-400) or amino groups (Amberlite IR-4-B, Deacidite). Resins of the first type in both classes, i. e. those with strongly dissociated functional groups, have been preferred for column separations.

Nucleotides contain both acidic and basic groups and should therefore be separable on either class of resin. In his first experiments (*52*) Cohn used a cation exchange resin (Dowex-50), but later (*53*) pointed out the advantages of anion exchange resins for nucleotide work. According to him the affinities of a structurally closely related series of compounds for an anion exchange resin should be determined by their net negative charge. In fact the orders of elution of the ribomononucleotides at different pH values were found to correspond well with the net charges calculated for those pH values from the data of Levene and Bass (*136*), except that the pyrimidine nucleotides tended to be more easily eluted than the less soluble purine nucleotides.

Perhaps the most satisfactory theory of ion exchange columns is that due to Mayer and Tompkins (*160*). They consider a column as consisting of a number of sections or "theoretical plates", in each of which the resin surface and the displacing solution are brought into perfect equilibrium. By applying the conditions that the substance passing down the column should be distributed between the adsorbed and dissolved phases of each plate in a constant ratio and that there should be material balance between adjacent plates, they are able to derive an expression describing the distribution of the substance throughout the column or alternatively its rate of flow from the bottom of the column.

It will be seen that according to this theory an ion exchange column is directly comparable to a single withdrawal countercurrent distribution. The possibility therefore exists of checking the purity of a substance by comparing its experimental and theoretical elution curves of concentration against volume of eluate. Indeed Cohn (*54*) has been able to show a close correspondence between the theoretical and experimental curves for the separation of uridylic and guanylic acids on Dowex-2. This test of purity is, however, not so easy to apply as in the case of countercurrent distribution, since the attainment of true equilibrium depends on sufficiently low rates of flow and small size of resin particles as well as on an eluting solvent of constant pH and ionic strength; there is therefore a greater chance that deviations from the theoretical curve result from imperfections of technique rather than impurity of the substance. On the other hand the number of theoretical plates in COHN's experiments was between one and two hundred, varying with the substance, and hence the ion exchange column is a much less tedious way of achieving high resolving power than a countercurrent distribution, and its successes in the nucleotide field are already considerable.

In analytical application the method has separated into two components "yeast adenylic acid" (43, 54) and later also uridylic (55), cytidylic (55, 147), and guanylic (54) acids from yeast. It is a relatively easy task to separate quantitatively adenosine, adenine, muscle adenylic acid, adenosine diphosphate (ADP) and triphosphate (ATP) (57), the substances being eluted in turn by solvents of increasing ionic strength (cf. 126). In this modification ion exchange chromatography is comparable to the familiar elution chromatography on alumina and the possibility of applying the simple theory to it is lost. A cation exchange column affords an easy separation of guanylic and uridylic acids from adenylic and cytidylic (164). Relatively large scale preparative isolations may also be carried out, for instance of cytidylic and uridylic acids anionically (56), and of cytidine and uridine cationically (81, 97). Evaporation of large volumes of effluent may sometimes be avoided through readsorption on a much smaller column (54).

3. Paper Chromatography.

Partition chromatography (193) has been applied in the nucleotide field using columns of starch (171, 172) and cellulose (40, 74) but most investigators have preferred the simpler and more rapid technique of paper strips. By this method many analyses of minute amounts of material can be made with little labour, once a suitable solvent system and method of detecting the spots of substances have been found.

The familiar aqueous butanol system has been widely favoured in one form or another (39, 42, 65, 102, 159) for the analysis of nucleoside mixtures, but does not cause migration of the nucleotides. Aqueous isobutyric acid buffered with ammonium isobutyrate gives a good spread of the ribonucleotides and permits their more definite identification through the characteristic changes in R_F with p_H of the solvent (157), which have the same fundamental cause as the variations with p_H in partition coefficient and in adsorption on ion exchange resins. CARTER (42, 57) has introduced the novel technique of allowing the end of the paper to be in contact with both an aqueous salt solution (e. g. 5% Na_2HPO_4) and an organic phase (e. g. isoamyl alcohol). This very useful system has sufficient resolving power to separate cleanly the isomeric adenylic acids "a" and "b".

Pyrimidine and purine derivatives are readily located on paper either visually by their fluorescence in ultraviolet light (42) or photographically through their ultraviolet absorption (159): the latter method has the advantage of giving a permanent and direct record of the experiment. After this stage the appropriate pieces of the paper may be cut out and the substances estimated quantitatively by the ultraviolet absorption of eluates from the paper (102). These methods are virtually incompatible with the use of collidine or phenols, which have desirable properties otherwise (109, 164). In some cases it may be helpful to use in the spray technique the colour reactions of different groups, for instance of phosphates (96, 157, 164), deoxy-sugars (39) or 1 : 2-glycols (39). Riboflavin derivatives show a characteristic green fluorescence in ultraviolet light (65, 86).

4. Use of Physical Properties.

The characteristic ultraviolet absorption spectra of the pyrimidines and purines afford the most convenient way of estimating and identifying the nitrogenous components of substances separated as described in the foregoing sections. The classical work of GULLAND and HOLIDAY on the structure of the nucleosides need not be recapitulated here (cf. 150), but attention may be drawn to some recent provision of useful data concerning both nucleosides (102, 169) and nucleotides (112a, 169).

A start has been made in the application of infrared spectroscopy (29, 38), and in due course the method should be available both for the detection of functional groups and for the "finger print identification" of nucleotides. The rate of progress in this direction is limited by difficulties in the preparation of suitable specimens.

In the meantime the examination of X-ray powder diagrams provides much firmer characterisation of the nucleotides than do the methods of melting points, optical rotations and crystal habits (36). FURBERG (87) has completed the X-ray structural analysis of cytidine. His results confirm the structure assigned on chemical grounds and in addition give valuable information about the interatomic distances and bond angles of the β-D-ribofuranoside ring.

5. Periodate Oxidation.

The periodate ion, IO_4^-, cleaves the carbon-carbon link of 1 : 2-glycols and of 2-hydroxyaldehydes as shown:

$$—CHOH—CHOH— \xrightarrow{IO_4^-} —CHO+OCH—+IO_3^-+H_2O$$

$$—CHOH—CHO \xrightarrow{IO_4^-} —CHO+CHO\cdot OH+IO_3^-$$

The reaction (110) is normally carried out in aqueous solution and proceeds in quantitative yield. It may therefore be conveniently applied to structural analysis in the nucleoside and nucleotide fields (*153*) by measuring the extent of reaction through an iodometric method requiring only small amounts of material. By this means, in combination with synthetical experiments, it has been proven that the purine ribonucleosides are 9-β-D-ribofuranosides (*70, 152*), that the pyrimidine ribonucleosides are 3-β-D-ribofuranosides (*70*) and that the deoxyribonucleosides are furanosides (*37*). Of greater interest, now that these problems have been solved, is the fact that only ribonucleotides in which both the 2' and 3' hydroxyls are unesterified will consume one mole of periodate. For ۔nstance the attack of ATP by periodate eliminates suggested structures involving a 3'-phosphate group (*154*). On the other hand the nucleotide fragment of vitamin B_{12} must bear the phosphate at 2' or 3' of the ribofuranose group (*40*). It will be noted that in the latter case the exploratory work with minute quantities was greatly aided by the technique of spraying a paper chromatogram with periodate solution and of detecting by Feulgen's reagent any dialdehydes formed (*39*). This device can clearly be of great help in many future investigations.

II. Ribonucleosides.

1. Isolation.

The purine ribosides are available from yeast nucleic acid through the standard method of Bredereck, Martini and Richter (*32*). Uridine and cytidine are conveniently separated from the liquors of this preparation by use of cation exchange resins (*81*). An alternative simple preparation of uridine (*148*) involves the destruction of the purine ribosides.

By direct acetone extraction of sponges Bergmann and Feeney (*28*) have isolated a new nucleoside, $C_{10}H_{14}N_2O_6$, $[\alpha]_D^{23} + 80°$; (c, 1,1 in 8% NaOH), which they name "spongothymidine". On vigorous acid hydrolysis it yields thymine (5-methyluracil), and its ultraviolet absorption spectrum is very similar to that of thymidine (XLI, p. 118). Acylation confirms the presence of three hydroxyl groups, and the uptake of only one mole of periodate without production of formic acid shows it to be a furanoside. On account of the high positive rotation Bergmann and Feeney suggest that the sugar may by xylofuranose rather than ribofuranose. The stability of the pyrimidine nucleosides makes it difficult to isolate the sugar from them unless the heterocyclic ring can first be hydrogenated (*140*).

In this connection it may be recalled that the occurrence of nucleosides derived from L-lyxose has been reported. Gulland and collaborators (*89*) have prepared from hydrolysates of yeast nucleic acid two substances initially regarded as the benziminazole derivatives of D-ribose and L-lyxose (I). Characterisation of sugars through these derivatives is complicated by epimerisation phenomena. In a second paper (*22*) it is made clear that the first substance is in fact *D*-arabobenziminazole, resulting from epimerisation of the true *D*-ribobenziminazole, with which the second substance isolated is probably identical (*21*). More recently SODI PALLARES and MARTINEZ GARZA (*179*) have isolated small quantities of 6:7-dimethyl-9-(*L*-1'-lyxityl)-*iso*-alloxazine, the *L*-lyxo analogue of *D*-riboflavin (LXVIII, p. 132), from human heart muscle and have identified it by comparison with a synthetic specimen (*114*). This discovery followed the isolation of *L*-lyxose itself from heart muscle (*180*). Although the sugar can hardly be derived from the flavin by degradation, it may well be a source of both the new flavin and of hitherto unknown purine and pyrimidine nucleosides.

(I.)

(II.) 1-α-*D*-Ribofuranosido-5:6-dimethylbenziminazole.

A nucleoside containing a base quite different from those of hitherto known nucleosides has been isolated as an acid degradation product of vitamin B_{12} (*34*, *40*). Synthesis has proved it to be 1-α-*D*-ribofuranosido-5:6-dimethylbenziminazole (II) (*34*).

2. Structure.

After the identities of the constituent base and sugar have been established the remaining problems of nucleoside structure are the position of attachment of the sugar to the base, the size of the glycoside ring and the configuration of the glycosidic centre. Strong evidence

(III.) [*X* = OH] Uridine.
(IV.) [*X* = NH₂] Cytidine.

(V.) [*X* = OH] Inosine.
(VI.) [*X* = NH₂] Adenosine.

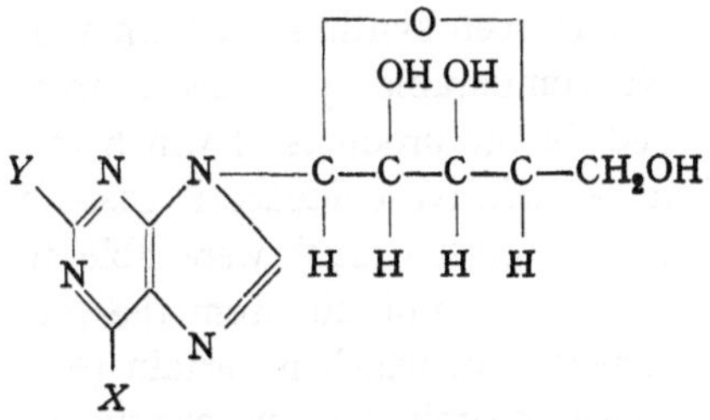

$$
\begin{array}{c}
\lceil\text{---O---}\rceil \\
\text{OH OH} \\
Y\text{---N}\quad\text{N---C---C---C---C---CH}_2\text{OH} \\
\text{H H H H}
\end{array}
$$

(VII.) [X = OH, Y = OH] Xanthosine.

(VIII.) [X = OH, Y = NH$_2$] Guanosine.

(IX.) [X = NH$_2$, Y = OH] Crotonoside.

regarding all three structural features is most readily obtained through periodate oxidation of the material and related synthetic substances.

Adenine hexopyranosides (X), in which the sugar is certainly attached to $N_{(9)}$ of the purine, are easily available through syntheses of type B discussed in the next section (p. 112). These compounds consume two moles of periodate producing formic acid and a dialdehyde (XI), similar to that obtained from a pentofuranoside (XII) by oxidation with one mole of periodate.

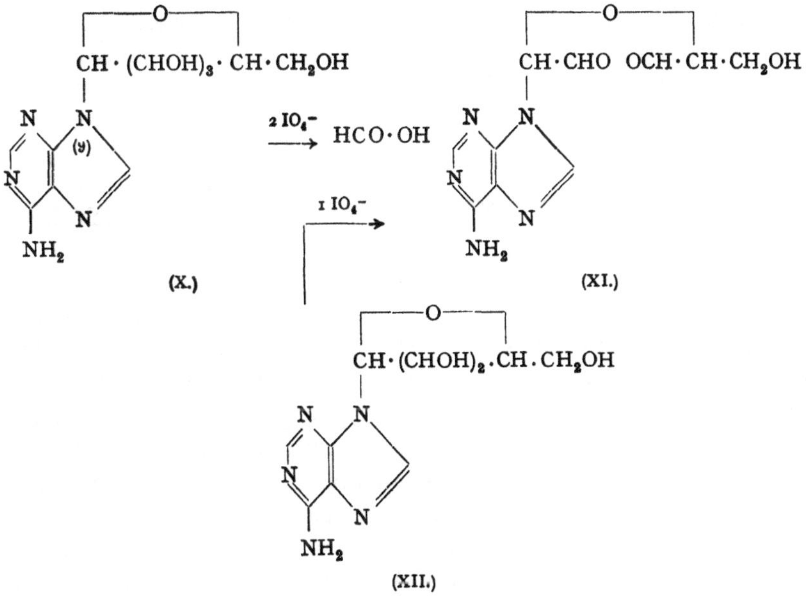

$$
\text{CH}\cdot(\text{CHOH})_3\cdot\text{CH}\cdot\text{CH}_2\text{OH}
$$

$$
\xrightarrow[]{2\,\text{IO}_4^-}\ \text{HCO}\cdot\text{OH}
$$

$$
\text{CH}\cdot\text{CHO}\quad\text{OCH}\cdot\text{CH}\cdot\text{CH}_2\text{OH}
$$

(X.)

$$
\xrightarrow[]{1\,\text{IO}_4^-}
$$

(XI.)

$$
\text{CH}\cdot(\text{CHOH})_2\cdot\text{CH}\cdot\text{CH}_2\text{OH}
$$

(XII.)

The dialdehydes from 9-D-mannopyranosido-adenine and from adenosine were found to be identical and the location of the ribose residue on $N_{(9)}$ of adenine was thus proven (*152*). The same conclusion holds for guanosine since it has been synthesised from the same intermediate as adenosine (*72*). No comparable synthesis of pyrimidine nucleosides has yet been developed, in the products of which the sugar is definitely attached to $N_{(3)}$, but degradative evidence is easier to obtain for these compounds. Levene and Tipson (*142*) were able to methylate 2′: 3′-diacetyl-5′-trityl uridine, and to obtain from the product by hydrolysis 1-methyluracil, the structure of which is certain (*27, 112*). Uridine and cytidine, which has been deaminated by nitrous acid to uridine, are therefore $N_{(3)}$-glycosides.

Early work of Levene and Tipson had established the furanoside character of the nucleosides by the laborious methylation procedure, but the size of the glycoside ring follows directly from the quantity of periodate used in an oxidation and this is easily determined by titration (*153*). A pentofuranoside [e. g. (XII)] has only one pair of adjacent hydroxyl groups and can only consume one mole of periodate. In a pentopyranoside on the other hand there are two such pairs, two moles of periodate are consumed and one mole of formic acid is liberated.

The dialdehyde (XI) has two centres of asymmetry, one of which is derived from the original glycosidic carbon. Comparison of the dialdehyde from the natural nucleoside with that from synthetic glycosides of known configuration is therefore a method of settling the third point, viz. the configuration of the glycosidic centre in the nucleoside. Glucopyranosides formed from α-acetobromoglucose have very probably the β-configuration (cf. *103*) and have served to show that all the commonly occurring natural pyrimidine and purine ribonucleosides have the β-configuration (*70, 72*).

The soundness of these methods of structural determination has been confirmed by the total X-ray analysis of cytidine (IV) which shows it to be 3-β-D-ribofuranosidocytosine (*87*).

Adenine thiomethylpentoside was isolated from yeast in 1912, but until very recently its structure was obscure (cf. *136*). This problem has now been completely solved. Suzuki and Mori (*182*) showed that the sugar contained a methylthio group and was a thiomethylaldopentose. Weygand, Trauth and Löwenfeld (*192*) prepared the osazone (XIV) of the thiomethylpentose and oxidised it with two moles of periodate whereupon mesoxalaldehyde diphenylhydrazone (XV) was produced in more than 50% yield. The sugar is therefore a 5-thiomethylpentose. Moreover application of a new and elegant method of sugar degradation (*191*) to 6-thiomethyl-D-glucose (XVII) yielded 5-thiomethyl-D-arabinose (XVI), which was different from the unknown sugar although the osazones were identical (*192*). The sugar of the nucleoside must therefore be 5-thiomethyl-D-ribose (XIII), in agreement with the conclusions arrived at from other evidence by Satoh and Makino (*175*).

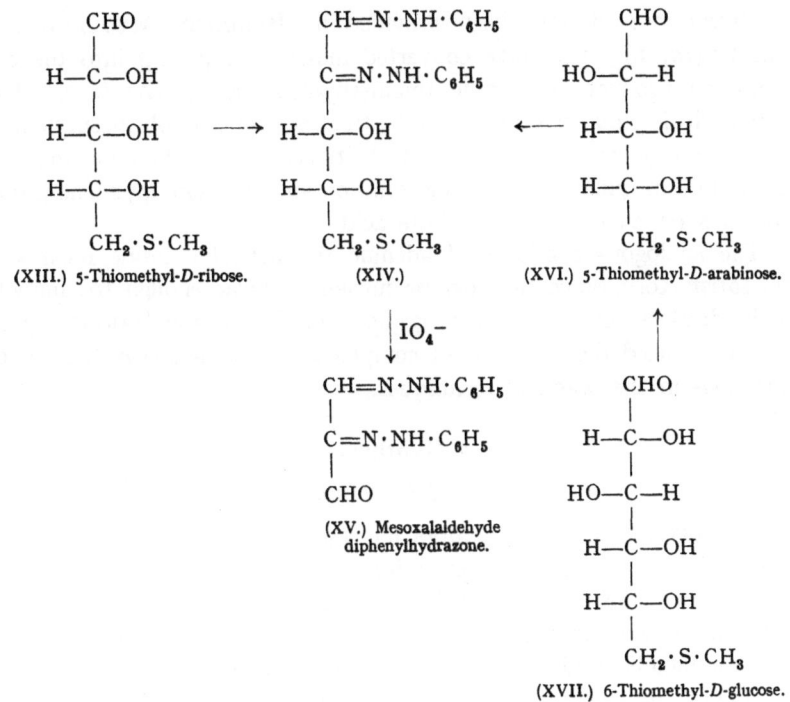

The nucleoside can hardly be other than a furanoside and it takes up the amount of periodate expected (*175*). Decisive confirmation of these deductions and the additional information that the glycosidic

(XVIII.) 5'-*p*-Toluenesulphonyl-2':3'-isopropylidene-inosine.

$K \cdot S \cdot CH_3 \longrightarrow$ $H_2SO_4/HOAc \longrightarrow$

(XIX.)

configuration is β have been provided by BADDILEY, WEYGAND and TRAUTH (*10, 18*), who have converted inosine (V, p. 104) into the deamination product of adenine thiomethylpentoside (XIX) (*123*). The 5′-methylthio group was introduced by treatment of 5′-*p*-toluenesulphonyl-2′ : 3′-isopropylidene inosine (XVIII) (*143a*) with potassium methyl-mercaptide in dimethylformamide solution and the isopropylidene group was then removed with cold dilute acid.

The analogous synthesis of adenine thiomethylpentoside itself was less satisfactory, since the adenine nucleus is basic enough to compete in the displacement of the tosyl group (*49*). SATOH and MAKINO (*175a*) have announced the independent completion of the same syntheses with variations in the practical techniques.

3. Synthesis.

a) Pyrimidine Nucleosides.

HILBERT and his collaborators (cf. *150*) were the first to elaborate a method for the synthesis of pyrimidine glycosides similar to the natural nucleosides. This consists of bringing 2 : 6-diethoxypyrimidine (XX) into reaction with an acetohalogen sugar and treating the product (XXI) with methanolic hydrogen chloride or methanolic ammonia to obtain an analogue of uridine (XXII) or cytidine (XXIII).

[R = acetylated glycosyl; R' = glycosyl.]

Application of this procedure to the syntheses of uridine and cytidine themselves by HOWARD, LYTHGOE and TODD (*107*) required the preparation of an acetohalogen-ribofuranose (XXVII). This was possible through the following series of reactions (*31, 71, 107*).

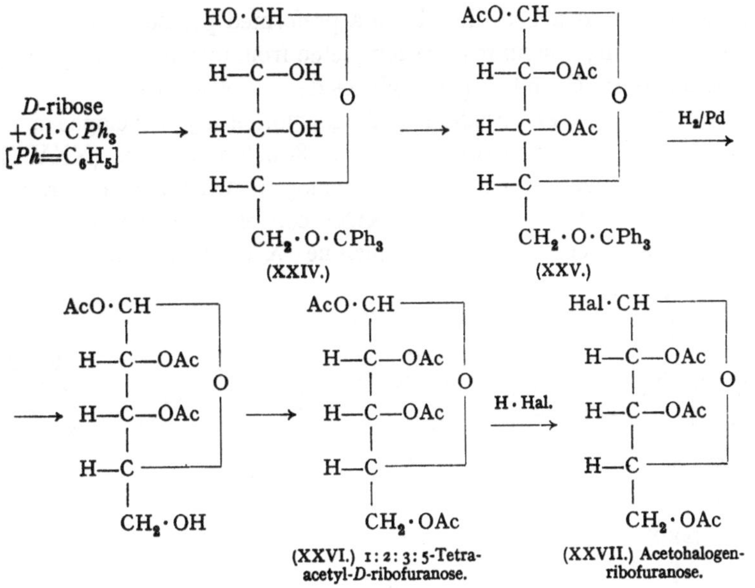

D-ribose
+Cl·CPh₃
[Ph=C₆H₅]

$$\text{D-ribose} + \text{Cl}\cdot\text{C}Ph_3 \longrightarrow$$

HO·CH ———
|
H—C—OH
| O
H—C—OH
|
H—C ———
|
CH₂·O·CPh₃

(XXIV.)

$$\longrightarrow$$

AcO·CH ———
|
H—C—OAc
| O
H—C—OAc
|
H—C ———
|
CH₂·O·CPh₃

(XXV.)

$$\xrightarrow{\text{H}_2/\text{Pd}}$$

AcO·CH ———
|
H—C—OAc
| O
H—C—OAc
|
H—C ———
|
CH₂·OH

(XXVI.)

$$\longrightarrow$$

AcO·CH ———
|
H—C—OAc
| O
H—C—OAc
|
H—C ———
|
CH₂·OAc

(XXVI.) 1:2:3:5-Tetra-
acetyl-D-ribofuranose.

$$\xrightarrow{\text{H}\cdot\text{Hal.}}$$

Hal·CH ———
|
H—C—OAc
| O
H—C—OAc
|
H—C ———
|
CH₂·OAc

(XXVII.) Acetohalogen-
ribofuranose.

Since this work an improved preparation of (XXV) (*24*) and a direct preparation of 1 : 2 : 3 : 5-tetraacetyl-D-ribofuranose (XXVI) from D-ribose (*195*) have been devised.

VISSER, GOODMAN and DITTMER (*190*) have succeeded in preparing synthetic thymine nucleosides for the first time from 2 : 4-diethoxy-5-methylpyrimidine and acetohalogen sugars. An important modification introduced by them is the continuous removal by distillation at low pressure of the ethyl halide formed during the reaction. All the compounds synthesised thus far are pyranosides (*165*).

b) Purine Nucleosides.

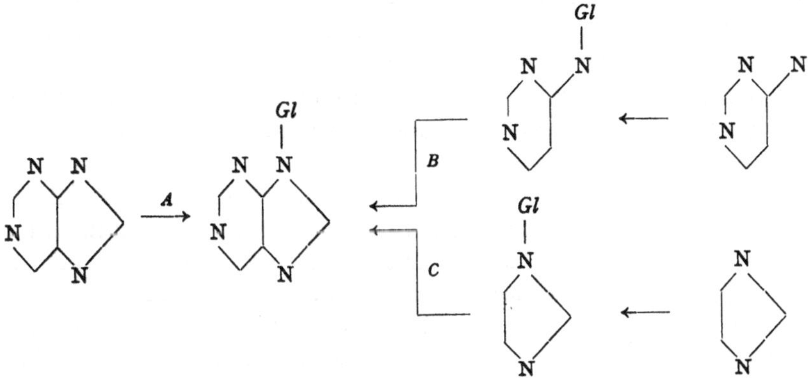

Three main types of purine nucleoside synthesis may usefully be distinguished and are represented schematically below. The sugar residue may be attached directly to a preformed purine nucleus as in *A* or the purine ring system may be completed from pyrimidine or glyoxaline glycosides as in *B* and *C* (*Gl* = glycoside or derivative).

Type A Syntheses. FISCHER and HELFERICH (*83*) developed a route of type *A*, in which the silver salt of 2 : 8-dichloroadenine (XXVIII) is brought into reaction with an acetohalogen sugar and the resulting 2 : 8-dichloroadenine glycoside (XXIX) converted as shown into the corresponding adenine (XXX) or guanine (XXXI) glycoside.

[*R* = acetylated glycosyl;   *R'* = glycosyl.]

The alternative synthesis (*101*) by route *B* of adenine glucoside identical with that of FISCHER and HELFERICH proved unambiguously that the sugar is attached to $N_{(9)}$ as shown. Further the provision of acetohalogenribofuranoses (*107*) and improvements in the preparation of 2 : 8-dichloroadenine (*70*) made possible the syntheses of adenosine (*71*) and guanosine (*72*) by DAVOLL, LYTHGOE and TODD. In analogous fashion have been prepared 9-α-D-arabofuranosido-adenine (*35*) and 9-β-D-xylofuranosido-adenine (*46*).

The direct use of the silver salt of adenine instead of 2 : 8-dichloro-adenine in these syntheses is not successful: the purine is then sufficiently basic to remove hydrogen halide from the carbohydrate reagent. DAVOLL and LOWY (*68*) have now neatly obviated this difficulty and so arrived at the most efficient currently available synthesis of purine nucleosides. In this the basicity of the amino groups of adenine or 2 : 6-diaminopurine is nullified by acetylation or benzoylation; also chloromercuric salts are used in preference to silver salts. As in the reactions with the silver salt of 2 : 8-dichloroadenine, alkylation occurs at $N_{(9)}$ rather than $N_{(7)}$. Regeneration of the amino from the acetamido group presents no difficulty. Methanolic ammonia at 0° liberates that at position 6, but the 2-acetamido group requires a sodium methoxide treatment. This differential stability of the acetamido substituents makes possible the synthesis of guanine glycosides shown below.

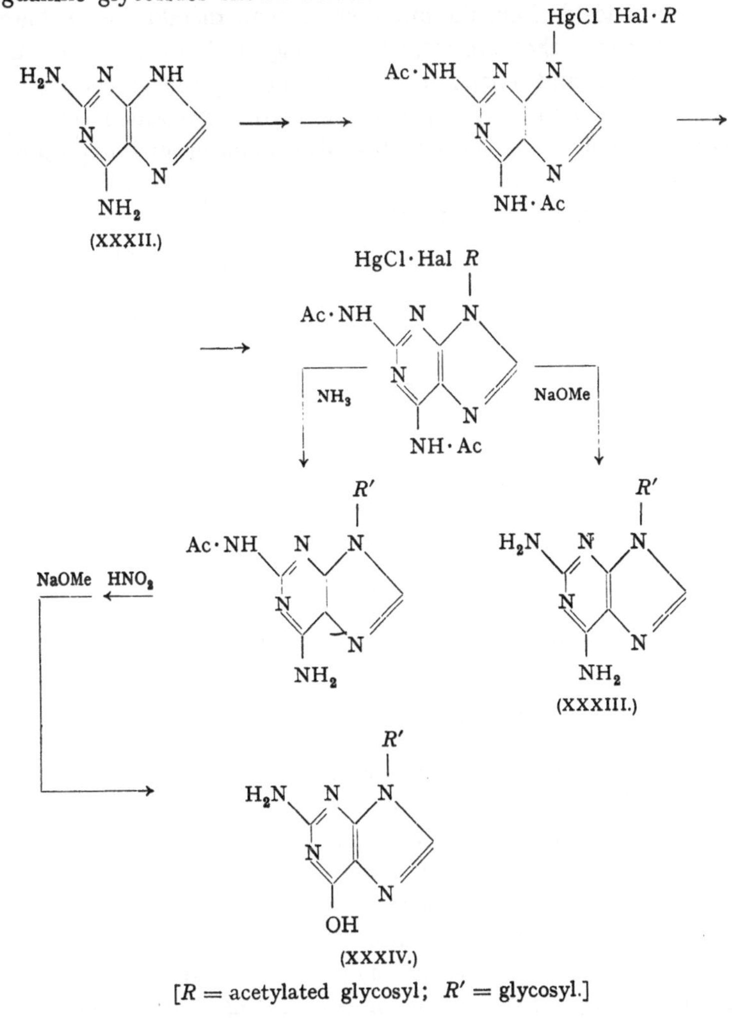

[R = acetylated glycosyl; R' = glycosyl.]

In this process acetochloro-D-ribofuranose gives an overall yield of 21% of guanosine (XXXIV; VIII) from the readily available 2 : 6-diaminopurine (XXXII), and in the corresponding simpler series of reactions an identical yield of adenosine from adenine. DAVOLL and

Lowy also indicate that selective deamination of the 2:6-diamino-9-glycosylpurines (XXXIII) to 9-glycosyl-isoguanines is possible. The synthesis of the naturally occurring isoguanine riboside, crotonoside (IX, p. 105) (47), is thus within reach.

Type B Syntheses. Syntheses of type *B* have the advantage over those of type *A* of unambiguity regarding the position of the sugar residue. But they necessarily involve more intermediates containing both heterocyclic nuclei and sugars, the most difficult sort to handle, and the overall yields from the initial sugar tend therefore to be lower. On balance however they are attractive enough to have been very fully exploited by TODD and collaborators (185). These syntheses involve four major steps, namely the attachment of the correct glycoside residue to a 4:6-diaminopyrimidine, the introduction of an amino group into position 5 of the pyrimidine ring, the completion of the glyoxaline ring, and any necessary alterations in the substituents of the pyrimidine ring.

[*Gl* = glycoside or derivative.]

Should the required nucleoside be a pyranoside the first step is relatively simple: it is only necessary to heat together the diaminopyrimidine and the sugar in alcoholic solution containing a trace of acid and to separate the glycoside by adsorption on alumina followed by aqueous elution (14). The preparation of similar furanosides, on the other hand, occasions some difficulty. The obvious adaptation of the pyranoside procedure by using a suitably protected sugar of fixed furanose structure is not very satisfactory in practice (4, 116). Recourse is therefore had to a novel method (116, 117) involving the condensation of an aldehydo sugar with the diaminopyrimidine, followed by removal of protecting groups from the 2', 3' and 4' hydroxyls but leaving that at 5'. Isomerisation to a cyclic structure then occurs and this structure must be furanoid so long as the 5′ hydroxyl is blocked. The initial steps of the synthesis of adenosine by this method (*118*) are shown below.

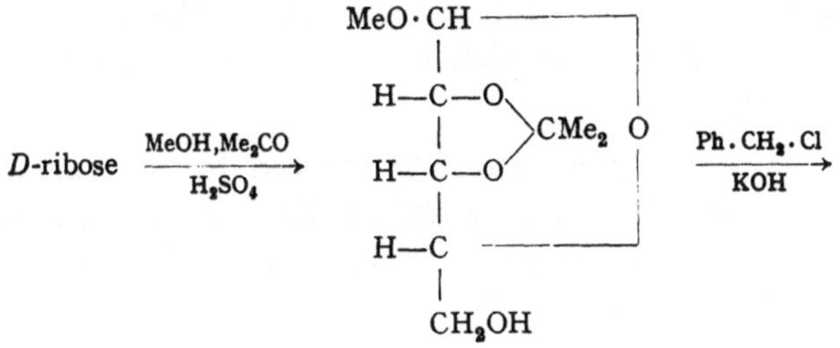

$$
D\text{-ribose} \xrightarrow[\text{H}_2\text{SO}_4]{\text{MeOH, Me}_2\text{CO}}
\begin{array}{l}
\text{MeO} \cdot \text{CH} \\
\text{H—C—O} \\
\text{H—C—O} \\
\text{H—C} \\
\text{CH}_2\text{OH}
\end{array}
\quad \text{CMe}_2 \quad \text{O}
\xrightarrow[\text{KOH}]{\text{Ph} \cdot \text{CH}_2 \cdot \text{Cl}}
$$

(XXXV.)

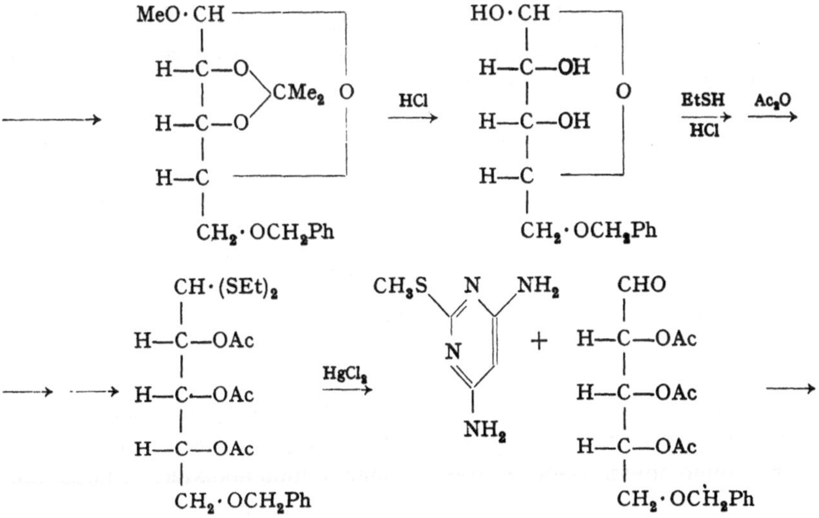

$$
\begin{array}{l}
\text{MeO} \cdot \text{CH} \\
\text{H—C—O} \\
\text{H—C—O} \\
\text{H—C} \\
\text{CH}_2 \cdot \text{OCH}_2\text{Ph}
\end{array}
\quad \text{CMe}_2 \quad \text{O}
\xrightarrow{\text{HCl}}
\begin{array}{l}
\text{HO} \cdot \text{CH} \\
\text{H—C—OH} \\
\text{H—C—OH} \\
\text{H—C} \\
\text{CH}_2 \cdot \text{OCH}_2\text{Ph}
\end{array}
\quad \text{O}
\xrightarrow[\text{HCl}]{\text{EtSH} \quad \text{Ac}_2\text{O}}
$$

$$
\begin{array}{l}
\text{CH} \cdot (\text{SEt})_2 \\
\text{H—C—OAc} \\
\text{H—C—OAc} \\
\text{H—C—OAc} \\
\text{CH}_2 \cdot \text{OCH}_2\text{Ph}
\end{array}
\xrightarrow{\text{HgCl}_2}
\begin{array}{l}
\text{CH}_3\text{S} \quad \text{N} \quad \text{NH}_2 \\
\text{N} \\
\text{NH}_2
\end{array}
+
\begin{array}{l}
\text{CHO} \\
\text{H—C—OAc} \\
\text{H—C—OAc} \\
\text{H—C—OAc} \\
\text{CH}_2 \cdot \text{OCH}_2\text{Ph}
\end{array}
\longrightarrow
$$

(XXXVI.)

$$
\begin{array}{l}
\text{CH}_3\text{S} \quad \text{N} \quad \text{N=CH} \cdot (\text{CHOAc})_3 \cdot \text{CH}_2\text{OCH}_2\text{Ph} \\
\text{N} \\
\text{NH}_2
\end{array}
\xrightarrow{\text{NH}_3}
$$

$$
\begin{array}{l}
\text{CH}_3\text{S} \quad \text{N} \quad \text{NH} \cdot \text{CH} \cdot (\text{CHOH})_2\text{CH} \cdot \text{CH}_2 \cdot \text{OCH}_2\text{Ph} \\
\text{N} \\
\text{NH}_2
\end{array}
$$

(with O bridge over CH·(CHOH)₂CH)

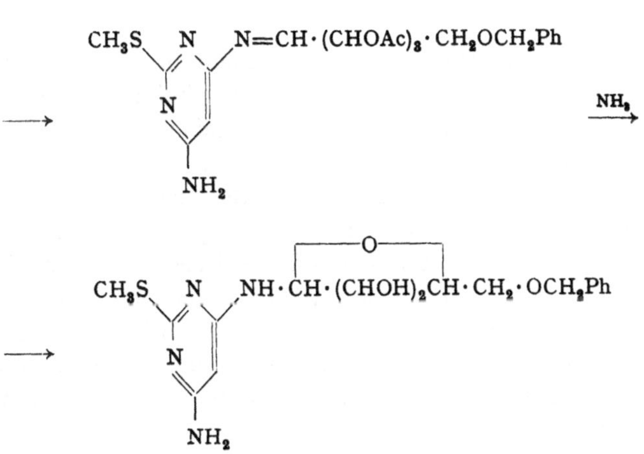

In effect 2:3:4-triacetyl-5-benzyl-D-ribose (XXXVI) is obtained from the useful ribofuranose derivative (XXXV) (*141*) through benzylation and WOLFROM's standard method for aldehydo sugar derivatives. The protecting benzyl group is retained throughout the later stages of the synthesis to avoid isomerisation of the labile pyrimidine furanosides into pyranosides (*104*).

The 5-amino group is readily introduced by nitrosation or, more generally, by coupling with a sufficiently reactive diazonium salt, followed in each case by reduction either catalytically or with ammonium sulfide or with zinc dust (*155*).

In his classical purine synthesis TRAUBE introduced the methinyl group of the glyoxaline ring by heating with formic acid. Clearly there are many possible substitutes for formic acid of a milder character, but the favoured one has been dithioformic acid or its salts, applied in the two stage process shown (*13*):

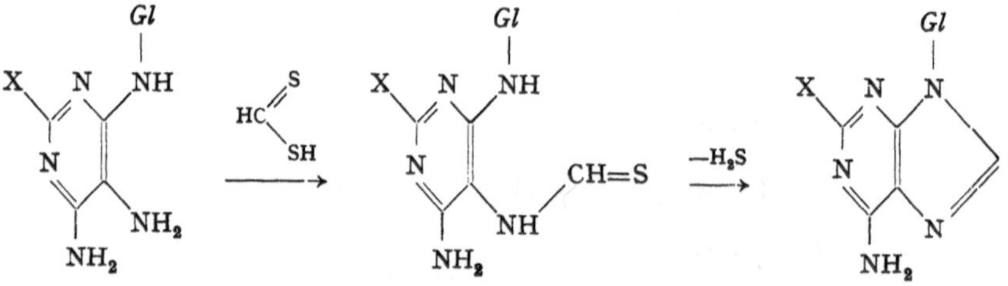

Heating the intermediate thioformamido compound in pyridine solution served for some early syntheses (*12, 15, 106,.115*) but later it was found advantageous to use alcoholic sodium alkoxides or potassium acetate in acetonitrile (*117, 119*) as cyclising agents. Cyclisation occurs preferentially at the substituted rather than at the unsubstituted amino group.

The first step in purine nucleoside syntheses of type *B*, the formation of a pyrimidine glycoside, imposes some limitations regarding the substituent X at position 2 of the pyrimidine ring. X cannot be a "tautomeric group", such as hydroxyl or amino, and in the case of the furanosides it cannot be hydrogen, but the methylthio substituent at 2 confers desirable properties. The final step of synthesis is therefore often to convert this into the required substituent, Y. Reductive fission to hydrogen is facile with RANEY nickel in alcoholic solution (*106, 118*). Oxidation with hydrogen peroxide gives a sulfone, which may be converted into a hydroxyl group by alkali or by sodium benzylate and hydrogenation (*3*). It should be possible to synthesise crotonoside (IX, p. 105) in this manner.

Many of the intermediate glycosidaminopyrimidines of type B syntheses mutarotate readily (*104*) and the α or β configuration of the final product is therefore as yet unpredictable. In one sense this is an advantage: for variation in the cyclisation conditions may lead to either product (*117*). As in the case of O-glycosides the optical rotation is some guide to the configuration (*105*).

Type C Syntheses. Hitherto syntheses of type C have suffered from inflexibility regarding the pyrimidine substituents besides ambiguity concerning the location of the sugar residue. On the other hand they derive increasing interest from the accumulation of evidence that purines may be built up in Nature from glyoxaline derivatives. BAXTER and SPRING (*25*, *26*) developed the synthesis of xanthine glycosides shown below and application of acetochlororibofuranose in it has lead to a synthesis of xanthosine (VII, p. 105) (*108*).

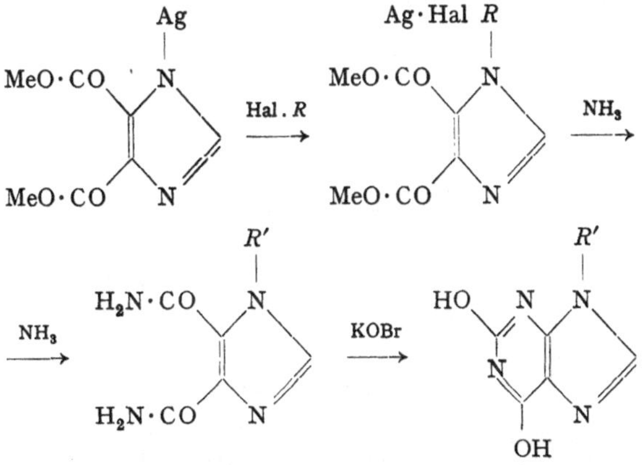

[R = acetylated glycosyl; R' = glycosyl.]

A variety of other purine syntheses from glyoxaline derivatives has been described (*59*, *60*, *99*) but nucleosides have not yet been prepared by these routes.

c) Nicotinamide Nucleosides.

FISCHER and RASKE (*84*) were able to form the quaternary pyridinium glucoside from pyridine and α-acetobromoglucose in presence of phenol, and after the isolation and degradation of diphosphopyridine nucleotide (DPN; L, p. 124), KARRER and coworkers (*113*) extended this method to the preparation of quaternary nicotinamide glycosides (XXXVII). Recently HAYNES and TODD (*98*) have improved the procedure, for instance by using acetonitrile as solvent, and, applying acetobromoribofuranose (XXVII, p. 109), have synthesised dihydronicotinamide-*D*-

riboturanoside (XXXVIII). Like the degradation product of DPN (*177*) the synthetic material was amorphous and no direct comparison was possible, but the synthetic material showed the appropriate activity as a bacterial growth factor.

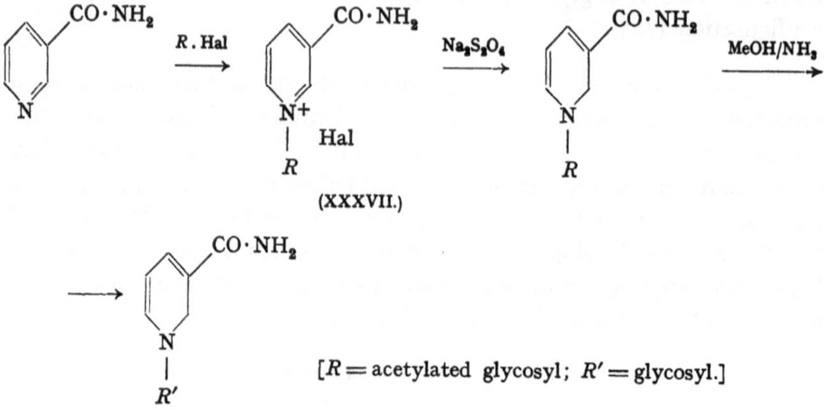

(XXXVII.)

[*R* = acetylated glycosyl; *R′* = glycosyl.]

(XXXVIII.)

d) Benziminazole Nucleosides.

The benziminazole glycosides related to vitamin B_{12} are formally similar to the purine nucleosides and thus analogous pathways of synthesis are available. Syntheses of type *A* (p. 110) from benziminazoles and acetohalogen sugars are not very satisfactory (*40*, *158*), and the route of choice is thus *B*, i. e. from aniline derivatives and their glycosides. In 1937 KUHN and STRÖBELE (*124*) prepared the glycosides of *o*-nitro-anilines, including 4 : 5-dimethyl-2-nitroaniline ribopyranoside, as intermediates in the synthesis of vitamin B_2 and its analogues, and these may be converted into benziminazole glycosides by reduction, thioformylation and cyclisation (*40*) in straightforward extension of the methods of purine glycoside synthesis. BRINK, FOLKERS and others (*34*) have gone further and prepared ribofuranosides from 5-tritylribose (XXIV, p. 109). For the cyclisation step they used formiminoether hydrochloride. As in the syntheses of the 9-*D*-ribopyranosido-2-methylthioadenines (*117*) the presence or absence of acetyl groups in the sugar residue determined the configuration of the product. The rôle of chelation of acetyl groups in such effects has been discussed previously (*38*, *117*, *119*). The nucleoside of positive optical rotation obtained without acetylation (II, p. 104) proved to be identical with the degradation product of vitamin B_{12}.

Other interesting mutarotations were observed by MAMALIS, PETROW and STURGEON (*158*) in their syntheses of a whole range of benziminazole pyranosides using cyclisation with ethylorthoformate.

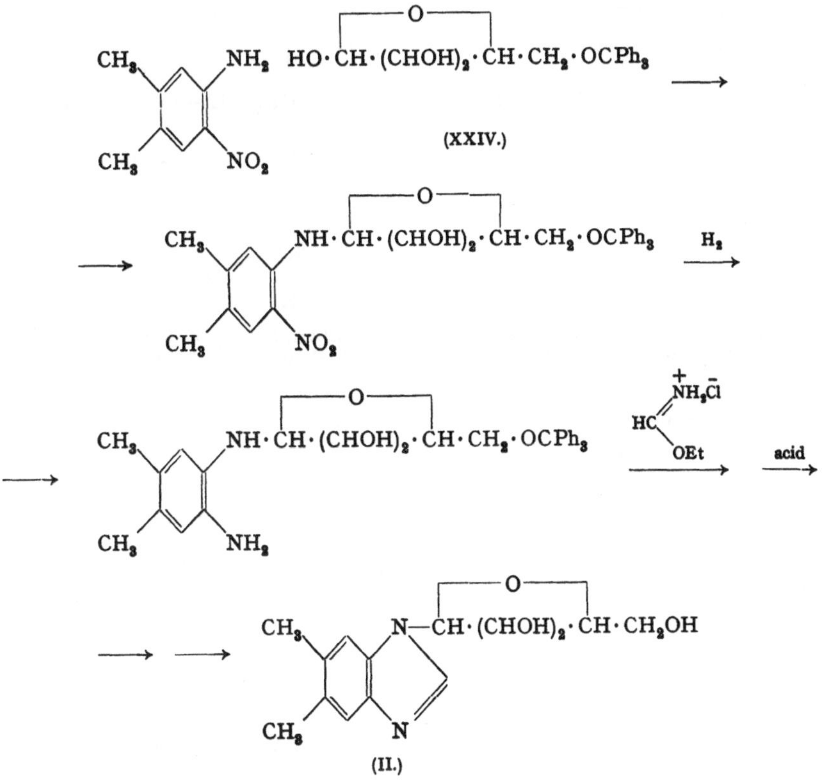

$$\text{(XXIV.)}$$

CH$_3$ ⎯ NH$_2$ HO·CH·(CHOH)$_2$·CH·CH$_2$·OCPh$_3$ (with O bridge) ⟶

CH$_3$ ⎯ NO$_2$

(XXIV.)

⟶   CH$_3$ ⎯ NH·CH·(CHOH)$_2$·CH·CH$_2$·OCPh$_3$ (with O bridge)   H$_2$ ⟶

CH$_3$ ⎯ NO$_2$

⟶   CH$_3$ ⎯ NH·CH·(CHOH)$_2$·CH·CH$_2$·OCPh$_3$ (with O bridge)

CH$_3$ ⎯ NH$_2$

HC(=NH$_2$Cl)OEt   acid ⟶

⟶ ⟶   CH$_3$ ⎯ N—CH·(CHOH)$_2$·CH·CH$_2$OH (with O bridge)

CH$_3$ ⎯ N

(II.)

III. Deoxyribonucleosides.

1. Isolation.

SCHINDLER (*176*) has described the isolation of the deoxyribosides of adenine (XLIII), guanine (XLIV), cytosine (XL) and thymine (XLI) from enzymic hydrolysates of thymonucleic acid by an improved process, which may advantageously be modified by the use of cation exchange (*37*). DEKKER and TODD (*74*) have shown that thymidine prepared in this way is contaminated with the hitherto unknown uracil deoxyriboside (XXXIX), which may be separated by a cellulose column. This material probably arises from deamination of cytosine residues during the preparation of commercial deoxyribonucleic acid. A preliminary account has appeared of the detection by ion-exchange columns of a substance, which appears to be 5-methyl-deoxycytidylic acid (*55*). Previous reports of the isolation of 5-methylcytosine from tuberculinic (*111*) and thymonucleic (*102*) acids thus receive some confirmation. Indeed DEKKER, ELMORE and TODD (*73*) have now isolated 5-methylcytosine deoxyriboside (XLII) as its picrate, m. p. 175—178° (dec.), distinct from the picrates of the other naturally occurring deoxyribosides.

$$\begin{array}{c} \text{┌──O──┐} \\ \text{│} \quad \underset{|}{\overset{}{\text{OH}}} \quad \text{│} \\ \text{CH·CH}_2\text{·C──C──CH}_2\text{OH} \\ \text{│} \quad \text{│ │} \\ \underset{\text{N}}{\overset{\text{O}}{}} \quad \text{H H} \\ \text{N} \\ \text{X} \end{array}$$

(XXXIX.) [X = OH] Uracil deoxyriboside,
(XL.) [X = NH₂] Cytosine deoxyriboside.

$$\begin{array}{c} \text{┌──O──┐} \\ \text{│} \quad \underset{|}{\overset{}{\text{OH}}} \quad \text{│} \\ \text{CH}_2\text{·CH}_2\text{──C──C·CH}_2\text{OH} \\ \text{│} \quad \text{│ │} \\ \underset{\text{N}}{\overset{\text{O}}{}} \quad \text{H H} \\ \text{N} \\ \text{CH}_3 \\ \text{X} \end{array}$$

(XLI.) [X = OH] Thymidine.
(XLII.) [X = NH₂] 5-Methylcytosine deoxy-riboside.

$$\begin{array}{c} \text{┌──O──┐} \\ \text{│} \quad \underset{|}{\overset{}{\text{OH}}} \quad \text{│} \\ \text{CH·CH}_2\text{──C──C──CH}_2\text{OH} \\ \text{│} \quad \text{│ │} \\ \text{N} \quad \text{N} \quad \text{H H} \\ \text{N} \\ \text{N} \\ \text{NH}_2 \end{array}$$

(XLIII.) Adenine deoxyriboside.

$$\begin{array}{c} \text{┌──O──┐} \\ \text{│} \quad \underset{|}{\overset{}{\text{OH}}} \quad \text{│} \\ \text{CH·CH}_2\text{·C──C·CH}_2\text{OH} \\ \text{│} \quad \text{│ │} \\ \text{H}_2\text{N} \quad \text{N} \quad \text{N} \quad \text{H H} \\ \text{N} \\ \text{N} \\ \text{OH} \end{array}$$

(XLIV.) Guanine deoxyriboside.

2. Structure.

It has been widely assumed that the deoxyribonucleosides are simply the 2'-deoxy derivatives of the corresponding ribonucleosides, but only a little experimental evidence had been collected in support (95) until recently. Unfortunately no method has yet been devised for the inter-conversion of the two series of compounds. Brown and Lythgoe (37) have now been able to prove the furanoside nature of the four common deoxyribosides by the periodate method. The configuration at the glycosidic carbon atom remains unknown.

3. Synthesis.

Despite considerable efforts to find better preparative methods (cf. e. g. 75, 167) 2-deoxy-D-ribose remains much less accessible than D-ribose. Consequently no progress has been made in the synthesis of deoxyribonucleosides comparable with that in the synthesis of ribo-nucleosides. Davoll and Lythgoe (69) have discussed the problem and have also described the preparation of crude 1-chloro-3 : 4-diacetyl-2-deoxy-D-ribose (XLV) and its condensation with the silver salt of theophylline to give the α and β forms of diacetyl-theophylline-7-2'-deoxyribopyranoside (XLVI). Initial experiments towards the synthesis of adenine deoxyribopyranoside through reaction between the new acetohalogen sugar and the silver salt of 2 : 8-dichloroadenine were rather

(XLV.) 1-Chloro-3:4-diacetyl-2-deoxy-D-ribose.

(XLVI.) Diacetyl-theophyline-7-2'-deoxyribopyranoside.

unpromising. The outlook is however not so discouraging since the reaction with the silver salt of benziminazole is successful (61), and DAVOLL and LOWY (68) have in the meantime described the much improved synthesis of purine ribonucleosides already discussed. The difficult problem remains of preparing the furanose analogue of (XLV).

IV. Isolation and Structure of Nucleotides.

No progress can be reported in the study of deoxyribonucleotides and this section is therefore confined to ribonucleotides.

1. Nucleotides derived from Phosphoric Acid.

a) Adenylic Acids.

Until 1949 only two simple adenylic acids, that is substances containing adenine, ribose and phosphate groups in equimolar amounts, were known. One, termed "muscle adenylic acid" owing to its part in the processes of muscular contraction, was regarded as adenosine-5'-phosphate and the other, "yeast adenylic acid", believed to be a characteristic fission product of yeast ribonucleic acid, as adenosine-3'-phosphate. In that year CARTER and COHN (43) announced the isolation of a third adenylic acid from yeast nucleic acid hydrolysates. Moreover

OH
|
O·P—OH
‖
O
|
O
|
OH
|
N   N——C—C—C—C—CH₂OH
        | | | |
        H H H H
NH₂

(XLVII.) Adenosine-2′-phosphate.

OH
|
O·P—OH
‖
O
|
O
|
OH
|
N   N——C—C—C—C—CH₂OH
        | | | |
        H H H H
NH₂

(XLVIII.) Adenosine-3′-phosphate.

O
|
OH OH
| |
N   N——C—C—C—C—CH₂·O·P—OH
        | | | | ‖
        H H H H   O
                            OH
NH₂

(XLIX.) Adenosine-5′-phosphate.

COHN and VOLKIN (58) have since shown that careful enzymatic degradation of calf liver ribonucleic acid leads to "muscle adenylic acid" in addition to the two other compounds (adenylic acids "a" and "b") which are obtained in the customary alkaline hydrolysis. The previous clarity of the picture was thus illusory. These important discoveries reveal new vistas in the structural chemistry of ribonucleic acid, but the first task must be to settle the structures of the three compounds.

Concerning the structure of "muscle adenylic acid" there is no difficulty. Its formulation as adenosine-5′-phosphate rests securely on its deamination to "muscle inosinic acid" and the well known work of LEVENE on that substance, on its uptake of one mole of periodate (153) showing the presence of two adjacent free hydroxyls, and on its synthesis from 2′ : 3′-isopropylidene adenosine (17).

CARTER and COHN noticed variations in the physical properties of "yeast adenylic acid", which suggested that it might·be, as they later proved that it in fact is, a mixture of two substances. Although it is possible to separate them by crystallisation, the fully satisfactory procedures are those developed by CARTER and COHN of paper chromatography (42) and anion exchange (54). Not only are the two compounds difficult to separate by older techniques but also they are interconvertible on boiling with 0,01 N-HCl during one hour (36). Further, although the optical rotations measured in formamide containing base are distinctly different, they are very similar in 10% HCl (145).

LEVENE and HARRIS (139) determined the structure of "yeast adenylic acid" by the following series of reactions. It was deaminated by nitrous acid to "yeast inosinic acid" and this was hydrolysed by its own acidity on keeping a 2,3% aqueous solution at 96° during one hour. From this solution was isolated the barium salt of a ribose phosphate, which after a purification process was compared with a salt obtained from guanylic acid through xanthylic acid. On the ground of their similarity in optical rotation these salts were regarded as identical and that from xanthylic acid had already (137, 138) been shown to be the barium salt of ribose-3-phosphate, since on hydrogenation it yielded a ribitol phosphate, the sodium salt of which had no detectable optical rotation in either water or borax solution. Now it is likely that LEVENE and HARRIS were dealing with the relatively less soluble adenylic acid "b" of CARTER and COHN, although unfortunately no constants are recorded for the material actually degraded; and likely that their chain of reasoning was correct, but the possibility of interconversion reactions in the acid media cannot be neglected. Indeed DOHERTY (78) has reported the degradation of both adenylic acids "a" and "b" to inactive ribitol phosphates by a similar scheme involving alcoholysis with benzyl alcohol and hydrogen chloride to a benzyl riboside phosphate. It has become therefore highly desirable that all three adenylic acids be subjected to a degradation process enabling the phosphate group to be allocated with certainty to the 2', 3' or 5' position, unless an unambiguous synthesis of either the 2' or 3' compound can be developed.

On the basis of his degradation DOHERTY (78) has suggested that the adenylic acids "a" and "b" are derived from two different nucleosides, the α- and β-9-D-ribofuranosidoadenines, rather than from a single nucleoside as tacitly assumed above. To the present author this seems improbable for the following reasons. Contrary to DOHERTY's suggestion the mutarotation of purine glycosides is not easily achieved (104). A greater difference in the optical rotations would be expected (cf. 105, 117) than is observed (see above, 145), although indeed the difference is not very large in the case of the α- and β-ribofuranosido-5 : 6-dimethylbenz-iminazoles (II, p. 104) (34). Both compounds are produced simultaneously in a synthetic procedure, which is unlikely to cause mutarotation (36, 37a). Finally KORNBERG and PRICER (121) have degraded TPN enzymatically to both "muscle adenylic acid" and adenylic acid "a" (see below).

In addition to adenosine-5'-phosphate COHN and VOLKIN (58) have reported the presence of another unexpected adenylic compound in the mixed enzymatic degradation products of ribonucleic acid. Whether this is a cyclic phosphate of adenosine remains to be seen.

b) Guanylic Acids.

COHN (54) has presented preliminary evidence that two guanylic acids, "a" and "b", analogous to the adenylic acids "a" and "b" exist in hydrolysates of rat liver ribonucleic acid.

c) Cytidylic and Uridylic Acids.

Improved procedures for the preparation of the pyrimidine nucleotides from yeast nucleic acid are based on fractionation of the phosphotung-states (149) and on anion exchange (56). The latter method yields some 7 g. of each compound from 100 g. of yeast nucleic acid in a relatively simple process.

Once again anion exchange chromatography has demonstrated heterogeneity of what were previously regarded as single nucleotides (55, 147). Indeed the two cytidylic acids with optical rotations of $+20,7°$ and $+49,4°$ may be separated by crystallisation of phosphotungstates and of brucine salts (147). The phenomenon of interconversion of the isomers during a short period of heating with 0,1 N-HCl was observed (55).

d) Nucleotide from Vitamin B_{12}.

Two hours heating with approximately N-HCl degrades vitamin B_{12c} to a nucleotide, isolable through partition chromatography on a cellulose column using wet butanol as solvent (40). Analytical figures and the results of more vigorous hydrolysis show it to be a monophosphate of the 1-α-D-ribofuranosido-5 : 6-dimethylbenziminazole already referred to in the sections on nucleosides (II, p. 104). Since it does not consume periodate the phosphate group must be at positions 2' or 3' of the glycoside. Whatever the structure of the vitamin itself may be it seems possible that, under the conditions of degradation used, isomerisation to a mixture of 2' and 3' phosphates may have occurred. A demonstration that the isolated substance is either a 2' or a 3' phosphate would therefore of itself provide no additional evidence regarding the structure of the vitamin.

e) Polynucleotides.

It is evident that, if the nucleic acids contain the nucleosides linked between their ribose residues by phosphate groups, it should be possible to isolate polynucleotides, which are dialkyl esters of phosphoric acid, by graded hydrolysis, and that the examination of such compounds would provide valuable information about the structure of nucleic acids. The development of powerful analytical procedures has now made this goal attainable. CARTER and COHN (44) have reported the separation of some nucleic acid fragments of unknown composition by anion exchange.

2. Monoalkyl Esters of Polyphosphoric Acids.

Several new prescriptions for the preparative isolation of ATP (LXXVI, p. 135) have appeared (77, 79, 122, 132). BAILEY (20) has provided valuable data on the stability of ATP salts on storage and has criticised the suggestion that the action of myosin on ATP produces a new isomeric ADP in addition to the well recognised compound (LXXIII). According to ALBAUM and OGUR (1) oat seedlings contain the pyrophosphate of an adenine pentoside, but the structure of this is obscure. The method of periodate titration, which has shown that the hydroxyls at 2' and 3' of ATP are unsubstituted (154), was not applied to it. On the other hand an adenosine triphosphate approaching 70% in purity extracted from mung beans (2) behaved like ATP towards periodate, although differing in other properties. LEPAGE and UMBREIT (135) had previously isolated an adenosine triphosphate from *Thiobacillus thiooxidans* isomeric with that from muscle and had concluded from the rate of acid hydrolysis of the ribose phosphate derived from it that it was adenosine-3'-triphosphate. This evidence would not, however, distinguish adenosine-3' from adenosine-2'-triphosphate.

Uridine-5'-pyrophosphate (41) is a fission fragment of UDPG (LII), which is discussed in the following sub-section.

3. Dialkyl Esters of Pyrophosphoric Acid.

a) Nicotinamide Nucleotides.

These compounds have become more accessible through improved isolation procedures (48, 100, 130, 133, 134, 181). The structure of DPN (L, p. 124) has long been accepted (cf. 151), but only very recently has the relation to it of TPN (LI) become clearer. The enzymatic conversion of DPN (L) into TPN, which had been reported by earlier workers, has been fully investigated by KORNBERG (120). An enzyme extractable from top ale yeast autolysates catalyses the reaction

$$DPN + ATP \rightarrow TPN + ADP.$$

Further evidence against the old suggestion that TPN contains a linear triphosphate group comes from the degradation of TPN with the nucleotide pyrophosphatase of potatoes (120a) carried out by KORNBERG and PRICER (121). Basic lead acetate precipitates the adenine nucleotide fragment from this hydrolysis leaving the nicotinamide nucleotide in solution. The latter is apparently identical with that from

DPN, but the adenine nucleotide fragment is a new substance composed of adenine, pentose and phosphate in the molar ratio $1:1:2$. From it is obtained by the action of a specific adenosine-5' phosphatase of potatoes the adenylic acid "a" of CARTER and COHN (43) in an overall yield of 61% from TPN. This is identified with certainty through ion exchange and paper chromatography and through enzymatic behaviour. As kidney particles degrade TPN to adenosine-5'-phosphate, adenylic acid "a" must be a phosphate of adenosine and not of 9-α-D-ribo-furanosidoadenine and, assuming it to be adenosine-2'-phosphate, TPN must be (LI).

(L.) [*R*=H] Diphosphopyridine nucleotide (DPN).
(LI.) [*R*=PO·(OH)₂] Triphosphopyridine nucleotide (TPN).

(LII.) Uridine diphosphate glucose (UDPG).

b) Flavin Nucleotides.

Further information concerning the new flavin dinucleotide (F–X) of SANADI and HUENNEKENS (*174*) from yeast will be awaited with interest [cf. section V, 2 c, p. 132 and (*86*)].

c) The Coenzyme of "Galactowaldenase".

Certain yeasts are adapted to the utilisation of galactose instead of glucose through the conversion of galactose-1-phosphate into glucose-1-phosphate by the enzyme "galactowaldenase". CAPUTTO, LELOIR, CARDINI and PALADINI (*41*) have shown that the corresponding coenzyme is widely distributed in animal tissue as well as in yeasts which cannot use galactose, and have accumulated considerable evidence as to its structure. The coenzyme is extracted from toluene-treated yeast by alcohol and purified by precipitation as the mercury salt and by charcoal adsorption. On paper chromatography using 77% alcohol the resulting material gives a spot (R_F 0,44), absorbing at 260 mμ and containing the coenzyme activity. The nature of the ultraviolet absorption suggests the presence of a uridine residue and on this assumption its intensity gives an estimate of the molarity of the solution used in further experiments. Electrometric titration reveals only two primary phosphoric acid dissociations, but graded hydrolysis with dilute acid produces a secondary acidic group and in a more vigorous treatment a further secondary acidic group and also a tertiary. The first stage also liberates a sugar, characterised as *D*-glucose by paper chromatography, the carbazole colour reaction and fermentation tests.

These results are explicable if the coenzyme is a dialkyl pyrophosphate, from which a labile glucose-1 residue is hydrolysed in the first step, the pyrophosphate link being broken in the second stage. Moreover the nucleotide end product has been identified as uridine-5'-phosphate (*129*) by direct comparison with a synthetic specimen (*161*). There is therefore good presumptive evidence for the formulation of the coenzyme as the glucose-1 ester of uridine-5'-pyrophosphate (LII). LELOIR and his collaborators have coined the name UDPG for it.

V. Synthesis of Nucleotides.

Nucleotides being composed of three different classes of substance, namely heterocyclic bases, carbohydrates and phosphoric acids, may in principle be synthesised in two ways, either from nitrogenous compounds and sugar phosphates or from nucleosides and phosphoric acid derivatives. The second method has been universally adopted to date. If ambiguity is to be avoided, it involves the protection by the classical methods of carbohydrate chemistry of all the hydroxyl groups of the glycoside except that to be phosphorylated. This is appropriately dealt with in discussion of particular syntheses, but a separate section is devoted first to the methods, which have only been developed in recent years, for the manipulation of phosphoric acid groups (cf. 5).

1. Methods of Phosphorylation.

The project of synthesising all the oligonucleotides poses four problems of phosphate chemistry; (a) the preparation of monoalkyl esters of phosphoric acid (e. g. the adenylic acids), (b) of dialkyl esters of phosphoric acid (i. e. hypothetical fragments of nucleic acid), (c) of monoalkyl esters of polyphosphoric acids (e. g. ADP, ATP), and (d) of dialkyl esters of pyrophosphoric acid (e. g. DPN, FAD). Adequate methods have been developed for (a) and (c), which involve only a single displacement on any phosphorus atom, whereas (b) and (d) involve two successive displacements on phosphorus. The latter more difficult problem still awaits satisfactory solution, which is however now in prospect.

Direct esterification with phosphoric acid is too crude a method for use in the nucleotide field, although polyphosphoric esters of vitamin B_1 have been prepared in this way (188, 189). Instead, hydrogen chloride is eliminated between the hydroxyl group of the sugar and a phosphoric acid chloride, in which chlorine is bound to phosphorus. The simplest reagent of this type for the preparation of monoalkyl phosphates would be chlorophosphonic acid, $Cl \cdot PO(OH)_2$. This compound is, not surprisingly, unknown but a mixture of phosphoryl chloride and 10% of water, which may be regarded as dichlorophosphinic acid, $Cl_2 \cdot PO \cdot OH$, has been used successfully by FORREST and TODD (86) and by FLEXSER and FARKAS (85) for the synthesis of riboflavin-5'-phosphate (LXX). The initial product was riboflavin-4':5'-phosphate (LXIX), but this cyclic phosphate could be opened up with 0,1 N-HCl at 100°. [According to REICH (170) the cyclic phosphates of catechol are similarly unstable in comparison with other diaryl phosphates.] The production of a cyclic phosphate in this instance and of polyphosphates from aneurin (173) emphasises the bifunctional nature of the reagent and suggests the possible use of it for solution of the second problem, the synthesis of dialkyl esters of phosphoric acid. But it is difficult to envisage clean syntheses of this type. Related to these experiments is the conversion in minute yield of nucleosides into nucleotides by phosphoryl chloride and aqueous baryta [cf. e. g. (90)].

Phosphorylating reagents containing a free hydroxyl group are inevitably crude mixtures and it is therefore preferable, when possible, to protect the hydroxyl with a group which may be removed easily later in the synthesis. The first group to be used successfully for this purpose was phenyl. Diphenyl chlorophosphonate (LIII) and phenyl dichlorophosphinate (LIV) are readily prepared from phenol and phosphoryl chloride as a mixture separated by distillation (33).

Ph·OH   POCl₃
(Ph=C₆H₅)

$$
\begin{array}{c}
\text{PhO} \diagdown \quad \diagup \text{O} \\
\text{P} \\
\text{PhO} \diagup \quad \diagdown \text{Cl}
\end{array}
\xrightarrow{ROH}
\begin{array}{c}
\text{PhO} \diagdown \quad \diagup \text{O} \\
\text{P} \\
\text{PhO} \diagup \quad \diagdown \text{OR}
\end{array}
\xrightarrow{2\,H_2}
\begin{array}{c}
\text{HO} \diagdown \quad \diagup \text{O} \\
\text{P} \\
\text{HO} \diagup \quad \diagdown \text{OR}
\end{array}
$$

(LIII.) Diphenyl chloro-phosphonate.   (LV.)

$$
\begin{array}{c}
\text{PhO} \diagdown \quad \diagup \text{O} \\
\text{P} \\
\text{Cl} \diagup \quad \diagdown \text{Cl}
\end{array}
\xrightarrow{2\,ROH}
\begin{array}{c}
\text{PhO} \diagdown \quad \diagup \text{O} \\
\text{P} \\
\text{RO} \diagup \quad \diagdown \text{OR}
\end{array}
\xrightarrow{H_2}
\begin{array}{c}
\text{HO} \diagdown \quad \diagup \text{O} \\
\text{P} \\
\text{RO} \diagup \quad \diagdown \text{OR}
\end{array}
$$

(LIV.) Phenyl dichloro-phosphinate.   (LVI.)

$\downarrow$ ROH

$$
\begin{array}{c}
\text{PhO} \diagdown \quad \diagup \text{O} \\
\text{P} \\
\text{RO} \diagup \quad \diagdown \text{Cl}
\end{array}
\xrightarrow{R'OH}
\begin{array}{c}
\text{PhO} \diagdown \quad \diagup \text{O} \\
\text{P} \\
\text{RO} \diagup \quad \diagdown \text{OR'}
\end{array}
\xrightarrow{H_2}
\begin{array}{c}
\text{HO} \diagdown \quad \diagup \text{O} \\
\text{P} \\
\text{RO} \diagup \quad \diagdown \text{OR'}
\end{array}
$$

(LVII.)   (LVIII.)

Either of these compounds may be used to phosphorylate hydroxyl groups of, for instance, nucleosides in pyridine solution to produce diphenylmonoalkylphosphates (LV) and monophenyldialkylphosphates (LVI) respectively. Further by using an inert solvent and only one mole of pyridine (19) it is possible to obtain the impure monophenyl monoalkyl chlorophosphonate (LVII) which could, in principle, be applied to the synthesis of monophenyldialkylphosphates containing two different alkyl groups (LVIII). Finally the protective phenyl group may be removed by hydrogenolysis in presence of ADAMS platinum catalyst (33) or by alkaline hydrolysis (30, 94). The principal successes of this method have been in the preparation of simple sugar phosphates [cf. e. g. (128)]. Difficulties encountered from time to time in the removal of phenyl groups restrict its application in the synthesis of more complex and less stable substances.

A more versatile protective group is benzyl. This is more labile than phenyl and more easily removed by hydrogenolysis, but phosphorylating reagents containing it are necessarily less stable and require greater care in manipulation. ATHERTON, OPENSHAW and TODD (7) found the first practicable preparation of dibenzyl chlorophosphonate (LX) in the application of the general method of McCOMBIE, SAUNDERS and

Stacey (*156*). From phosphorus trichloride, benzyl alcohol and a base is obtained (*6, 7, 9*) a substance which may be regarded as dibenzyl phosphite or as its tautomer, dibenzyl phosphonate (LIX), and oxidation of this with chlorine in carbon tetrachloride or with sulfuryl chloride (*6*) leads to dibenzyl chlorophosphonate. Alternatively the halogenphosphonate may be obtained by a novel reaction involving treatment with an inert base and a polyhalogen compound, for example carbon tetrachloride or trichlorobromomethane (*8, 9*).

Used in the same fashion as diphenyl chlorophosphonate, dibenzyl chlorophosphonate is an excellent reagent for the phosphorylation of alcohols and amines. Both benzyl groups may be removed from the intermediate trialkylphosphate (LXI) by hydrogenolysis with palladium

catalysts, but in addition there exists the important possibility of removing a single benzyl group selectively. One method is that of "quaternisation", in which a tertiary base attacks the benzyl group yielding the quaternary benzylammonium salt of the dialkyl phosphate (*11, 50*). On account of this reaction the preceding stage may often be improved by using a sterically hindered base, such as 2 : 6-lutidine, instead of pyridine for neutralisation of the hydrogen chloride eliminated between the chlorophosphonate and the hydroxyl group (*6*).

CLARK and TODD (51) have discovered another method in which the debenzylating agent is a salt, for instance lithium chloride. In this example an equilibrium is set up between the trialkyl phosphate and lithium chloride on the one hand and the lithium salt of the dialkyl phosphate and benzyl chloride on the other. Precipitation of this lithium salt from the solution leads to quantitative reaction. It will be observed that in both cases the monodebenzylated ester is produced as an anion (LXII) and that therefore a second debenzylation would produce a doubly charged anion. The considerable difference between the first two dissociation constants of phosphoric acids thus favours the halt of the reaction at the desired half-way stage. The great importance of these discoveries will become apparent in the discussion of the synthesis of nucleotides derived from polyphosphoric acids. Synthetic application will also surely come of monobenzyl phosphite (monobenzyl phosphonate) (LXIII) prepared analogously from dibenzylphosphite (11, 51).

Tetrabenzyl pyrophosphate (LXIV) is produced when dibenzyl chlorophosphonate is treated with aqueous alkali under certain conditions (9, 76). The reaction is as follows:

(LXIV.) Tetrabenzyl pyrophosphate.

The reaction between salts of dialkyl phosphoric acids and chlorophosphonates is the standard method for the synthesis of polyphosphates. In more complex cases silver salts are used in anhydrous media. TOY (187) has introduced the technique of using pyridine salts in inert media. An alternative route would lie in the use of tribenzylpyro phosphate, prepared by debenzylation of tetrabenzylpyro phosphate (11).

Tetrabenzylpyro phosphate may also be used as a mild phosphorylating reagent by virtue of its capacity to liberate dibenzyl phosphoric acid in presence of an alcohol or amine (9). In other words the mixed anhydrides of dialkyl or diaryl phosphoric acids and any strong acid, not merely hydrogen chloride, are potential phosphorylating agents. Some current work in Cambridge is based on this principle.

Finally one other method of phosphorylation may be mentioned, the attack of phosphate anions on ethylene oxide derivatives (7, 127). The reaction is however sluggish and of limited potential application.

2. Nucleotides derived from Phosphoric Acid.

a) 5'-Phosphates.

Given the methods discussed in the previous section the problems in the synthesis simple nucleotides are to protect other reactive groups and to find a suitable solvent for the phosphorylation. Protection of the 2' and 3' hydroxyls for the synthesis of 5'-phosphates occasions no difficulty. LEVENE and TIPSON (143) showed that uridine reacts with acetone in presence of zinc chloride giving mono*iso*propylidene uridine,

(VI.) Adenosine.

(LXV.) Isopropylidene-adenosine.

(LXVI.) 2':3'-Isopropylidene-adenosine-5'-dibenzylphosphate.

(XLIX.) Adenosine-5'-phosphate.

(LXVII.) Adenosine-5'-benzylphosphate.

which contains a free primary hydroxyl since with sodium iodide its monotosyl derivative gives an iodo *iso*propylidene uridine under the conditions of OLDHAM and RUTHERFORD's test (*166*). In like fashion may be prepared the *iso*propylidene derivatives of the other three common ribonucleosides, e. g. (LXV) (*144*, *161*). Phosphorylation of these compounds with phosphoryl chloride or diphenyl chlorophosphonate is not very satisfactory, but BADDILEY and TODD (*17*) were able to obtain a good yield of 2′: 3′-*iso*propylidene-adenosine-5′-dibenzylphosphate (LXVI) by using dibenzyl chlorophosphonate in pyridine at temperatures low enough to avoid N-phosphorylation.

Hydrogenolysis of the benzyl groups followed by hydrolysis of the *iso*propylidene residue with dilute acid then gives adenosine-5′-phosphate (XLIX). Application of this procedure to the cases of uridine and cytidine is straightforward (*161*), but there are special difficulties in the case of guanosine. Unlike the other compounds *iso*propylidene guanosine is virtually insoluble in pyridine and there may be a greater tendency to N-phosphorylation in this instance. However an alternative procedure has been devised by MICHELSON and TODD (*161*). In this the *iso*propylidene guanosine is dissolved in dimethylformamide before dilution with pyridine and treatment with phosphoryl chloride at − 10°. After hydrolyses with baryta and sulfuric acid guanosine-5′ phosphate is obtained in moderate yield.

b) 2′- and 3′-Phosphates.

Unfortunately no method is yet available for the simultaneous protection of the 5′ and 3′ (or 2′) hydroxyls. It had indeed been supposed that, in analogy with a number of instances in the carbohydrate field, benzaldehyde would condense with ribonucleosides forming a six membered ring between the 3′ and 5′ positions. Some evidence was brought forward in favour of this assumption (*30*, *91*) and various nucleotides synthesised from the benzylidene nucleosides were described as 2′-phosphates (*92*, *93*, *94*, *161*). In a thorough re-examination of the situation BROWN, HAYNES and TODD (*36*) have shown that the evidence mentioned was either erroneous or misinterpreted and that all the synthetic compounds described as 2′-phosphates are in fact 5′-phosphates. The failure to realise the identity of the products of the two synthetic routes through *iso*propylidene and benzylidene nucleosides arose from their insufficient characterisation by the customary methods and serves to emphasise the value of ion exchange chromatography and X-ray powder diagrams. These results are most easily explained by the assignment of the 2′: 3′ structure to the benzylidene nucleosides (cf. also *145a*).

The syntheses, which were carried out via these intermediates, of adenosine-, uridine- and cytidine-3′-phosphates (*161*) thus lose their

unambiguity. At the time of writing it is only possible to cover the
5'-hydroxyl with a trityl group and to produce from the monotrityl-
nucleosides mixtures of the 2'- and 3'-phosphates, separable by anion
exchange chromatography (37a).

c) Dialkyl Esters of Phosphoric Acid.

FORREST and TODD (86) have reinvestigated the preparation of
riboflavin-5'-phosphate (FMN) (LXX). The principal obstacle is the
extremely low solubility of riboflavin derivatives, which hinders their
purification and attack by reagents. In these circumstances a direct

$$CH_2 \cdot (CHOH)_3 \cdot CH_2OH$$

POCl$_3$/H$_2$O $\longrightarrow$

(LXVIII.) Riboflavin.

$$CH_2 \cdot (CHOH)_2 \cdot CH \cdot CH_2$$

(LXIX.) Riboflavin-4': 5'-phosphate.

$$CH_2 \cdot (CHOH)_3 \cdot CH_2$$

(LXX.) Riboflavin-5'-phosphate.

approach is preferable to a complex synthetic scheme involving the manipulation of intermediates. Addition of phosphoryl chloride to a dilute pyridine solution of riboflavin (LXVIII) produces complex substances probably containing riboflavin residues linked through phosphorus. But in presence of about sufficient water to hydrolyse one chlorine from phosphoryl chloride, giving perhaps essentially dichlorophosphinic acid, the reaction is simpler. Purification of the material precipitated from the reaction medium by the chromatopile technique (*163*) with *n*-propanol/pyridine/water as solvent gives riboflavin-4′ : 5′-phosphate (LXIX). The same compound appears to be produced from natural flavin-adenine dinucleotide by the action of bases (*86*). Riboflavin-5′-phosphate (LXX) is obtained from this cyclic dialkyl phosphate by dilute acid hydrolysis and chromatopile purification. The assigned structures are secured by the observation that (LXIX) takes up one mole and (LXX) two moles of periodate; in neither case is formaldehyde liberated. The synthetic compound (LXX) shows the same behaviour as natural FMN in paper chromatography and the method therefore makes FMN much more accessible and provides the first convincing evidence as to the structure of FMN.

Another cyclic phosphate was apparently prepared by E. FISCHER (*82*) in his pioneer work on the phosphorylation of nucleosides. He treated theophylline glucopyranoside with phosphoryl chloride in pyridine at — 20° and isolated as the major product a crystalline monobasic phosphate of theophylline glucoside. FISCHER concluded that the phosphoric acid residue was esterified with two of the sugar hydroxyls. The 3′- and 6′-hydroxyls of glucopyranosyl are in suitable stereochemical relation to form a cyclic phosphate, which would be resistant to attack by periodate.

A different type of dialkyl phosphate is that in which two nucleosides are esterified with the same phosphoric acid group, as is presumed to occur in the nucleic acids. GULLAND and SMITH (*94*) have prepared a compound of this type from benzylidene uridine and phenyl dichlorophosphinate, $PhO \cdot PO \cdot Cl_2$, the protecting groups being removed by alkali and acid. This compound must now be formulated as diuridine-5′-phosphate.

3. Mononucleotides derived from Polyphosphoric Acids.

The syntheses have been achieved of both the important members of this class, namely adenosine-5′-diphosphate and -triphosphate. Before the discovery of the methods of selective debenzylation already discussed under methods of phosphorylation, BADDILEY and TODD (*17*) found that a very mild acid hydrolysis of 2′ : 3′-*iso*propylidene-adenosine-5′-dibenzyl-phosphate (LXVI, p. 130) removes the *iso*propylidene and one benzyl group. Adenosine-5'-benzyl-phosphate (LXVII, p. 130) results in high yield and forms an ideal starting material for extension of the phosphate system. Silver chloride can be eliminated between a suspension of its silver salt (LXXI) in anhydrous acetic acid and dibenzyl chlorophosphonate (LX) giving adenosine-5'-tribenzyl-diphosphate (LXXII), from which ADP (LXXIII) is obtained on hydrogenation.

$$
\underset{\text{(LXXI.)}}{\overset{\displaystyle O}{\underset{\displaystyle OCH_2 \cdot Ph}{Ad\!-\!O\!-\!\overset{\|}{P}\!-\!O \cdot Ag}}} + \underset{\text{(LX.)}}{\overset{\displaystyle O}{\underset{\displaystyle OCH_2 \cdot Ph}{Cl \cdot \overset{\|}{P} \cdot OCH_2 \cdot Ph}}} \longrightarrow \underset{\text{(LXXII.) Adenosine-5'-tribenzyl-diphosphate.}}{\overset{\displaystyle O \qquad\qquad O}{\underset{\displaystyle OCH_2 \cdot Ph \ \ OCH_2 \cdot Ph}{Ad\!-\!O\!-\!\overset{\|}{P}\!-\!\!-\!O\!-\!\!-\!\overset{\|}{P} \cdot OCH_2 \cdot Ph}}} \longrightarrow
$$

$$
\xrightarrow{\ 3\,H_2\ } \quad \underset{\text{(LXXIII.) ADP.}}{\overset{\displaystyle O \qquad O}{\underset{\displaystyle OH \quad\ OH}{Ad\!-\!O\!-\!\overset{\|}{P}\!-\!O\!-\!\overset{\|}{P}\!-\!OH}}} \qquad\qquad \underset{\text{(LXXIV.)}}{\overset{\displaystyle O \qquad\qquad O}{\underset{\displaystyle OCH_2 \cdot Ph \ \ OCH_2 \cdot Ph}{Ad \cdot O\!-\!\overset{\|}{P}\!-\!\!-\!O\!-\!\!-\!\overset{\|}{P}\!-\!O \cdot Ag}}}
$$

$$
\underset{\text{(LXXVI.) ATP.}}{\overset{\displaystyle O \qquad O \qquad O}{\underset{\displaystyle OH \quad\ OH \quad\ OH}{Ad \cdot O \cdot \overset{\|}{P}\!-\!O\!-\!\overset{\|}{P}\!-\!O\!-\!\overset{\|}{P} \cdot OH}}} \xleftarrow{\ 4\,H_2\ } \underset{\text{(LXXV.)}}{\overset{\displaystyle O \qquad\quad O \qquad\quad O}{\underset{\displaystyle Ph \cdot CH_2O \qquad OCH_3 \cdot Ph \ \ OCH_2 \cdot Ph}{Ad \cdot O \cdot \overset{\|}{P}\!-\!O\!-\!\overset{\|}{P}\!-\!\!-\!O\!-\!\!-\!\overset{\|}{P} \cdot OCH_2 \cdot Ph}}}
$$

[Ad = Adenosine-5'—]

In later experiments (16) it was found preferable to replace acetic acid by a mixture of phenol and methyl cyanide, which gives a homogeneous solution of the silver salt. In this case the yield of ADP from (LXXI) was 55%. BADDILEY, MICHELSON and TODD (16) have been able to carry the process further. After partial debenzylation with N-methyl-morpholine a second silver salt (LXXIV) is obtained and this in turn may be brought into reaction with dibenzyl chlorophosphonate and the product (LXXV) hydrogenated to yield ATP (LXXVI). Both the synthetic nucleotides have been compared with the natural substances after isolation as their acridine salts.

A second, simpler and more effective synthesis of ATP has been described by MICHELSON and TODD (*162*). This consists of the reaction between dibenzyl chlorophosphonate and the disilver salt of adenosine-5'-phosphate, followed by hydrogenation. This would be expected to give an isomeric ATP (LXXVIII) but in fact the only recognisable product is identical with the natural material and that prepared by the previous synthesis. To provide a sure understanding of this surprising result is not possible at the present time. The various benzylated intermediates are not crystalline and are extremely labile towards loss of both benzyl and phosphate groups. Thus, although analytical figures are in fair agreement with the structures shown, these are by no means certain. It could therefore be argued that the first synthesis leads to (LXXVIII) through an isomeric adenosine-5'-dibenzyl pyrophosphate. Against this hypothesis electrometric titration (*146*) favours the accepted structure (LXXVI) of ATP. Alternatively partial debenzylation may occur during the initial step of the second synthesis: in this case the yield is surprisingly high.

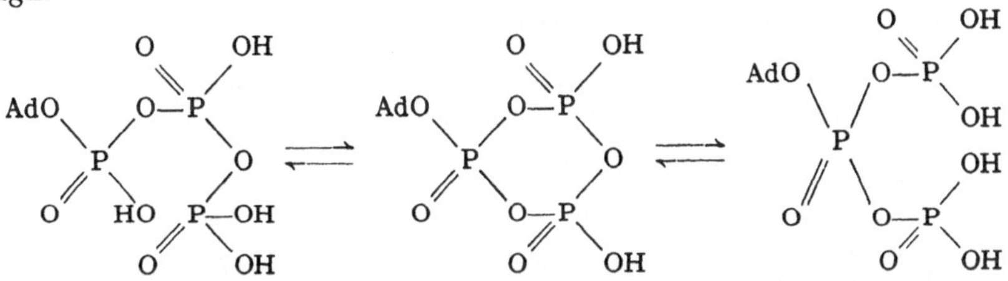

(LXXVI.) ATP.   (LXXVII.) Intermediate.   (LXXVIII.) Isomeric ATP.

[Ad = Adenosine-5'-]

The most plausible explanation is that advanced by MICHELSON and TODD (*162*), namely that the two isomeric forms of ATP are in equilibrium through the cyclic intermediate (LXXVII). It is to be hoped that it will prove possible to test this interesting possibility through tracer studies, since it may have wider implications in the biochemical sphere.

4. Dinucleotides derived from Pyrophosphoric Acids.

No success in this field has been reported, but the progress elsewhere makes the synthesis of coenzyme dinucleotides now a practical possibility. Two types of approach may be envisaged. In one a mononucleotide derivative of pyrophosphoric acid would be esterified with a nucleoside, in the other two mononucleotides would be condensed together. The acid chloride or acid anhydride derivative of a mononucleotide required in the second case would have a further application in the synthesis of polynucleotides classed as dialkyl esters of phosphoric acid. The preparation of such compounds as adenosine-5'-benzyl chlorophosphonate is therefore a central problem of current work in Cambridge on nucleotide synthesis.

References.

1. ALBAUM, H. G. and M. OGUR: An Adenine-pentose-pyrophosphate from Plant Tissues. Arch. Biochemistry 15, 158 (1947).
2. ALBAUM, H. G., M. OGUR and A. HIRSHFELD: The Isolation of Adenosine Triphosphate from Plant Tissue. Arch. Biochemistry 27, 130 (1950).
3. ANDREWS, K. J. M., N. ANAND, A. R. TODD and A. TOPHAM: Experiments on the Synthesis of Purine Nucleosides. XXVI. 9-D-Glucopyranosido*iso*-guanine. J. chem. Soc. (London) 1949, 2490.
4. ANDREWS, K. J. M., G. W. KENNER and A. R. TODD: Experiments on the Synthesis of Purine Nucleosides. XXIV. 9-D-Galactosido-2-methylthioadenines. J. chem. Soc. (London) 1949, 2302.
5. ATHERTON, F. R.: Some Aspects of the Organic Chemistry of Phosphorus Oxyacids. Quart. Rev. 3, 146 (1949).
6. ATHERTON, F. R., G. A. HOWARD and A. R. TODD: Studies on Phosphorylation. IV. Further Studies on the Use of Dibenzyl Chlorophosphonate and the Examination of Certain Alternative Phosphorylation Methods. J. chem. Soc. (London) 1948, 1106.
7. ATHERTON, F. R., H. T. OPENSHAW and A. R. TODD: Studies on Phosphorylation. I. Dibenzyl Chlorophosphonate as a Phosphorylating Agent. J. chem. Soc. (London) 1945, 382.
8. — — — Studies on Phosphorylation. II. The Reaction of Dialkyl Phosphites with Polyhalogen Compounds in Presence of Bases. A New Method for the Phosphorylation of Amines. J. chem. Soc. (London) 1945, 660.
9. ATHERTON, F. R. and A. R. TODD: Studies on Phosphorylation. III. Further Observations on the Reaction of Phosphites with Polyhalogen Compounds in Presence of Bases and its Application to the Phosphorylation of Alcohols. J. chem. Soc. (London) 1947, 674.
10. BADDILEY, J.: Adenine Thiomethylpentoside. A Proof of Structure and Synthesis. J. chem. Soc. (London) 1951 (in press).
11. BADDILEY, J., V. M. CLARK, J. J. MICHALSKI and A. R. TODD: Studies on Phosphorylation. V. The Reaction of Tertiary Bases with Esters of Phosphorous, Phosphoric and Pyrophosphoric Acids. A New Method of Selective Debenzylation. J. chem. Soc. (London) 1949, 815.
12. BADDILEY, J., G. W. KENNER, B. LYTHGOE and A. R. TODD: Experiments on the Synthesis of Purine Nucleosides. X. A Synthesis of 9-D-Ribopyranosido-adenine. J. chem. Soc. (London) 1944, 657.
13. BADDILEY, J., B. LYTHGOE, D. McNEIL and A. R. TODD: Experiments on the Synthesis of Purine Nucleosides I. Model Experiments on the Synthesis of 9-Alkylpurines. J. chem. Soc. (London) 1943, 383.
14. BADDILEY, J., B. LYTHGOE and A. R. TODD: Experiments on the Synthesis of Purine Nucleosides. III. 4-Glycosidaminopyrimidines. J. chem. Soc. (London) 1943, 571.
15. — — — Experiments on the Synthesis of Purine Nucleosides. VI. The Synthesis of 9-D-Xylosido-2-methyladenine and of 6-D-Xylosidamino-2-methylpurine. J. chem. Soc. (London) 1944, 318.

16. BADDILEY, J., A. M. MICHELSON and A. R. TODD: Nucleotides. II. A Synthesis of Adenosine Triphosphate. J. chem. Soc. (London) 1949, 582.

17. BADDILEY, J. and A. R. TODD: Nucleotides. I. Muscle Adenylic Acid and Adenosine Diphosphate. J. chem. Soc. (London) 1947, 648.

18. BADDILEY, J., F. WEYGAND and O. TRAUTH: The Structure and Synthesis of Adenine Thiomethyl Pentoside: Synthesis of 5'-Methylthioinosine. Nature (London) 167, 359 (1951).

19. BAER, E. and M. KATES: Synthesis of Enantiomeric α-Lecithins. J. Amer. chem. Soc. 72, 942 (1950).

20. BAILEY, K.: The Enzymic Degradation of Adenosinetriphosphate. Biochemic. J. 45, 479 (1949).

21. BARKER, G. R.: Private communication.

22. BARKER, G. R., K. R. FARRAR and J. M. GULLAND: The Constitution of Yeast Ribonucleic Acid. X. Further Studies on the Nature of the Carbohydrate Radicals. J. chem. Soc. (London) 1947, 21.

23. BARKER, G. R. and J. M. GULLAND: The Constitution of Yeast Ribonucleic Acid. V. Synthesis of Yeast Adenylic Acid. J. chem. Soc. (London) 1942, 231.

24. BARKER, G. R. and M. V. LOCK: The Chemistry of Ribose and its Derivatives. II. The Constitution of an Anhydro-D-Ribose. J. chem. Soc. (London) 1950, 23.

25. BAXTER, R. A., A. C. MCLEAN and F. S. SPRING: Application of the HOFMANN Reaction to the Synthesis of Heterocyclic Compounds. V. The Synthesis of 9-D-Mannopyranosidoxanthine and of 9-D-Ribopyranosidoxanthine. J. chem. Soc. (London) 1948, 523.

26. BAXTER, R. A. and F. S. SPRING: Application of the HOFMANN Reaction to the Synthesis of Heterocyclic Compounds. IV. The Synthesis of 9-D-Xylopyranosidoxanthine. J. chem. Soc. (London) 1947, 378.

27. BEHREND, R. u. R. THURM: Über die Konstitution der Alkylderivate des Methyluracils und der S-Methylharnsäure. Liebigs Ann. Chem. 323, 160 (1902).

28. BERGMANN, W. and R. J. FEENEY: The Isolation of a New Thymine Pentoside from Sponges. J. Amer. chem. Soc. 72, 2809 (1950).

29. BLOUT, E. R. and M. FIELDS: Absorption Spectra. VII. The Infrared Spectra of Some Nucleic Acids, Nucleotides and Nucleosides. J. biol. Chemistry 178, 335 (1949).

30. BREDERECK, H. u. E. BERGER: Nucleinsäuren. XVII. Nucleotidsynthesen: Synthese der Uridylsäure. Ber. dtsch. chem. Ges. 73, 1124 (1940).

31. BREDERECK, H., M. KÖTHNIG u. E. BERGER: Über die d-Ribose (Darstellung einer kristallisierten Anhydroribose). Ber. dtsch. chem. Ges. 73, 956 (1940).

32. BREDERECK, H., A. MARTINI u. F. RICHTER: Nucleinsäuren. XIX. Fermentative und chemische Darstellung der Nucleoside aus Hefenucleinsäure. Ber. dtsch. chem. Ges. 74, 694 (1941).

33. BRIGL, P. u. H. MÜLLER: Zur Synthese von Phosphorsäureester. I. Ber. dtsch. chem. Ges. 72, 2121 (1939).

34. BRINK, N. G., F. W. HOLLY, C. H. SHUNK, E. W. PEEL, J. J. CAHILL and K. FOLKERS: Vitamin B$_{12}$. IX. 1-α-D-Ribofuranosido-5:6-dimethylbenziminazole. A Degradation Product of Vitamin B$_{12}$. J. Amer. chem. Soc. 72, 1866 (1950).

35. BRISTOW, N. W. and B. LYTHGOE: Experiments on the Synthesis of Purine Nucleosides. XXV. 1:2:3:5-Tetracetyl D-Arabofuranose and the D-Arabofuranosides of Theophylline and Adenine. J. chem. Soc. (London) 1949, 2306.

36. BROWN, D. M., L. J. HAYNES and A. R. TODD: Nucleotides. VI. The Structure of the Synthetic Nucleotides Prepared from the Benzylidene Ribonucleosides. J. chem. Soc. (London) 1950, 2299.

37. Brown, D. M. and B. Lythgoe: Deoxyribonucleosides and Related Compounds. II. A Proof of the Furanose Structure of the Natural 2-Deoxyribonucleosides. J. chem. Soc. (London) 1950, 1990.

37a. Brown, D. M. and A. R. Todd: Unpublished.

38. Brownlie, I. A., G. B. B. M. Sutherland and A. R. Todd: Infra-red Spectroscopic Measurements of the Substituted Pyrimidines. I. The Presence of Hydrogen Bonding in 4-Triacetyl-D-Xylosidaminopyrimidines. J. chem. Soc. (London) 1948, 2265.

39. Buchanan, J. G., C. A. Dekker and A. G. Long: The Detection of Glycosides and Non-reducing Carbohydrate Derivatives in Paper Partition Chromatography. J. chem. Soc. (London) 1950, 3162.

40. Buchanan, J. G., A. W. Johnson, J. A. Mills and A. R. Todd: Chemistry of the Vitamin B_{12} Group. I. Acid Hydrolysis Studies. Isolation of a Phosphorus-containing Degradation Product. J. chem. Soc. (London) 1950, 2845.

41. Caputto, R., L. F. Leloir, C. E. Cardini and A. C. Paladini: Isolation of the Coenzyme of the Galactose Phosphate Glucose Phosphate Transformation. J. biol. Chemistry 184, 333 (1950).

42. Carter, C. E.: Paper Chromatography of Purine and Pyrimidine Derivatives of Yeast Ribonucleic Acid. J. Amer. chem. Soc. 72, 1466 (1950).

43. Carter, C. E. and W. E. Cohn: Separation of Three Naturally Occurring Adenine Ribonucleotides by Paper Chromatography and Ion Exchange. Federat. Proc. (Amer. Soc. exp. Biol.) 8, 190 (1949).

44. – – The Enzymatic Degradation of Ribonucleic Acid by Crystalline Ribonuclease. J. Amer. chem. Soc. 72, 2604 (1950).

45. Cavalieri, L. F., A. Bendich, J. F. Tinker and G. B. Brown: Ultraviolet Absorption Spectra of Purines, Pyrimidines and Triazolopyrimidines. J. Amer. chem. Soc. 70, 3875 (1948).

46. Chang, P. and B. Lythgoe: Experiments on the Synthesis of Purine Nucleosides. XXVII. 1:2:3:5-Tetracetyl D-Xylofuranose and the D-Xylofuranosides of Theophylline and Adenine. J. chem. Soc. (London) 1950, 1992.

47. Cherbuliez, E. et K. Bernhard: Recherches sur la graine de croton. I. Sur le crotonoside (2-oxy-6-amino-purine-d-riboside). Helv. chim. Acta 15, 464 (1932).

48. Clark, H. W., A. L. Dounce and E. Stotz: Extraction and Purification of Diphosphopyridine Nucleotide. J. biol. Chemistry 187, 459 (1949).

49. Clark, V. M., G. W. Kenner, A. R. Todd and F. J. Weymouth: Unpublished.

50. Clark, V. M. and A. R. Todd: Studies on Phosphorylation. VI. The Reaction Between Organic Bases and Esters of the Oxyacids of Phosphorus. An Interpretation Based on a Comparison of Certain Aspects of the Chemistry of Sulphur and Phosphorus. J. chem. Soc. (London) 1950, 2023.

51. – – Studies on Phosphorylation. VII. The Action of Salts on Neutral Benzyl Esters of the Oxyacids of Phosphorus. A New Method of Selective Debenzylation. J. chem. Soc. (London) 1950, 2031.

52. Cohn, W. E.: The Separation of Purine and Pyrimidine Bases and of Nucleotides by Ion Exchange. Science (New York) 109, 377 (1949).

53. – Separation of Mononucleotides by Anion Exchange Chromatography. J. Amer. chem. Soc. 71, 2275 (1949).

54. – The Anion-exchange Separation of Ribonucleotides. J. Amer. chem. Soc. 72, 1471 (1950).

55. – Heterogeneity in Pyrimidine Nucleotides. J. Amer. chem. Soc. 72, 2811 (1950).

56. COHN, W. E. and C. E. CARTER: The Preparation of Uridylic and Cytidylic Acids from Yeast Ribonucleic Acid by an Ion-exchange Method. J. Amer. chem. Soc. **72**, 2606 (1950).

57. — — The Separation of Adenosine Polyphosphates by Ion Exchange and Paper Chromatography. J. Amer. chem. Soc. **72**, 4273 (1950).

58. COHN, W. E. and E. VOLKIN: Nucleoside-5'-phosphates from Ribonucleic Acid. Nature (London) **167**, 483 (1951).

59. COOK, A. H. and E. SMITH: Studies in the Azole Series. XXIII. A New Synthesis of 6-Aminopurines. J. chem. Soc. (London) **1949**, 3001.

60. COOK, A. H. and G. H. THOMAS: Studies in the Azole Series. XXX. New Syntheses of 2- and 8-Aminopurines. J. chem. Soc. (London) **1950**, 1888.

61. COOLEY, G., B. ELLIS, P. MAMALIS, V. PETROW and B. STURGEON: The Chemistry of Anti-pernicious Anaemia Factors. V. The Interrelationship and Structure of the α-, β- and γ-Components. J. Pharm. Pharmacol. **2**, 579 (1950).

62. CRAIG, L. C.: Extraction. Analyt. Chemistry **23**, 41 (1951).

63. CRAIG, L. C., C. GOLUMBIC, H. MIGHTON and E. TITUS: Distribution Studies. III. The Use of Buffers in Counter Current Distribution. J. biol. Chemistry **161**, 321 (1945).

64. CRAIG, L. C. and D. CRAIG: Extraction and Distribution. In: A. WEISSBERGER, Technique of Organic Chemistry **3**, 171 (1950).

65. CRAMMER, J. L.: Paper Chromatography of Flavin Nucleotides. Nature (London) **161**, 349 (1948).

66. DANIELLI, J. F. and R. BROWN (editors): Nucleic Acid. Symposia, Soc. experim. Biology **1** (1947).

67. DAVIES, C. W.: Ion Exchange Resins in Chromatography. Research (London) **3**, 447 (1950).

68. DAVOLL, J. and B. A. LOWY: A New Synthesis of Purine Nucleosides. The Synthesis of Adenosine, Guanosine and 2.6-Diamino-9-β-D-ribofuranosyl-purine. J. Amer. chem. Soc. **73**, 1650 (1951).

69. DAVOLL, J. and B. LYTHGOE: Deoxyribonucleosides and Related Compounds. I. Synthetic Applications of Some 1-Halogeno-2-deoxy-sugar Derivatives. J. chem. Soc. (London) **1949**, 2526.

70. DAVOLL, J., B. LYTHGOE and A. R. TODD: Experiments on the Synthesis of Purine Nucleosides. XII. The Configuration at the Glycosidic Centre in Natural and Synthetic Pyrimidine and Purine Nucleosides. J. chem. Soc. (London) **1946**, 833.

71. — — — Experiments on the Synthesis of Purine Nucleosides. XIX. A Synthesis of Adenosine. J. chem. Soc. (London) **1948**, 967.

72. — — — Experiments on the Synthesis of Purine Nucleosides. XX. A Synthesis of Guanosine. J. chem. Soc. (London) **1948**, 1685.

73. DEKKER, C. A., D. T. ELMORE and A. R. TODD: Unpublished.

74. DEKKER, C. A. and A. R. TODD: Uracil Deoxyriboside. Nature (London) **166**, 557 (1950).

75. DERIAZ, R. E., W. G. OVEREND, M. STACEY, E. G. TEECE and L. F. WIGGINS: Deoxy-sugars. V. A Reinvestigation of the Glycal Method for the Synthesis of 2-Deoxy-D-and L-Ribose. J. chem. Soc. (London) **1949**, 1879.

76. DEUTSCH, A. and O. FERNÖ: Use of Dibenzylphosphoryl Chloride for Phosphorylations. Nature (London) **156**, 604 (1945).

77. DEY, B. B., H. C. FRIEDMANN and C. SIVARAMAN: Preparation of Adenosine Triphosphate from Bull Frogs. Current Sci. **18**, 4 (1949).

78. DOHERTY, D. G.: Studies on the Structure of Isomeric Nucleotides. Abstr. Amer. chem. Soc. Meeting (Chicago) 56 C (1950).

79. DOUNCE, A. L., A. ROTHSTEIN, G. J. BEYER, R. MEIER and R. M. FREER: Detailed Procedure for the Preparation of Highly Purified Adenosine Triphosphate. J. biol. Chemistry 174, 361 (1948).

80. DUNCAN, J. F. and B. A. J. LISTER: Ion Exchange. Quart. Rev. 2, 307 (1948).

81. ELMORE, D. T.: The Isolation of Uridine and Cytidine from Yeast Ribonucleic Acid. J. chem. Soc. (London) 1950, 2084.

82. FISCHER, E.: Über Phosphorsäureester des Methylglucosids und Theophyllinglucosids. Ber. dtsch. chem. Ges. 47, 3193 (1914).

83. FISCHER, E. u. B. HELFERICH: Synthetische Glucoside der Purine. Ber. dtsch. chem. Ges. 47, 210 (1914).

84. FISCHER, E. u. K. RASKE: Verbindung von Acetobromglucose und Pyridin. Ber. dtsch. chem. Ges. 43, 1750 (1910).

85. FLEXSER, L. A. and W. G. FARKAS: Quoted in ref. (86).

86. FORREST, H. S. and A. R. TODD: Nucleotides. V. Riboflavin-5'-phosphate. J. chem. Soc. (London) 1950, 3295.

87. FURBERG, S.: An X-ray Study of the Stereochemistry of the Nucleosides. Acta chem. Scand. 4, 751 (1950).

88. GREENE, R. D. and A. BLACK: The Preparation of Pure d-Riboflavin from Natural Sources. J. Amer. chem. Soc. 59, 1920 (1937).

89. GULLAND, J. M. and G. R. BARKER: The Constitution of Yeast Ribonucleic Acid. VI. The Nature of the Carbohydrate Radicals. J. chem. Soc. (London) 1943, 625.

90. GULLAND, J. M. and G. I. HOBDAY: The Constitution of Yeast Ribonucleic Acid. IV. Syntheses of Uridylic and Guanylic Acids, Uridine-5-phosphate and Guanosine-5-phosphate. J. chem. Soc. (London) 1940, 746.

91. GULLAND, J. M. and W. G. OVEREND: Benzylideneguanosine. J. chem. Soc. (London) 1948, 1380.

92. GULLAND, J. M. and H. SMITH: The Constitution of Yeast Ribonucleic Acid. XI. Synthesis of Uridine-2'-phosphate. J. chem. Soc. (London) 1947, 338.

93. — — The Constitution of Yeast Ribonucleic Acid. XII. Synthesis of Cytidine-2'-phosphate. J. chem. Soc. (London) 1948, 1527.

94. — — The Constitution of Yeast Ribonucleic Acid. XIII. The Synthesis of Di-2'-uridine Phosphate. J. chem. Soc. (London) 1948, 1532.

95. GULLAND, J. M. and L. F. STORY: Constitution of Purine Nucleosides. VI. Adenine Deoxyriboside, Adenine Glucoside, and a Route to the Synthesis of the Naturally Occurring Nucleosides. J. chem. Soc. (London) 1938, 259.

96. HANES, C. S. and F. A. ISHERWOOD: Separation of the Phosphoric Esters on the Filter Paper Chromatogram. Nature (London) 164, 1107 (1949).

97. HARRIS, R. J. C. and J. F. THOMAS: The Isolation of Uridine from Yeast Ribonucleic Acid. J. chem. Soc. (London) 1948, 1936.

98. HAYNES, L. J. and A. R. TODD: Codehydrogenases. I. The Synthesis of Dihydronicotinamide-D-ribofuranoside [N-D-Ribofuranosidyl-1:2 (or 6)-dihydronicotinamide]. J. chem. Soc. (London) 1950, 303.

99. HEILBRON, I.: Presidential Address. Concerning Amino-acids, Peptides and Purines. J. chem. Soc. (London) 1949, 2099.

100. HOGEBOOM, G. H. and G. T. BARRY: Purification of Diphosphopyridine Nucleotide by Counter-current Distribution. J. biol. Chemistry 176, 935 (1948).

101. HOLLAND, A., B. LYTHGOE and A. R. TODD: Experiments on the Synthesis of Purine Nucleosides. XVIII. A Synthesis of 9-D-Glucopyranosidoadenine. J. chem. Soc. (London) 1948, 965.

102. HOTCHKISS, R. D.: The Quantitative Separation of Purines, Pyrimidines and Nucleosides by Paper Chromatography. J. biol. Chemistry 175, 315 (1948).

103. HOWARD, G. A.: The Replacement of the Halogen Group of Acetohalogeno-sugars with Retention of Configuration. J. chem. Soc. (London) 1950, 1045.

104. HOWARD, G. A., G. W. KENNER, B. LYTHGOE and A. R. TODD: Experiments on the Synthesis of Purine Nucleosides. XIV. An Interpretation of Some Interconversion Reactions of N-Glycosides. J. chem. Soc. (London) 1946, 855.

105. — — — — Experiments on the Synthesis of Purine Nucleosides. XV. The Configuration of Some Synthetic Purine and Pyrimidine Glycosides. J. chem. Soc. (London) 1946, 861.

106. HOWARD, G. A., B. LYTHGOE and A. R. TODD: Experiments on the Synthesis of Purine Nucleosides. XI. The Synthesis of 9-D-Xylopyranosido-2-methyl-thioadenine and its Conversion into 9-D-Xylopyranosidoadenine. J. chem. Soc. (London) 1945, 556.

107. — — — A Synthesis of Cytidine. J. chem. Soc. (London) 1947, 1052.

108. HOWARD, G. A., A. C. McLEAN, G. T. NEWBOLD, F. S. SPRING and A. R. TODD: Application of the HOFMANN Reaction to the Synthesis of Heterocyclic Compounds. VI. Experiments on the Synthesis of Purine Nucleosides. XXI. A Synthesis of Xanthosine. J. chem. Soc. (London) 1949, 232.

109. HUMMEL, J. P. and O. LINDBERG: Studies on the Mechanism of Aerobic Phosphorylation. J. biol. Chemistry 180, 1 (1949).

110. JACKSON, E. L.: Periodic Acid Oxidation. Organic Reactions (New York) 2, 341 (1944).

111. JOHNSON, T. B. and R. D. COGHILL: Researches on Pyrimidines. CIII. The Discovery of 5-Methylcytosine in Tuberculinic Acid, the Nucleic Acid of the Tubercle Bacillus. J. Amer. chem. Soc. 47, 2838 (1925).

112. JOHNSON, T. B. and F. W. HEYL: Researches on Pyrimidines: Some Condensation Products of Substituted Pseudothiourea: Synthesis of 1-Methyluracil. Amer. chem. J. 37, 628 (1907).

112a. KALCKAR, H. M.: Differential Spectrophotometry of Purine Compounds by Means of Specific Enzymes. II. The Determination of Adenine Compounds. J. biol. Chemistry 167, 445 (1947).

113. KARRER, P., B. H. RINGIER, J. BÜCHI, H. FRITZSCHE u. U. SOLMSSEN: Modellversuche betreffend die wasserstoffübertragenden Wirkungsgruppen der Cofermente. Helv. chim. Acta 20, 55 (1937).

114. KARRER, P., H. SALOMON, K. SCHÖPP, F. BENZ u. B. BECKER: Flavinsynthesen. VI. Synthesen von drei weiteren Stereoisomeren des Lactoflavins. Helv. chim. Acta 18, 908 (1935).

115. KENNER, G. W., B. LYTHGOE and A. R. TODD: Experiments on the Synthesis of Purine Nucleosides. IX. A Synthesis of 9-D-Xylopyranosidoadenine. J. chem. Soc. (London) 1944, 652.

116. — — — Experiments on the Synthesis of Purine Nucleosides. XVII. The Preparation of 4-Glycofuranosidaminopyrimidines, and a Synthesis of 9-L-Arabofuranosido-2-methylthioadenine. J. chem. Soc. (London) 1948, 957.

117. KENNER, G. W., H. J. RODDA and A. R. TODD: Experiments on the Synthesis of Purine Nucleosides. XXII. The Synthesis of the α- and β-Forms of 9-Triacetyl-D-ribopyranosido-2-methylthioadenine and Further Studies on the Synthesis of 9-Glycofuranosido Purines. J. chem. Soc. (London) 1949, 1613.

118. KENNER, G. W., C. W. TAYLOR and A. R. TODD: Experiments on the Synthesis of Purine Nucleosides. XXIII. A New Synthesis of Adenosine. J. chem. Soc. (London) 1949, 1620.

119. KENNER, G. W. and A. R. TODD: Experiments on the Synthesis of Purine Nucleosides. XIII. An Improved Method for the Cyclisation of 4-Glycosidamino-5-thioformamidopyrimidines. J. chem. Soc. (London) 1946, 852.

120. KORNBERG, A.: Enzymic Synthesis of Triphosphopyridine Nucleotide. J. biol. Chemistry **182**, 805 (1950).

120a. KORNBERG, A. and W. E. PRICER: Nucleotide Pyrophosphatase. J. biol. Chemistry **182**, 763 (1950).

121. — — On the Structure of Triphosphopyridine Nucleotide. J. biol. Chemistry **186**, 557 (1950).

122. KRISHNAN, P. S. and W. L. NELSON: Some Observations on the Isolation of Adenosine Triphosphate from Skeletal Muscle. Arch. Biochemistry **19**, 65 (1948).

123. KUHN, R. u. K. HENKEL: Über die Senkung der Körpertemperatur durch Adenylthiomethylpentose. Hoppe-Seyler's Z. physiol. Chem. **269**, 41 (1941).

124. KUHN, R. u. R. STRÖBELE: Über O-Nitranilin-glucoside. Ber. dtsch. chem. Ges. **70**, 773 (1937).

125. KUNIN, R.: Ion Exchange. Analyt. Chemistry **23**, 45 (1951).

126. KUNIN, R. and R. J. MYERS: The Anion Exchange Equilibria in an Anion Exchange Resin. J. Amer. chem. Soc. **69**, 2874 (1947).

127. LAMPSON, G. P. and H. A. LARDY: Phosphoric Esters of Biological Importance. III. The Synthesis of Propanediol Phosphate. J. biol. Chemistry **181**, 697 (1949).

128. LELOIR, L. F.: Sugar Phosphates. Fortschr. Chem. organ. Naturstoffe **8**, 47 (1951).

129. — Private communication to A. R. TODD.

130. LEPAGE, G. A.: Preparation of Diphosphopyridine Nucleotide. J. biol. Chemistry **168**, 623 (1947).

131. — Adenosine Diphosphate. Biochemical Prepar. **1**, 1 (1949).

132. — Adenosine Triphosphate. Biochemical Prepar. **1**, 5 (1949).

133. — Diphosphopyridine Nucleotide. Biochemical Prepar. **1**, 28 (1949).

134. LEPAGE, G. A. and G. C. MUELLER: Preparation of Triphosphopyridine Nucleotide. J. biol. Chemistry **180**, 975 (1949).

135. LEPAGE, G. A. and W. W. UMBREIT: The Occurrence of Adenosine-3-triphosphate in Autotrophic Bacteria. J. biol. Chemistry **148**, 255 (1943).

1.′6. LEVENE, P. A. and L. W. BASS: Nucleic Acids. Amer. chem. Soc. Monographs no. 56. New York: The Chemical Catalog Co. 1931.

137. LEVENE, P. A. and S. A. HARRIS: The Ribose Phosphoric Acid from Xanthylic Acid. J. biol. Chemistry **95**, 755 (1932).

138. — — The Ribose Phosphoric Acid from Xanthylic Acid. II. J. biol. Chemistry **98**, 9 (1932).

139. — — The Ribose Phosphoric Acid. from Yeast Adenylic Acid. J. biol. Chemistry **101**, 419 (1933).

140. LEVENE, P. A. u. F. B. LA FORGE: Über die Hefenucleinsäure. V. Die Struktur der Pyridin-Nucleoside. Ber. dtsch. chem. Ges. **45**, 608 (1912).

141. LEVENE, P. A. and E. T. STILLER: The Synthesis of Ribose-5-phosphoric Acid. J. biol. Chemistry **104**, 299 (1934).

142. LEVENE, P. A. and R. S. TIPSON: N-Methyl Uridine and its Bearing on the Structure of Uridine. J. biol. Chemistry **104**, 385 (1934).

143. — — The Partial Synthesis of Ribose Nucleotides. I. Uridine 5-Phosphoric Acid. J. biol. Chemistry **106**, 113 (1934).

143a. — — The Partial Synthesis of Ribose Nucleotides. II. Muscle Inosinic Acid. J. biol. Chemistry **111**, 313 (1935).

144. — — Phosphorylation of Monoacetone Adenosine and of Diacetyl Adenosine. J. biol. Chemistry **121**, 131 (1937).

145. LIPKIN, D. and G. C. McELHENY: The Specific Rotation of Yeast Adenylic Acid in Anhydrous Formamide. J. Amer. chem. Soc. **72**, 2287 (1950).

145a. — — An Isomeric Benzylideneguanosine. Nature (London) **167**, 238 (1951).

146. LOHMANN, K.: Untersuchungen zur Konstitution der Adenylpyrophosphorsäure. Biochem. Z. **254**, 381 (1932).

147. LORING, H. S., N. G. LUTHY, H. W. BORTNER and L. W. LEVY: The Isolation of an Isomeric Cytidylic Acid from Hydrolysates of Yeast Ribonucleic Acid. J. Amer. chem. Soc. **72**, 2811 (1950).

148. LORING, H. S. and J. McT. PLOESER: Deamination of Cytidine in Acid Solution and the Preparation of Uridine and Cytidine by Acid Hydrolysis of Yeast Nucleic Acid. J. biol. Chemistry **178**, 439 (1949).

149. LORING, H. S., P. M. ROLL and J. G. PIERCE: Preparation of Cytidylic Acid and Diammonium Uridylate from Yeast Nucleic Acid. J. biol. Chemistry **174**, 729 (1948).

150. LYTHGOE, B.: Chemistry of Nucleosides and Nucleotides. Nucleotides Related to Nucleic Acids. Annu. Rep. Progr. Chem. **41**, 200 (1944).

151. — Chemistry of Adenine Nucleotide Coenzymes. Annu. Rep. Progr. Chem. **42**, 175 (1945).

152. LYTHGOE, B., H. SMITH and A. R. TODD: Experiments on the Synthesis of Purine Nucleosides. XVI. 9-β-D-Mannopyranosidoadenine. A Proof of the Location of the Sugar Residue in Adenosine. J. chem. Soc. (London) **1947**, 355.

153. LYTHGOE, B. and A. R. TODD: Experiments on the Synthesis of Purine Nucleosides. VIII. The Determination of the Lactol Ring Structure of Purine Glycosides by Periodate Oxidation. J. chem. Soc. (London) **1944**, 592.

154. — — Structure of Adenosine di- and tri-Phosphate. Nature (London) **155**, 695 (1945).

155. LYTHGOE, B., A. R. TODD and A. TOPHAM: Experiments on the Synthesis of Rurine Nucleosides. V. The Coupling of Pyrimidine Derivatives with Diazonium Salts. A Method for the Preparation of 5-Aminopyrimidines. J. chem. Soc. (London) **1944**, 315.

156. McCOMBIE, H., B. C. SAUNDERS and G. J. STACEY: Esters Containing Phosphorus. I. J. chem. Soc. (London) **1945**, 380.

157. MAGASANIK, B., E. VISCHER, R. DONIGER, D. ELSON and E. CHARGAFF: The Separation and Estimation of Ribonucleotides in Minute Quantities. J. biol. Chemistry **186**, 37 (1950).

158. MAMALIS, P., V. PETROW and B. STURGEON: The Chemistry of Antipernicious Anaemia Factors. IV. Benziminazole Glycosides. J. Pharm. Pharmacol. **2**, 491 (1950).

159. MARKHAM, R. and J. D. SMITH: Chromatographic Studies of Nucleic Acids. I. A technique for the Identification and Estimation of Purine and Pyrimidine Bases, Nucleosides and Related Substances. Biochemic. J. **45**, 294 (1949).

160. MAYER, S. W. and E. R. TOMPKINS: Ion Exchange as a Separation Method. IV. A Theoretical Analysis of the Column Separations Process. J. Amer. chem. Soc. **69**, 2866 (1947).

161. MICHELSON, A. M. and A. R. TODD: Nucleotides. III. Mononucleotides Derived from Adenosine, Guanosine, Cytidine and Uridine. J. chem. Soc. (London) **1949**, 2476.

162. — — Nucleotides. IV. A Novel Synthesis of Adenosine Triphosphate. J. chem. Soc. (London) **1949**, 2487.

163. MITCHELL, H. K. and F. A. HASKINS: A Filter Paper "Chromatopile". Science (New York) **110**, 278 (1949).

164. MONTREUIL, J. et P. BOULANGER: Chromatographie quantitative sur papier des ribonucléotides. C. R. hebd. Séances Acad. Sci. **231**, 247 (1950).

165. NEWMARK, M. Z., I. GOODMAN and K. DITTMER: The Lactal Ring Structure of Some Synthetic Pyrimidine Nucleosides. J. Amer. chem. Soc. **71**, 3847 (1949).

165a. Nucleic Acids. [Symposium]: cf. Reference (*66*).

166. OLDHAM, J. W. H. and J. K. RUTHERFORD: A Method for the Identification and Estimation of the 6-Hydroxyl Group in Glucose. J. Amer. chem. Soc. **54**, 366 (1932).

167. OVEREND, W. G., M. STACEY and L. F. WIGGINS: Deoxy-sugars. IV. A Synthesis of 2-Deoxy-*D*-ribose from *D*-Erythrose. J. chem. Soc. (London) **1949**, 1358.

167a. Partition Chromatography: cf. Reference (*193*).

168. PLAUT, G. W. E., S. A. KUBY and H. A. LARDY: Systems for the Separation of Phosphoric Esters by Solvent Distribution. J. biol. Chemistry **184**, 243 (1950).

169. PLOESER, J. McT. and H. S. LORING: The Ultra-violet Spectra of the Pyrimidine Ribonucleosides and Ribonucleotides. J. biol. Chemistry **178**, 431 (1949).

170. REICH, W. S.: A New Method of Phosphorylation. Nature (London) **157**, 133 (1946).

171. REICHARD, P.: Partition Chromatography on Starch of Ribonucleosides. Nature (London) **162**, 662 (1948).

172. — Preparation of Ribonucleosides from Small Amounts of Ribonucleic Acid. J. biol. Chemistry **179**, 763 (1949).

173. ROUX, H., Y. TEYSSEIRE et G. DUCHESNE: Sur l'obtention de nouveaux esters polyphosphoriques de l'aneurine. I. Préparation des esters polyphosphoriques de l'aneurine. Bull. Soc. Chim. biol. **30**, 592 (1948).

174. SANADI, D. R. and F. M. HUENNEKENS: Isolation of Flavin-adenine Dinucleotide. Abstr. Amer. chem. Soc. Meeting (Detroit) 60 C (1950).

175. SATOH, K. and K. MAKINO: Structure of Adenylthiomethylpentose. Nature (London) **165**, 769 (1950).

175a. — — Structure of Adenylthiomethylpentose. Nature (London) **167**, 238 (1951).

176. SCHINDLER, O.: Verbesserte Isolierungsmethode der Desoxyribonucleoside. Helv. chim. Acta **32**, 979 (1949).

177. SCHLENK, F.: Nicotinamide Riboside. Arch. Biochemistry **3**, 93 (1943).

178. — Chemistry and Enzymology of Nucleic Acids. Adv. Enzymology (New York) **9**, 455 (1949).

179. SODI PALLARES, E. and H. MARTINEZ GARZA: Isolation of *L*-Lyxoflavine from the Human Myocardium. Arch. Biochemistry **22**, 63 (1949).

180. SODI PALLARES, E., F. VÉLEZ OROZCO and J. RIVERO CAVALLO: Note on a Pentose Found in the Heart Muscle. Arch. inst. cardiol. (Mexico) **17**, 575 (1947) [Chem. Abstr. **41**, 7480 (1947)].

181. SUMNER, J. B., P. S. KRISHNAN and E. B. SISLER: An Improved Method for the Preparation of Coenzyme. I. Arch. Biochemistry **12**, 19 (1947).

182. SUZUKI, U. u. T. MORI: Über einen durch Hydrolyse von Adenylthiozucker der Hefe entstehenden schwefelhaltigen Zucker. Biochem. Z. **162**, 413 (1925).

183. TINKER, J. F. and G. B. BROWN: The Characterisation of Purines and Pyrimidines by the Method of Countercurrent Distribution. J. biol. Chemistry **173**, 585 (1948).

184. TIPSON, R. S.: The Chemistry of Nucleic Acids. Adv. Carbohydrate Chem. **1**, 193 (1945).

185. TODD, A. R.: Synthesis in the Study of Nucleotides. J. chem. Soc. (London) **1946**, 647.

186. TOMPKINS, E. R.: Laboratory Application of Ion Exchange Techniques. J. chem. Educat. **26**, 32, 92 (1949).

187. TOY, A. D. F.: Preparation and Properties of Some Unsymmetrical Tetralkyl Pyrophosphates. J. Amer. chem. Soc. **72**, 2065 (1950).

188. VELLUZ, L., G. AMIARD et J. BARTOS: Un nouveau dérivé de phosphorylation de la vitamin B_1, l'acide thiamine triphosphorique. Bull. Soc. chim. France **1948**, 871.

189. VISCONTINI, M., G. BONETTI u. P. KARRER: Zur Herstellung der Cocarboxylase und des Aneurin-triphosphorsäureesters. Helv. chim. Acta **32**, 1478 (1949).

190. VISSER, D. W., I. GOODMAN and K. DITTMER: The Synthesis of Thymine Nucleosides. J. Amer. chem. Soc. **70**, 1926 (1948).

191. WEYGAND, F. u. R. LÖWENFELD: Ein neuer Zuckerabbau. Ber. dtsch. chem. Ges. **83**, 559 (1950).

192. WEYGAND, F., O. TRAUTH u. R. LÖWENFELD: Konstitutionsaufklärung des Thiozuckers der Adenylthiomethylpentose. Ber. dtsch. chem. Ges. **83**, 563 (1950).

193. WILLIAMS, R. T. (editor): Partition Chromatography. Biochemical Soc. Symposia 3 (1949).

194. WILLIAMSON, B. and L. C. CRAIG: Distribution Studies. V. Calculation of Theoretical Curves. J. biol. Chemistry **168**, 687 (1947).

195. ZINNER, H.: Darstellung der β-Tetracetyl-ribofuranose. Ber. dtsch. chem. Ges. **83**, 153 (1950).

(Received, March 5, 1951.)

Die Veilchenriechstoffe

Von H. Schinz, Zürich.

Inhaltsübersicht.

Einleitung.

Begriff der Veilchenriechstoffe.

Zu den „natürlichen" Veilchenriechstoffen zählt man die wohlriechenden Bestandteile der Veilchenblüten und -blätter, sowie das Iron, den Geruchsträger der Iriswurzelstöcke, welche wegen ihres veilchenartigen Geruchs, botanisch aber unkorrekt, auch „Veilchenwurzeln" genannt werden.

Für die Parfümindustrie ist von den natürlichen Veilchenriechstoffen vor allem das Iron von Bedeutung, aber auch Veilchenblätteröl wird viel gebraucht, während das Blütenöl wegen seines außerordentlich hohen Preises spärlich Verwendung findet. Noch wichtiger sind für die Riechstoffindustrie die künstlichen Veilchenriechstoffe α- und β-Jonon, welche von Tiemann und Krüger 1890 entdeckt wurden.

Die Jonone, obschon Kunstprodukte, müssen bei einer Besprechung, welche eigentlich nur den natürlichen Veilchenriechstoffen gelten sollte, ebenfalls Erwähnung finden, weil zwischen der „Geschichte" der Jonone und derjenigen des Irons eine Art Wechselwirkung besteht. Den Anlaß zur Auffindung der Jonone gaben nämlich die ersten wissenschaftlichen Arbeiten von Tiemann und Krüger über das Iron. Anderseits hat die Kenntnis der Struktur und der Reaktionen der Jonone in einer späteren Phase die Untersuchungen am Iron vereinfacht.

Es ist übrigens bemerkenswert, daß die Jonone nicht etwa die natürlichen Veilchenriechstoffe verdrängt, sondern im Gegenteil Anlaß zu deren Produktionssteigerung gegeben haben, da besonders günstige Geruchswirkungen mit Mischungen von Natur- und Kunstprodukten erzielt werden.

Nachdem die Jonone längst in großem Maßstab industriell hergestellt wurden, konnte neuerdings gezeigt werden, daß sie selbst sowie einige ihrer Abkömmlinge in geringen Mengen im ätherischen Öl verschiedener

Pflanzenarten und sogar in tierischen Sekreten vorkommen. Dies ist ein weiterer Grund, in einer Arbeit über natürliche Veilchenriechstoffe auch die Jonone kurz zu behandeln.

Allgemeiner Verlauf der Arbeiten über die Veilchenriechstoffe.

Die Chemie der Veilchenriechstoffe beginnt mit den ersten Untersuchungen von Tiemann und Krüger über das Iron (S. 156). Diese befassen sich mit der Isolierung dieses Ketons aus dem Irisöl und einigen chemischen Reaktionen desselben. Aus den Analysen wurde für das Iron die Bruttoformel $C_{13}H_{20}O$ gefolgert, die sich erst viel später als falsch erwies. Beim Versuch, ein Keton dieser Zusammensetzung auf einfache Art durch Kondensation von Citral mit Aceton herzustellen, wurde durch Zufall das Jonon entdeckt. Ein Kolben, welcher von dem bei der genannten Reaktion erhaltenen Citrylidenaceton enthielt, entwickelte beim Reinigen mit konz. Schwefelsäure Veilchengeruch. Tiemann und Krüger zeigten, daß der wohlriechende Stoff durch Cyclisation des Citrylidenacetons unter dem Einfluß der Säure entstanden war. Das neue, mit dem Iron nicht identische Keton wurde Jonon, das Citrylidenaceton Pseudojonon genannt. Damit war zum ersten Mal die Synthese eines Veilchenriechstoffes verwirklicht. Tiemann und Krüger fanden, daß das Jonon in einer α- und einer β-Form auftreten kann und bestimmten die Konstitution der beiden Jonone durch Abbau.

Der Grund dafür, daß Tiemann und Krüger als Objekt ihrer Untersuchungen über das Veilchenaroma nicht Veilchenblüten-, sondern Irisöl wählten, liegt darin, daß das Irisöl leichter zu beschaffen und viel billiger ist. Eine Analyse des teuren Veilchenblütenöls hätte das Arbeiten mit sehr kleinen Mengen Material erfordert. Dazu waren die Methoden jener Zeit noch nicht fein genug. Die genannten Autoren gingen von der Annahme aus, daß die Geruchsträger des Iriswurzel- und des Veilchenblütenöls identisch oder nahe verwandt seien. Dabei stützten sie sich allerdings auf bloße äußerliche Vergleiche; chemische Anhaltspunkte waren nicht vorhanden.

Der große Erfolg der Jonone bewirkte, daß die Arbeiten über Veilchenriechstoffe in der darauffolgenden Periode mehr technisch als wissenschaftlich orientiert waren. Die Möglichkeit praktischer Auswertung ließ eine Reihe von Fabrikations-, Trennungs- und Reinigungsverfahren für die Jonon-isomeren sowie für die in der Seitenkette methylierten Homologen entstehen. Diese Methoden sind in zahlreichen Patenten niedergelegt. An wissenschaftlichen Untersuchungen hat aber die Zeitspanne bis 1919 außer einer mißglückten Ironsynthese nichts Nennenswertes aufzuweisen.

Auch die folgenden zehn Jahre waren nicht besonders fruchtbar. Die aus dieser Zeit stammenden analytischen Arbeiten von Ruzicka sowie die synthetischen Versuche von DIELS (S. 159) auf dem Gebiet des Irons fußen immer noch auf der Formel von TIEMANN und KRÜGER, die sich später als falsch erwies.

Zu Beginn der Dreißigerjahre setzten die wichtigen Arbeiten über das Iron von RUZICKA und Mitarbeitern ein.

Die richtige Bruttoformel für das Iriswurzelketon $C_{14}H_{22}O$ wurde gefunden. Der oxydative Abbau führte diese Autoren zuerst jedoch zur Annahme einer unrichtigen Strukturformel. Eine Reihe von Versuchen zur Synthese des Irons bewegte sich demzufolge in falscher Richtung. Bis zur Überwindung dieses Irrweges verstrich wiederum geraume Zeit. Die richtige Konstitutionsformel des α-, β- und γ-Irons erschien in einer aus dem Jahr 1946 datierten Arbeit. Die Synthese des α-Irons wurde gleichzeitig von SCHINZ, RUZICKA, SEIDEL und TAVEL (*137*) sowie von NAVES und Mitarbeiter (*90*) veröffentlicht. In der Folgezeit erschien eine große Zahl von Arbeiten der beiden genannten Forschergruppen über Konstitution und Synthese der verschiedenen Iron-isomeren. Heute darf das Ironproblem, mit Ausnahme einiger stereochemischer Fragen und der Synthese des γ-Irons als gelöst betrachtet werden. Es liegen bereits einige zusammenfassende Abhandlungen von SCHINZ (*134—136*), RUZICKA (*109*), STOLL (*147*) sowie NAVES (*78*) über die Irone vor, die in Zeitschriften des engeren Fachgebietes erschienen sind.

Inzwischen wurde auch das Veilchenblätteröl verschiedentlich untersucht. Nonadienal und Nonadienol wurden isoliert und durch Synthese dargestellt. An diesen Arbeiten sind WALBAUM und ROSENTHAL (*167*), SPÄTH und KESZTLER (*145*) sowie RUZICKA und SCHINZ (S. 189) beteiligt. Die letzteren Autoren untersuchten auch das Veilchenblütenöl und isolierten daraus neben Nonadienol und Nonadienal ein wohlriechendes Keton $C_{13}H_{20}O$ unbekannter Struktur.

I. Die Jonone.

In diesem Kapitel werden von den in den älteren Arbeiten über die Jonone beschriebenen Reaktionen nur diejenigen erwähnt, welche beim Iron ebenfalls angewandt wurden. Von der neueren Literatur (von etwa 1930 an) werden dagegen alle wichtigeren Untersuchungen an den Jononen berücksichtigt, auch wenn sie nicht auf das Iron übertragen wurden.

Konstitution der Jonone.

Nach TIEMANN und KRÜGER bildet sich durch Cyclisation des Pseudojonons mit zirka 60%iger Schwefelsäure ein Gemisch von α- und β-Jonon (*161*), mit konz. Schwefelsäure β- (*160*), mit Phosphorsäure α-Jonon (*35*). Die Cyclisation wurde später von HIBBERT und CANNON (*40*) sowie von ROYALS (*104a*) eingehender untersucht.

Die Konstitution der beiden, über kristallisierte Derivate gereinigten Ketone (I) und (IV)* bewiesen TIEMANN und KRÜGER durch Abbau des α-Jonons zu Isogeronsäure (II) und β,β-Dimethyladipinsäure (III) bzw. des β-Jonons zu Geronsäure (V) und α,α-Dimethyladipinsäure (VI) (*159*).

(I.) α-Jonon.   (II.) Isogeronsäure.   (III.) β,β-Dimethyl-adipinsäure.

(IV.) β-Jonon.   (V.) Geronsäure.   (VI.) α,α-Dimethyl-adipinsäure.

Durch Behandlung mit Jodwasserstoffsäure und Phosphor gehen beide Jonone unter Abspaltung von Wasser in den Kohlenwasserstoff Jonen über. TIEMANN und KRÜGER nahmen für diesen die Formel (VII) an (*161*). Nach BOGERT (*11*) besitzt die Verbindung aber wahrscheinlich die Formel (VIII) mit einem Benzolring. RUZICKA und RUDOLPH (*114*) gewannen durch Dehydrieren von Jonen mit Schwefel 1,6-Dimethyl-naphthalin (IX).

(VII.)   (VIII.)   (IX.) 1,6-Dimethyl-naphthalin.

TIEMANN und KRÜGER führten das Jonen durch Oxydation mit Chromsäure und Kaliumpermanganat über verschiedene Zwischenstufen in die sogenannte Joniregen-tricarbonsäure (X) und durch weiteren Abbau in Dimethylhomophthalsäure (XI) über, die mit einem synthetischen Kontrollpräparat identifiziert wurde (*161*).

(X.)   (XI.) Dimethylhomophthalsäure.

* Unsere Schreibweise der ungesättigten Seitenkette in den Strukturformeln der Jonone und Irone — eigentlich diejenige der *trans*-Form — ist willkürlich. Wir hätten ebensogut diejenige der *cis*-Form wählen können, da über die gegenseitige Lage der Substituenten an der Doppelbindung des Butenonylrestes bisher keine wirklich sicheren Anhaltspunkte vorliegen.

An der Doppelbindung der Seitenkette ist bei den Jononen theoretisch *cis-trans*-Isomerie möglich. Da keiner der Bearbeiter der Jonone je 2 α- oder 2 β-Jonone erwähnt, ist anzunehmen, daß sich immer nur die eine der beiden stereoisomeren Formen bildet. NAVES und BACHMANN (*84*) halten sie für die *cis*-Form.

Hydrierung der Jonone.

SKITA (*141*) stellte aus α- und β-Jonon durch Hydrieren in Gegenwart von kolloidalem Palladium Dihydro-α- und Dihydro-β-jonon (XII und XIII) sowie Tetrahydro-jonon (XV) dar.

| (XII.) Dihydro-α-jonon. | (XIII.) Dihydro-β-jonon. | (XIV.) Kerngesättigtes Dihydro-jonon, *cis* und *trans*. | (XV.) Tetrahydro-jonon, *cis* und *trans*. |

Diese Hydrojonone sowie die entsprechenden Alkohole wurden später von PALFRAY, SABETAY und KANDEL (*94*), KANDEL (*43*) sowie von NAVES und BACHMANN (*83*) unter Verwendung von Nickelkatalysatoren gewonnen. Dihydro-α-jonon erhält man nach NAVES und BACHMANN auch durch Reduktion von α-Jonon mit Natrium und Alkohol und Rückoxydation des erhaltenen Dihydro-jonols zum Keton.

Das im Kern gesättigte Dihydrojonon (XIV) hatte SKITA (*141*) durch Kondensation von Dihydro-cyclocitral mit Aceton dargestellt. Dieses Präparat war wahrscheinlich sterisch nicht einheitlich. Reines *trans*-Dihydrojonon beschrieben später PRELOG und FRICK (*97*), COLOMBI (*17*) sowie BÄCHLI (*3*). NAVES und ARDIZIO (*82*) beobachteten die Bildung der entsprechenden *cis*-Verbindung als Nebenprodukt bei der Hydrierung von β-Jonon zu Dihydro-β-jonon. BÄCHLI erhielt dieses Isomere in größerer Menge bei der Hydrierung von β-Jonon-propylenacetal und isolierte es in reiner Form.

Reines Tetrahydrojonon erhält man am besten durch Hydrieren von α- oder β-Jonon — in Eisessiglösung in Gegenwart von Platinoxyd — zu Tetrahydrojonol und Reoxydation zum entsprechenden Keton (*13*). Das so erhaltene Präparat besitzt *cis*-Form. *trans*-Tetrahydrojonon wird durch Hydrieren des *trans*-Dihydrojonons erhalten (*97*, *17*). Die Zuteilung der *cis*- bzw. *trans*-Form erfolgt unter Anwendung der Regel von AUWERS und SKITA.

In der Seitenkette methylierte Jonone.

Kondensiert man Citral mit Methyl-äthylketon statt mit Aceton, so erhält man ein methyliertes Pseudojonon. Da nebeneinander Reaktion an der zur CO-Gruppe benachbarten Methyl- und an der Methylengruppe stattfindet, entsteht ein Gemisch von zwei isomeren Produkten, von denen das eine einen geraden, das andere einen verzweigten Butenonrest enthält.

Cherbuliez und Hegar (*16*) stellten die beiden Methylpseudojonone durch Kondensation von Citral mit 1-Brombutanon-(2) bzw. 2-Brom-butanon-(3) in reiner Form her. Beets (*9*) erhielt das Isomere mit ver-zweigtem Butenonrest unter Anwendung von 2-Chloro-butanon-(3) in besserer Ausbeute.

Die Methyl-pseudojonone gehen bei der Cyclisation in die technisch wichtigen Methyljonone über. Methyl-α- und Methyl-β-jonon kommen in je einer Form mit normaler und mit iso-Seitenkette vor (XVI und XVII).

(XVI.)   (XVII.)

Dagegen entsteht nach Köster (*51*) bei der Kondensation von Cyclo-citral (α oder β) mit Methyl-äthylketon ausschließlich Methyljonon mit normaler Butenonkette. Der gleiche Autor beschreibt ein Trennungs-verfahren für die normalen und iso-Formen mit Bisulfit. Ferner macht er auf die Möglichkeit einer *cis-trans*-Umlagerung in der Seitenkette bei der Kondensation von Citral mit Ketonen aufmerksam.

Andere Untersuchungen an den Jononen.

Sobotka, Bloch, Cahnmann, Feldbau und Rosen (*142*) trennten (±)-α-Jonon mittels *L*-Methylhydrazid in die optischen Antipoden.

Naves und Bachmann (*84*) isolierten aus dem technischen Jonon-gemisch ein tricyclisches Isomeres der Jonone.

Henbest (*37*) sowie etwas später Büchi, Seitz und Jeger (*14*) stellten durch Einwirkung von N-Bromsuccinimid und nachfolgende Abspaltung von Bromwasserstoff aus β-Jonon Dehydro-β-jonon her.

Köster (*49*) machte Untersuchungen über gegenseitige Umwand-lungen von α- und β-Jonon. β-Jonon wird aus dem α-Isomeren durch Isomerisieren mit konz. Schwefelsäure oder alkoholischer Lauge in der Kälte erhalten. Young, Cristol, Andrews und Lindenbaum (*170*) erwähnen die teilweise Bildung von α-Jonon bei der Spaltung von β-Jonon-semicarbazon mittels Phthalsäureanhydrid; mit verdünnter Schwefelsäure soll diese Isomerisierung vermieden werden.

Von Interesse sind auch die von Karrer und Stürzinger (*46*) [vgl. Naves (*92*)] hergestellten Mono- und Di-epoxyde beider Jonone, sowie die von Karrer und Benz (*44*) beobachtete anormale Reaktion des β-Jonons mit Acetylen.

STOLL, RUZICKA und SEIDEL (*154*), sowie STOLL, BOLLE und RUZICKA stellten Oxyde dar, welche sich von Tetrahydrojonon ableiten (*148—150*).

Die Reaktionen des β-Jonons zur Kettenverlängerung (Reaktion mit β-Methyl-crotonsäureester, Glycidester-synthese usw.) seien hier nur angetönt, da sie zur Chemie des Vitamins A gehören und den Rahmen dieser Zusammenstellung überschreiten.

Über die physikalischen Eigenschaften, Raman- und Ultraviolett-Spektren der Jonone und Jononabkömmlinge berichten ausführlich NAVES und Mitarbeiter (*83, 79, 88, 90*), über Ultraviolett-, Infrarot- und Raman-Spektren anderseits die Arbeitsgruppe RUZICKA (*124, 31*).

Vorkommen von Jononen und ihren Abkömmlingen in Naturprodukten.

In der älteren Literatur begegnet man häufig der Ansicht, das Jonon, ein synthetisches Produkt, sei mit dem riechenden Prinzip der Veilchenblüten identisch. Es ist zu betonen, daß es sich hier um eine bloße Vermutung handelt. Veilchenblütenöl wurde zum ersten Mal von RUZICKA und SCHINZ (188) eingehend analysiert, wobei geringe Mengen eines Ketons isoliert wurden, das zwar ebenfalls die Bruttoformel $C_{13}H_{20}O$ besitzt, sich aber im Geruch von den Jononen unterscheidet. Seine Konstitution ist bisher unbekannt.

Dagegen ist das Vorkommen von Jononen und Jonon-abkömmlingen in einigen anderen Naturprodukten — Pflanzenfamilien außerhalb der Violaceen und zum Teil auch in tierischen Sekreten — bezeugt. Diese Angaben stammen alle aus dem Zeitraum der letzten 25 Jahre. Die nachgewiesenen Mengen sind immer sehr gering.

β-Jonon kommt nach PENFOLD und PHILLIPS (*95*) sowie SABETAY (*131*) im Blütenöl von *Boronia megastigma* NEES vor. NAVES und PARRY (*91*) fanden in einem Öl der gleichen Pflanze β-Jonon sowie (+)-α- und ($\pm$)-α-Jonon. TISCHER (*162*) gibt an, Jonon aus *Trentepohlia Jolithus* [L.] WALLR. (Veilchenmoos) aufgefunden zu haben. Ferner isolierte NAVES (*74*) (+)-α-Jonon, (+)-β-Jonon und cis-(2,3)-Dihydrojonon aus *Costus*wurzelöl. BOHNSACK (*63*) fand β-Jonon im Himbeersaft.

(+)-Dihydro-γ-jonon wurde von RUZICKA, SEIDEL und PFEIFFER (*125*) in geringer Ausbeute aus den flüchtigen Anteilen des grauen Ambra isoliert. Der Hauptbestandteil des Ambras ist das Ambreïn (XVIII), ein kristallisierter, tricyclischer Triterpenalkohol (*113*). Nach RUZICKA, DÜRST und JEGER (*112, 42*) liefert dieser bei der Oxydation mit Kaliumpermanganat als Spaltprodukte optisch aktives Dihydro-γ-jonon (XIX) und ein bicyclisches Lacton (XX).

Die Konstitution des ersten der beiden Produkte ergab sich einerseits aus der Bildung des bekannten Tetrahydrojonons (XXI) durch katalytische Hydrierung und anderseits aus dem Resultat des Ozonabbaues, wobei das Diketon $C_{12}H_{20}O_2$ (XXII) und Formaldehyd entstanden.

Wahrscheinlich bildet sich auch das natürliche Dihydro-γ-jonon des
Ambras durch Oxydationsprozesse des Ambreïns.

(XVIII.) Ambreïn.   (XIX.)   (XX.)
                                    Dihydro-γ-jonon.

(XXI.)   (XXII.)

Ruzicka, Büchi und Jeger (111) erhielten Dihydro-γ-jonon (XXV)
synthetisch durch Anlagerung von Chlorwasserstoff an Dihydro-α-jonon
(XXIII—XXIV) und Wiederabspaltung des Halogenwasserstoffes durch
Erhitzen mit Silberstearat. Aus dem erhaltenen Gemisch von Dihydro-α-
und Dihydro-γ-jonon wurden die beiden Isomeren durch Chromatographie der Semicarbazone getrennt.

(XXIII.)   (XXIV.)   (XXV.) Dihydro-γ-jonon.

In den flüchtigen Bestandteilen des Ambras fanden Ruzicka und
Seidel (122) ferner ein Oxyd $C_{13}H_{22}O$, für das sie die allerdings nicht
ganz sichere Formel (XXVI) in Erwägung ziehen; außerdem wurde ein
Keton $C_{13}H_{20}O$ isoliert, in dem vielleicht das γ-Jonon (XXVII) vorliegt.

(XXVI.)   (XXVII.) γ-Jonon.

Eine weitere Gruppe von Substanzen, die mit den Jononen verwandt
sind, wurde von Prelog und Mitarbeitern (99, 102) aus dem Harn trächtiger Stuten isoliert. Folgende Individuen wurden identifiziert: zwei diastereomere 5-Oxy-tetrahydrojonole (XXVIII), ein 5-Oxo-cis(2,3)-tetra-
hydrojonol (XXIX) und ein 5-Oxy-cis(2,3)-tetrahydrojonon (XXX),
ferner zwei diastereomere 5-Oxo-tetrahydrojonone (XXXI). Die beiden
diastereomeren Glykole (XXVIII) kommen nach Lederer, Prelog
und Schneider (55) auch im Castoreum vor.

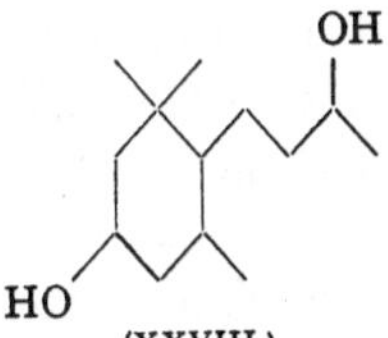

(XXVIII.) 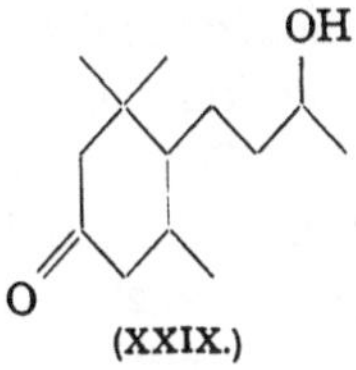   (XXIX.) 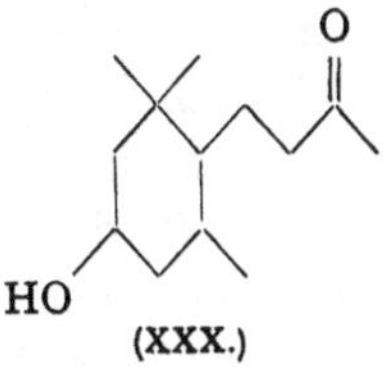   (XXX.)

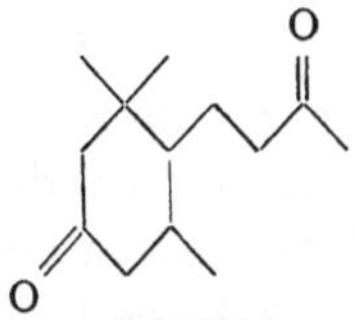

(XXXI.) 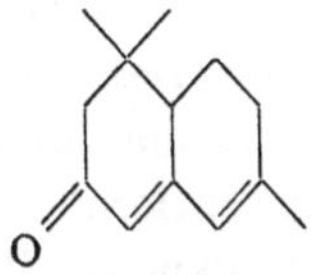   (XXXII.)

Der Stutenharn enthält auch ein bicyclisches Keton (XXXII), welches mit den genannten monocyclischen Verbindungen verwandt ist (101).

Die aromatischen Verbindungen (XXXIII) und (XXXIV), die ebenfalls im Stutenharn vorkommen (98, 99), sind wahrscheinlich durch Dehydrierung von Jononderivaten entstanden. Dabei findet Retropinakolinumlagerung einer der geminalen Methylgruppen statt. Solche Übergänge konnten von Büchi, Seitz und Jeger (14) [vgl. Karrer und Ochsner (45)] auch in vitro ausgeführt werden.

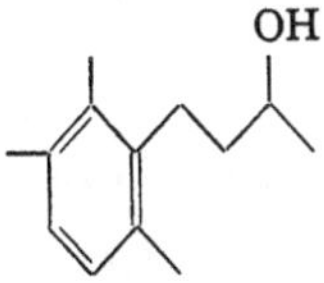

(XXXIII.) 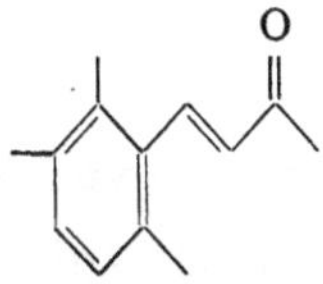   (XXXIV.)

Mit den oben genannten Verbindungen verwandt sind die Jononderivate mit einer Oxy- oder Oxogruppe in 4-Stellung (XXXV), (XXXVI) und (XXXVII), welche bei der biochemischen Oxydation von β-Jonon im Tierkörper (Kaninchen) entstehen (10, 100).

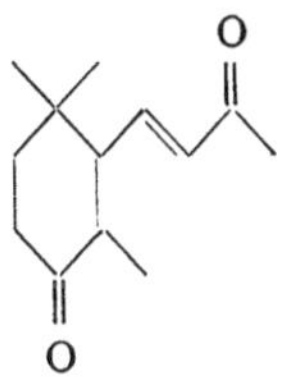

(XXXV.) 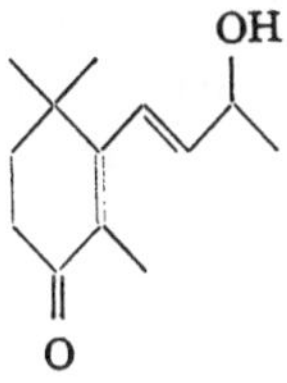   (XXXVI.) 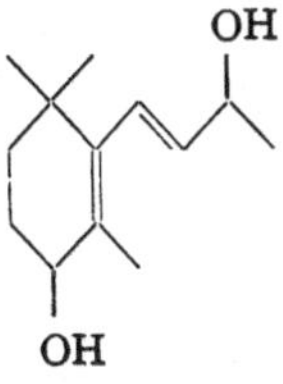   (XXXVII.)

II. Die Irone.

Gewinnung des Irons.

Zur Gewinnung des Irons verwendet man die Wurzelknollen (Rhizome) von *Iris florentina*, *I. germanica* und *I. pallida*. Die genannten Schwertlilienarten werden besonders in der Toscana, aber auch in anderen Gegen-den Mittelitaliens sowie in Südfrankreich angebaut. Die gewaschenen und getrockneten Wurzelstöcke kommen teilweise direkt als „Veilchenwurzeln" oder gepulvert als „Veilchenwurzelpulver" in den Handel.

Der Riechstoff wird den Wurzelknollen durch Destillation mit Wasserdampf entzogen, wobei man 0,1—0.2% „Irisöl" gewinnt. Eine andere Methode besteht in der Extraktion mit Lösungsmitteln, besonders Petroläther, welche 0,4—0,5% salbenartige „concrète" liefert.

Das nach beiden Verfahren gewonnene Produkt besteht zum größten Teil aus Myristinsäure und anderen Pflanzensäuren, die durch Schütteln mit Soda oder Lauge entfernt werden. Der ölige Neutralteil, dessen Menge ca. 0,05% der getrockneten Wurzel beträgt, enthält außer Iron eine Reihe ebenfalls riechender Begleitsubstanzen (Aldehyde, Terpenalkohole, Ester), die zum Teil durch fraktionierte Destillation abgetrennt werden. Tiemann und Krüger (161) wandten zur Trennung wiederholte Destillation mit Wasserdampf an, verseiften die Ester und oxydierten die Aldehyde mit Silberoxyd zu den entsprechenden Säuren. Das Iron wurde schließlich als nicht kristallines Phenylhydrazon abgetrennt und aus diesem mit Schwefelsäure wieder abgeschieden. Solche Präparate zeigen im allgemeinen $[\alpha]_D = + 40$ bis $+ 45°$. Heute läßt sich die Reinigung des Irons leichter mittels des bekannten Reagens P von Girard und Sandulescu erzielen.

Arbeiten von Tiemann und Krüger.

Tiemann und Krüger (161) fanden für das Iron Analysenwerte, aus welchen sie die Formel $C_{13}H_{20}O$ ableiteten. Behandlung mit Bromlauge zeigte, daß ein Methylketon vorlag. Da der oxydative Abbau mit Kaliumpermanganat keine Bruchstücke lieferte, aus denen Schlüsse auf die Konstitution möglich waren, führten die Autoren das Keton mittels Jodwasserstoffsäure und Phosphor unter Abspaltung von Wasser in den Kohlenwasserstoff Iren über, der für einen weiteren Abbau geeigneter schien. Durch Oxydation erhielten sie aus Iren Joniregen-tricarbonsäure (XL) und Dimethyl-homophthalsäure (XLI), d. h. die gleichen Produkte, die auch aus Jonen entstehen.

(XXXVIII.)   (XXXIX.)   (XL.)   (XLI.)

Dadurch waren die geruchlich verwandten Ketone Iron und Jonon in enge gegenseitige Beziehung gebracht. Tiemann und Krüger schlugen auf Grund dieses Resultates für das Iren Formel (XXXIX), für das Iron

Formel (XXXVIII) vor. Iron und Jonon einerseits, Iren und Jonen anderseits wurden als Isomere mit verschiedener Lage der Doppelbindung betrachtet.

Erste Synthesen auf Grund der Formel von Tiemann und Krüger.

Die Iron-Formel von Tiemann und Krüger (*161*) wurde während 40 Jahren als richtig angenommen.

Im Jahre 1909 publizierten Merling und Welde (*59*) die Synthese eines Ketons, das, wie sie glaubten, die Konstitution (XXXVIII) besitzen sollte. Durch Kondensation von Isopropyliden-acetessigester mit Natriumacetessigester wurde der sogenannte „Isophoron-carbonsäureester" (XLII) hergestellt. Dieser lieferte mit Phosphorpentachlorid den zweifach ungesättigten, chlorierten Ester (XLIII). Die entsprechende Säure sollte nach diesen Autoren durch Reduktion mit Natrium und Alkohol neben (XLIV) und (XLV) hauptsächlich Δ^4-Cyclogeraniumsäure (XLVI) ergeben. Nach Angaben einer Patentschrift (*58*) ist (XLVI) sogar das einzige Reaktionsprodukt. Die vermeintliche Säure (XLVI) wurde zum entsprechenden Aldehyd (XLVII) reduziert und dieser mit Aceton kondensiert.

(XLII.)   (XLIII.)

(XLIV.)   (XLV.)   (XLVI.)   (XLVII.)   (XLVIII.)

Ruzicka und Brugger (*110*) zeigten aber in einer 1941 publizierten Arbeit, daß das Keton von Merling und Welde lediglich aus α-Jonon bestand. Der Beweis erfolgte durch Abbau des Ketons zu Isogeronsäure sowie durch Identifikation der bei der Reduktion des Enolchlorids entstehenden Säure als α-Cyclogeraniumsäure.

Ruzicka (*105*) unternahm 1919 einen Versuch zur Herstellung eines Ketons, dessen Struktur der Tiemannschen Formel entspricht, das aber eine zusätzliche Methylgruppe besitzt. Der durch Einwirkung von Bromisobuttersäureester auf γ-Acetylbuttersäureester erhaltene ungesättigte Ester (XLIX) ergab durch Kondensation nach Dieckmann, Methylierung des erhaltenen Ketoesters (L) und Ketonspaltung das Tetramethylcyclohexenon (LI). Daraus hätte der Aldehyd (LII) und schließlich das Keton (LIII) hergestellt werden müssen. Die Methylgruppe in 6-Stellung sollte die Lage der Doppelbindung bei den nachfolgenden Reaktionen fixieren. Diese wurden aber überhaupt nicht ausgeführt, weil die Ausbeute an (LI) zu gering war und eine Variante zu dessen Darstellung [Thermolyse des Anhydrids von (XLIX)] das isomere ungesättigte Keton (LIV) geliefert hatte.

Es ist bemerkenswert, daß Ruzicka schon damals die Synthese einer Verbindung mit zusätzlicher Methylgruppe am richtigen Ort beabsichtigte, ohne zu wissen, daß es sich um das Ironskelett handelte.

1935 beschrieb Verley (*163*) die Synthese eines Ketons der Tiemannschen Formel (LIX), welche von Rhodinal (LV) ausging:

Die beschriebenen Reaktionen lassen sich aber nach Ruzicka und Brugger (*110*) nicht reproduzieren.

Tetrahydro-iron.

Da die von Tiemann und Krüger (*161*) aufgestellte Ironformel sich aus den Abbauresultaten des Irens ableitete, d. h. eines Produktes, bei dessen Herstellung Umlagerungen nicht ausgeschlossen sind, stellte Ruzicka 1919 durch katalytische Hydrierung in Gegenwart von Platinschwarz Tetrahydro-iron her (*106*). Damit lag zum ersten Mal der sichere Beweis vor, daß das Iron wirklich zwei Doppelbindungen besitzt, denn Tiemann und Krüger hatten sich nur auf die Molekularrefraction gestützt.

Wenn die Formel von Tiemann und Krüger richtig gewesen wäre, hätte das Tetrahydro-iron, abgesehen von der optischen Aktivität, mit dem Tetrahydro-jonon übereinstimmen müssen. Es erwies sich aber als gänzlich verschieden, während die Tetrahydroketone aus α- und β-Jonon untereinander identisch waren. Am auffallendsten war, daß Tetrahydro-iron 14° höher siedete. Ruzicka glaubte, unter Vorbehalt der Richtigkeit der Tiemannschen Formel, diesen Unterschied durch Stereoisomerie erklären zu können: Beim Tetrahydro-jonon befänden sich Butanonkette und Methylgruppe in *cis-*, beim Tetrahydro-iron in *trans-*Stellung. Der Unterschied der Siedepunkte war allerdings größer, als er sonst bei Diastereomeren zu sein pflegt. Skepsis gegenüber der Formel von Tiemann schien schon deshalb angebracht.

Diensynthesen auf Grund der Tiemannschen Formel.

Die Kondensation von Diels und Alder hätte die Synthese eines Ketons der Tiemannschen Formel auf folgende Art ermöglichen sollen:

(LX.) (LXI.) (LXII.) (LXIII.)

Die Reaktion wurde einerseits von den genannten Autoren selbst (*21,22*), anderseits von Ruzicka und Mitarbeitern (*60, 61*) ausgeführt [vgl. auch Naves (*80, 81*)]. Das als Ausgangsmaterial dienende Dimethylbutadien (LX) wird aber im Verlauf der Reaktion vollständig zu (LXII) isomerisiert, welches bei der Kondensation mit Crotonaldehyd nicht den Aldehyd (LXI), sondern das Isomere (LXIII) liefert. Das Trimethyl-butadien (LXIV), bei dem eine solche Isomerisierung wegen seines Baues nicht möglich ist, liefert den Aldehyd (LXV) und daraus ein „Methyliron" (LXVI). Das Produkt war jedoch geruchlich uninteressant.

(LXIV.) (LXV.) (LXVI.)

Richtige Bruttoformel des Irons.

Erst 1933 wurde die richtige Bruttoformel des Irons gefunden. Ruzicka Seidel und Schinz (*126*) zeigten an Hand genauer Analysen einiger Derivate, daß das Iris-keton die Zusammensetzung $C_{14}H_{22}O$ und nicht $C_{13}H_{20}O$ besitzt, wie man während 40 Jahren allgemein angenommen hatte. Nun wurde die Formel von Tiemann verlassen und neue Versuche zur Konstitutionsaufklärung des Irons unternommen.

Zuerst wurde das Iren von neuem untersucht. Durch Dehydrieren mit Selen erhielten die genannten Autoren 1,2,6-Trimethyl-naphthalin (LXVIII), das durch Vergleich mit einem synthetischen Kontrollpräparat mittels des Pikrats und Styphnates identifiziert wurde. Daraus folgt,

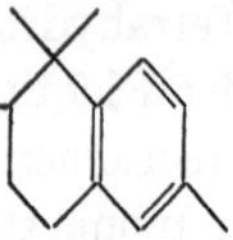

(LXVII.) 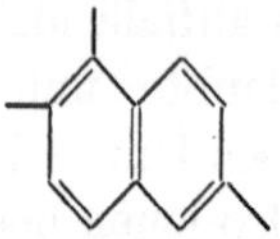   (LXVIII.)

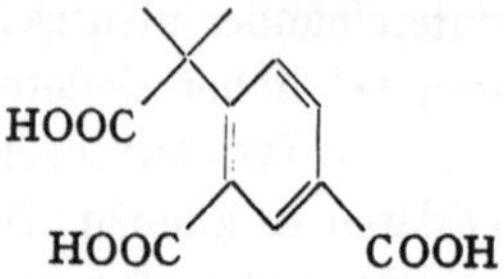

HOOC

HOOC   COOH

(LXIX.) Joniregen-tricarbonsäure.

daß Iren nicht, wie TIEMANN und KRÜGER glaubten, ein Isomeres, sondern ein Homologes des Jonens ist und die Formel (LXVII) eines 1,1,2,6-Tetra-methyltetralins besitzt. In der Joniregen-tricarbonsäure (LXIX), welche TIEMANN und KRÜGER durch oxydativen Abbau erhalten hatten, ist die zusätzliche Methylgruppe verschwunden. Dies läßt sich dadurch er-klären, daß als Zwischenprodukt ein Methylketon und eine α-Ketosäure entstehen, welch letztere bei der Weiteroxydation ein Kohlenstoffatom verliert.

Ein synthetisches 1,1,2,6-Tetramethyl-tetralin stellten später BOGERT und APFELBAUM (12) ausgehend von m-Bromtoluol (LXX) dar. Die physikalischen Konstanten dieses Kohlenwasserstoffs stimmten mit denjenigen der Irens annähernd überein und die Dehydrierung führte zum 1,2,6-Trimethyl-naphthalin. Das Produkt wurde auch nach der Methode von TIEMANN und KRÜGER oxydativ abgebaut.

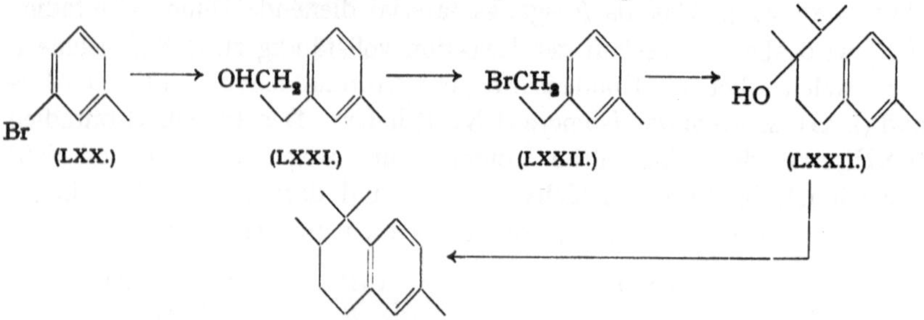

Br   (LXX.)   OHCH₂   (LXXI.)   BrCH₂   (LXXII.)   HO   (LXXII.)

(LXXIV.)

Da sich Jonen aus den Jononen und Iren aus Iron unter den gleichen Bedingungen bilden, schien für Iron die Formel (LXXV) oder (LXXVI) wahrscheinlich. Man hätte dann allerdings durch oxydativen Abbau Methyl-isogeronsäure bzw. Methyl-geronsäure erhalten müssen.

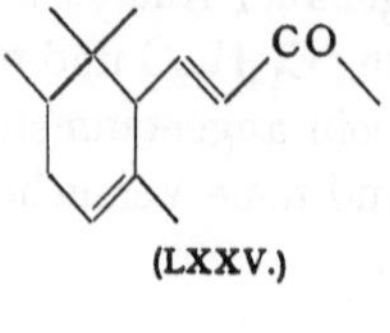

(LXXV.) 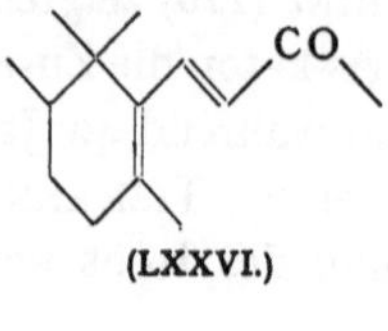   (LXXVI.)

RUZICKA, SEIDEL und SCHINZ (*126*) stellten ferner durch Reinigung über kristallisierte Derivate zum ersten Mal einheitliche Iron-präparate her. Sie verwendeten dazu das Thiosemicarbazon und das noch besser geeignete Phenyl-semicarbazon. Durch fraktionierte Kristallisation konnten Präparate mit ganz verschiedenen Schmelzpunkten gewonnen werden. Aus diesen Derivaten wurden Ketonpräparate mit verschiedenem Drehvermögen erhalten. Dabei zeigte sich gleichzeitig auch die leichte Isomerisierbarkeit des Irons. Gewisse Keton-präparate wurden z. B. schon beim Erwärmen mit Oxalsäurelösung verändert und lieferten nachher Derivate mit anderen Schmelzpunkten als das Ausgangsmaterial.

Synthese eines 6-Methyl-jonons (Gemisch von α und β).

Eine 1940 publizierte Arbeit von RUZICKA und SCHINZ (*115*) beschreibt die Darstellung eines 6-Methyl-jonons. 2,3-Dimethylhepten-(2)-on-(6) (LXXIX), das aus dem Bromid (LXXVII) über den β-Ketoester (LXXVIII) gewonnen worden war, wurde nach REFORMATZKY in den Oxyester (LXXX) übergeführt. Wasserabspaltung und Verseifung lieferte ε-Methyl-geraniumsäure (LXXXI), welche zu 6-Methyl-cyclogeraniumsäure (LXXXII) isomerisiert wurde. Der daraus durch Reduktion gewonnene Aldehyd (LXXXIII) gab bei der Kondensation mit Aceton ein 6-Methyl-jonon, das aus einem Gemisch von α- und viel β-Form zu bestehen schien. Die teilweise Wanderung der Doppelbindung aus der α- in die β-Stellung mußte während der Bildung der Stufen (LXXXIII) und (LXXXIV) eingetreten sein.

(LXXVII.) (LXXVIII.) (LXXIX.) (LXXX.)

(LXXXI.) ε-Methyl-geraniumsäure. (LXXXII.) 6-Methyl-cyclogeraniumsäure. (LXXXIII.) (LXXXIV.)

Wie wir jetzt wissen, lag hier das erste künstliche Irongemisch vor. Es wurde jedoch nicht als solches erkannt, da der Geruch infolge des hohen Gehaltes an β-Form — welche im Naturprodukt nicht vorkommt — eine süßliche, mehr an Jonon als an Iron erinnernde Nuance zeigte.

Die gleiche Synthese hatten auch KILBY und KIPPING (*47*) versucht, waren aber auf einer Zwischenstufe stehen geblieben.

Oxydativer Abbau des Irons mit Ozon und Chromsäure; Hypothese der 7-Ring-Struktur des Irons.

Bis vor zehn Jahren waren keine Oxydationsprodukte des Irons in der Literatur beschrieben, aus denen sich die Konstitution des Ketons ableiten ließ. Erst 1942 gaben RUZICKA, SEIDEL, SCHINZ und PFEIFFER (*127*) eine geeignete Abbaumethode bekannt, wonach sie das Iron in Eisessig ozonisierten und die Ozonidlösung einer Nachoxydation mit Chromsäure unterwarfen. Dazu wurde zuerst ein über das nicht kristallisierte p-Sulfophenylhydrazon gereinigtes Keton verwendet. Die sauren Anteile der Oxydationsprodukte wurden in die Methylester übergeführt und diese — nach Entfernung von Aldehyd- und Ketosäureestern als p-Nitrophenylhydrazone — der fraktionierten Destillation unterworfen. Nach Verseifung der einzelnen Fraktionen wurden drei kristallisierte Dicarbonsäuren isoliert:

1. $C_8H_{14}O_4$, optisch inaktiv, eine Estergruppe schwer verseifbar.
2. $C_9H_{16}O_4$, (+)-drehend, eine Estergruppe schwer verseifbar.
3. $C_{10}H_{18}O_4$, (+)-drehend, beide Estergruppen leicht verseifbar.

Die Säure $C_8H_{14}O_4$ erwies sich als ($\pm$)-α,α,β-Trimethylglutarsäure (LXXXVIII). Sie wurde durch Vergleich mit einem synthetischen Präparat identifiziert.

Die Säure $C_9H_{16}O_4$ lieferte beim Erhitzen auf 300° ein Keton $C_8H_{14}O$; dieses gab eine Mono-benzylidenverbindung, aus welcher durch Einwirkung von Ozon eine kristallisierte (+)-drehende Dicarbonsäure $C_8H_{14}O_4$ entstand. Letztere lieferte beim Erhitzen ein Anhydrid. Die Säure C_8 ist wahrscheinlich (+)-α,α,β-Trimethylglutarsäure (LXXXVIII) und die Ausgangssäure C_9 (+)-α,α,β-Trimethyladipinsäure (LXXXV).

Die Säure $C_{10}H_{18}O_4$ gab beim Erhitzen auf 300° ein Keton $C_9H_{16}O$, aus dessen Mono-benzylidenverbindung bei der Ozonisation eine kristallisierte (+)-Dicarbonsäure $C_9H_{16}O_4$ erhalten wurde. Diese ging ihrerseits beim Erhitzen in ein Keton $C_8H_{14}O$ über, welches bei der Umsetzung mit Benzaldehyd eine Dibenzylidenverbindung lieferte. Letztere gab beim Ozonisieren eine kristallisierte (+)-Dicarbonsäure $C_7H_{12}O_4$, die als (+)-Trimethyl-bernsteinsäure (XCV) identifiziert werden konnte. Der Vergleich geschah allerdings mit einer synthetischen ($\pm$)-Säure. Für die Dicarbonsäure $C_{10}H_{18}O_4$ läßt sich darnach die Formel (LXXXIX) der $(+)$-β,β,γ-Trimethyl-pimelinsäure, und für die anderen Stufen des Abbaues die Formeln (XC) bis (XCIV) ableiten.

Der gleiche Abbau wurde mit zwei Iron-präparaten aus Phenyl-semicarbazonen der Schmelzpunkte 177—179° bzw. 155—160° unternommen. Da man hier mit kleineren Mengen arbeitete, wurden die Dicarbonsäuren nicht kristallin erhalten. Es konnte aber bei der ersten Probe die Entstehung der Säuren $C_{10}H_{18}O_4$ und $C_9H_{16}O_4$, im zweiten diejenige der Säure $C_9H_{16}O_4$, mit Hilfe der aus ihnen entstehenden Ketone nachgewiesen werden.

Da sich unter den Abbausäuren des Irons sowohl $(+)$-drehende, als auch optisch inaktive Produkte finden, mußte man annehmen, daß das natürliche, optisch aktive Iron ein inaktives Isomeres als Beimengung enthalte. Es konnte zwar auch später nie ein solches in reiner Form isoliert werden, jedoch erhielten RUZICKA, SEIDEL und FIRMENICH (*124*) aus den Mutterlaugen des $(+)$-Tetrahydro-iron-semicarbazons ein optisch inaktives isomeres Produkt.

Die Bildung der β,β,γ-Trimethyl-pimelinsäure (LXXXIX), bei der sich sieben Kohlenstoffatome in gerader Kette befinden, ließ sich nun aber aus einem Iron mit der Struktur (XCVI) oder (XCVII) nicht erklären. RUZICKA, SEIDEL, SCHINZ und PFEIFFER (*127*) schlugen deshalb eine Formel mit einem siebengliedrigen Ring vor. Dabei wurden zwei Möglichkeiten, (XCVIII) und (XCIX), in Betracht gezogen, die beide mit den Ergebnissen des oxydativen Abbaues in Einklang stehen.

Die Bildung des Irens bei der Behandlung mit Jodwasserstoffsäure und Phosphor versuchte man durch die Annahme zu erklären, daß unter den energischen Reaktionsbedingungen eine Umlagerung stattgefunden habe. Dabei könnte ein Keton der Formel (XCVI) bzw. (XCVII) eventuell als Zwischenstufe eine Rolle spielen.

Ultraviolett-Spektrum des natürlichen Irons; Isomerisierungen.

Um zwischen den hypothetischen Formeln (XCVIII) und (XCIX) zu entscheiden, bestimmten Ruzicka, Seidel und Firmenich (*124*) an gereinigtem Iron (Phenylsemicarbazon, Schmp. 177—179°) und zum Vergleich an reinem α- und β-Jonon die physikalischen Daten, darunter auch die Ultraviolett-Absorptionsspektren.

Die Molekularrefraktionen des Irons und des α-Jonons zeigten Exaltationen von 1,1 bis 1,2 bzw. 1,3 bis 1,4. Die Ultraviolett-Absorptionskurven der beiden Ketone sind fast identisch und streben beide dem bei circa 230 mμ liegenden, für eine aliphatische α,β-ungesättigte Ketongruppierung typischen Maximum, log ε = 4,2, zu. Das β-Jonon zeigt eine wesentlich höhere Exaltation von 2,5 bis 2,6 und das Maximum der Absorption liegt bei 298 mμ, log ε = 4,0. Die Kurven des α-Jonons und des Irons zeigen bei 298 mμ nur kleine Inflexionen.

Daraus wurde der Schluß gezogen, daß das Iron nicht die Struktur (XCIX), sondern (XCVIII) besitze.

Die gleichen Autoren konnten ferner zeigen, daß beim Kochen des Irons mit 20%iger Schwefelsäure eine allmähliche Umlagerung der α- in die β-Form stattfindet. Der ungefähre Gehalt an β-Form solcher durch Isomerisierung erhaltener Gemische ergibt sich aus der Ultraviolett-Absorptionskurve.

Synthesen „ironartiger" Ketone mit 7-Ring-struktur.

M. Stoll und Scherrer (*155*) prüften die Frage, ob die 7-Ring-struktur an sich schon eine Geruchsverbesserung erzeuge. Ein zu diesem Zweck hergestelltes Keton der Formel (C), mit unbestimmter Lage der Ring-Doppelbindung, zeigte geruchlich keinerlei Ähnlichkeit mit Iron.

(C.)   (CI.)

Auch ein isomeres Keton (CI) von Ruzicka, Seidel, Schinz und Pfeiffer (*129*) bot kein Interesse. Als Ausgangsmaterial zur Darstellung dieses sowie der im folgenden erwähnten Produkte diente das 1,1,2-Trimethyl-suberon (CIV).

Von Firmenich (*26*) stammt ein Versuch zur Darstellung eines „β-Irons" (CII); er blieb jedoch auf einer Zwischenstufe stehen. Sprecher (*146*) erstrebte das gleiche Ziel auf einem anderen Weg; das Schluß-produkt besaß aber eine andere Konstitution. Dagegen gelang ihm die Herstellung eines „Tetrahydro-irons" (CIII).

(CII.)   (CIII.)

Ein „α-Iron" mit 7-Ring-struktur (CXIV) erhielt schließlich Boss-HARD (*13*). Der Synthesengang, dessen Vorstufen (CIV) bis (CIX) schon von SPRECHER benutzt worden waren, erfolgte nach folgendem Schema:

(CIV.)   (CV.)   (CVI.)
1,1,2-Trimethyl-
suberon.

Das Keton besaß gleiche physikalische Daten und das gleiche Ultra-violett-Absorptionsspektrum wie das Iron; die Doppelbindung war also im Verlauf der beiden letzten Reaktionen gewandert. Das Phenylsemi-carbazon zeigte zufällig den gleichen Schmelzpunkt wie das höchst-schmelzende Präparat aus Iron. Geruchlich war jedoch die synthetische Verbindung vom natürlichen Iron stark verschieden.

Auch BARBIER (*5—7*) sowie NAVES (*65*) beschäftigten sich mit 7-Ring-produkten, die zur Darstellung von Iron oder ironähnlichen Verbindungen Verwendung finden sollten.

Abbau des Tetrahydro-irons; Beweis der 6-Ring-struktur.

Schließlich führte der Abbau des Tetrahydro-irons RUZICKA, SEIDEL und BRUGGER (*123*) zur Ableitung der richtigen Strukturformel des Irons:

a) *Säure* $C_{12}H_{22}O_2$.

Tetrahydro-iron (über Semicarbazon, Schmp. 203—204° gereinigt) (CXV) wurde einem Abbau nach Wieland (CXV—CXVIII) unterworfen. Dabei entstand eine Monocarbonsäure $C_{12}H_{22}O_2$ (CXVIII), von der eine inaktive und eine (+)-Form isoliert wurden. Die Ester beider Säuren waren leicht verseifbar.

$$R—CH_2—CH_2—CO—CH_3 \longrightarrow R—CH_2—CH_2—\overset{\displaystyle OH}{\underset{\displaystyle CH_3}{C}}—CH_3 \longrightarrow$$
$$R = C_{10}H_{19}$$
$$\text{(CXV.)} \qquad\qquad \text{(CXVI.)}$$

$$\longrightarrow R—CH_2—CH=\overset{\displaystyle CH_3}{\underset{\displaystyle CH_3}{C}} \longrightarrow \qquad R—CH_2—COOH$$
$$\text{(CXVII.)} \qquad\qquad C_{12}H_{22}O_2$$
$$\text{(CXVIII.)}$$

$$R—CH_2—CH_2—COOH \longrightarrow R—CH_2—CH_2—CH_2OH \longrightarrow R—CH_2—CH=CH_2$$
$$\text{(CXIX.)} \qquad\qquad\qquad \text{(CXX.)} \qquad\qquad\qquad \text{(CXXI.)}$$

Die (+)-Säure $C_{12}H_{22}O_2$ wurde auch noch auf einem anderen Weg erhalten: Tetrahydro-iron wurde mit Bromlauge zur sogenannten „Tetrahydro-ironsäure" $C_{13}H_{22}O_2$ (CXIX) oxydiert. Der aus dem entsprechenden Ester nach Bouveault-Blanc erhaltene Alkohol (CXX) wurde nach Tschugaeff in den Kohlenwasserstoff (CXXI) übergeführt. Bei der Ozonisation entstand aus diesem die (+)-Säure (CXVIII).

b) *Abbau der Säure $C_{12}H_{22}O_2$ zu einem Tetramethyl-cyclohexanon.*

Die (±)-Säure $C_{12}H_{22}O_2$ (CXVIII) wurde mittels der modifizierten Curtiusschen Methode in ein tertiäres Amin $C_{11}H_{23}N$ verwandelt. Der über die quaternäre Ammoniumbase hergestellte Kohlenwasserstoff $C_{11}H_{20}$ lieferte bei der Ozonisation ein Keton $C_{10}H_{18}O$, welches mit 1,1,3,6-Tetramethyl-cyclohexanon-(2) (CXXIV) identisch war. Das Amin und der Kohlenwasserstoff besitzen demnach die Formeln (CXXII) bzw. (CXXIII).

(CXXII.)   (CXXIII.)   (CXXIV.) 1,1,3,6-Tetra-methyl-cyclohexanon-(2).

Das Keton (CXXIV) wurde von Schäppi und Seidel (*133*) synthetisch folgendermaßen erhalten:

(CXXV.)   (CXXVI.)   (CXXVII.)   → (CXXIV.)

Daraus geht hervor, daß das Iron einen 6-Ring besitzt. Des weiteren ergibt sich die Lage der Seitenkette.

c) *Säure* $C_{11}H_{20}O_2$.

Die ($\pm$)-Säure $C_{12}H_{22}O_2$ (CXXVIII) lieferte anderseits durch einen weiteren WIELANDschen Abbau (CXXVIII—CXXXI) eine ($\pm$)-Säure $C_{11}H_{20}O_2$, die (+)-Säure $C_{12}H_{22}O_2$ gab eine (+)-Säure $C_{11}H_{20}O_2$.

$$R\text{—}CH_2\text{—}COOH \rightarrow R\text{—}CH_2\text{—}\underset{CH_3}{\overset{OH}{C}}\text{—}CH_3 \rightarrow R\text{—}CH{=}C\underset{CH_3}{\overset{CH_3}{<}} \rightarrow R\text{—}COOH$$

$C_{12}H_{22}O_2$ (CXXIX.)   (CXXX.) $C_{11}H_{20}O_2$
(CXXVIII.)   (CXXXI.)

Die Ester der beiden Säuren $C_{11}H_{20}O_2$ (CXXXI) sind ebenso schwer verseifbar wie Dihydro-cyclogeraniumsäureester (CXXXII). Dies spricht ebenfalls für das Vorliegen eines 6-Ringes (CXXXIII), da die isomere Verbindung mit einem 7-Ring (CXXXIV) leichter verseifbar sein müßte, weil in diesem Falle die COOR-Gruppe nur von einer Seite her sterisch gehindert ist.

(CXXXII.)   (CXXXIII.)   (CXXXIV.)

Die Säure $C_{11}H_{20}O_2$ wurde außerdem auf folgendem Weg erhalten. Tetrahydro-iron gab bei der BECKMANNschen Umlagerung ein Gemisch von zwei Isoximen (CXXXV und CXXXVII). Bei der Verseifung mit Salzsäure entstand die kristalline „Tetrahydro-ironsäure" $C_{13}H_{24}O_2$ (CXXXVI) und ein Amin $C_{12}H_{25}N$ (CXXXVIII). Die quaternäre Ammoniumbase des letzteren lieferte beim Erhitzen einen Kohlenwasserstoff $C_{12}H_{22}$ (CXXXIX). Beim Ozonisieren desselben entstand die (+)-Säure $C_{11}H_{20}O_2$ (CXL).

$$R\text{—}CH_2\text{—}CH_2\text{—}CO\text{—}NH\text{—}CH_3 \rightarrow R\text{—}CH_2\text{—}CH_2\text{—}COOH$$
(CXXXV.)   (CXXXVI.)

$$R\text{—}CH_2\text{—}CH_2\text{—}NH\text{—}CO\text{—}CH_3 \rightarrow R\text{—}CH_2\text{—}CH_2\text{—}NH_2 \rightarrow$$
(CXXXVII.)   (CXXXVIII.)

$$\rightarrow R\text{—}CH{=}CH_2 \rightarrow R\text{—}COOH$$
(CXXXIX.)   (CXL.)

Ein Tetrahydro-iron mit geringer optischer Drehung gab mit Brom-
lauge eine nicht kristalline „Tetrahydro-ironsäure". Das durch weiteren
Abbau erhaltene Amin $C_{12}H_{25}N$ war verschieden von demjenigen aus der
kristallinen Säure. Der durch Zersetzung der quaternären Base des
Ammoniumsalzes des Amins erhaltene Kohlenwasserstoff $C_{12}H_{22}$ gab beim
Ozonisieren eine nicht-kristalline Säure $C_{11}H_{20}O_2$. Es liegen hier Produkte
vor, die sich von den entsprechenden kristallinen Verbindungen durch
Stereoisomerie unterscheiden.

d) *Abbau der Säure $C_{11}H_{20}O_2$ zu einem Trimethyl-cyclopentanon.*

Die (+)-Säure $C_{11}H_{20}O_2$ wurde über das Azid mit Hilfe des modifizierten Curtiusschen Abbaues in ein Amin $C_{10}H_{21}N$ übergeführt. Die
Pyrolyse des Phosphats dieser Base ergab einen Kohlenwasserstoff
$C_{10}H_{18}$. Mit Ozon wurde daraus eine Ketosäure $C_{10}H_{18}O_3$ und aus dieser
mit Bromlauge eine Dicarbonsäure $C_9H_{16}O_4$ gewonnen. Diese war identisch mit β,β,γ-Trimethyl-pimelinsäure (CXLVI). Beim Erhitzen auf
300° bildete sich das bekannte 1,1,5-Trimethyl-cyclopentanon-(3)
(CXLVII). Die erwähnten Zwischenprodukte entsprechen deshalb den
Substanzen (CXLII) bis (CXLV).

(CXLI.)   (CXLII.)   (CXLIII.)   (CXLIV.)

(CXLV.)   (CXLVI.)   (CXLVII.)
                 β,β,γ-Trimethyl-   1,1,5-Trimethyl-
                 pimelinsäure.   cyclopentanon-(3).

Man muß annehmen, daß unter den energischen Bedingungen der
Phosphatzersetzung eine Verschiebung der Doppelbindung des Kohlenwasserstoffes stattfindet (CXLIII → CXLIV).

Aus diesen Resultaten geht ebenfalls hervor, daß das Iron einen Ring
mit sechs Kohlenstoffatomen besitzt.

e) *Dehydrierung verschiedener Abbauprodukte des Irons.*

Die Tetrahydro-ironsäure $C_{13}H_{24}O_2$ (CXLVIII), die Abbausäure
$C_{12}H_{22}O_2$ (CXLIX) und das Tetrahydro-iran (CL) lieferten bei der Dehydrierung mit Selen das 1,2,3,4-Tetramethyl-benzol (CLI); dieses wurde
durch Oxydation mit Kaliumpermanganat in Mellophansäure (CLII)
übergeführt.

Die Abbausäure $C_{11}H_{20}O_2$ (CLIII) ging in das 1,2,4-Trimethylbenzol (CLIV) über, das zu Trimellitsäure (CLV) oxydiert wurde.

Diese Resultate können nicht unbedingt als Beweise der 6-Ring-struktur des Irons gelten, da unter den energischen Bedingungen der Dehydrierung Umlagerungen möglich wären.

(CXLVIII.)   (CXLIX.)   (CL.)

(CLI.)   (CLII.) Mellophansäure.

(CLIII.)   (CLIV.)   (CLV.) Trimellitsäure.

γ-Iron.

Der Abbau des Tetrahydro-irons hatte also zu mehreren Produkten geführt, welche über die 6-Ring-struktur des Irons keinen Zweifel mehr übrig ließen. Nun mußte aber eine Erklärung für die Bildung der β,β,γ-Tri-methyl-pimelinsäure (CLXVI) gefunden werden, die man aus Iron mit Ozon und Chromsäure erhalten hatte. Die zuerst auf Grund der De-hydrierung zu Iren in Erwägung gezogene Formel (CLVI) für das Iron kam nicht in Frage. Die Entstehung der Trimethyl-pimelinsäure läßt sich dagegen verstehen, wenn man für das Iron eine semicyclische Doppel-bindung im Sinne von Formel (CLVII) annimmt. Der Abbau dieser Ver-bindung führt primär zur β-Ketosäure (CLVIII). Durch Säurespaltung erhält man daraus die genannte Dicarbonsäure (CLIX). Die Keton-spaltung liefert dagegen das Keton (CLX). Dieses Keton konnte in der Tat aus den neutralen Oxydationsprodukten vom Abbau des Irons mit Ozon und Chromsäure isoliert und identifiziert werden. Diese Verbindung ist auch identisch mit dem Keton $C_9H_{16}O$, das durch Pyrolyse der β,β,γ-Trimethyl-pimelinsäure entsteht.

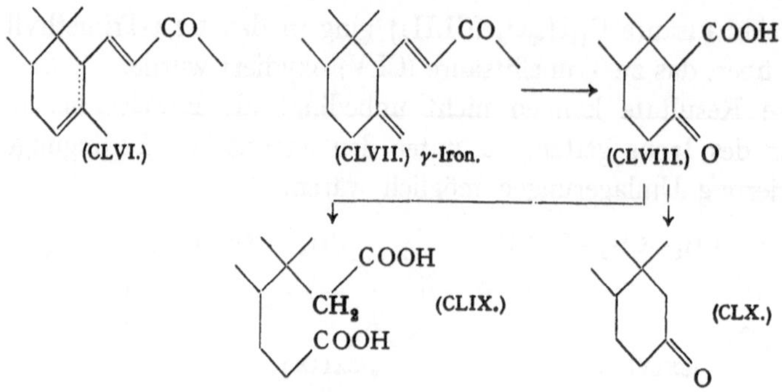

(CLVI.) (CLVII.) γ-Iron. (CLVIII.) O

(CLIX.) (CLX.)

Die Anwesenheit der semicyclischen Doppelbindung wurde auch durch reichliche Bildung von Formaldehyd bei der Ozonisation bewiesen. Da der größte Teil hiervon schon während dem Einleiten des Gasstroms entweicht, wurde er früher übersehen.

Ruzicka, Seidel und Schinz nannten das Iron mit der semicyclischen Doppelbindung „γ-Iron".

Weitere Nachweise des γ-Irons.

Die Anwesenheit eines Ketons mit semicyclischer Doppelbindung folgt nach Ruzicka und Seidel (_121_) auch aus einigen Abbauprodukten des Dihydro-irons und Dihydro-irans. Bei der partiellen Hydrierung des Irons in Gegenwart von Nickelkatalysator wird zuerst die Doppelbindung der Seitenkette abgesättigt. Das so erhaltene Dihydro-iron liefert bei der Reduktion nach Wolff-Kishner den Kohlenwasserstoff Dihydro-iran.

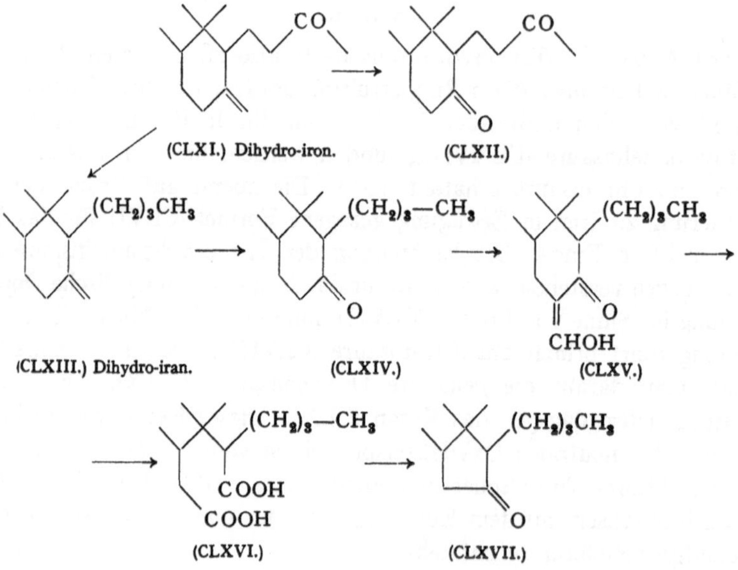

(CLXI.) Dihydro-iron. (CLXII.)

(CLXIII.) Dihydro-iran. (CLXIV.) (CLXV.)

(CLXVI.) (CLXVII.)

Aus Dihydro-iron (CLXI) erhält man bei der Ozonisation und nachfolgenden Oxydation mit Chromsäure aus den neutralen Reaktionsprodukten ein cyclisches Diketon $C_{13}H_{22}O_2$ (CLXII).

Das Dihydro-iran (CLXIII) liefert ein cyclisches Monoketon $C_{13}H_{24}O$ (CLXIV), das über die Oxymethylenverbindung (CLXV) in eine Dicarbonsäure $C_{13}H_{24}O_4$ (CLXVI) und letztere durch Erhitzen des Thoriumsalzes in ein cyclisches Keton $C_{12}H_{22}O$ der wahrscheinlichen Formel (CLXVII) übergeführt wurde.

Die Bildung der Produkte (CLXII) und (CLXIV) ist ein weiterer Beweis für das Vorliegen der semicyclischen Doppelbindung im γ-Iron.

Argumente von NAVES und Mitarbeitern zugunsten der 6-Ring-struktur des Irons; RAMAN-Spektren.

Inzwischen waren auch NAVES und BACHMANN (64, 85—88) mit Hilfe ausschließlich physikalischer Methoden zur Auffassung gelangt, daß das Iron einen 6-Ring besitze. Durch Vergleich einer Reihe von physikalischen Daten des Irons und des Dihydro-irols mit denen des Jonons und Dihydro-jonols wiesen diese Autoren auf die nahe Verwandtschaft von Iron und Jonon hin.

NAVES und BACHMANN verglichen den RAMAN-Effekt von Dihydroirol (aus Iron durch Reduktion mit Natrium und Alkohol erhalten) einerseits mit demjenigen von 1,1,5-Trimethylcyclohexen-(4), dem sogenannten Methyl-α-cyclo-geraniolen ($>C=CH-$), anderseits mit demjenigen von 1,1,3-Trimethyl-cyclohepten-(5 bzw. 6) ($-CH=CH_2-$). Dihydroirol und Trimethyl-cyclohexen zeigten die gleichen typischen Spektrallinien bei einer Frequenz von 1674 cm^{-1}, während die analoge, für die Ring-Doppelbindung ($-CH=CH-$) charakteristische Linie beim Trimethyl-cyclohepten bei 1642 cm^{-1} lag. Daraus schlossen diese Forscher, daß Dihydro-irol eine $>C=CH-$ und nicht eine $-CH=CH-$-Doppelbindung enthalte; deshalb sollte auch das Iron einen 6-Ring besitzen, und zwar sollte dieser Verbindung die Konstitution eines 6-Methyl-α-jonons zukommen.

Nach GÜNTHARD und RUZICKA (32) hätte aber, auch wenn die Linie bei 1642 cm^{-1} sichtbar gewesen wäre, daraus nicht unbedingt auf das Vorliegen eines 7-Ringes geschlossen werden dürfen, weil die Linie der semicyclischen Doppelbindung $>C=CH_2$ (γ-Iron) an der gleichen Stelle liegt. Diese Möglichkeit wurde von NAVES und BACHMANN nicht in Betracht gezogen.

Anwesenheit von α-Iron im natürlichen Irongemisch.

Die von RUZICKA, SEIDEL und SCHINZ (126) erhaltenen Phenylsemicarbazone verschiedener Schmelzpunkte wiesen auf das Vorhandensein von mehreren Isomeren hin. Man wußte aber damals nocht nicht, ob es sich dabei um Doppelbindungs- oder Stereoisomerie handle.

Der Gehalt des natürlichen Irisketons an γ-Iron beträgt nach Ruzicka, Seidel, Schinz und Pfeiffer (*128*) 70—75%. β-Iron, das sich leicht spektroskopisch nachweisen läßt, ist im genuinen Iron nicht in nennenswerter Menge vorhanden. Nach Ruzicka und Mitarbeitern bestehen die restlichen 25—30% des Irisketons aus dem α-Isomeren.

Die Anwesenheit von α-Iron ergibt sich u. a. aus dem Resultat der Ozonisation von Dihydro-iron und Dihydro-iran (*121*). In beiden Fällen waren nämlich außer den aus der γ-Verbindung stammenden Ketonen saure Anteile erhalten worden, die sich aus der α-Form gebildet haben mußten.

So erhielt man aus Dihydro-iron eine Keto-dicarbonsäure $C_{13}H_{22}O_5$ (wahrscheinlich CLXX), welche sich aus einem Keton der Formel (CLXVIII) (Dihydro-α-Iron) über die Diketo-monocarbonsäure CLXIX gebildet haben konnte. Sie wurde nach Clemmensen zu einer Dicarbonsäure $C_{13}H_{24}O_4$ (CLXXI) reduziert. Diese wurde über das Thoriumsalz in ein Keton $C_{12}H_{22}O$ (CLXXII) übergeführt. Das Oxymethylenketon (CLXXIII) lieferte bei der Ozonisation eine Dicarbonsäure der wahrscheinlichen Konstitution (CLXXIV).

(CLXVIII.) Dihydro-α-iron.   (CLXIX.)   (CLXX.)   (CLXXI.)

(CLXXII.)   (CLXXIII.)   (CLXXIV.)

Beim Abbau des Dihydro-irans entstand außer den neutralen Anteilen eine Ketosäure der wahrscheinlichen Formel (CLXXVI), die sich leicht zur Dicarbonsäure (CLXXVII) weiter oxydieren ließ. Das Dihydro-iran mußte also ebenfalls einen gewissen Gehalt an α-Form (CLXXV) aufweisen.

(CLXXV.) Dihydro-α-iran.   (CLXXVI.)   (CLXXVII.)

Diese Versuche zeigen die Anwesenheit von α-Iron im genuinen Iron-gemisch. Damit sind die genannten Autoren, allerdings nur für ca. 25 bis 30% des natürlichen Öls, zu der früher auf Grund der Dehydrierung angenommenen alten Strukturformel zurückgekehrt. Nach NAVES sollen gewisse Sorten natürlichen Irons überhaupt fast vollständig aus α-Iron bestehen. Da NAVES die genaue Bezugsquelle dieses Ketons jedoch nicht erwähnt, ist seine Angabe unkontrollierbar.

α-, β- und γ-Iron aus genuinen und isomerisierten Gemischen von Natur-Iron.

Die Frage der Zusammensetzung des natürlichen Irons und die Methodik ihrer Bestimmung haben zu Kontroversen zwischen den beiden Arbeitsgruppen RUZICKA und NAVES geführt.

Nach RUZICKA und Mitarbeitern (*128, 130, 139, 33, 138*) ergibt sich die Menge des γ-Irons im natürlichen Iron oder in durch Isomerisieren erhaltenen Gemischen von Iron-isomeren durch Bestimmung des beim Ozonisieren gebildeten Formaldehyds als Dimedon-derivat. Der Formaldehyd entweicht zum großen Teil mit dem Gasstrom, kann aber leicht in Wasser aufgefangen werden. Mittels Versuchen an Modellsubstanzen wurde festgestellt, daß auf diese Weise ca. 35% der theoretischen Menge Formaldehyd erfaßt werden. Diesem Umstand wird durch Anwendung des Korrekturfaktors 100/35 Rechnung getragen. Iron aus verschiedenen Handelssorten von Iriswurzelöl sowie aus Rhizomen selbst extrahiertes Keton zeigten einen Gehalt von 70—75% γ-Iron.

Nach NAVES und Mitarbeitern (*85, 64, 66, 75, 77*) besteht das natür-liche Iron aus toscanischem Irisöl praktisch vollständig aus α-Iron. Erst später von NAVES untersuchte Proben anderer Provenienz enthielten auch γ-Iron, dessen Menge zwischen 5 und 55% schwankte (*66, 71, 77*). Die Bestimmung der beim Ozonisieren gebildeten Menge Formaldehyd erfolgte nach DOEUVRE.

Nach RUZICKA und Mitarbeitern (*130*) zeigen α- und γ-Iron die gleichen Absorptionskurven im Ultraviolett: $\lambda_{max} = 230$ mμ, log $\varepsilon = 4,2$. Der Gehalt an β-Iron ($\lambda_{max} = 298$ mμ, log $\varepsilon = 4,0$) in Gemischen der drei Irone läßt sich aus den Extinktionskoeffizienten bei 298 mμ berechnen. Weniger genau ist die Bestimmung mit Hilfe der Molekularrefraktion. β- und γ-Iron lassen sich also direkt bestimmen. Der Gehalt an α-Iron ergibt sich aus der Differenz: $(\alpha + \beta + \gamma) - (\beta + \gamma)$.

Reines γ-Iron erhält man durch Hydrolyse des Phenylsemicarbazons vom Schmelzpunkt 177—179° mit Phthalsäureanhydrid.

α-Iron in fast reiner Form bereitet NAVES aus seinen angeblich γ-freien Ironmustern (Phenylsemicarbazon, Schmp. 162—163°) durch Reinigung mit GIRARD-Reagens. Nach RUZICKA und Mitarbeitern ist die Isolierung reinen α-Irons aus den stark γ-haltigen natürlichen Gemischen schwierig.

Sie erhielten ein vielleicht teilweise invertiertes α-Iron künstlich durch Einwirkung schwacher Säuren, z. B. Oxalsäure, auf das natürliche Irongemisch (*130*). Dabei bildet sich aber immer auch eine gewisse Menge β-Iron. Aus einem mit Ameisensäure isomerisierten Gemisch von natürlichem Iron isolierten BÄCHLI, SEIDEL, SCHINZ und RUZICKA (*4*) über das Phenylsemicarbazon ein (+)-α-Iron, durch Isomerisieren mit alkoholischer Lauge ein (—)-α-Iron. Im letzten Fall wurde außerdem (+)-β-Iron gewonnen.

(CLXXVIII.) α-Iron.   (CLXXIX.) β-Iron.   (CLXXX.) γ-Iron.

Nach KÖSTER (*50*) wird (+)-β-Iron aus dem natürlichen Irongemisch durch Einwirkung von konz. Schwefelsäure oder alkoholischen Alkalien hergestellt. Die Isomerisierung vollzieht sich schon in der Kälte mit großer Leichtigkeit. Das (+)-β-Iron wird über das Semicarbazon in reiner Form gewonnen. Das Semicarbazon und das Phenylsemicarbazon des β-Irons färben sich am Lichte in kurzer Zeit gelb. Während γ-Iron mit Säuren und Alkalien in homogener Lösung rasch zu α- und β-Iron isomerisiert wird, ist es gegen verdünnte wäßrige Säuren und Alkalien in der Kälte ziemlich beständig (*139*).

Konstanten und Derivate der drei stellungsisomeren Irone sind in einer Tabelle zusammengestellt (S. 179).

Synthese von ($\pm$)-α- und ($\pm$)-β-Iron.

Das 1940 von RUZICKA und SCHINZ (*115*) beschriebene synthetische Gemisch von 6-Methyljononen war damals wegen des hohen Gehaltes an β-Form nicht als Iron erkannt worden.

Einige Jahre später stellten SCHINZ, RUZICKA, SEIDEL und TAVEL (*137*) einerseits, NAVES, GRAMPOLOFF und BACHMANN (*90, 89*) anderseits ein aus praktisch reinem 6-Methyl-α-jonon bestehendes Keton dar. Die Synthesen der beiden Arbeitsgruppen sind praktisch identisch und wurden gleichzeitig im Jahre 1947 publiziert.

2,3-Dimethyl-hepten-(2)-on-(6) (CLXXXI) wird mit Acetylen in das Carbinol (CLXXXII) und dieses durch partielle Hydrierung in Methyllinalool (CLXXXIII) übergeführt. Das daraus durch Allyl-umlagerung erhaltene Methylgeraniol (CLXXXIV) liefert bei Behandlung mit Aluminiumisopropylat in Aceton-Benzol das Pseudoiron (CLXXXVI). Dieses besteht aus verschiedenen Stereoisomeren. Von den vier möglichen Formen konnten drei durch Derivate verschiedener Schmelzpunkte nachgewiesen werden (*140, 90*). Das Pseudoiron geht unter dem Einfluß von Phosphorsäure in das cyclische Isomere (CLXXXVIII) über, welches zur Hauptsache aus α-Iron besteht. Mit konz. Schwefelsäure entsteht dagegen hauptsächlich β-Iron.

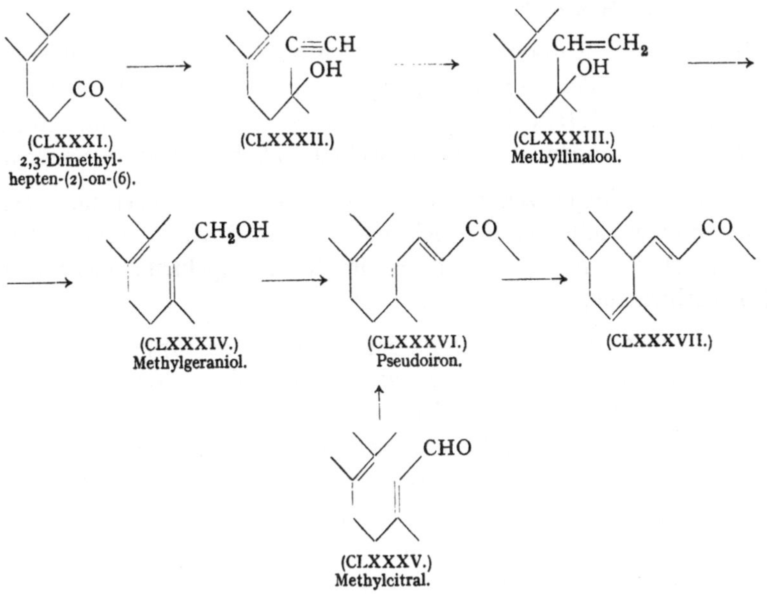

(CLXXXI.)
2,3-Dimethyl-hepten-(2)-on-(6).

(CLXXXII.)

(CLXXXIII.)
Methyllinalool.

(CLXXXIV.)
Methylgeraniol.

(CLXXXVI.)
Pseudoiron.

(CLXXXVII.)

(CLXXXV.)
Methylcitral.

NAVES, GRAMPOLOFF und BACHMANN benutzen zum Teil die folgende Variante: (CLXXXIII) wird mit Chromsäure zu Methylcitral (CLXXXV) oxydiert und dieses mit Aceton zu (CLXXXVI) kondensiert. Die Cyclisation zu 6-Methyl-α-jonon wird mit 62,5%iger Schwefelsäure ausgeführt. Die Oxydation des Methyl-linalools mit Chromsäure wurde von STOLL und COMMARMONT (*152*) sowie von NAVES (*66*) besonders bearbeitet.

Wenn die beiden Synthesen einerseits praktisch gleich sind, werden anderseits die erhaltenen Produkte von den beiden Forschergruppen verschieden beurteilt. Nach SCHINZ, RUZICKA, SEIDEL und TAVEL liegt in dem von ihnen dargestellten Produkt die (±)-Form eines im natürlichen Irongemisch enthaltenen Bestandteils vor und das synthetische Produkt ist geruchlich dem Naturprodukt gleichwertig. Nach NAVES, GRAMPOLOFF und BACHMANN besteht aber keine Identität zwischen den beiden Verbindungen; im besonderen sind sie auch geruchlich ganz verschieden.

Infrarot-Spektren.

GÜNTHARD und RUZICKA (*31, 32*) zeigten an einer Reihe von Beispielen die Übereinstimmung der Wellenlänge von gleichartigen Frequenzen im RAMAN-Effekt und im Infrarot. Für die bei Substanzen der Irongruppe typischen Frequenzen machten sie folgende Zuordnungen:

$$v\text{-(CC)} \quad 1675 \text{ cm}^{-1} \qquad 1675 \text{ cm}^{-1} \qquad 1645 \pm 5 \text{ cm}^{-1}$$
$$\delta\text{-(CH)} \quad 810\text{---}840 \text{ cm}^{-1} \qquad \text{---} \qquad 890 \text{ cm}^{-1}$$

Nach diesen Autoren besitzt die Infrarot-Technik gegenüber der RAMAN-Technik den Vorteil, daß bei ihr neben der v-(CC)-Schwingung stets auch eine δ-(CH)-Schwingung sichtbar ist. Ferner sind beim Infrarot-Spektrum die Banden deutlicher ausgebildet als beim RAMAN-Spektrum.

Die v-(CC)-Frequenzen der RAMAN-Spektren von NAVES und BACH-MANN stimmen mit den Beobachtungen von GÜNTHARD und RUZICKA im Infrarot überein, dagegen fehlen im RAMAN-Spektrum meistens die δ-(CH)-Schwingungen.

Nach GÜNTHARD und RUZICKA darf aus einer bei 1675 cm^{-1} liegenden (CC)-Schwingung nur dann auf das Vorliegen des α-Typs geschlossen werden, wenn im Spektrum gleichzeitig auch die zugehörigen δ-(CH)-Schwingung bei ca. 812 cm^{-1} vorhanden ist; diese ist im Infrarot deutlich, im RAMAN-Spektrum dagegen unscharf oder überhaupt nicht ausgebildet. Die Frequenz 1670—1675 cm^{-1} ist überhaupt wenig typisch, da sie nicht nur beim·α-, sondern auch beim β- und — mit geringerer Intensität — sogar beim γ-Typ (hier neben der 1642er Frequenz) auftritt.

Unter Benutzung der δ-(CH)-Schwingungen bei 890 cm^{-1} und mit Hilfe einer verbesserten Aufnahmetechnik wurden von den letztgenannten Autoren quantitative Bestimmungen an künstlichen Gemischen von Dihydro-α- und Dihydro-γ-jononderivaten ausgeführt und die Methode hierauf auf Iron-präparate übertragen. Die so gefundenen Werte für γ-Iron stimmten mit den Resultaten der Formaldehydmethode praktisch überein.

Bestimmung und Interpretation der RAMAN- und Infrarot-Spektren führten wiederum zu Meinungsverschiedenheiten der beiden Arbeitsgruppen (33, 88, 78).

Stereoisomerie-Möglichkeiten der Irone und Hydroirone.

Beim γ- und α-Iron (CLXXXVIII) bzw. (CLXXXIX) ist *cis-trans*-Isomerie der Substituenten in den Stellungen 2 und 6 im Kern und überdies eine solche in bezug auf die Substituenten an der Doppelbindung der Seitenkette möglich. Es ergeben sich somit die vier Racemformen a) *cis* (2,6), *cis* ($2^1,2^2$); b) *trans* (2,6), *cis* ($2^1,2^2$); c) *cis* (2,6), *trans* ($2^1,2^2$); d) *trans* (2,6), *trans* ($2^1,2^2$).

Beim β-Iron (CXC) fällt die Stereoisomerie am Ring weg; es sind deshalb nur die beiden Racemate *cis* ($2^1,2^2$) und *trans* ($2^1,2^2$) möglich.

Für die optisch aktiven Irone verdoppelt sich die Zahl der möglichen Isomeren auf je 8 α- und γ- sowie 4 β-Formen. Im ganzen sind also 20 aktive und 10 (±)-Irone möglich.

(CLXXXVIII.) γ-Iron.   (CLXXXIX.) α-Iron.   (CXC.) β-Iron.

NAVES (73) schlug für die Irone eine spezielle Nomenklatur vor, die sich an die bei den Mentholen übliche anlehnt. Er bezeichnet a) kurz als „Irone", b) als „Isoirone", c) als „Neoirone" und d) als „Neo-isoirone".

Bei den drei stellungsisomeren Dihydro-ironen (CXCI), (CXCII) und (CXCIII) mit gesättigter Seitenkette sind zwei racemische γ, zwei α- und eine β-Form möglich. Das im Ring gesättigte Dihydroiron (CXCIV) kann in acht und das Tetrahydro-iron (CXCV) in vier (±)-Formen auftreten. Die Zahl der entsprechenden möglichen optisch aktiven Verbindungen ist doppelt so groß.

(CXCI.)   (CXCII.)   (CXCIII.)

(CXCIV.)   (CXCV.)

Stereoisomerie-(2,6) der Irone.

Im Jahre 1948 beschrieb NAVES (67) ein neues (±)-6-Methyl-α-jonon. Der Unterschied in der Darstellung gegenüber der früheren Synthese liegt darin, daß die Cyclisation des Pseudo-methyljonons statt mit 62,5%iger Schwefelsäure oder Phosphorsäure diesmal mit Bortrifluorid ausgeführt wurde. Dabei bildete sich nach NAVES vorwiegend das *cis*-(2,6)-Isomere (Phenylsemicarbazon, Schmp. ca. 160°), während nach der früheren Methode die *trans*-Form (Phenylsemicarbazon, Schmp. 175°) erhalten wird. Nur das (±)-*cis*-(2,6)-6-Methyl-α-jonon soll geruchlich dem (+)-α-Iron des natürlichen Iriswurzelketons ebenbürtig sein, die *trans*-Verbindung dagegen wie die in der Seitenkette methylierten Jonone riechen. Sowohl das natürliche α- als auch das γ-Iron besitzen nach NAVES *cis*-(2,6)-Struktur.

Dieser Punkt gab wiederum Anlaß zu regem Hin und Her der Meinungen der beiden Arbeitsgruppen NAVES und RUZICKA.

Nach SEIDEL, SCHINZ und RUZICKA (140) erhält man weder bei der Cyclisation mit Phosphorsäure noch mit Bortrifluorid sterisch einheitliches α-Iron. In beiden Fällen wurden Derivate mit verschiedenen Schmelzpunkten (zwischen circa 155° und 172°) isoliert. Die Cyclisation mit Bortrifluorid liefert allerdings ein Gemisch mit etwas anderem Isomerenverhältnis. Einen größeren Einfluß auf das Verteilungsverhältnis von *cis*- und *trans*-Form übt aber die Art des verwendeten Pseudoirons aus. Höhersiedende Anteile desselben geben mehr *cis*-(2,6)-α-Iron. Vielleicht spielt die räumliche Lage der Substituenten an der Doppelbindung in Konjugation zur CO-Gruppe eine Rolle bei der Ausrichtung der Substituenten an den C-Atomen 2 und 6 des Cyclisationsproduktes. Ferner sind nach Seidel, Schinz und Ruzicka die Geruchsunterschiede zwischen *cis*-reichen und *cis*-armen Gemischen bei weitem nicht so groß wie Naves meint. Hingegen spielt für den Geruch die Anwesenheit geringer Mengen von β-Iron und von nicht-ketonischen Verunreinigungen eine große Rolle.

Später fand Naves (*68, 69*) bei eingehenderer Untersuchung, daß auch seine Produkte weniger einheitlich waren, als er ursprünglich geglaubt hatte. Er fand im natürlichen Iron-gemisch auch eine bestimmte Menge *trans*-(2,6)-Iron (sogenanntes „Iso-iron") und isolierte anderseits aus dem durch Cyclisation mit Phosphorsäure oder 62%iger Schwefelsäure gewonnenen „(±)-α-Isoiron" aus (±)-*cis*-(2,6)-Iron bestehende Anteile.

Die Zuteilung der räumlichen Lage der Substituenten wurde von Naves (*67*) mit Hilfe der Auwers-Skitaschen Regel an den in der Seitenkette gesättigten Dihydroironen vorgenommen (vgl. S. 181).

Stereoisomerie-(2^1,2^2) der Irone.

Naves und Bachmann (*85, 68, 72*) schlossen auf Grund der für die Doppelbindung in Konjugation zur Carbonylgruppe charakteristischen Raman-Frequenzen sowie der Parachorwerte auf *cis*(2^1,2^2)-Form beim natürlichen und synthetischen α-Iron. Diese Annahme würde auch die leichte Bildung von Iren erklären. Die genannten Autoren isolierten nun aus dem natürlichen Irongemisch noch ein anderes α-Iron, das nach dem Raman-Spektrum *trans*-(2^1,2^2)-Form besitzt. Bei der partiellen Hydrierung entsteht daraus ebenfalls Dihydro-α-iron. Die neue Verbindung, die durch ein sehr schwer lösliches Phenylsemicarbazon vom Schmelzpunkt 181° charakterisiert wird, wurde „Neo-α-iron" genannt.

Der Gehalt natürlicher Iron-gemische an Neo-α-iron steht in direkter Beziehung zum Gehalt an γ-Iron (*70*). Die Frage, ob sich das Neo-α-iron aus dem γ-Iron bildet, scheint noch nicht ganz abgeklärt.

Das γ-Iron (Phenylsemicarbazon, Schmp. 178—179°) besitzt nach Naves und Bachmann ebenfalls *trans*-(2^1,2^2)-Form. Sie nennen es deshalb „Neo-γ-iron".

Bei der Synthese von (±)-α-Iron (*68*) bilden sich ebenfalls geringe Mengen des Neo-α-ketons. Auch Seidel, Schinz und Ruzicka (*140*) fanden bei einem Cyclisationsversuch von hochsiedendem Pseudoiron-anteilen mit Bortrifluorid ein Keton mit hochschmelzendem, schwerlöslichem Phenylsemicarbazon, das vielleicht dem „Neo-α-iron" entspricht.

Vergleich der natürlichen und synthetischen Irone.

Zum Vergleich der verschiedenen natürlichen optisch aktiven und der synthetischen racemischen Irone sind ihre physikalischen Konstanten und die Schmelzpunkte ihrer Derivate in *Tabelle 1* zusammengestellt; es werden dort stets die Angaben der beiden Forschergruppen RUZICKA und NAVES berücksichtigt. Bei Produkten, welche mehrmals hergestellt wurden, sind im allgemeinen nur die reinsten Präparate aufgeführt. Wahrscheinlich sind nicht alle erwähnten Derivate einheitliche Individuen, da die Trennungen oft unvollkommen sind. Ferner neigen z. B. die 2,4-Dinitrophenylhydrazone zu Dimorphismus. Die Schmelzpunkte von NAVES sind korrigiert, diejenigen von RUZICKA unkorrigiert.

Der Vergleich der optisch aktiven Verbindungen und der Racemate ist beim β-Iron am leichtesten, da hier die 2,6-Stereoisomerie wegfällt. Da die Schmelzpunkte aller Derivate der (+)- und der (±)-Ketone übereinstimmen und die Präparate bei der Mischprobe keine Depressionen zeigen, darf man annehmen, daß die Ketone auch an der Doppelbindung in der Seitenkette gleiche sterische Form aufweisen.

Was das α-Iron betrifft, so zeigte ein Phenylsemicarbazon von durch Isomerisieren von Natur-Iron erhaltenem α-Iron von RUZICKA, SEIDEL, SCHINZ und TAVEL vom Schmelzpunkt zirka 160° mit einem gleichschmelzenden Präparat aus (±)-α-Iron (Cyclisation mit Phosphorsäure) bei der Mischprobe keine Depression. NAVES findet anderseits Identität (abgesehen vom Drehvermögen) seiner optisch aktiven mit den synthetischen „Ironen", „Isoironen" und „Neo-ironen".

Tabelle 1. Charakterisierung der verschiedenen Irone durch physikalische Konstanten und Schmelzpunkte von Derivaten.

PS = Phenylsemicarbazon; DNP = 2,4-Dinitrophenylhydrazon; TS = Thiosemicarbazon; S = Semicarbazon; BP = p-Brom-phenylhydrazon; NP = p-Nitrophenylhydrazon.

(+)-γ-*Irone.*

RUZICKA:

d_4^{15} = 0,939; n_D^{15} = 1,505; $[α]_D$ = + 22° (*126*)
d_4^{20} = 0,9368; n_D^{20} = 1,500; $[α]_D$ = + 8° (*4*)
PS 178—179° (*126*); DNP 130—131° (*4*); TS 127,5—128,5° (*4*)

NAVES:

a) „Neo-γ-iron" d_4^{20} = 0,9355; n_D^{20} = 1,5019; $[α]_D$ = + 7,6° (*71*)
  PS 178—179° (*66, 68, 71**); TS 169,5—170°*; DNP 146—146,5° (*68, 71*);
   BP 178—179°**
b) ein anderes γ-Iron:
  PS 174—177° (*77*); DNP 126,5—127° (*77*)

---

 * Siehe auch Helv. chim. Acta **32**, 599 (*1949*).
 ** Helv. chim. Acta **32**, 618 (*1949*).

$(+)$-α-*Irone*.

Ruzicka:

$d_4^{20} = 0,9358$; $n_D^{20} = 1,5013$; $[\alpha]_D = + 114°$ (4)
PS 160—161° (130); 153—154° (4); TS 180° (126); DNP nicht konstant (4)

Naves:

a) cis: $d_4^{20} = 0,9349$; $n_D^{20} = 1,5003$; $[\alpha]_D = + 226°$ (72)*
   PS 162,5—163° $(85, 66)$; 157,5—158° (72); TS 182,5—183°*; DNP 125,5 bis
   126° $(85, 66)$; BP 169—170° (73)**
b) trans: PS 174,5—175,5° (68)

$(—)$-α-*Irone*.

Ruzicka:

Isomerisieren von natürlichem Iron mit alkoholischer Kalilauge:
$d_4^{20} = 0,9318$; $n_D^{20} = 1,4973$; $[\alpha]_D = — 79,5°$ (4)
PS 164—165° (4); DNP 103—104° (4)

Naves:

„Neo-α-Iron":
$d_4^{20} = 0,9347$; $n_D^{20} = 1,5013$; $[\alpha]_D = — 8,2°$ $(70, 71)$
PS 181—182° $(68, 70)$

---

   * Siehe auch Helv. chim. Acta **32**, 599 (1949).
   ** Siehe auch Helv. chim. Acta **32**, 618 (1949).

$(+)$-β-*Irone*.

Köster (50):

Isomerisieren von natürlichem Iron mit konz. Schwefelsäure:
$d_4^{18} = 0,9472$; $n_D^{25} = 1,5160$; $[\alpha]_D = + 48°$
S 166—167°; TS 164—165°

Ruzicka:

a) Isomerisieren von natürlichem Iron mit alkoholischer Kalilauge:
   $d_4^{21} = 0,9434$; $n_D^{21} = 1,5178$; $[\alpha]_D = + 41°$ (130)
   S 167—168° (130); TS 165—166° (130); PS 160—161° (130)
b) Isomerisieren mit konz. Schwefelsäure:
   $d_4^{20} = 0,9444$; $n_D^{20} = 1,5178$; $[\alpha]_D = + 11,3°$ (4)
   DN 131—132° (4)

Naves:

Isomerisieren mit konz. Schwefelsäure:
$d_4^{20} = 0,9456$; $n_D^{20} = 1,5180$; $[\alpha]_D = + 20,0°$*
PS 167—168° (66); S 167,5—168°; 169—169,5° (66); TS 167,5—168°*; DNP 135
   bis 136° (66)

---

   * Helv. chim. Acta **32**, 599 (1949).

$(\pm)$-α-*Irone*.

Ruzicka:

$d_4^{19} = 0,9345$; $n_D^{19} = 1,5001$ (Cyclisation mit Phosphorsäure) $(137, 130, 140)$
$d_4^{19} = 0,9413$; $n_D^{19} = 1,5021$ (Cyclisation mit Phosphorsäure) $(140, 130)$
$d_4^{20} = 0,9358$; $n_D^{20} = 1,5005$ (Cyclisation mit Bortrifluorid) (140)

a) PS 166—167° (137, 130); 171—172° (140); 172—173° (140)
DNP 120—121° (140); 100—101° (140); 104—105° (140)

b) PS 160—162° (130); zirka 160° (140); DNP 148—149° (140); 151—152° (140)

c) PS 178—179° (140)

Naves:

„Iso-α-iron" (Cyclisation mit Phosphorsäure oder 62%iger Schwefelsäure):
$d_4^{20} = 0{,}9355$; $n_D^{20} = 1{,}4970$ (90)
$d_4^{20} = 0{,}9346$; $n_D^{20} = 1{,}5013$ (77)
PS 174,5—175,5° (90, 66); DNP 103—103,5° (66)
„α-Iron" (Cyclisation mit Bortrifluorid):
$d_4^{20} = 0{,}9368$ (67); $n_D^{20} = 1{,}5019$ (67)
PS 164,5—165 (67); DNP 125,5—126° (72)
„Neo-α-iron" (Cyclisation mit Bortrifluorid):
$d_4^{20} = 0{,}9349$; $n_D^{20} = 1{,}5009$ (68)*
PS 181—182° (68)*; TS 188,5—189°*; DNP 153,5—154° (67, 68, 72); BP 164
bis 165°**

---

* Siehe auch Helv. chim. Acta **32**, 599 (1949).
** Helv. chim. Acta **32**, 618 (1949).

$$(\pm)\text{-}\beta\text{-}Iron.$$

Ruzicka:

Cyclisation mit konz. Schwefelsäure:
$d_4^{19} = 0{,}9474$; $n_D^{19} = 1{,}5180$ (130)
PS 164—166° (167°) (130); DNP 131—132° (140); TS 166—167° (140); S 167
bis 168° (140)

Naves (66):

$(\pm)$-α-Iron mit Schwefelsäure isomerisiert:
$d_4^{20} = 0{,}9465$; $n_D^{20} = 1{,}5183$
PS 167—168°; DNP 135—136°

Dihydro-irone.

Die Dihydro-irone spielen eine wichtige Rolle beim Studium der 2,6-Stereoisomerie der Irone. Nach Naves (66, 67, 69) besitzt das aus $(\pm)$-α-Iron (Phenylsemicarbazon, Schmp. 174—175°) durch partielle Hydrierung gewonnene Dihydroderivat eine kleinere Dichte und einen tieferen Brechungsindex als das aus natürlichem $(+)$-α-Iron, sowie aus dem anderen stereoisomeren $(\pm)$-α-Iron (beide Phenylsemicarbazone, Schmp. 164,5—165°) hergestellte Dihydroiron. Deshalb teilt dieser Autor auf Grund der Regel von Auwers und Skita dem ersten der genannten Dihydro-ketone (Semicarbazon, Schmp. 143—144°) *trans*-Form, dem zweiten und dritten (Semicarbazon beider Präparate, Schmp. 172,5—173°) *cis*-Form zu. Diese Zuteilung wird an den Dihydroketonen und nicht an den unhydrierten Ausgangsverbindungen vorgenommen, weil sich die Derivate der ersteren besser voneinander trennen lassen und überdies bei ihnen die (2^1,2^2)-Stereoisomerie wegfällt.

Der experimentelle Befund wird von Seidel, Schinz und Ruzicka (*140*) im wesentlichen bestätigt (Schmelzpunkt der (±)-Dihydro-ketonsemi-carbazone 146—147° bzw. 171—172°), es ist aber nicht sicher, ob die Zuordnung von Naves und Bachmann richtig ist. Beckett, Pitzer und Spritzer (*8*) zeigten nämlich, daß die Regel von Auwers und Skita bei 1,3-dialkylierten Cyclohexanverbindungen keine Gültigkeit hat. Wir wissen allerdings nicht, ob dies für den ungesättigten Ring des Irons ebenfalls zutrifft, da die Voraussetzung — Sesselform des Rings — in diesem Falle nicht erfüllt ist.

Bächli, Seidel, Schinz und Ruzicka (*4*) erhielten aus (+)-α-Iron ein Dihydroketon (Semicarbazon, Schmp. 171—172°). Aus (—)-α-Iron gewannen sie ein Dihydroderivat (Semicarbazon, Schmp. 143,5—144,5°). Diese Präparate gaben bei der Mischprobe mit den entsprechenden (±)-Präparaten keine Schmelzpunktserniedrigungen. Naves und Bach-mann (*85, 86, 66*) gaben ein weiteres Dihydroderivat von (+)-α-Iron an (Semicarbazon, Schmp. 203—203,5°), welches sie aus einem Natur-iron erhielten, das nach ihrer Meinung praktisch vollständig aus α-Iron bestand. Diese Dihydroverbindung ist bis heute nicht abgeklärt.

Dihydro-γ-iron wurde von Naves und Bachmann durch Hydrieren eines γ-haltigen natürlichen Irons zur Dihydro-Stufe und Trennung mittels des Semicarbazons gewonnen. Bächli, Seidel, Schinz und Ruzicka trennten ein durch Oxydation von Dihydro-irol (aus natürlichem Iron durch Reduktion mit Natrium und Alkohol gewonnen) erhaltenes Ketongemisch auf gleiche Art. Dihydro-β-iron aus (+)-β-Iron wurde ebenfalls von beiden Arbeitsgruppen hergestellt. Die Hydrierung des β-Irons zum Dihydro-keton verläuft aber weniger einheitlich als die des α-Isomeren.

Bächli (*3*) stellte ein im Kern gesättigtes racemisches Dihydro-iron dar. Das Propylenketal von (±)-β-Iron wurde in Gegenwart von Raney-Nickel partiell hydriert. Auf dem entstandenen Gemisch von Dihydro-ketonen trennte man durch fraktionierte Hydrolyse der Verbindung mit Girard-Reagens *P* ein α,β-ungesättigtes Dihydro-iron ab. Da aus β-Jonon unter den gleichen Bedingungen *cis*-Dihydro-jonon entsteht, besitzt dieses Dihydro-iron wahrscheinlich *cis*-(2,3)-Struktur. In bezug auf die Lage der Substituenten an den Kohlenstoffatomen 2 und 6 scheint das Produkt nicht ganz einheitlich. Ein *trans*-(2,3)-Dihydro-iron erhielt Bächli auf folgendem Weg: 6-Methyl-cyclogeraniumsäure wurde nach Rosenmund zum entsprechenden Aldehyd reduziert. Die Kondensation desselben mit Aceton führte zu einem Dihydro-iron, welches in Analogie mit den Verhältnissen beim Dihydro-jonon *trans*-(2,3)-Form besitzen muß.

Die Eigenschaften der verschiedenen Dihydro-irone sind in Tabelle 2 zusammengestellt.

Tabelle 2. Charakterisierung der verschiedenen Dihydro-irone.
DH = Dihydroverbindung; S = Semicarbazon; DNP = 2,4-Dinitrophenylhydrazon.

Optisch-aktive Dihydro-irone.

DH aus Gemisch von (+)-α- und (+)-γ-Iron:
RUZICKA: S 180—181°; 170—171° (*121*)

DH aus (+)-α-Iron:
RUZICKA: S 171—172° (*4*); DNP 128,5—129° (*4*)
NAVES: S 203—203,5° (*85, 66*); 172—173° (*67*); DNP 106—106,5° (*85, 66*); 130
bis 131° (*67*)

DH aus (—)-α-Iron:
RUZICKA: S 143,5—144,5° (*4*); DNP 114,5—115° (*4*)

DH aus (+)-γ-Iron:
RUZICKA: S 199—200° (*4*); DNP 108—109° (*4*)
NAVES: S 199,5—200° (*66, 71*); DNP 109—110° (*71*)

DH aus (+)-β-Iron:
RUZICKA: S 161—162° (*4*); DNP 104—105° (*4*)
NAVES: S 157—158° (*69*); DNP 103—104° (*69*)

Racemische Dihydro-irone.

DH-α-iron:
RUZICKA (*140*):
a) d_4^{20} = 0,9282; n_D^{20} = 1,4803
   S 146—147°; DNP 115—116°
b) d_4^{20} = 0,9315; n_D^{20} = 1,4828
   S 171—172°; DNP 128—129°

NAVES:
a) d_4^{20} = 0,9260; n_D^{20} = 1,4804 (*69*)
   S 143,5—144° (*67*); DNP 116—117° (*67*)
b) d_4^{20} = 0,9305; n_D^{20} = 1,4826 (*69*)
   S 172,5—173° (*67*); DNP 130—131° (*67*)

DH-β-Iron:
RUZICKA (*140*):
S 162—162,5°; DNP 104—105°

Tetrahydro-irone.

(+)-Tetrahydro-iron wurde von RUZICKA und Mitarbeitern (*106, 124, 123*) durch Hydrierung von Iron mit Hilfe von Platinkatalysatoren in Essigester- und Eisessiglösung gewonnen. Bei der Hydrierung in Eisessig bis zur Brom-beständigkeit bildet sich immer viel (oder quantitativ) Tetrahydro-irol, welches zum Keton rückoxydiert wird. NAVES (*90, 69*) führte Hydrierungen von Iron zu Tetrahydro-irol mit Platinoxyd in Eisessig bei 60° durch, ferner solche von Dihydro-iron zu Tetrahydro-irol in Gegenwart von RANEY-Nickel bei 100—110° unter 100 Atm. Druck.

Das (+)-Tetrahydro-iron aus natürlichem Iron-Gemisch gibt in der Hauptsache ein Semicarbazon, Schmp. 203—204°, und daneben tiefer schmelzende Präparate, vor allem ein solches vom Schmp. 160—161°.

Das NAVESsche Tetrahydro-iron aus Natur-iron (69) sowie aus reinem γ-Iron (71) lieferte fast ausschließlich das hochschmelzende Semicarbazon. KÖSTER (50), wie auch NAVES (69) stellten aus durch Isomerisieren von natürlichem Iron erhaltenem β-Iron das Tetrahydro-keton dar (Semicarbazon, ebenfalls Schmp. circa 200°).

Auch bei der Hydrierung der (±)-Irone erhielten SEIDEL, SCHINZ und RUZICKA (140) neben einem Semicarbazon, Schmp. ca. 200°, ein solches vom Schmp. ca. 160°, und zwar lieferte α-Iron etwa zehnmal mehr des zweiten Präparates als des ersten; bei der Hydrierung von β-Iron war das Verhältnis umgekehrt. Das durch Cyclisation mit Bortrifluorid hergestellte (±)-α-Iron ergab die beiden Tetrahydroderivate im Verhältnis 2 : 3. Die Semicarbazone vom Schmp. ca. 200° und ca. 160° der optisch aktiven Tetrahydro-ketone geben bei der Mischprobe mit den entsprechenden racemischen Präparaten keine Depressionen. Dagegen zeigen die entsprechenden 2,4-Dinitrophenylhydrazone Schmelzpunktserniedrigungen. Vielleicht sind sie teilweise isomerisiert.

NAVES (69) erhielt durch Hydrieren bei 100—110° unter 100 Atm. Druck aus dem einen stereoisomeren racemischen Dihydro-iron (Semicarbazon, Schmp. 143,5—144°) 60% Tetrahydro-keton (Semicarbazon, Schmp. 161—162°) und 40% des Präparates (Semicarbazon, Schmp. 199 bis 200°). Zur Erklärung dieses Phänomens nimmt er Stereomutation an. Diese wird vollständig, wenn die Hydrierung mit Platinoxyd in Eisessig bei 60° vorgenommen wird. SEIDEL, SCHINZ und RUZICKA (140) erhielten beim Hydrieren des Dihydroketons (Semicarbazon, Schmp. 171—172°) fast ausschließlich das Tetrahydroketon (Semicarbazon, Schmp. ca. 200°). Nach diesem Resultat scheint es wahrscheinlich, daß das Tetrahydro-iron (Semicarbazon, Schmp. ca. 200°) die gleiche sterische Form (2,6) besitzt, wie das Dihydro-iron (Semicarbazon, Schmp. 172°).

Das im Kern gesättigte cis-(2,3)-Dihydro-iron von BÄCHLI (3) ergab zur Hauptsache das Tetrahydroderivat (Semicarbazon, Schmp. ca. 200°). Das trans-(2,3)-Isomere lieferte dagegen ein neues Tetrahydro-keton; das bei 162—163° schmelzende Semicarbazon war mit dem oben beschriebenen Präparat (Schmp. 160°) nicht identisch.

Die verschiedenen Tetrahydro-irone und ihre Derivate sind in Tabelle 3 zusammengestellt.

Tabelle 3. Charakterisierung der verschiedenen Tetrahydro-irone. $S =$ Semicarbazon; $NP = p$-Nitrophenylhydrazon; $DNP = $ 2,4-Dinitrophenylhydrazon.

Optisch aktive Tetrahydro-irone.

RUZICKA (140):
$d_4^{25} = 0,9173$; $n_D^{25} = 1,4721$; $[\alpha]_D = + 35°$ (124)
a) S 203—204°; DNP 116—117°; NP 139—140°
b) S 160—161°; DNP 76—77°

c) S 162—164°; DNP 135—136°
d) S 154—155°; DNP 81—83°

NAVES (69):
$d_4^{20} = 0,9218$; $n_D^{20} = 1,4743$; $[\alpha]_D = + 39,7°$
S 199—200°; DNP 130—130,5°; NP 142—143°

KÖSTER (50):
S 203°; DNP 110—112°

Racemische Tetrahydro-irone aus synthetischen Ironen.

RUZICKA:
a) S 198—199°; DNP 128—129°; NP 142—143° (*140*)
b) S 160—161°; DNP 136—137°; 88—89° (*140*)
c) S 162—163°; DNP 110,5—111,5° (*3*)

NAVES (69):
a) $d_4^{20} = 0,9223$; $n_D^{20} = 1,4742$
   S 199—200°; DNP 130—130,5°; NP 142—143°
b) $d_4^{20} = 0,9143$; $n_D^{20} = 1,4715$
   S 161—162°; DNP 115—116°; *134—135*°

Weitere Ironsynthesen.

Nach CAROLL (*15*) liefert Linalool mit Acetessigester in Gegenwart geringer Mengen eines alkalischen Katalysators in einer einzigen Stufe Dihydro-pseudojonon. TAVEL (*158*) ersetzte den Acetessigester durch dessen α-Monochlorderivat und erhielt auf diese Weise α-Chlor-dihydro-pseudojonon und daraus durch Erhitzen mit Pyridin Pseudojonon. Methyl-linalool (CXCVI) ergab auf diese Weise das Chlorketon (CXCVII) und daraus Pseudoiron (CXCVIII), das zu Iron cyclisiert wurde.

(CXCVI.)　　　　　　　(CXCVII.)　　　　　(CXCVIII.) Pseudoiron
Methyl-linalool.

GRÜTTER (*29*) verwendete als Ausgangsmaterial für eine neuartige Ironsynthese das dem 2,3-Dimethyl-hepten-(2)-on-(6) (CXCIX) isomere Keton, 2-Methyl-3-methylen-heptanon-(6) (CCI), welches aus Thujon gewonnen werden kann. Daraus erhielt er nach der üblichen Methode eine mit dem Pseudoiron (CC) isomere Verbindung (CCII), die sich ebenso wie jenes zu Iron cyclisieren ließ.

(CXCIX.)　　　　　(CC.)　　　　　(CCI.)　　　　　(CCII.)
2-Methyl-3-methylen-
heptanon-(6).

Favre (25) stellte aus 2,3-Dimethyl-hepten-(2)-on-(6) sowie aus 2-Methyl-3-methylen-heptanon-(6) mit Bromessigester nach Reformatzky, Acetylierung des erhaltenen Oxyesters und Pyrolyse des Acetats (164) Methyl-geraniumsäureester (CCIII) bzw. sein Isomeres (CCIV) dar, welche sich beide zum gleichen 6-Methyl-α-cyclogeraniumsäure-ester (CCV) cyclisieren lassen. Die Cyclisation kann auch mit der entsprechenden freien Säure erfolgen. Ester (CCV) wurde mit Lithium-aluminiumhydrid zu 6-Methyl-α-cyclogeraniol (CCVI) reduziert und dieses nach Lauchenauer und Schinz (54) zu 6-Methyl-α-cyclocitral (CCVII) oxydiert. Letzteres gab bei der Kondensation mit Aceton in Gegenwart von Natriumamid oder Natriumhydrid Iron (CCVIII), das zum größten Teil aus der β-Form besteht. Es war also, wie seinerzeit bei der ersten Synthese des 6-Methyl-jonons von Ruzicka und Schinz, unter dem Einfluß des Alkalis Wanderung der Doppelbindung eingetreten.

(CCIII.) Methyl-geraniumsäureester.   (CCIV.)   (CCV.) 6-Methyl-α-cyclogeraniumsäureester.   (CCVI.) 6-Methyl-α-cyclogeraniol.

(CCVII.) 6-Methyl-α-cyclocitral.   (CCVIII.) Iron.

Rouvé und Stoll (104) stellten ε-Methylcitral unter Verwendung der Reaktion von Arens und van Dorp (1) her:

(CCIX.)   (CCX.)   (CCXI.) ε-Methylcitral.

Andere Arbeiten mit Iron.

Ruzicka, Seidel, Schinz und Tavel (130) stellten durch Reduktion von natürlichem Irongemisch nach Meerwein Irol dar, welches, wie das Ausgangsmaterial, aus der α- und γ-Form bestand, die über die Allophanate getrennt wurden.

Winter, Schinz und Stoll (169) bereiteten ein 4-Methyljonon, Rouvé und Stoll (103) ein 5-Methyljonon. Diese Synthesen erfolgten ausgehend von den Ketonen (CCXII) bzw. (CCXIV) nach der gleichen Methode, die Schinz, Ruzicka, Seidel und Tavel zur Synthese des (±)-α-Irons benutzt hatten. Das Keton (CCXIII) war praktisch reines α-, (CCXV) enthielt dagegen mehr als die Hälfte β-Verbindung. Naves, Grampoloff und Bachmann (*90*) bereiteten durch Kondensation von Methylcitral mit Methyläthylketon und Cyclisation ein Gemisch von „Dimethyljononen"; die Produkte besitzen kein geruchliches Interesse.

(CCXII.)   (CCXIII.)   (CCXIV.)   (CCXV.)

Dihydro-irol aus natürlichem Irongemisch (α) wurde von Naves und Bachmann (*85*) sowie von Seidel, Schinz und Ruzicka (*139*) durch Reduktion mit Natrium und Alkohol dargestellt.

Tetrahydro-irol wird durch direkte Hydrierung von Iron mit 3 Mol, Wasserstoff gewonnen (*90, 140*). Naves stellte aus racemischem cis- und trans-Dihydro-α-iron die entsprechenden Tetrahydro-irole in reiner Form her (*69*).

Bächli (*3*) führte durch Abbau mit Kaliumpermanganat (±)-α-Iron (CCXVI) in β-Methylisogeronsäure (CCXVII), (±)-β-Iron (CCXVIII) in β-Methyl-geronsäure (CCXIX) über. Die beiden Ketosäuren wurden durch die Semicarbazone charakterisiert.

(CCXVI.) α-Iron.   (CCXVII.) β-Methyl-isogeronsäure.   (CCXVIII.) β-Iron.   (CCXIX.) β-Methyl-geronsäure.

III. Die Riechstoffe des Veilchenblätter- und Veilchenblütenöls.

Veilchenblätteröl.

Das Veilchenblätteröl wird durch Wasserdampfdestillation der Veilchenblätter selbst oder eines aus diesen mit flüchtigen Lösungsmitteln gewonnenen Extraktes hergestellt. 1000 kg Blätter liefern ca. 20 g flüchtiges Öl. Es besteht zu circa einem Drittel aus Nonadien-(2,6)-al-(1). Dieser Aldehyd wurde zum ersten Mal von Walbaum und Rosenthal (*167*) in Form des Semicarbazons isoliert. Durch Überführung in *n*-Nonansäure wurde gezeigt, daß er geradkettige Struktur besitzt. Die Lage der Doppelbindung wurde später durch andere Autoren bestimmt. Späth und Kesztler (*145*) fanden bei der Oxydation des Semicarbazons mit Permanganat als Spaltstücke Oxal- und Bernsteinsäure. Bei der Ozonisation entstand ferner Propionaldehyd. Die Oxydation des mit 1 Mol Wasserstoff partiell hydrierten Semicarbazons lieferte *n*-Heptylsäure. Auch Ruzicka und Schinz (*116*) erhielten durch Ozonisation des Semicarbazons Propionaldehyd und, nach Nachoxydation der nicht-flüchtigen Teile mit Permanganat, Bernsteinsäure; außerdem wurde Glyoxal nachgewiesen. Von beiden Forschergruppen wurde überdies das Vorliegen der geraden Kette nochmals bestätigt. Aus diesen Resultaten ergab sich für den Aldehyd die Formel von Nonadien-(2,6)-al-(1) (CCXX).

$$CH_3—CH_2—CH=CH—CH_2—CH_2—CH=CH—CHO$$

(CCXX.) Nonadien-(2,6)-al-(1).

Ruzicka und Schinz (*118*) isolierten aus dem Veilchenblätteröl auch den primären Alkohol Nonadien-(2,6)-ol-(1) (CCXXIV), dessen Prozentgehalt im Vergleich zu demjenigen des Aldehyds jedoch sehr gering ist. Den gleichen Alkohol fanden später Takei und Ono (*156*) im Gurkensaft.

Ruzicka und Schinz isolierten aus Veilchenblätteröl noch *n*-Hexanol, je ein optisch aktives Heptenol und Octenol. Diese Stoffe sind für die Erzeugung des Veilchenblättergeruches nicht so wichtig wie das Nonadienol und das Nonadienal; es dürfte ihnen für den Gesamteffekt aber doch eine gewisse Bedeutung zukommen. Einige andere, von diesen Autoren ebenfalls isolierte Substanzen, wie Salicylsäuremethylester und aliphatische Monocarbonsäuren, sind dagegen ohne praktische Bedeutung.

Veilchenblütenöl.

Das Veilchenblütenöl ist das teuerste aller im Handel vorkommenden ätherischen Öle. Es wurde deshalb selten untersucht. Eine Prüfung von v. Soden (*143*) aus dem Jahre 1904 beschränkte sich auf die Bestimmung einiger physikalischer Daten. Sabetay und Trabaud (*132*) isolierten aus einem Öl von Parmaveilchen Eugenol.

Erst Ruzicka und Schinz (*119*) unterzogen das Veilchenblütenöl einer eingehenden Analyse. Sie isolierten in der Hauptsache die gleichen Substanzen wie aus dem Blätteröl, d. h. Nonadienal (CCXX), Nonadienol (CCXXIV) und die genannten Alkohole C_6—C_8. Der Gehalt an Nonadienal ist im Blütenöl niedriger, derjenige an Nonadienol höher als im Blätteröl. Außerdem wurde eine geringe Menge eines optisch aktiven Ketons $C_{13}H_{20}O$ aufgefunden. Dieses zeigt den wahren Geruch der Veilchenblüten, der sich von demjenigen der Jonone und der Irone deutlich unterscheidet. Es besitzt also die gleiche Bruttozusammensetzung wie die Jonone, seine Konstitution ist jedoch unbekannt. Ruzicka und Schinz bezeichneten das neue Keton, das sie durch das *p*-Bromphenylhydrazon charakterisierten, mit dem Namen „Parmon".

Synthesen von Nonadienol und Nonadienal.

RUZICKA und SCHINZ (*117*) stellten 1934 ein Nonadienol und ein Nonadienal synthetisch dar. Durch Reduktion von Sorbinsäureester (CCXXI) mit Natrium und Alkohol wurde Hexen-(3)-ol-(1) (CCXXII) gewonnen. Das diesem Alkohol entsprechende Bromid oder ein Gemisch von Chlorid und Jodid wurde über die Magnesiumverbindung nach GRIGNARD mit Acrolein zum Carbinol (CCXXIII) umgesetzt, welches bei der Allyl-umlagerung Nonadienol (CCXXIV) liefert. Letzteres wurde mit Chromsäure zum Aldehyd oxidiert.

$$CH_3-CH{=}CH-CH{=}CH-COOR \longrightarrow CH_3-CH_2-CH{=}CH-CH_2-CH_2OH \longrightarrow$$

(CCXXI.) Sorbinsäureester.   (CCXXII.) Hexen-(3)-ol-(1).

$$\longrightarrow CH_3-CH_2-CH{=}CH-CH_2-CH_2-CHOH-CH{=}CH_2 \longrightarrow$$

(CCXXIII.)

$$\longrightarrow CH_3-CH_2-CH{=}CH-CH_2-CH_2-CH{=}CH-CH_2OH$$

(CCXXIV.) Nonadien-(2,6)-ol-(1).

(CCXXV.)

Das so erhaltene Nonadienol und Nonadienal sind mit den entsprechenden Naturprodukten nicht identisch, sondern stereoisomer. Wiederholung der Synthese (*120*) mit einem natürlichen Hexenol aus japanischem Pfefferminzöl führte dagegen zu Produkten, die mit den natürlichen Verbindungen übereinstimmten. Diese Variante wurde auch von TAKEI, ONO, KUROIWA, TAKAHATA und SIMA (*157*) ausgeführt.

RUZICKA, SCHINZ und SUSZ (*120*) konnten mit Hilfe der RAMAN-Spektren zeigen, daß das natürliche Hexen-(3)-ol-(1) *cis*-, das durch Reduktion von Sorbinsäureester gewonnene Produkt *trans*-Form besitzt. Die für die Doppelbindung charakteristische Linie liegt im ersten Fall bei einer Frequenz von 1654 cm^{-1}, im zweiten bei 1670 cm^{-1}. Daraus folgt, daß das nach der ersten Synthese gewonnene Nonadienol *trans-$\Delta^{6,7}$*, das Naturprodukt und das der zweiten Synthese *cis-$\Delta^{6,7}$*-Form besitzt. Nur die *cis-$\Delta^{6,7}$*-Produkte zeigen den guten Veilchenblättergeruch.

STOLL und ROUVÉ (*153*) stellten das zur Synthese von *cis-$\Delta^{6,7}$*-Nonadienol nötige *cis*-Hexen-(3)-ol-(1) durch partielle Reduktion des entsprechenden Hexynols (CCXXV) in Gegenwart von kolloidalem Palladium dar. Das Hexynol erhielten sie auf folgendem Weg:

$$CH_3-CH_2-CO-CH_3 \longrightarrow CH_3-CH_2-CCl_2-CH_3 \longrightarrow CH_3-CH-C{\equiv}CH \longrightarrow$$

$$\longrightarrow CH_3-CH_2-C{\equiv}CMgBr \longrightarrow CH_3-CH_2-C{\equiv}C-CH_2-CH_2OH \qquad (CCXXV.)$$

Eine neue Methode von NORMANT (*93*) geht zur Darstellung von Hexen-(3)-ol-(1) von einem Handelsprodukt, dem Tetrahydrofuran (CCXXVI) aus. Das daraus erhaltene Dichlorid (CCXXVII) wird mit Äthylmagnesiumbromid ins 1-Äthyl-2-chlor-tetrahydrofuran (CCXXVIII) übergeführt. Dieses liefert bei Behandlung mit Natrium und nachfolgender Hydrolyse Hexen-(3)-ol-(1) (CCXXIX) in guter Ausbeute. Nach einer

Untersuchung von Stoll und Commarmont (*151*) besteht das so erhaltene Produkt jedoch zum größten Teil aus der *trans*-Form.

$$\text{(Tetrahydrofuran)} \longrightarrow \text{(2,3-Dichlortetrahydrofuran)} \longrightarrow \text{(1-Äthyl-2-chlortetrahydrofuran)} \longrightarrow \begin{array}{c} CH\!-\!CH_2 \\ \| \quad | \\ C_2H_5\!-\!CH \quad CH_2OH \end{array}$$

(CCXXVI.)	(CCXXVII.)	(CCXXVIII.)	(CCXXIX.)
Tetra-hydrofuran.	2,3-Dichlor-tetrahydrofuran.	1-Äthyl-2-chlor-tetrahydrofuran.	Hexen-(3)-ol-(1).

Crombie und Harper (*19*) stellten *trans*-Hexen-(3)-ol-(1) durch Reduktion der β,γ-Hexensäure mit Lithiumaluminiumhydrid her und bestimmten dessen Infrarot-Spektrum sowie dasjenige des natürlichen *cis*-Hexenols. Sondheimer (*144*) gewann *cis*-Hexenol durch partielle Hydrierung von Hexynol in Gegenwart von Palladium-calciumcarbonat und *trans*-Hexenol durch partielle Reduktion von Hexynol mit Natrium in flüssigem Ammoniak. Das Hexynol stellten sie, wie Stoll und Rouvé, aus Äthylacetylen, letzteres jedoch durch Kondensation von Äthyljodid mit Natriumacetylid her.

1947 publizierte Hunsdiecker (*41*) eine nach einem neuen Prinzip verlaufende Synthese des Nonadienols und Nonadienals. 1-Brom-penten-(2) (CCXXX) wurde mit 1,4-Dibrompenten-(2)' (CCXXXI) unter besonderen Versuchsbedingungen (*27*) zu 1-Brom-nonadien-(2,6) (CCXXXII) kondensiert und dieses über das Benzoat in Nonadienol übergeführt. Durch Oxydation wurde der Aldehyd hergestellt. Da das 1-Brom-penten-(2) durch Einwirkung von Äthylmagnesiumbromid auf Acrolein und Bromieren des erhaltenen sekundären Carbinols unter Allyl-umlagerung gewonnen worden war, mußte es nach Grédy (*28*) *trans*-Form besitzen und folglich ein *trans*-$\Delta^{6,7}$-Nonadienol liefern. Was das Dibrombuten betrifft, so geben die Autoren nicht an, welches der beiden Stereoisomeren sie verwendeten.

$$\overset{Mg}{CH_3\!-\!CH_2\!-\!CH\!=\!CH\!-\!CH_2Br} + Br\!-\!CH_2\!-\!CH\!=\!CH\!-\!CH_2Br \longrightarrow$$

(CCXXX.) 1-Brompenten-(2). (CCXXXI.) 1,4-Dibrombuten-(2).

$$\longrightarrow CH_3\!-\!CH_2\!-\!CH\!=\!CH\!-\!CH_2\!-\!CH_2\!-\!CH\!=\!CH\!-\!CH_2Br$$

(CCXXXII.) 1-Brom-nonadien-(2,6).

IV. Geruch und Konstitution.

Empfindung und Charakterisierung der Gerüche.

Die Empfindung von Gerüchen ist diejenige Sinneswahrnehmung, welche sich am schwierigsten definieren läßt und am meisten vom Individuum abhängige Schwankungen aufweist. Es ist deshalb begreiflich, daß Physiologen und Chemiker immer wieder nach Gesetzmäßigkeiten suchten, um die Mannigfaltigkeit der Erscheinungen zu klassieren und dadurch besser verständlich zu machen.

Daß eine Geruchsempfindung zustande komme, muß der Stoff, welcher wahrgenommen werden soll, eine gewisse Flüchtigkeit besitzen und außerdem wasser- und lipoidlöslich sein, um zur Nasenschleimhaut des Beobachters gelangen zu können. Dort wirken die Substanzpartikel auf die Enden der Geruchsnerven (57). Die meisten Autoren nehmen eine Reizwirkung physikalischer Art, z. B. die Erzeugung elektrischer Ströme an, welche die Geruchsempfindung hervorrufen sollen, andere halten chemische Einwirkungen für wahrscheinlicher. Nach RUZICKA (107) sollen z. B. chemische Reaktionen mit speziellen, vom menschlichen Organismus erzeugten Substanzen, den sogenannten „Osmoceptoren", stattfinden.

Ebenso wenig wie jeder farbige Stoff ein Farbstoff, ist nicht jeder riechende Stoff ein Riechstoff. In Analogie zu den für die Farbstoffe charakteristischen chromophoren Gruppen bezeichnet man bei den Riechstoffen gewisse chemische Radikale als osmophore Gruppen. Es sind dies vor allem die Reste —OH, —CHO, —CO— und —COOR. Ein Riechstoff muß mindestens eine osmophore Gruppe besitzen.

Dem Geruchssinn verwandt ist die Empfindung des Geschmackes; hier spielen sich die Reizwirkungen an den Zungen- und Gaumennerven ab. Nach PLANCK (96) ist der Geruchssinn unvergleichlich empfindlicher als der Geschmackssinn.

Zahlreich sind die Versuche der Physiologen, Psychologen und Chemiker, Ordnung in die ungeheure Fülle der verschiedenen Gerüche und Geruchsnuancen zu bringen. Man stellte sogenannte „Grundgerüche" auf, aus denen sich alle anderen Gerüche zusammensetzen lassen sollten. So spricht ZWAARDEMAKER (171) von neun, HENNING (38) von sechs, CROCKER und HENDERSON (18) von vier Grundgerüchen.

Diese Bestrebungen scheinen zum Teil einem übertriebenen Hang nach steifer Klassierung entsprungen. Das System von CROCKER und HENDERSON mit den in je 9 Intensitätsgraden wahrnehmbaren vier Grundgerüchen: blumig, sauer, brenzlig, caprylig und den sich daraus ergebenen $9 \times 9 \times 9 \times 9 = 6561$ (!) verschiedenen Geruchsnuancen geht in dieser Beziehung zweifellos etwas zu weit.

Gleiche Gerüche bei Substanzen verschiedener chemischer Konstitution.

Die Frage über die Beziehungen zwischen Konstitution und Geruch hat die Riechstoffchemiker von jeher beschäftigt. Immer und immer wieder stießen sich dabei die Systematiker an der Tatsache, daß manche Substanzen ganz verschiedenen chemischen Baues, z. B. die Nitromoschusse und das Muscon des natürlichen Moschus, den gleichen Geruch aufweisen.

Gleiche Gerüche bei Stoffen verschiedener chemischer Konstitution erscheinen uns aber heute nicht mehr erstaunlich, seit wir wissen, daß z. B. zwei Farbstoffe verschiedener Struktur, wie etwa ein Azofarbstoff und ein Blütenfarbstoff, das gleiche Rot oder Gelb zeigen, oder daß das *p,p'*-Dioxy-diäthylstilben (das sogenannte „Stilböstrol") gleiche physiologische Wirkung ausübt wie die Follikelhormone Oestron und Oestradiol.

Die „Gleichheit von Gerüchen" ist übrigens kein absoluter Begriff. Ein geübtes Riechorgan nimmt feinere Unterschiede wahr als ein ungeübtes.

Eine interessante Beobachtung zur Unterscheidung gleich oder ähnlich riechender Substanzen verschiedener Konstitution stammt von Guillot (*30*). Es sind dies die sogenannten „partiellen Anosmien". Es gibt Leute, die z. B. den Geruch des Nitromoschus nicht wahrnehmen, während sie auf Muscon normal ansprechen. Dieses Phänomen zeigt, daß Nitromoschus und Muscon auf verschiedene Nerven oder Gruppen von Nerven der Nasenschleimhaut wirken. Bei Leuten mit partiellen Anosmien ist nur die eine Gruppe dieser Nerven reaktionsfähig.

Geruch und Konstitution innerhalb der einzelnen Körperklassen.

Klare Zusammenhänge zwischen Geruch und Konstitution findet man jedoch, sobald man die Untersuchungen auf eine einzelne chemische Körperklasse beschränkt.

Bei den Nitromoschussen bewirkt z. B. Ersatz des Pseudobutyl- durch den *n*-Butylrest oder asymmetrische Anordnung der drei Nitrogruppen Verschwinden des Moschusgeruches (*48*) (*108*).

Wenn wir die Ketone mit großer Ringgliederzahl betrachten, so nehmen wir bei Zunahme der Ringgliederzahl einen allmählichen Übergang des Camphergeruchs (Cyclooctan) in den Moschusgeruch wahr, der bei C_{15} bis C_{17} ein Optimum erreicht. Gegen C_{20} hin klingt jeglicher Geruch ab. Ferner konstatieren wir innerhalb dieser Stoffklasse, daß die Absättigung einer Doppelbindung (Zibeton → Dihydrozibeton) keinen wesentlichen Einfluß auf den Geruch ausübt. Auch bei Einführung von seitlichen Methylgruppen wird keine starke Änderung des Geruches beobachtet.

$$CH_3-C=CH-CH_2-CH_2-C=CH-CH_2OH$$
$$||$$
$$CH_3CH_3$$

Geraniol.

$$CH_3-C=CH-CH_2-CH_2-CH-CH_2OH$$
$$||$$
$$CH_3CH_3$$

Citronellol.

Für eine zusammenfassende Darstellung dieses Gebietes vgl. Lederer (*54a*).

In anderen Körperklassen kann dagegen eine Doppelbindung für die Geruchsnuance von Wichtigkeit sein. Dies ist z. B. bei den aliphatischen Monoterpenalkoholen und den entsprechenden Aldehyden der Fall. Das doppelt ungesättigte Geraniol und seine α,β-Dihydroverbindung, das Citronellol, sind beides wichtige, aber stark voneinander verschiedene Riechstoffe. Das gleiche gilt für die entsprechenden Aldehyde Citral und Citronellal. Durch Ringschluß verschlechtert sich bei Geraniol und bei Citral der Geruch. Cyclogeraniol und Cyclocitral riechen campherartig.

Geruch und Konstitution bei den Veilchenriechstoffen.

a) *Sogenannte „Veilchenketone"*.

Während bei Geraniol und Citral der Ringschluß eine Geruchsverschlechterung zur Folge hat, ist es beim Citrylidenaceton (Pseudo-jonon) gerade umgekehrt: das geruchlich indifferente aliphatische Keton geht durch Cyclisation in die nach Veilchen riechenden Jonone über. Die β-Form riecht stärker, in Verdünnung süßlicher als die α-Form. Die γ-Form ist bis jetzt nicht bekannt. Das gesättigte Tetrahydro-jonon (*cis-* und *trans*-Form) zeigt Zedernholzgeruch. Die durch Absättigen der Seitenkette erhältlichen Dihydro-jonone (α- und β-Form) besitzen eine halb veilchenartige, halb zedernholzähnliche Nuance. Dem Dihydro-γ-jonon (XXV, S. 154), das in den flüchtigen Bestandteilen des grauen Ambras vorkommt (*54a*), ist eine besonders angenehme Form dieses Übergangsgeruches eigen.

Bei Ersatz der Gruppe $=CH-CO-CH_3$ durch $=CH-CHOH-CH_3$, $=CH-CHO$ oder $=CH-COOR$ verschwindet der Veilchengeruch. Modifikation der Seitenkette in der Art von $=CH-CO-CH_2-CH_3$ oder

$$=\overset{\overset{\textstyle CH_3}{\textstyle \cdot}}{C}-CO.CH_3$$

durch Eintritt einer Methylgruppe erzeugt eine Nuancenverschiebung, die allgemein als günstig empfunden wird: die Methyljonone sind besonders geschätzte Riechstoffe und kommen unter verschiedenen Phantasienamen, wie „Iraldeïn", „Iralia" usw. in den Handel.

Einen noch tiefer greifenden Einfluß vermag eine zusätzliche Methylgruppe im Kern auszuüben, vorausgesetzt, daß sie an der richtigen Stelle sitzt. 6-Methyl-α- und 6-Methyl-β-jonon sind nichts anderes als α- bzw. β-Iron (CLXXVIII—CLXXIX, S. 174). Der Geruch der Irone ist kräftiger und feiner als derjenige der Jonone und wird dementsprechend höher geschätzt.

Das isomere 4-Methyl- und das 5-Methyl-jonon (*169, 103*) sind dagegen ohne parfümistisches Interesse. Das erste der beiden Produkte (α-Form) riecht zwar kräftig, jedoch wie Jonon ohne die Note des Irons; das zweite (Gemisch von α und β) zeigt einen nur schwachen, zwischen dem der Jonone und des Irons liegenden Geruch.

α- und γ-Iron (CLXXVIII, bzw. CLXXX, S. 174) riechen ähnlich, doch wird die Nuance der α-Form allgemein als kräftiger und angenehmer beurteilt. Die Nuancenverschiebung vom α- zum β-Iron entspricht etwa derjenigen beim Übergang vom α- zum β-Jonon. Auch in bezug auf die bei der Reduktion zum Dihydro- und Tetrahydroketon auftretende Geruchsänderung herrscht bei den Ironen Parallelität mit den Erscheinungen bei den Jononen.

Über den Einfluß der feineren Strukturunterschiede — relative Lage der Substituenten an den Kohlenstoffatomen 2 und 6, geometrische Isomerie an der Doppelbindung der Seitenkette — auf den Geruch läßt sich vorläufig noch kein abschließendes Urteil fällen, da die verschiedenen Forschergruppen nicht in allen Punkten einig sind. NAVES (67) findet, daß für das Zustandekommen des typischen Irongeruchs cis-Lage der Butenon-kette und der Methylgruppe in 6-Stellung unerläßlich sei. Nach SEIDEL, SCHINZ und RUZICKA (140) ist jedoch der Unterschied zwischen dem cis-(2,6)- und trans-(2,6)-Isomeren nicht so groß, wie NAVES meint.

b) *Nonadienal und Nonadienol.*

Auch beim Nonadien-(2,6)-ol-(1) (CCXXIV, S. 189), dem sogenannten „Veilchenblätteralkohol", liegen interessante Beobachtungen über die Abhängigkeit des Geruchs von der Struktur vor (120). Bei dieser Substanz ist die *cis-trans*-Isomerie an der Doppelbindung zwischen den Kohlenstoffatomen 6 und 7 von großer Bedeutung. Nur die *cis*-Form zeigt den blumigen, mild-ausgeglichenen Geruch, während das *trans*-Isomere zwar ebenfalls stark, aber unangenehm scharf riecht und als Riechstoff völlig unbrauchbar ist.

Das gleiche gilt für das Nonadien-(2,6)-al-(1) (CCXX, S. 188).

Dieses Phänomen zeigt sich übrigens auch beim Hexen-(3)-ol-(1), dem sogenannten „Blätteralkohol", der für die Darstellung des synthetischen Nonadienols dient. Er kommt in natürlicher Form verestert im japanischen Pfefferminzöl und frei in den grünen Teeblättern vor. Die Doppelbindung dieses Alkohols entspricht derjenigen zwischen den Kohlenstoffatomen 6 und 7 des Nonadienols. Nur *cis*-Hexenol zeigt den angenehmen, vollen Blättergeruch, während das *trans*-Isomere widerlich stechend riecht.

Die Doppelbindung zwischen den Kohlenstoffatomen 2 und 3 des Nonadienols besitzt beim synthetischen und wahrscheinlich auch beim natürlichen Produkt *trans*-Lage. Die zugehörige *cis*-Form ist bisher nicht bekannt.

Das mit dem Nonadienol isomere 7-Methyl-octadien-(2,6)-ol-(1) („Apogeraniol", CCXXXIII) riecht weniger stark als Nonadien-(2,6)-ol-(1) und besitzt nicht die Nuance der Veilchenblätter (168).

$$CH_3 > C=CH-CH_2-CH_2-CH=CH-CH_2OH$$
$$CH_3$$

(CCXXXIII.) Apogeraniol.

V. Zur Biogenese der Veilchenriechstoffe.

Hypothese über die Biogenese der Terpenverbindungen im allgemeinen.

Terpenverbindungen finden sich vor allem in den Blüten, Blättern und zum Teil auch in den Wurzeln von Pflanzen und bisweilen sogar in tierischen Organen. Über die Art ihrer Bildung und die Rolle, welche ihnen hier zukommt, weiß man bis heute noch nichts Sicheres. Es existieren einige Erklärungsversuche, über deren völlig hypothetischen Charakter man sich aber klar sein muß.

Da alle Terpenverbindungen formal aus Isoprenresten zusammengesetzt sind, wurde oft vermutet, daß diese Stoffe in den Pflanzen entweder aus Isopren selbst oder aus sauerstoffhaltigen Abkömmlingen desselben aufgebaut werden.

Verschiedene Forscher haben in vitro solche von Isopren ausgehende Synthesen verwirklicht.

So erhielt WAGNER-JAUREGG (*165*, *166*) durch Kondensation von Isopren mittels Eisessig-Schwefelsäure ein Substanzgemisch, aus dem nach Verseifung Geraniol, α-Terpineol, 1,4- und 1,8-Cineol isoliert werden konnten. FAVORSKY und LEBEDEVA (*24*) gewannen durch Schütteln von Dimethyl-vinyl-carbinol (sogen. „tert. Isoprenalkohol") mit 20%iger Schwefelsäure Linalool, Geraniol und Terpinhydrat. LENNARTZ (*56*) stellte aus Isopren durch katalytische Hydratation und Kondensation in Eisessig ein Gemisch von Acetaten her, das nach Destillation und Verseifung primären Isoprenalkohol („Prenol"), Linalool, Geraniol und zwei andere primäre, sowie einen tertiären Monoterpenalkohol unbekannter Konstitution, ferner Farnesol und einen aliphatischen Diterpenalkohol lieferte.

Das Isopren könnte, wie FAVORSKY meint, aus Leucin stammen, das sich durch Abbau von Proteinen bildet. KREMERS (*52*), ASCHAN (*2*), LENNARTZ (*56*) und andere Autoren nehmen dagegen an, daß sich durch Kondensation von Aceton und Acetaldehyd das Aldol (CCXXXIV) und daraus durch Wasserabspaltung β-Methyl-crotonaldehyd („Isopren-aldehyd", CCXXXV) bilde. Aldolkondensation von zwei Molekülen des letzteren führt zu Dehydrocitral (CCXXXVI). Durch biologische Dehydrierungen, Reduktionen, Oxydationen, Cyclisationen usw. oder durch Kombination

(CCXXXIV.)

(CCXXXV.) „Isopren-aldehyd".   (CCXXXVI.) Dehydrocitral.

13*

von verschiedenen solcher Reaktionen könnten sich aus dem Dehydrocitral die verschiedenartigsten Terpenverbindungen bilden. Die Ausgangsstoffe Aceton und Acetaldehyd sind als intermediäre Stoffwechselprodukte bekannt und finden sich in den Pflanzenzellen. Für die Kondensation von Aceton und Acetaldehyd in der obengenannten Weise muß man eine besonders katalytische Wirkung von Enzymen annehmen, da sie sonst das Aldol (CCXXXVII) liefert. Eine Kondensation zum Aldol (CCXXXIV) konnte in vitro mit Hilfe von Glykokoll erzielt werden (53). Statt mit Acet-

$$CH_3—CHOH—CH_2—CO—CH_3$$
(CCXXXVII.)

aldehyd könnte die Kondensation des Acetons auch mit Brenztraubensäure vor sich gehen (39), welche ebenfalls ein normales Stoffwechselprodukt der Pflanze ist. Die so entstehenden Umsetzungsprodukte würden durch Verlust von Kohlendioxyd die gleichen Verbindungen liefern, die man durch Kondensation von Aceton mit Acetaldehyd direkt erhält.

Außer Aldolkondensationen könnten nach DIELS (20) bei der Biogenese der Terpene auch Dienkondensationen in irgendeiner Form zur Wirkung kommen. Dieser Autor stellte in vitro Limonen und Camphoren aus Isopren bzw. Myrcen nach der von ihm und ALDER aufgefundenen Reaktion unter sogenannten „physiologischen Bedingungen" her.

EMDE (23) sowie HALL (36) vermuten ferner, daß Terpene durch Abbau von Zuckern und anderen Kohlehydraten entstehen. Nach EMDE soll dabei der Lävulinsäure, einem Abbauprodukt der Zucker, als Aufbauelement der Terpene eine besonders wichtige Rolle zukommen. Diese Auffassung ist insofern bemerkenswert, weil sie als Baustein eine geradlinige Verbindung annimmt (34), die erst durch geeignete Anordnung Ketten mit seitlichen Methylgruppen in regelmäßigen Abständen liefern soll.

Hypothesen über die Biogenese der Veilchenriechstoffe.

In Fortsetzung der Theorie der Aldolkondensation ließe sich auch die Entstehung der Veilchenketone erklären. Die Bildung von Pseudojonon aus Citral und Aceton oder von Jonon aus Cyclocitral und Aceton, wie sie im Laboratorium ausgeführt wird, ist nichts anderes als eine Aldolkondensation mit gleichzeitiger Dehydratation. In diesem Falle handelt es sich um eine normale Kondensation (Reaktion an der Aldehydgruppe), im Gegensatz zur Bildung von Aldol (CCXXXIV) aus Aceton und Acetaldehyd (Reaktion der Ketogruppe).

Für die Entstehung des Irons, welches eine zusätzliche Methylgruppe in Stellung 6 des Cyclohexenringes besitzt, müßte man nach dieser Anschauungsweise als erste Stufe eine Kondensation von Propionaldehyd (statt Acetaldehyd) mit Aceton annehmen. Dadurch erhielte man, nach

Abspaltung von Wasser aus dem entstandenen Aldol (CCXXXVIII) den α,β-Dimethylcrotonaldehyd (CCXXXIX). Dieser würde durch weitere Aldolkondensation mit einem Mol β-Methylcrotonaldehyd Dehydromethylcitral (CCXL) liefern, welches auf oben geschilderte Weise Iron liefern würde.

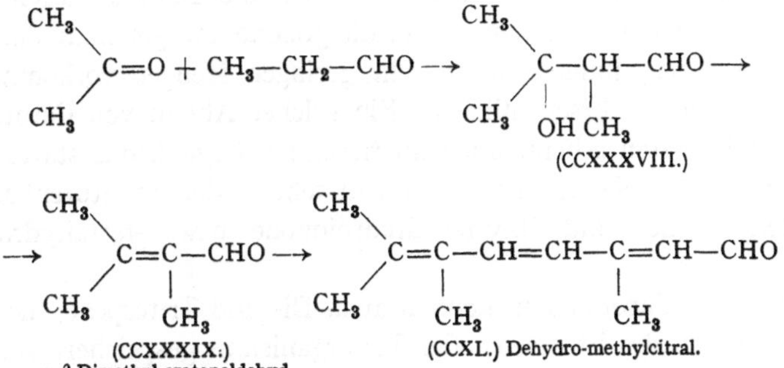

$$\underset{CH_3}{\overset{CH_3}{>}}C=O + CH_3-CH_2-CHO \rightarrow \underset{\underset{OH\ CH_3}{|\ \ |}}{\underset{CH_3}{\overset{CH_3}{>}}C-CH-CHO} \rightarrow$$

(CCXXXVIII.)

$$\rightarrow \underset{CH_3}{\overset{CH_3}{>}}\underset{\underset{CH_3}{|}}{C=C}-CHO \rightarrow \underset{CH_3}{\overset{CH_3}{>}}C=\underset{\underset{CH_3}{|}}{C}-CH=CH-\underset{\underset{CH_3}{|}}{C}=CH-CHO$$

(CCXXXIX.)
α,β-Dimethyl-crotonaldehyd.

(CCXL.) Dehydro-methylcitral.

Das Iron ist die einzige bisher bekannte Terpenverbindung mit dieser supplementären Methylgruppe. Doch ist ein solcher Stoff aus einer anderen Gruppe von chemischen Verbindungen bekannt, nämlich das Ergosterin (CCXLI), ein Steroid der Hefepilze (109). Auch hier befindet sich eine Methylgruppe an dem der Isopropylgruppe benachbarten Kohlenstoffatom.

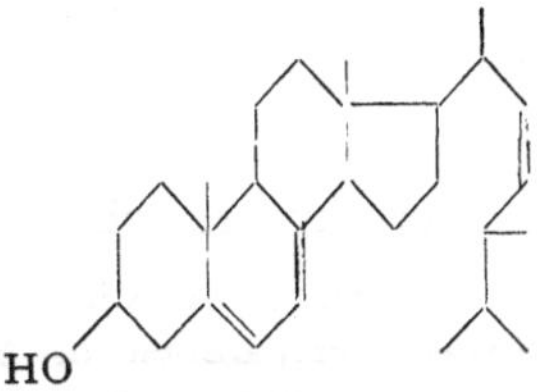

HO

(CCXLI.) Ergosterin.

Es läßt sich leicht auch für die Bildung des Nonadien-(2,6)-ol-(1) und des entsprechenden Aldehyds ein analoges Bildungsschema aufstellen. Die Kondensation von Propion- und Acetaldehyd zum Aldol (CCXLII) müßte wiederum unter der Einwirkung eines besonderen Enzyms erfolgen, da sie normalerweise das Aldol (CCXLIII) liefert. Das aus (CCXLII) entstehende Penten-(2)-al-(1) (CCXLIV) würde durch weitere Kondensation mit Acetaldehyd — unter den genannten besonderen Bedingungen

$$CH_3-CH_2-CHOH-CH_2-CHO \rightarrow CH_3-CH_2-CH=CH-CHO \rightarrow$$
(CCXLII.)   (CCXLIV.) Penten-(2)-al-(1).

$$\rightarrow CH_3 \quad CH_2-CH=CH-CH-CH-CHO$$
(CCXLV.) Heptadienal.

$$\downarrow$$

$$CH_3-CHOH-\underset{\underset{CH_3}{|}}{CH}-CHO \qquad CH_3-CH_2-CH=CH-CH=CH-CH=CH-CHO$$
(CCXLIII.)   (CCXLVI.) Nonatrienal.

— ein Heptadienal (CCXLV) und dieses durch Wiederholung der gleichen Reaktion schließlich Nonatrienal (CCXLVI) liefern. Letzteres könnte durch partielle biologische Hydrierung in Nonadienal und unter gleichzeitiger Reduktion der Aldehydgruppe in Nonadienol übergehen.

Für die Veilchenketone ist allerdings noch eine prinzipiell andere Möglichkeit in Betracht zu ziehen, nämlich die Bildung durch Abbau von Carotinoiden in der Pflanze. Für die Jonone und Jononabkömmlinge, die in der Natur immer nur in sehr geringen Mengen vorkommen, ist diese sogar viel wahrscheinlicher. Ein solcher Abbau von Carotinoiden zu Jononderivaten scheint auch im tierischen Organismus stattzufinden und dürfte u. a. die Erklärung zur Entstehung der im Stutenharn aufgefundenen Oxo- und Oxy-tetrahydrojonone bzw. -tetrahydro-jonole geben.

Außer den Carotinoiden können auch Di- und Triterpene, die sich in anderen Teilen des Pflanzen- oder Tierorganismus gespeichert vorfinden, durch Abbau zu Jononderivaten führen. Ein Beispiel hierfür ist das im grauen Ambra vorkommende Dihydro-γ-jonon, dessen Entstehung aus dem Triterpenalkohol Ambreïn, welcher die Hauptmenge des rohen Ambras ausmacht, augenfällig ist.

Ob man auch für das Iron eine solche Entstehungsweise annehmen kann, wissen wir nicht. Wir kennen bisher kein Carotinoid mit zusätzlicher Methylgruppe am β-Jononkern; es läßt sich vorläufig auch nicht beurteilen, ob vielleicht eine Methylierung in der Pflanze stattfindet.

Literaturverzeichnis.

1. Arens, J. F. and D. A. van Dorp: A New Method for the Synthesis of α,β-Unsaturated Aldehydes. Preparation of β-Methylcinnamic Aldehyde, Citral and β-Ionylidene-acetaldehyde. Recueil Trav. chim. Pays-Bas **67**, 973 (1948).

2. Aschan, O.: Hypothetisches bezüglich der biologischen Entstehung der Harzsäuren. Chemiker-Ztg. **49**, 689 (1925).

3. Bächli, P.: Doppelbindungs- und Stereo-Isomerie bei den Ironen. Dissert. Eidgen. Techn. Hochschule Zürich. 1950.

4. Bächli, P., C. F. Seidel, H. Schinz u. L. Ruzicka: Veilchenriechstoffe. 31. Mitt. Weitere Versuche mit natürlichem Iron. Helv. chim. Acta **32**, 1744 (1949).

5. Barbier, H.: Extension de cycles dans la série hydroaromatique. I. Essais avec la 1,1,3-triméthyl-cyclohexyl-méthylamine-5 (dihydro-isophoryl-méthylamine). Helv. chim. Acta **23**, 519 (1940).

6. — Extension de cycles dans la série hydroaromatique. II. Essais avec la 1,1,3-triméthyl-cyclohexyl-méthylamine-2 (dihydro-cyclogéranyl-méthylamine). Helv. chim. Acta **23**, 524 (1940).

7. — Les produits d'oxydation du 1,1,4-triméthyl-cycloheptène. Helv. chim. Acta **23**, 1477 (1940).

8. Beckett, Ch. W., K. S. Pitzer and R. Spritzer: The Thermodynamic Properties and Molecular Structure of Cyclohexane, Methylcyclohexane,

Ethylcyclohexane and the Seven Dimethylcyclohexanes. J. Amer. chem. Soc. **69**, 2488 (1947).

9. BEETS, M. G. J.: Contribution to the Chemistry of Ionones. II. A Convenient Laboratory Synthesis of the Isomethylionones. Recueil Trav. chim. Pays-Bas **69**, 307 (1950).

10. BIELIG, H. J. u. A. HAYOSIDA: Über das Verhalten des β-Jonons im Tierkörper. Biochemische Hydrierungen VIII. Hoppe-Seyler's Z. physiol. Chem. **266**, 99 (1940).

11. BOGERT, M. T.: The Structure of Vitamine A and the Synthesis of Ionenes. Science (New York) **76**, 475 (1932).

12. BOGERT, M. T. and P. M. APFELBAUM: The Synthesis of 1,1,2,6-Tetramethyltetralin and the Constitution of Irene. J. Amer. chem. Soc. **60**, 930 (1938).

13. BOSSHARD, A.: Synthetische Arbeiten auf dem Gebiete der Veilchenriechstoffe. Dissert. Eidgen. Techn. Hochschule Zürich. 1946.

14. BÜCHI, G., K. SEITZ u. O. JEGER: Veilchenriechstoffe. 29. Mitt. Über die Dehydrierung von β- und α-Jonon. Helv. chim. Acta **32**, 39 (1949).

15. CAROLL, M. F.: Addition of α,β-Unsaturated Alcohols to the Active Methylene Group. I. The Action of Ethyl-aceto-acetate on Linalool and Geraniol. J. chem. Soc. (London) **1940**, 704.

16. CHERBULIEZ, E. et H. HEGAR: Sur la synthèse de méthyl-pseudo-ionones de structure définie. Helv. chim. Acta **15**, 191 (1932).

17. COLOMBI, L.: Stereoisomeria dei metilchetoni α,β-non saturi e sintesi del biciclosesquilavandulolo. Dissert. Eidgen. Techn. Hochschule Zürich. 1949.

18. CROCKER, E. C. and L. F. HENDERSON: Analysis and Classification of Odors. An Effort to Develop a Workable Method. Amer. Perfumer **22**, 325 (1927).

19. CROMBIE, L. and ST. H. HARPER: "Leaf Alcohol" and the Stereochemistry of the cis- and the trans-n-Hex-3-en-1-ols and n-Pent-3-en-1-ols. J. chem. Soc. (London) **1950**, 873.

20. DIELS, O.: Bedeutung der Diensynthese für Bildung, Aufbau und Erforschung von Naturstoffen. Fortschr. Chem. organ. Naturstoffe **3**, 1 (1939).

21. DIELS, O. u. K. ALDER: Synthesen in der Hydroaromatischen Reihe. I. Anlagerung von „Dien"-Kohlenwasserstoffen. Liebigs Ann. Chem. **460**, 98 (1927).

22. — — Synthesn in der Hydromatischen Reihe. III Synthese von Terpenen, Camphern, hydroaromatischen und heterocyclischen Systemen. Liebigs Ann. Chem. **470**, 62 (1929).

23. EMDE, H.: Mitteilungen zur Biosynthese (VI, VII, VIII). Helv. chim. Acta **14**, 881 (1931).

24. FAVORSKY, A. E. et A. J. LEBEDEVA: Synthèses dans le domaine des terpènes en partant de l'acétylène. Bull. Soc. chim. France [5] **6**, 1347 (1939).

25. FAVRE, H.: Synthèse d'une d,l-γ Irone. Thèse, École Polytechn. Féd. Zürich, (à paraître).

26. FIRMENICH, G.: Recherches synthétiques dans le domaine de l'irone. Synthèse du géranyl-géraniol. Thèse, École Polytechn. Féd. Zürich. 1940.

27. GILMAN, H. et J. H. M. McGLYMPHY: La préparation indépendante des bromures d'allylmagnésium. Bull. Soc. chim. France [4] **43**, 1322 (1928).

28. GRÉDY, B.: Effet Raman et Chimie organique. Comparaison des spectres Raman de quelques alcools éthyléniques et de plusieurs de leurs dérivés »cis« et »trans«. Bull. Soc. chim. France [5] **3**, 1101 (1936).

29. GRÜTTER, H.: I. Zur Bildung der Allo-cycloterpene. II. Eine neue Synthese des Lavandulols. III. Eine neue Synthese des Irons. Dissert. Eidgen. Techn. Hochschule Zürich. 1950.

30. Guillot, M.: Physiologie des sensations. Anosmies partielles et odeurs fondamentales. C. R. hebd. Séances Acad. Sci. **226**, 1307 (1948).

31. Günthard, H. u. L. Ruzicka: Veilchenriechstoffe. 27. Mitt. Infrarotspektren in der Iron- und Jononreihe. Helv. chim. Acta **31**, 642 (1948).

32. — — Veilchenriechstoffe. 33. Mitt. Infrarot- und Raman-Spektren in der Ironreihe. Helv. chim. Acta **32**, 2125 (1949).

33. Günthard, H., L. Ruzicka, H. Schinz u. C. F. Seidel: Veilchenriechstoffe. 34. Mitt. Über das Mengenverhältnis von γ- und α-Iron im Irisöl. Helv. chim. Acta **32**, 2198 (1949).

34. Haagen-Smit, A. J.: The Chemistry, Origin and Function of Essential Oil in Plant Life. The Essential Oils I, 17 (Günther, E.). New York. 1948.

35. Haarmann & Reimer, Holzminden: Verfahren zur Darstellung von Jonon. D. R. P. 129027; Friedländer, Fortschr. der Teerfarbenfabrikation, Bd. VI, S. 1246.

36. Hall, J. A.: Relationships in Phytochemistry: Condensation of Sugar Derivatives. Chem. Reviews **20²**, 305 (1937).

37. Henbest, H. B.: Synthesis of Dehydro-β-ionone. Nature (London) **161**, 481 (1948).

38. Henning, H.: Der Geruch. Frankfurt. 1916.

39. Hesse, G., H. Eilbracht u. F. Reicheneder: Das krystallisierte Caliotropis-Harz. III. Mitt. über afrikanische Pfeilgifte. Liebigs Ann. Chem. **546**, 233 (1941).

40. Hibbert, H. and L. T. Cannon: Condensation of Citral With Ketones and Synthesis of Some New Ionones. J. Amer. chem. Soc. **46**, 119 (1924).

41. Hunsdiecker, H.: Eine neue Synthese eines Nonadien-(2,6)-ols-(1) und Nonadien-(2,6)-als-(1). Ber. dtsch. chem. Ges. **80**, 137 (1947).

42. Jeger, O., O. Dürst u. L. Ruzicka: Zur Kenntnis der Triterpene. 117. Mitt. Die Konstitution des Ambreïns. Helv. chim. Acta **30**, 1859 (1947).

43. Kandel, J.: Hydrogénation catalytique de la β-ionone: dihydro-β-ionone, dihydro-β-ionol; quelques dérivés de l'α et du β-ionol. C. R. hebd. Séances Acad. Sci. **205**, 994 (1937).

44. Karrer, P. u. J. Benz: Anlagerung von Acetylen an β-Jonon. Helv. chim. Acta **31**, 390 (1948).

45. Karrer, P. u. P. Ochsner: Dehydrierungen von β- (und α)-Jonon mittels Bromsuccinimid. Helv. chim. Acta **31**, 2093 (1948).

46. Karrer, P. u. H. Stürzinger: Mono- und Diepoxyde des α- und β-Jonons und Umwandlungsprodukte dieser Verbindungen. Helv. chim. Acta **29**, 1829 (1946).

47. Kilby, B. A. and F. B. Kipping: Synthesis of Irone. J. chem. Soc. (London) **1939**, 435.

48. Kohn, G.: Die Riechstoffe. Braunschweig. 1924.

49. Köster, H.: Über die Doppelbindung des Jononrings. III. Über Riechstoffe der Jonongruppe. Ber. dtsch. chem. Ges. **77**, 553 (1944).

50. — Über β-Iron. IV. Über Riechstoffe der Jonongruppe. Ber. dtsch. chem. Ges. **77**, 559 (1944).

51. — Über die Kondensation des Citrals mit Ketonen. V. Über Riechstoffe der Jonongruppe. Ber. dtsch. chem. Ges. **80**, 248 (1947).

52. Kremers, R. E.: Biogenesis of Oil of Peppermint. J. biol. Chemistry **50**, 31 (1922).

53. Kusin, A. u. N. Newrajewa: Über die Wege der biochemischen Synthese von Kohlenstoffketten vom Isoprentypus. Biokhimiya **6**, 261 (1941) [Chem. Zbl. **1942** II, 41].

54. LAUCHENAUER, A. u. H. SCHINZ: Eine neue Modifikation der Reaktion nach OPPENAUER. Helv. chim. Acta **32**, 1265 (1949).

54a. LEDERER, E.: Odeurs et parfums des animaux. Fortschr. Chem. organ. Naturstofte **6**, 87 (1950).

55. LEDERER, E., V. PRELOG et R. SCHNEIDER: Recherches sur des extraits d'organes. 14ème comm. Sur la présence de dérivés de l'ionone dans le castoréum. Helv. chim. Acta **31**, 2133 (1948).

56. LENNARTZ, TH.: Synthese von aliphatischen Terpenalkoholen aus Isopren. Ber. dtsch. chem. Ges. **76**, 831 (1943).

57. MARTINET, J.: Propriétés organoleptiques et constitution chimique. Traité de Chimie Organique **11**, 595 (V. GRIGNARD). 1936.

58. MERLING, G.: Vertahren zur Darstellung von Δ^4-Cyclogeraniumsäure [1,3,3-Trimethyl-cyclohexen-(4)-carbonsäure-(2)]. D. R. P. 175587 (1905); FRIEDLÄNDER, Fortschr. der Teerfarbenfabrikation, Bd. VIII, S. 1298.

59. MERLING, G. u. R. WELDE: Synthese von Veilchenriechstoffen. Liebigs Ann. Chem. **366**, 119 (1909).

60. NAEF, M. et Cie., Genf: Herstellung eines Alkyl-cyclohexenaldehyds. S. P. 136907 (1928) [Chem. Zbl. **1930** II, 2053].

61. — Herstellung von Alkylcyclohexenaldehyden, Alkylcyclohexencarbonsäuren sowie deren Ester. F. P. 672025 (1929) [Chem. Zbl. **1930** I, 2796].

62. NAVES, Y. R.: Études sur les matières végétales volatiles. XLVI. Sur le dédoublement de la d,l-α-ionone. Helv. chim. Acta **30**, 769 (1947).

63. — (im Namen von H. BOHNSACK): Sur la présence d'ionones dans les produits végétaux. Helv. chim. Acta **30**, 956 (1947).

64. — Études sur les matières végétales volatiles. LII. Sur la constitution et sur la synthèse de l'irone. Helv. chim. Acta **30**, 2221 (1947); pli cacheté (10. VI. 1943).

65. — Études sur les matières végétales volatiles. LIX. Cyclisation de l'acide α,α,β-triméthylsubérique en triméthyl-1,1,2-cycloheptanone-7. Helv. chim. Acta **31**, 156 (1948).

66. — Études sur les matières végétales volatiles. LXII. Sur la constitution et sur la synthèse d'irones. Helv. chim. Acta **31**, 893 (1948).

67. — Études sur les matières végétales volatiles. LXIV. Synthèse de la cis-(2,6)-d,l-α-irone. Helv. chim. Acta **31**, 1103 (1948).

68. — Études sur les matières végétales volatiles. LXVII. Sur la séparation et sur l'isolement des α- et γ-irones et sur une α-irone jusqu'à présent inédite. Helv. chim. Acta **31**, 1280 (1948).

69. — Études sur les matières végétales volatiles. LXIX. Sur les tetrahydroirones. Helv. chim. Acta **31**, 1871 (1948).

70. — Études sur les matières végétales volatiles. LXX. Présence de néo-α-irone dans des essences d'iris. Helv. chim. Acta **31**, 1876 (1948).

71. — Études sur les matières végétales volatiles. LXXIII. Sur la cis-(2,6)-d-γ-irone, dont la phényl-4-semicarbazone F. 178—179°. Helv. chim. Acta **31**, 2047 (1948).

72 — Études sur les matières végétales volatiles. LXXXI. Sur les cis-(2,6)-α-irones. Helv. chim. Acta **32**, 611 (1949).

73. — Études sur les matières végétales volatiles. LXXXIV. Sur la nomenclature triviale des irones. Helv. chim. Acta **32**, 969 (1949).

74. — Études sur les matières végétales volatiles. LXXXVI. Présence d'un mélange d'ionones et de dihydro-ionones (isoirone de HAARMANN et REIMER) dans l'huile essentielle de racine de costus. Helv. chim. Acta **32**, 1064 (1949).

75. Naves, Y. R.: Études sur les matières végétales volatiles. LXXXVII. Sur l'évaluation des γ-irones et de leurs dérivés par ozonolyse. Helv. chim. Acta **32**, 1151 (1949).

76. — Études sur les matières végétales volatiles. XCI. Sur les conversions du méthyl-3-linalol en méthyl-3-citrals et des nérolidols en farnésals. Helv. chim. Acta **32**, 1798 (1949).

77. — Études sur les matières végétales volatiles. XCVI. Sur les évaluations des γ-irones et de leurs dérivés par ozonolyse et sur des irones isomères. Helv. chim. Acta **32**, 2186 (1949).

78. — The Structures and Syntheses of Irones. Proc. Scient. Sect. Toilet Good Assoc. No. 11 (1949).

79. Naves, Y. R. et P. Ardizio: Études sur les matières végétales volatiles. LXVIII. Absorption de dérivés des ionones et des irones dans l'ultraviolet moyen. Helv. chim. Acta **31**, 1427 (1948).

80. — — Études sur les matières végétales volatiles. LXXV. Synthèse d'aldéhydes (tetraméthyl-cyclohexényl)-formiques en vue de la préparation d'isoirones. Helv. chim. Acta **31**, 2252 (1948).

81. — — Études sur les matières végétales volatiles. LXXVI. Synthèses d'isomères des irones. Helv. chim. Acta **31**, 2256 (1948 .

82. — — Études sur les matières végétales volatiles. LXXVII. Sur les dihydroionones. Helv. chim. Acta **32**, 206 (1949).

83. Naves, Y. R. et P. Bachmann: Études sur les matières végétales volatiles. XXVI. Contribution à la connaissance des ionones. Helv. chim. Acta **26**, 2151 (1943).

84. — — Études sur les matières végétales volatiles. XXIX. Isolement d'un isomère tricyclique des ionones. Helv. chim. Acta **27**, 645 (1944).

85. — — Études sur les matières végétales volatiles. LIII. Contribution à la connaissance de l'irone. Helv. chim. Acta **30**, 2222 (1947); pli cacheté (10. VI. 1943).

86. — — Études sur les matières végétales volatiles. LIV. Absorption spectrale et spectres Raman de l'irone, du dihydroirol et de substances voisines. Helv. chim. Acta **30**, 2233 (1947); pli cacheté (10. VI. 1943).

87. — — Études sur les matières végétales volatiles. LV. Effet Raman dans la série des méthyl-3-linalol, méthyl-3-citrals et méthyl-6-ionones. Helv. chim. Acta **30**, 2241 (1947); pli cacheté (19. VI. 1944).

88. — — Études sur les matières végétales volatiles. LXXIX. Spectres Raman et structures des irones et de leurs dérivés. Helv. chim. Acta **32**, 394 (1949).

89. Naves, Y. R. et A. V. Grampoloff: Études sur les matières végétales volatiles. XCVIII. Synthèse de méthyl-6-ionones par l'intermédiaire des méthyl-6-citrals. Études sur les matières végétales volatiles. XCIX. Synthèse des méthyl-6-citrals. Helv. chim. Acta **32**, 2552 (1949); plis cachetés (24. II. et 6. III. 1944).

90. Naves, Y. R., A. V. Grampoloff et P. Bachmann: Études sur les matières végétales volatiles. L. Études dans les séries des méthyl-3-linalols, des méthyl-3-citrals et des méthyl-6-ionones. Helv. chim. Acta **30**, 1599 (1947).

91. Naves, Y. R. et G. R. Parry: Études sur les matières végétales volatiles. XLV. Présence d'ionones dans l'essence concrète de *Boronia megastigma* Nees. Helv. chim. Acta **30**, 419 (1947).

92. Naves, Y. R., O. Schwarzkopf et A. D. Lewis: Études sur les matières végétales volatiles. XLVIII. Sur les époxydes des ionones. Helv. chim. Acta **30**, 880 (1947).

93. NORMANT, H.: Méthode générale de préparation des alcools primaires β-éthyléniques. C. R. hebd. Séances Acad. Sci. **226**, 733 (1948).

94. PALFRAY, L., S. SABETAY et J. KANDEL: Hydrogénation catalytique de l'α-ionone: ionol, dihydroionol, tétrahydroionol, dihydroionone, tetrahydroionone. C. R. hebd. Séances Acad. Sci. **203**, 1376 (1936).

95. PENFOLD, A. R. u. L. W. PHILLIPS: Öl von *Boronia megastigma.* Proc. Roy. Soc. W. Australia **14**, 1 (1927/28) [Referat: SCHIMMEL, Berichte **1929**, 6].

96. PLANCK, R.: Der gegenwärtige Stand der Klassifikation und objektiven Bewertung von Geschmacks- und Geruchsempfindungen. Die Chemie **57**, 154 (1944).

97. PRELOG, V. u. H. FRICK: Veilchenriechstoffe. XXV. Mitt. Über die beiden diastereomeren Tetrahydrojonane. Helv. chim. Acta **31**, 417 (1948).

98. PRELOG, V., J FÜHRER, R. HAGENBACH u. H. FRICK: Untersuchungen über Organextrakte. 11. Mitt. Über die Konstitution des Alkohols $C_{13}H_{20}O$ aus dem Harn trächtiger Stuten. Helv. chim. Acta **30**, 113 (1947).

99. PRELOG, V., J. FÜHRER, R. HAGENBACH u. R. SCHNEIDER: Untersuchungen über Organextrakte. 13. Mitt. Über die lsolierung von Jononderivaten aus dem Harn trächtiger Stuten. Helv. chim. Acta **31**, 1799 (1948).

100. PRELOG, V. u. H. L. MEIER: Untersuchungen über Organextrakte und Harn. 18. Mitt. Über die biochemische Oxydation von β-Jonon im Tierkörper. Helv. chim. Acta **33**, 1276 (1950).

101. PRELOG, V. u. B. VATERLAUS: Untersuchungen über Organextrakte und Harn. 16. Mitt. Über die Konstitution des ungesättigten Ketons $C_{13}H_{18}O$ aus dem Harn trächtiger Stuten. Helv. chim. Acta **32**, 2082 (1949).

102. — — Untersuchungen über Organextrakte und Harn. 19. Mitt. Über die Konstitution der Oxyketone E und G aus dem Harn von trächtigen Stuten. Helv. chim. Acta **33**, 1725 (1950).

103. ROUVÉ, A. et M. STOLL: Produits à odeur de violette. 20ème comm. Synthèse de la 5-méthyl-ionone. Helv. chim. Acta **30**, 2216 (1947).

104. — — Une nouvelle préparation du triméthyl-2,3,6-octadiène-2,6-al-8 (ε-méthyl-citral). Helv. chim. Acta **33**, 2019 (1950).

104a. ROYALS, E. E.: Cyclization of Pseudoionone by Acid Reagents. Ind. Engng. Chem. **38**, 546 (1946).

105. RUZICKA, L.: Über die Herstellung von Polymethyl-cyclohexenonen des Irontypus. Helv. chim. Acta **2**, 144 (1919).

106. — Über die Beziehungen zwischen den Jononen und Iron. Helv. chim. Acta **2**, 352 (1919).

107. — Die Grundlagen der Geruchschemie. Chemiker-Ztg. **44**, 93, 129 (1920).

108. — Sur les produits à odeur de musc et les cycles à grand nombre de chaînons. Bull. Soc. chim. France [4] **43**, 1145 (1928).

109. — Sur l'irone et autres substances à odeur de violette. Ind. Parf. **3**, 374 (1948).

110. RUZICKA, L. u. W. BRUGGER: Veilchenriechstoffe. 10. Mitt. Über die vermeintliche Ironsynthese von MERLING und WELDE. J. prakt. Chem. **158**, 125 (1941).

111. RUZICKA, L., G. BÜCHI u. O. JEGER: Veilchenriechstoffe. 24. Mitt. Synthese des Dihydro-γ-ionons, eines Abbauproduktes des Ambreïns. Helv. chim. Acta **31**, 293 (1948).

112. RUZICKA, L., O. DÜRST u. O. JEGER: Zur Kenntnis der Triterpene. 113. Mitt. Überführung des Triterpens Ambreïn in ein Abbauprodukt des Diterpens Manool. Helv. chim. Acta **30**, 353 (1947).

113. Ruzicka, L. u. F. Lardon: Zur Kenntnis der Triterpene. 105. Mitt. Über das Ambreïn, einen Bestandteil des grauen Ambra. Helv. chim. Acta **29**, 912 (1946).

114. Ruzicka, L. u. E. A. Rudolph: Beiträge zur Kenntnis der Dehydrierung mit Schwefel und des dehydrierenden Abbaus mit Braunstein und Schwefelsäure. Helv. chim. Acta **10**, 915 (1927).

115. Ruzicka, L. u. H. Schinz: Veilchenriechstoffe. 9. Mitt. Über die Synthese der kernmethylierten Jononhomologen 1,1,3,6-Tetramethyl-2-(buten-2^1-ylon-2^3-cyclohexen-(2 bzw. 3.) Helv. chim. Acta **23**, 959 (1940).

116. — — Veilchenriechstoffe. 4. Mitt. Über das Veilchenblätteröl. Zur Konstitution des Veilchenblätteraldehyds, Nonadien-(2,6)-al-(1). Helv. chim. Acta **17**, 1592 (1934).

117. — — Veilchenriechstoffe. 5. Mitt. Synthese des Veilchenblätteraldehyds, Nonadien-(2,6)-al-(1) bzw. eines Stereoisomeren desselben. Helv. chim. Acta **17**, 1602 (1934).

118. — — Veilchenriechstoffe. 6. Mitt. Die nichtaldehydischen Bestandteile des Veilchenblätteröls. Helv. chim. Acta **18**, 381 (1935).

119. — — Veilchenriechstoffe. 13. Mitt. Über das ätherische Öl der Veilchenblüten. Helv. chim. Acta **25**, 760 (1942).

120. Ruzicka, L., H. Schinz u. B. P. Susz: Veilchenriechstoffe. 14. Mitt. Zur Stereoisomerie von Hexen-(3)-ol-(1), Nonadien-(2,6)-ol-(1) und Nonadien-(2,6)-al-(1). Helv. chim. Acta **27**, 1561 (1944).

121. Ruzicka, L. u. C. F. Seidel: Veilchenriechstoffe. 22. Mitt. Über den Abbau von Dihydro-iron und Dihydro-iran. Helv. chim. Acta **31**, 160 (1948).

122. — — Über die flüchtigen Anteile des grauen Ambra. 2. Mitt. Über ein Oxyd $C_{13}H_{22}O$, einen Oxyaldehyd $C_{17}H_{30}O_2$ und ein Keton $C_{13}H_{20}O$. Helv. chim. Acta **33**, 1285 (1950).

123. Ruzicka, L., C. F. Seidel u. W. Brugger: Veilchenriechstoffe. 17. Mitt. Über den Abbau des Tetrahydro-irons. Helv. chim. Acta **30**, 2168 (1947).

124. Ruzicka, L., C. F. Seidel u. G. Firmenich: Veilchenriechstoffe. 11. Mitt. Physikalische Konstanten und kristallisierte Derivate des Irons vor und nach dem Kochen mit verd. Schwefelsäure. Helv. chim. Acta **24**, 1434 (1941).

125. Ruzicka, L., C. F. Seidel u. M. Pfeiffer: Über die flüchtigen Bestandteile des grauen Ambra. 1. Mitt. Isolierung von Dihydro-γ-ionon. Helv. chim. Acta **31**, 827 (1948).

126. Ruzicka, L., C. F. Seidel u. H. Schinz: Veilchenriechstoffe. 3. Mitt. Über die Bruttoformel und einige Umsetzungen des Irons. Helv. chim. Acta **16**, 1143 (1933).

127. Ruzicka, L., C. F. Seidel, H. Schinz u. M. Pfeiffer: Veilchenriechstoffe. 12. Mitt. Über den Abbau des Irons mit Ozon und Chromsäure. Helv. chim. Acta **25**, 188 (1942).

128. — — — — Veilchenriechstoffe. 15. Mitt. Die Konstitution des Irons. Helv. chim. Acta **30**, 1807 (1947); pli cacheté (28. VI. 1946).

129. — — — — Veilchenriechstoffe. 26. Mitt. Synthese des 1,1,7-Trimethyl-3-(buten-3^1-ylon-3^3)-cyclohepten-(2), eines 7-gliedrigen Analogons des β-Irons. Helv. chim. Acta **31**, 422 (1948).

130. Ruzicka, L., C. F. Seidel, H. Schinz u. Ch. Tavel: Veilchenriechstoffe. 23. Mitt. Einige ergänzende Versuche zur Konstitutionsaufklärung und Synthese des Irons. Helv. chim. Acta **31**, 257 (1948).

131. Sabetay, S.: Sur la présence de la β-ionone dans un produit naturel. Bull. Soc. chim. France [4] **45**, 1169 (1929).

132. SABETAY, S. et L. TRABAUD: Sur la teneur en eugénol libre de l'essence de fleur de violette de Parme. C. R. hebd. Séances Acad. Sci. **209**, 843 (1939).

133. SCHÄPPI, G. u. C. F. SEIDEL: Veilchenriechstoffe. 18. Mitt. Einige Synthesen aus dem Gebiet der Abbauprodukte von Tetrahydro-iron. Helv. chim. Acta **30**, 2199 (1947).

134. SCHINZ, H.: Sur la constitution de l'irone et la synthèse de la d,l-α-irone. Ind. Parf. **3**, 37 (1948).

135. — Struttura dell'irone e sintesi del d,l-α-irone. Riv. ital. Essenze, Profumi Piante officin. **30**, 148 (1948).

136. — The Constitution of Irone and the Synthesis of d,l-α-Irone. Perfum. essent. Oil Rec. **39**, 107 (1948).

137. SCHINZ, H., L. RUZICKA, C. F. SEIDEL et CH. TAVEL: Produits à odeur de violette. 16ème comm. Synthèse de la d,l-α-irone. Helv. chim. Acta **30**, 1810 (1947); pli cacheté (28. VI. 1946).

138. SCHINZ, H., C. F. SEIDEL u. L. RUZICKA: Veilchenriechstoffe. 35. Mitt. Bemerkungen zur Publikation von Y. R. NAVES über die Bestimmung des γ-Irons. Helv. chim. Acta **32**, 2560 (1949).

139. SEIDEL, C. F., H. SCHINZ u. L. RUZICKA: Veilchenriechstoffe. 30. Mitt. Die Bestimmung der semicyclischen Methylengruppe bei Substanzen der Iron- und Jononreihe. Helv. chim. Acta **32**, 1739 (1949).

140. — — — Veilchenriechstoffe. 32. Mitt. Über synthetisches α- und β-Iron, ihre Dihydro- und Tetrahydroderivate. Helv. chim. Acta **32**, 2102 (1949).

141. SKITA, A.: Hydrierungen mit Platinmetallen als Katalysator. VI. Ber. dtsch. chem. Ges. **45**, 3312 (1912).

142. SOBOTKA, H., E. BLOCH, H. CAHNMANN, E. FELDBAU and E. ROSEN: Studies on Ionone. II. Optical Resolution of d,l-α-Ionone. J. Amer. chem. Soc. **65**, 2061 (1943).

143. SODEN, H. v.: Über ätherische Öle, welche durch Extrahieren frischer Blüten mit flüchtigen Lösungsmitteln gewonnen werden (ätherische Blütenextraktöle). J. prakt. Chem. **69**, 256 (1904).

144. SONDHEIMER, F.: Studies of Compounds Related to Natural Perfumes. I. Concerning cis- and trans-Hex-3-en-1-ols. J. chem. Soc. (London) **1950**, 877.

145. SPÄTH, E. u. F. KESZTLER: Die Konstitution des Veilchenblätteraldehyds. Ber. dtsch. chem. Ges. **67**, 1496 (1934).

146. SPRECHER, H. v.: Synthetische Versuche auf dem Gebiete des Irons. Dissert. Eidgen. Techn. Hochschule Zürich. 1943.

147. STOLL, M.: L'évolution récente de la Chimie des produits à odeur de violette. Chim. et Ind. **63**, 131 (1950).

148. STOLL, M., P. BOLLE et L. RUZICKA: Synthèses d'époxydes hydroaromatiques. IV. Epoxy-3,4-tetrahydro-ionone et ses produits de transformation. Helv. chim. Acta **33**, 1502 (1950).

149. — — — Synthèses d'époxydes hydroaromatiques. V. Essai en vue de la préparation de l'anhydride du triméthyl-1,1,3-butanonyl-2²-cyclohexanol-4. Lactone de l'acide dihydroxy-3,4-dihydro-homocyclogéranique. Helv. chim. Acta **33**, 1510 (1950).

150. — — — Synthèses d'époxydes hydroaromatiques. VI. Anhydride de l'hydroxy-4-triméthyl-1,1,3-butanonyl-2²-cyclohexane. Helv. chim. Acta **33**, 1515 (1950).

151. STOLL, M. et A. COMMARMONT: Au sujet de l'hexène-3-ol-1 de H. NORMANT. Helv. chim. Acta **32**, 597 (1949).

152. — — Transformation du linalol et du γ-méthyl-linalol en citral et ε-méthyl-citral. Helv. chim. Acta **32**, 1354 (1949).

153. Stoll, M. et A. Rouvé: Synthèse du *cis-β,γ*-hexénol (hexénol naturel). Helv. chim. Acta **21**, 1542 (1938).

154. Stoll, M., L. Ruzicka et C. F. Seidel: Synthèses d'époxydes hydroaromatiques. III. Anhydride de la tétrahydro-ionol-3-one-2[3]. Helv. chim. Acta **33**, 1245 (1950).

155. Stoll, M. et W. Scherrer: Produits à odeur de violette. VIII. Synthèse du 1,1,6-Trimethyl-3-(butène-3[1]-ylone-3[3])-cycloheptène. Helv. chim. Acta **23**, 941 (1940).

156. Takei, S. u. M. Ono: Blätteröl. III. Riechstoffe der Gurken (*Cucumis sativus* L.). Chem. Zbl. **1939** II, 3705.

157. Takei, S., M. Ono, Y. Kuroiwa, T. Takahata u. T. Sima: Blätteröl. II. Einige aus Blätteralkohol synthetisch gewonnene Riechstoffe. Chem. Zbl. **1938** II, 3696.

158. Tavel, Ch.: Nouvelles synthèses de la pseudoionone et de la pseudoirone. Helv. chim. Acta **33**, 1266 (1950). Pli cacheté (28. X. 1948).

159. Tiemann, F.: Über die Veilchenketone und die in Beziehung dazu stehenden Verbindungen der Citral- (Geranial-) Reihe. Ber. dtsch. chem. Ges. **31**, 808 (1898).

160. — Über die Zerlegung des Jonons in zwei Spielarten, α- und β-Jonon. Ber. dtsch. chem. Ges. **31**, 867 (1898).

161. Tiemann, F. u. P. Krüger: Über Veilchenaroma. Ber. dtsch. chem. Ges. **26**, 2675 (1893).

162. Tischer, J.: Carotinoide der Süßwasseralgen. II. Über die Carotinoide und die Bildung von Jonon in Trentepohlia nebst Bemerkungen über den Gehalt dieser Algen an Erythrit. Hoppe-Seyler's Z. physiol. Chem. **243**, 103 (1936).

163. Verley, A.: Sur la synthèse de l'irone de Tiemann. Bull. Soc. chim. France [5] **2**, 1205 (1935).

164. Vodoz, Ch. A. et H. Schinz: Quelques observations sur la préparation de l'acide géranique synthétique et de ses isomères cycliques. Helv. chim. Acta **33**, 1313 (1950).

165. Wagner-Jauregg, Th.: Die Dimerisation des Isoprens. Liebigs Ann. Chem. **488**, 176 (1931).

166. — Synthesen von Terpenen aus Isopren. Liebigs Ann. Chem. **496**, 52 (1932).

167. Walbaum, H. u. A. Rosenthal: Über das ätherische Öl der grünen Veilchenblätter. J. prakt. Chem. [2] **124**, 55 (1930).

168. Willimann, L.: Einige Reaktionen an acetalisierten Ketoestern und über ein Apogeraniol. Dissert. Eidgen. Techn. Hochschule Zürich. 1949.

169. Winter, M., H. Schinz et M. Stoll: Produits à odeur de violette. 19ème comm. Synthèse de la 4-méthyl-ionone. Helv. chim. Acta **30**, 2213 (1947).

170. Young, W. G., St. J. Cristol, L. J. Andrews and S. L. Lindenbaum: Polyenes. II. The Purification of β-Ionone. J. Amer. chem. Soc. **66**, 855 (1944).

171. Zwaardemaker, H.: Die Physiologie des Geruchs. Frankfurt. 1895.

(Eingelaufen am 27. Dezember 1950.)

Neuere Entwicklungen auf dem Gebiete der Flechtenstoffe.

Von Y. Asahina, Tokyo.

Einleitung.

Seit dem Erscheinen der zusammenfassenden Übersicht des Verfassers in diesen *Fortschritten* (*2*) (1939) wurden zahlreiche neue Angaben auf dem Gebiete der Flechtenstoffe bekannt, so daß es als gerechtfertigt erscheint, den vorliegenden Beitrag zu veröffentlichen.

Überblickt man die erwähnte neuere Entwicklung, so ist vor allem die wesentliche Erweiterung der Körperklassen der Diphenylenoxyd- und der Xanthon-Derivate hervorzuheben. Früher stand die Usninsäure als Vertreter einer besonderen Gruppe vereinzelt da. Nachdem aber ihre von ROBERTSON (*30, 37, 41*) und von SCHÖPF (*83, 84*) ermittelte Struktur nun als endgültig angesehen werden darf, reihen wir diese Säure in die Unterklasse der Diphenylenoxyde ein.

Durch die Untersuchungen des irischen Forschers NOLAN (*31, 32, 38, 43, 58, 59, 64, 66—69, 96*) wurden zahlreiche chlorhaltige Depsidone aufgeklärt. Bei einer weiteren Aufarbeitung der Krustenflechten darf man mit Bestimmtheit eine künftige, wesentliche Ausdehnung dieses Gebietes erwarten.

In Indien haben SESHADRI und Mitarbeiter (*74—78, 87, 88*) mit einem Reichtum von experimentellem Material zur Förderung der Flechtenchemie beigetragen. Ihre überraschende Beobachtung, daß das Erythrin optisch aktiv ist, wurde bisher von keiner Seite in Frage gestellt. Ferner hat SESHADRI (*87*) eine geistvolle Theorie der Biogenese von Flechtenstoffen aufgestellt (p. 234).

Was praktische Anwendungen betrifft, hat die Usninsäure als ein besonders gegen Tuberkelbazillen wirksames Antibioticum Aufmerksamkeit erregt, obzwar sich leider ihre Heilkraft in vivo noch nicht mit Sicherheit feststellen ließ.

Die im früheren Übersichtsartikel des Verfassers (2) befolgte systematische Anordnung ist im wesentlichen beibehalten worden; die dort behandelten Verbindungen werden untenstehend nur kurz erwähnt, falls keine Neubearbeitung derselben vorliegt.

I. Verbindungen der Fettreihe.

1. Fettsäuren und Laktone.

Über die Protolichesterinsäure (und Derivate), Nephromopsinsäure, Nephrosterinsäure und Nephrosteransäure, Caperatsäure und Roccellsäure liegen keine neueren Angaben vor.

Rangiformsäure.

Die Rangiformsäure wurde zuerst von PATERNÒ (*71*) in *Cladonia rangiformis* HOFFM. aufgefunden. Sie bildet, aus Benzol umgelöst, farblose, glänzende Blättchen vom Schmp. 104—106°. Nach HESSE (*45*) besitzt die Rangiformsäure die Zusammensetzung $C_{21}H_{36}O_6$ und ist ein Monomethylester der Tricarbonsäure Norrangiformsäure (Schmp. 119°). In neuerer Zeit haben ASAHINA und SASAKI (*17*) dieselbe Säure aus *Cladonia mitis* SANDST. isoliert und die Zusammensetzung als $C_{21}H_{38}O_6$ richtiggestellt. Früher hatte HESSE (*47, 48, 50*) aus *Cl. sylvatica* (L.) HARM. eine Säure vom Schmp. 100° (Sylvatsäure genannt) isoliert. Da damals *Cl. mitis* von *Cl. sylvatica* nicht scharf unterschieden war, hatte HESSE als Ausgangsmaterial ein Gemisch der beiden Pflanzen benutzt, und demnach dürfte seine Sylvatsäure nicht ganz reine Rangiformsäure gewesen sein.

Nun haben ASAHINA und SASAKI unter der Annahme, daß die Norrangiformsäure eine Desoxy-norcaperatsäure sei, Heptadecan-1,2,3-tricarbonsäure (Formel I; $R = C_{14}H_{29}$) synthetisiert, jedoch ist es ihnen nicht gelungen, die letztere in die stereochemisch einheitlichen Komponenten zu spalten. Endlich hat AOKI (*1*) die synthetische Tricarbonsäure durch Erhitzen im Vakuum in ein Anhydrid-Gemisch übergeführt, welches sich durch Chloroform in zwei Bestandteile spalten ließ (Schmp. 82—83° bzw. 67—69°).

Bei der gleichen Operation liefert die natürliche Norrangiformsäure ebenfalls zwei Anhydride (Schmp. 82—83° bzw. 66—69°), die sich mit den aus synthetischen Säuren erhaltenen als identisch erwiesen. Dadurch wurde es klar, daß die Norrangiformsäure, deren Monomethylester die

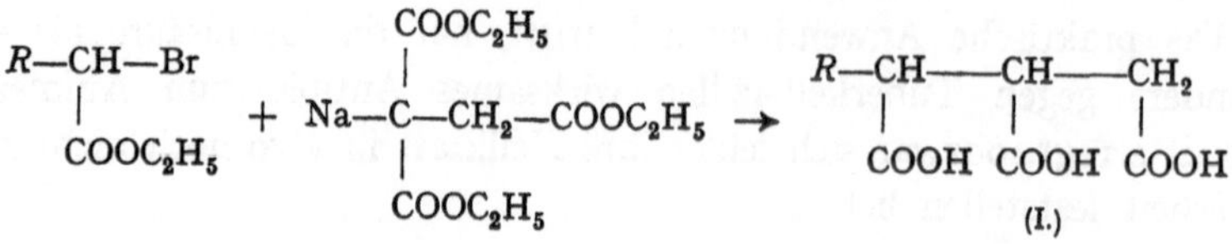

$$R-CH-Br \atop COOC_2H_5 + Na-C-CH_2-COOC_2H_5 \atop COOC_2H_5 \atop COOC_2H_5 \to R-CH-CH-CH_2 \atop COOH \ COOH \ COOH$$

(I.)

$$HOOC-CH-C=CH_2 \atop CH_3(CH_2)_{12}-CH \quad CO \atop O$$

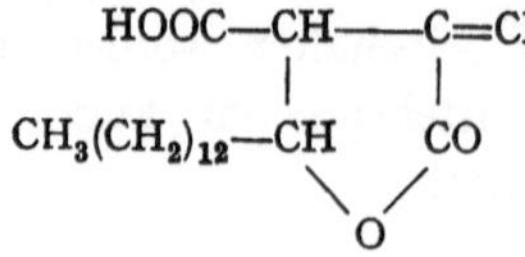

Protolichesterinsäure.

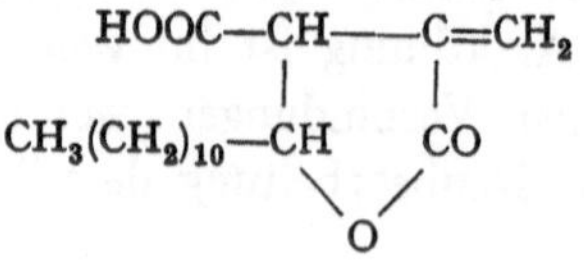

$$HOOC-CH-C=CH_2 \atop CH_3(CH_2)_{10}-CH \quad CO \atop O$$

Nephrosterinsäure.
(Mit gesättigter Doppelbindung: Nephrosteransäure.)

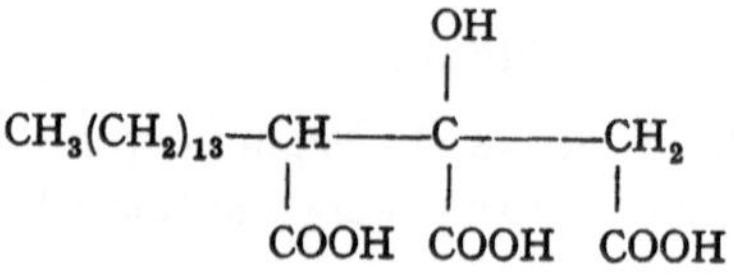

$$OH \atop CH_3(CH_2)_{13}-CH-C-CH_2 \atop COOH \ COOH \ COOH$$

Nor-caperatsäure.

$$CH_3(CH_2)_{11}-CH-CH-CH_3 \atop COOH \ COOH$$

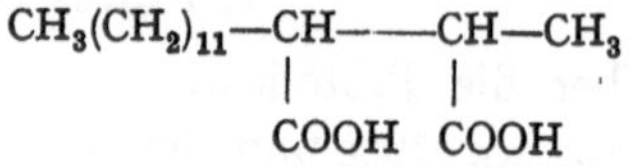

Roccellsäure.

Rangiformsäure bildet, eine Form der Heptadecan-1,2,3-tricarbonsäure ist (Desoxy-norcaperatsäure).

2. Triterpenoide*.

Zeorin.

Das Zeorin scheint in zwei Formen aufzutreten. Sowohl Paterno (*70*) als auch Hesse (*46, 51*) geben dafür den Schmelzpunkt 230—231° an, während Zopf (*110*) sowie Asahina und Akagi (*5*) einen weit höheren Schmelzpunkt fanden (247—254° bzw. 253°). Später hat Takeda** bei weiterem Umlösen des Asahina-Akagischen Präparates aus Benzol einen Schmelzpunkt 236° erhalten. Eine neuere, einschlägige Untersuchung von Asahina und Yosioka (*28*) läßt sich wie folgt zusammenfassen.

Bei mildem Acetylieren liefert Zeorin ein Monoacetat, während bei energischem Eingriff Anhydro-zeorinacetat gebildet wird. Das letztere liefert beim Verseifen mit kochender alkoholischer Salzsäure Zeorinin, welches von Zopf (*110, 112*) direkt aus Zeorin durch Einwirkung von alkoholischer Salzsäure gewonnen wurde. Bei der katalytischen Hydrierung liefert Zeorinin den gesättigten Alkohol Desoxyzeorin, der beim Oxydieren in ein Keton, Desoxyzeorinon, übergeführt wird. Wenn auch das letztere kein Oxim bildet, so läßt es sich durch Reduktion

---

* Diese Gruppe wurde früher (*3*) als „Neutrale, gegen Alkali indifferente Substanzen" bezeichnet. — Über das Gebiet der Triterpene hat jüngst Jeger (*55*) in diesen Fortschritten zusammenfassend berichtet.

** Privatmitteilung von K. Takeda.

mit Natrium und Alkohol in das Desoxyzeorin rückverwandeln. Folglich enthält das Zeorin je eine tertiäre und eine sekundäre Hydroxyl-Gruppe. Wird Zeorininacetat mit Chromsäure oxidiert, so entsteht ein Produkt $C_{32}H_{52}O_4$, welches ein Oxim liefert. Diese Beobachtung ist wohl darauf zurückzuführen, daß eine Methylen-Gruppe des Zeorininacetats zu CO oxidiert und zugleich 1 Molekül Wasser an die Doppelbindung angelagert wurde.

Durch Einwirkung von Wasserstoffsuperoxyd auf Zeorinin oder dessen Acetat erhält man Zeorinin-Oxyd $C_{30}H_{50}O_2$ (bzw. das Acetat). Beim Kochen mit alkoholischer Salzsäure liefern die beiden Oxyde Iso-dehydro-zeorinin, $C_{30}H_{48}O$, welches wahrscheinlich durch hydrolytische Öfnung des Oxydrings und darauffolgende Abspaltung von 2 Molekülen Wasser zustande kommt. Bei der katalytischen Hydrierung geht es in Zeorinin über.

Leukotylin

ist, wie früher erwähnt (2), ein Bestandteil der japanischen Flechte *Parmelia laucotyliza* NYL. Es besitzt die Zusammensetzung $C_{30}H_{52}O_3$ und drei Hydroxyl-Gruppen. Durch Erhitzen mit Selen wird es zu Agathalin abgebaut [ASAHINA und AKAGI (5)].

Ursolsäure.

Die Ursolsäure bildet ein spezifisches Stoffwechselprodukt der *Pirolaceen, Ericaceen* und einiger anderer Phanerogamen; ihr Vorkommen in Flechten war früher unbekannt. Nachdem aber Zeorin und Leukotylin sich als Triterpenoide erwiesen hatten, so ist es weniger überraschend, daß die Ursolsäure von BREADEN, KEANE und NOLAN (31) in *Cladonia sylvatica* (L.) HARM. entdeckt worden ist. Es scheint uns indessen als fraglich, ob das Ausgangsmaterial der genannten Forscher wirklich reine *Cladonia sylvatica* war. Nach ZOPF (111) enthält *Cl. sylvatica* D-Usninsäure, Fumarprotocetrarsäure und einen farblosen Körper, der allem Anschein nach keine Perlatolinsäure ist. Demgegenüber fand ZOPF (116) in *Cl. impexa* (unter *Cl. laxinscula* und *Cl. condensata*) zweifellos L-Usninsäure und eine farblose Säure, die später von ASAHINA (4) sowie von BREADEN, KEANE und NOLAN (31) als Perlatolinsäure erkannt wurde. Die letztgenannten Forscher geben an, daß ihr Ausgangsmaterial (angeblich *Cl. sylvatica*) eine schwach linksdrehende Usninsäure, Perlatolinsäure, Ursolsäure und daneben eine kleine Menge Fumarprotocetrarsäure enthält. Vermutlich haben sie ein Gemisch von *Cl. impexa* und *Cl. sylvatica* in den Händen gehabt.

Ein Vorkommen von racemischer Usninsäure in Flechten wurde bisher nicht beobachtet.

3. Zuckeralkohole.

Seit dem Erscheinen des erwähnten Berichtes (2) wurde das Vorkommen folgender Zuckeralkohole in Flechten sichergestellt:

D-Arabit in *Parmelia latissima* Fée (69), in *Cladonia impexa* (L.) Harm. (31) und in *Ramalina tayloriana* Zahlbr. (88).

Mannit in *Buellia canescens* DNotrs (66), in *Lecanora sordida* Th. Fr. (57), in *Pertusaria concreta* Nyl. f. *Westringi* Nyl. (32) und in *Lecanora parella* Ach. (64).

II. Verbindungen der Benzolreihe.

1. Pulvinsäure-Derivate.

Vulpinsäure, Pinastrinsäure und Calycin (o-Oxy-pulvinsäureanhydrid) wurden früher behandelt (2).

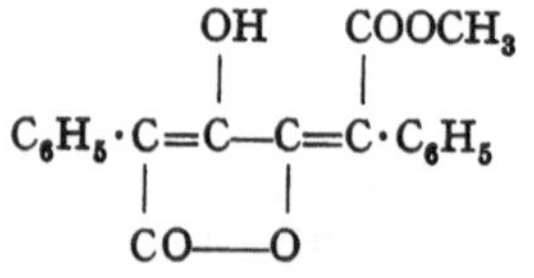

$$
\begin{array}{cc}
\text{OH} \quad \text{COOCH}_3 & \text{OH} \quad \text{COOCH}_3 \\
| \qquad | & | \qquad | \\
C_6H_5 \cdot C{=}C{-}C{=}C \cdot C_6H_5 & (p)CH_3O \cdot C_6H_4 \cdot C{=}C{-}C{=}C \cdot C_6H_5 \\
| \qquad | & | \qquad | \\
CO{-}O & CO{-}O
\end{array}
$$

Vulpinsäure. 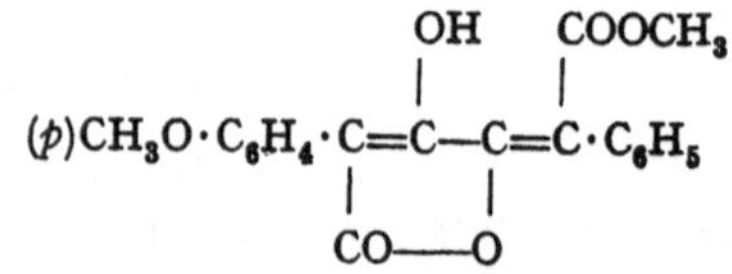   Pinastrinsäure.

2. Diphenylenoxyd-Derivate*.

Didymsäure.

Diese Säure wurde von Asahina zuerst in *Cladonia didyma* Fée mikrochemisch aufgefunden und später in verschiedenen rotfrüchtigen Cladonien nachgewiesen. Nogami (65) hat die Zusammensetzung $C_{22}H_{26}O_5$ festgestellt und je eine Carboxyl-, Methoxyl- und Hydroxyl-Gruppe ermittelt. Ferner hat dieser Autor, ausgehend von Didymsäure, durch trockene Destillation Decarboxy-didymsäure $C_{21}H_{26}O_3$, durch Kochen mit Jodwasserstoff Decarboxy-nor-didymsäure und beim Oxydieren mit Permanganat n-Capronsäure erhalten.

Die Didymsäure zeichnet sich durch die blaue Chlorkalk-Reaktion aus, welche beim Strepsilin lange bekannt war. Diese eigentümliche Reaktion wurde auch beim 2,7-Dioxy-4,5-diphenylenoxyd beobachtet (6). Wie nun Shibata (91) gezeigt hat, liefert die Didymsäure (II) bei der Kalischmelze unter Verlust der Carboxyl-Gruppe und des Äther-Methyls eine Verbindung $C_{20}H_{26}O_4$, welche vier Hydroxyle aufweist. Die zwei hinzugetretenen Hydroxyle werden wohl durch Sprengung einer Oxyd-Sauerstoffbrücke gebildet sein. Ferner ergab das Ultraviolett-Spektrum der Didymsäure und ihrer Derivate eine vollkommene Übereinstimmung mit demjenigen von 2,7-Dioxy-4,5-dialkyl-diphenylenoxyden. Ein end-

---

* Früher: „Gruppe II. Usninsäure."

gültiger Strukturbeweis der Didymsäure wurde von Shibata (*91*) nach dem folgenden Schema 1 herbeigeführt.

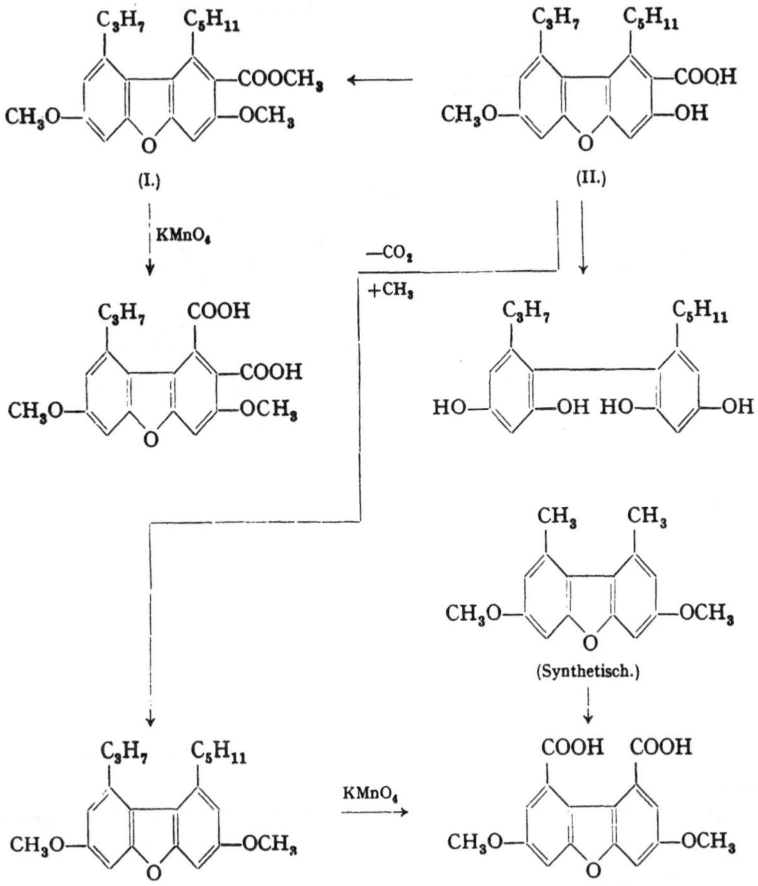

Dimethylester, Schmp. 189°. (Keine Schmp.-Depression in der Mischprobe.)

Schema 1. Strukturbeweis für die Didymsäure (Shibata).

Strepsilin.

Zopf (*114*) hat gezeigt, daß die blaugrüne Chlorkalk-Reaktion der Flechte *Cladonia strepsilis* Wain. durch den Gehalt an Strepsilin bedingt ist. Shibata (*92*) stellte die Zusammensetzung des Strepsilins $C_{15}H_{10}O_5$ fest und wies zwei Hydroxyle sowie eine Lakton-Bindung nach (III). Auch hat er die Lichtabsorptionskurve des Strepsilins mit denen der Didymsäure und der 3,7-Dioxy-diphenylenoxyd-Derivate als übereinstimmend gefunden. Der endgültige Konstitutionsbeweis wurde gleichfalls von Shibata (*92*) erbracht (Schema 2, S. 214).

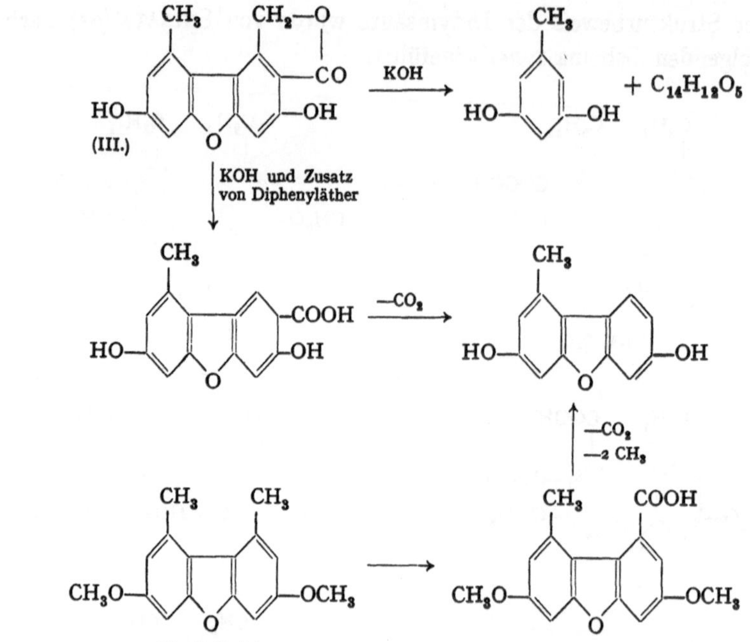

Schema 2. Strukturbeweis für Strepsilin (SHIBATA).

Usninsäure.

Wie ASAHINA und Mitarbeiter (*25, 26, 106*) gefunden haben, liefern Mono- und Diacetyl-usninsäure beim Erhitzen mit absolutem Alkohol Mono- bzw. Diacetyl-usninsäure-äthoxylat. Das letztere wird einerseits durch Erhitzen mit Essigsäure in Diacetyl-acetousnetinsäure-äthyl-ester (V) und anderseits beim Erhitzen mit Wasser im Rohr in Diacetyl-decarbusninsäure (VI) abgebaut. Demnach kommt dem Äthoxylat die Konstitution (IV) zu.

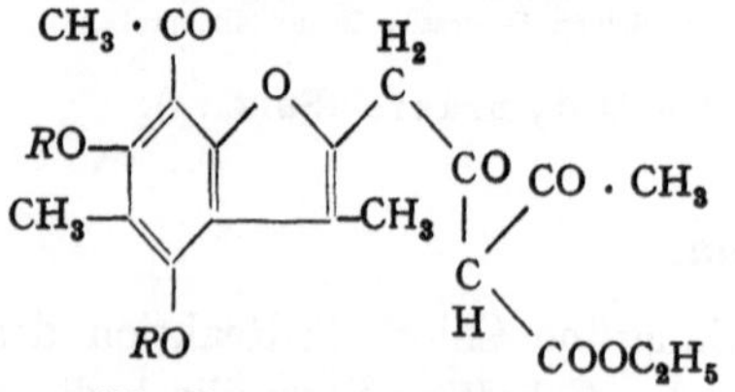

(IV.) Diacetyl-usninsäure-äthoxylat.

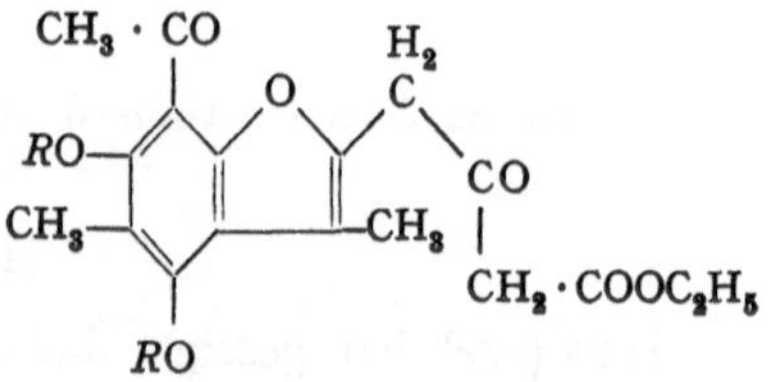

(V.)

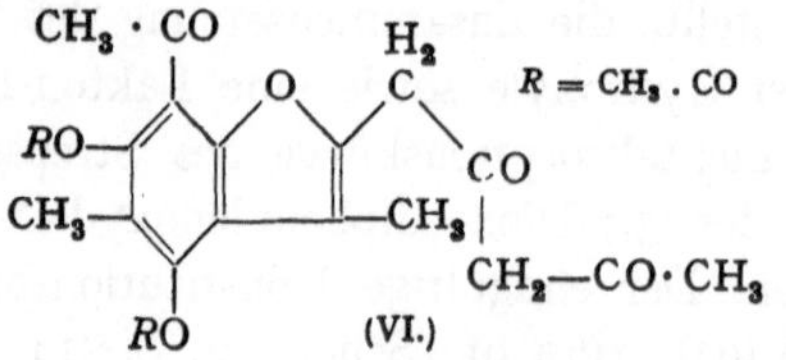

(VI.)

$R = CH_3 \cdot CO$

Die Äthoxylat-Bildung ist somit nicht als eine Lakton-Spaltung, sondern als die Loslösung einer Keto-Bindung aufzufassen. Diesem Anspruch wird die CURD-ROBERTSONsche Formel (VII) (37) am besten gerecht.

Durch den Ozon-Abbau der Diacetyl-usninsäure erhielten SCHÖPF und ROSS (83, 84) neben Aceton-oxalester eine Verbindung $C_{16}H_{16}O_7$ vom Schmp. 132°, der sie die Struktur (VIII, $R = CO . CH_3$) zuschrieben. Dabei haben sie auch beobachtet, daß das entacetylierte Produkt (VIII, $R = H$; Schmp. 233°) beim nochmaligen Acetylieren ein bei 131—132° schmelzendes Produkt liefert, welches trotz des fast übereinstimmenden Schmelzpunktes vom ursprünglichen Abbauprodukt verschieden ist und als eine umlaktonisierte Verbindung (IX, $R = CO . CH_3$) aufgefaßt wurde. ASAHINA und OKAZAKI (16) haben die Ozon-Spaltung der Diacetyl-usninsäure nachgeprüft und erhielten außer den erwähnten Produkten auch den Ester (X). Sie stellten dabei die Strukturisomerie von (VIII) und (IX) in Abrede, da diese Substanzen durch Impfen ineinander übergeführt werden konnten.

SCHÖPF und ROSS (83, 84) haben darauf aufmerksam gemacht, daß die Usninsäure-Formel (VII) eine gewisse Schwierigkeit bietet, da diese Säure bereits unter milden Bedingungen racemisierbar ist. Da am asymmetrischen Zentrum (Kohlenstoffatom 3) kein bewegliches Atom vorhanden ist, wurde die Racemisierbarkeit „auf ein besonders leicht erfolgendes Durchschwingen des Moleküls" zurückgeführt, wofür jedoch ähnliche Beispiele nicht vorliegen. Zur Erklärung der leichten Spaltbarkeit zwischen $C_{(3)}$ und $C_{(11)}$ (z. B. unter Bildung des Usninsäure-äthoxylats) zogen SCHÖPF und ROSS die Doppelbindungsregel von O. SCHMIDT (81) heran, wonach bei einer α,β-ungesättigten Substanz die Bindungskraft zwischen den γ- und δ-Kohlenstoffatomen geschwächt sei. Bei Zufuhr von genügender Energie würde demnach die Bindung zwischen $C_{(3)}$ und $C_{(11)}$ derart gelockert, daß der 3-Methylgruppe ein Durchschwingen ermöglicht würde, ohne daß eine Aufspaltung stattfindet.

Usnonsäure. Seinerzeit hatte WIDMAN (*104*) die *D,L*-Usninsäure mit Kaliumpermanganat in *D,L*-Usnonsäure übergeführt, die sich von der ersteren durch einen Mehrgehalt von 1 Atom Sauerstoff unterscheidet. ASAHINA und Mitarbeiter (*25, 26, 106*) haben aus *D*-Usninsäure die *D*-Usnonsäure erhalten, welche durch milde Reduktion in *D*-Usninsäure rückverwandelt werden kann. Wird nun die *D*-Usnonsäure mit Alkohol gekocht, so entsteht *D*-Isooxyacetusnetinsäure-äthylester, der bei der Reduktion inaktiven Acetusnetinsäure-ester liefert. Da bei der Bildung der letzteren Verbindung die Asymmetrie am $C_{(3)}$ verlorengegangen ist, muß durch Eintritt eines Sauerstoffatoms im linksstehenden Ring ein neues asymmetrisches Zentrum gebildet worden sein.

SCHÖPF und ROSS (*83, 84*) haben die Konstitution der Usnonsäure durch das Symbol (XI) ausgedrückt:

(XI.) Usnonsäure.

Dihydro-usninsäure und Tetrahydro-desoxy-usninsäure (*24*). Durch katalytische Hydrierung wird die Diacetyl-usninsäure zunächst in Dihydro-diacetyl-usninsäure und dann in Tetrahydró-diacetyl-desoxy-usninsäure übergeführt. Die Dihydro-usninsäure, welcher die Konstitution (XII) zukommt, ist viel beständiger als Usninsäure. Unter den Bedingungen, bei denen die Usninsäure Acetusnetinsäure-ester, sowie Diacetylusninsäure-äthoxylat liefert, erleidet die Dihydro-usninsäure keine Ringsprengung. Wird Dihydro-usninsäure mit Permanganat oxydiert und das (nicht kristallisierende) Reaktionsprodukt im Vakuum trocken-destilliert, so erhält man 2,6-Dioxy-3-methyl-acetophenon (XIII). Wird Dihydro-usninsäure unter Zusatz von Chlorcalcium destilliert, so bildet sich neben Acetylaceton das 6-Oxy-3,5-dimethyl-7-acetyl-cumaranon (XIV) [ZOPF (*114*)]. Offenbar ist das Produkt (XIII) aus dem rechtsseitigen Kern der Dihydro-usninsäure entstanden. Das Cumaranon (XIV) müßte hingegen dem linksseitigen Ring entstammen, in dem das Sauerstoffatom in der 4-Stellung durch intermolekulare Disproportionierung entfernt wurde (*10*). *D*-Usninsäure liefert beim

Destillieren unter Zusatz von Chlorcalcium Decarbusninsäure neben etwas *rac.* Usninsäure. Hier geht durch Abspaltung von Kohlensäure die Cumaron-Bildung rasch vonstatten, wodurch die Seitenkette in der 2-Stellung stabilisiert wird (*10*) (Schema 3).

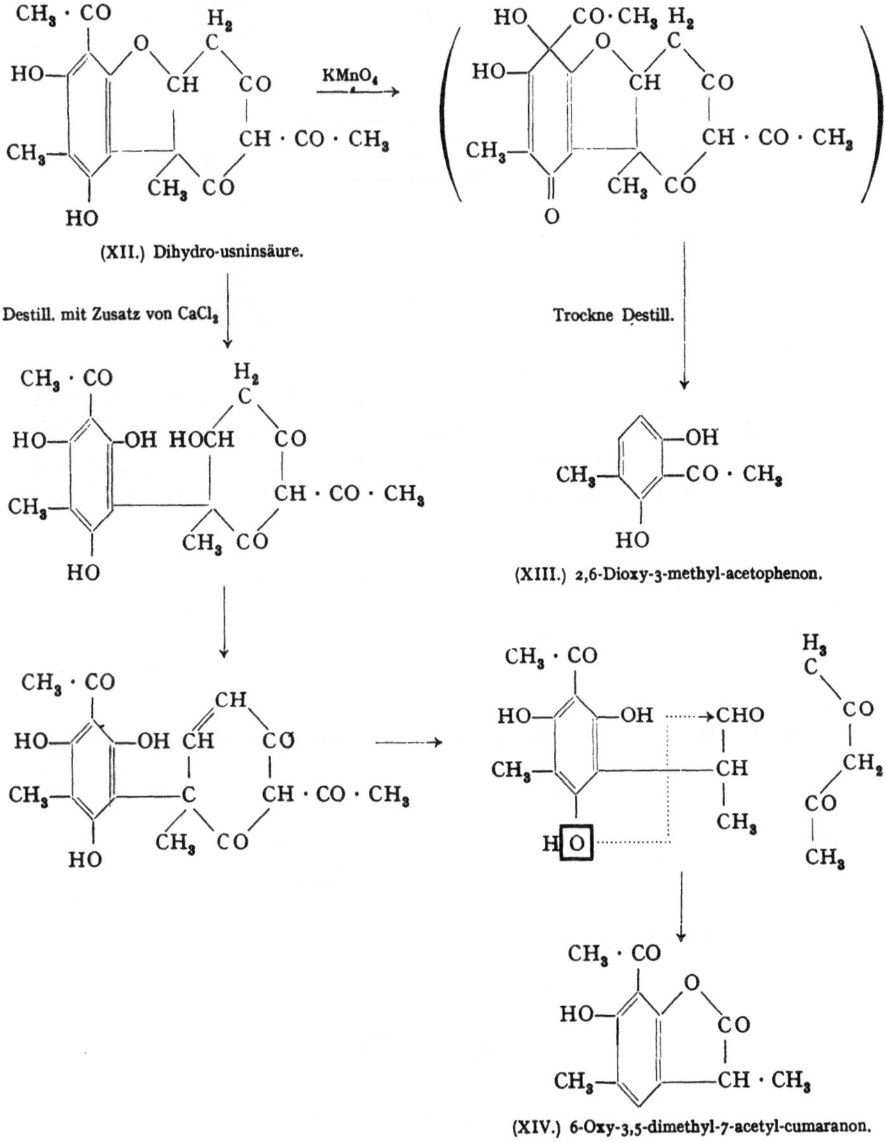
KMnO4
(XII.) Dihydro-usninsäure.
Destill. mit Zusatz von CaCl2
Trockne Destill.
(XIII.) 2,6-Dioxy-3-methyl-acetophenon.
(XIV.) 6-Oxy-3,5-dimethyl-7-acetyl-cumaranon.

Schema 3. Abbau und Struktur der Dihydro-usninsäure.

Usnolsäure und Decarbusnol. Durch Einwirkung von konzentrierter Schwefelsäure erhält man aus Usninsäure die Usnolsäure [PATERNÒ (*72*),

Stenhouse (99)] und aus Decarbusninsäure Decarbusnol [Widman (105)].
Die Usnolsäure wird beim Behandeln mit Anilin in Decarbusninsäure-
anilid und beim Decarboxylieren in Decarbusnol übergeführt (23):

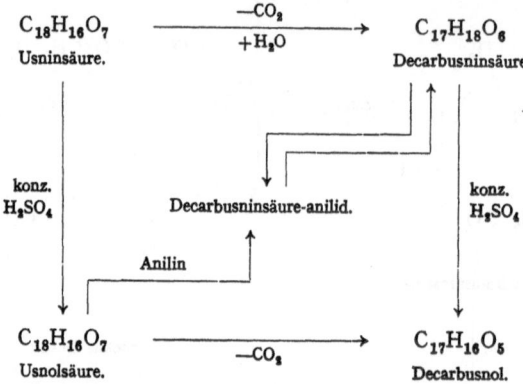

$C_{18}H_{16}O_7$ — Usninsäure. $\xrightarrow[+H_2O]{-CO_2}$ $C_{17}H_{18}O_6$ — Decarbusninsäure.

konz. H_2SO_4 Decarbusninsäure-anilid.   konz. H_2SO_4

Anilin

$C_{18}H_{16}O_7$ — Usnolsäure. $\xrightarrow{-CO_2}$ $C_{17}H_{16}O_5$ — Decarbusnol.

Sowohl Usnolsäure als auch Decarbusnol bilden mit o-Phenylendiamin
kein gefärbtes Kondensationsprodukt (sie sind demnach keine 1,3-Di-
ketone), sie geben aber mit Ehrlichschem Reagens eine prachtvolle
blaue Färbung, die von Asahina zum folgenden Nachweis der Usninsäure benutzt wurde.

Eine Spur von Usninsäure-Kristallen wird auf dem Objektträger mit einem
Tropfen 80%iger Schwefelsäure (enthaltend 0,1% p-Dimethylaminobenzaldehyd)
versetzt und zu gelindem Sieden erhitzt. Dann wird das so erhaltene dunkelbraune
Produkt mit wenigen Tropfen Alkohol verdünnt, wobei eine indigoblaue Färbung
erscheint.

Da das Diacetyl-usninsäure-äthoxylat auch beim Lösen in konzentrierter Schwefelsäure Usnolsäure-äthylester gibt, so ist die Bildung
der Usnolsäure sowie des Decarbusnols so zu deuten, daß die enolisierte
1,3-Diketon-Seitenkette an irgendeiner Stelle des Cumaron-Kerns Cyclisierung erfuhr. So haben Robertson und Mitarbeiter (41) dem Decarbusnol die Konstitution (XV, $R = H$) und der Usnolsäure die Struktur
(XV, $R = COOH$) zugeteilt.

(XV.)

Demgegenüber fanden ASAHINA und OKAZAKI (*15*), daß Decarbusnol bei der Einwirkung von Ozon neben Formaldehyd ein Triketon $C_{16}H_{14}O_6$ (XVII) liefert — ein Zeichen, daß es eine semicyclische Methylen-Gruppe besitzt. Ferner haben sie bei der Destillation vom Tetrahydro-decarbusnol mit Kalk 3-Methyl-4-äthyl-cyclopentadien-on(1), und bei derselben Operation aus dem dihydrierten Triketon das 3-Methyl-cyclopenten-on (1) erhalten. Diese Umwandlungen lassen sich mit der Konstitution (XVI, $R = H$) für Decarbusnol erklären (Schema 4).

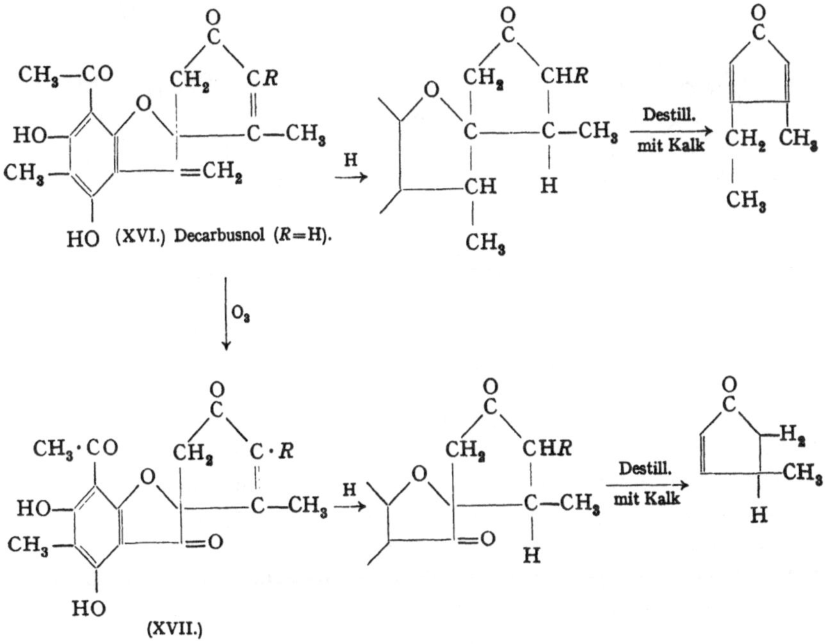

Schema 4. Abbau des Decarbusnols.

3. Xanthon-Derivate.

Lichexanthon.

ASAHINA und NOGAMI (*12*) isolierten aus *Parmelia formosana* eine kristalline Substanz $C_{16}H_{14}O_5$, die alle Eigenschaften eines Xanthonderivates zeigte. Ferner wurden in derselben ein Hydroxyl und zwei Methoxyle nachgewiesen, so daß diese Verbindung als ein Oxy-dimethoxy-methyl-xanthon (XVIII) anzusprechen ist. Da bei der Kalischmelze des Lichexanthons Orcin erhalten wurde, ist es wohl durch Kombinierung von je einem Mol Orcin und Phloroglucin entstanden. Diese Vermutung wurde durch die Synthese bestätigt (Schema 5, S. 220).

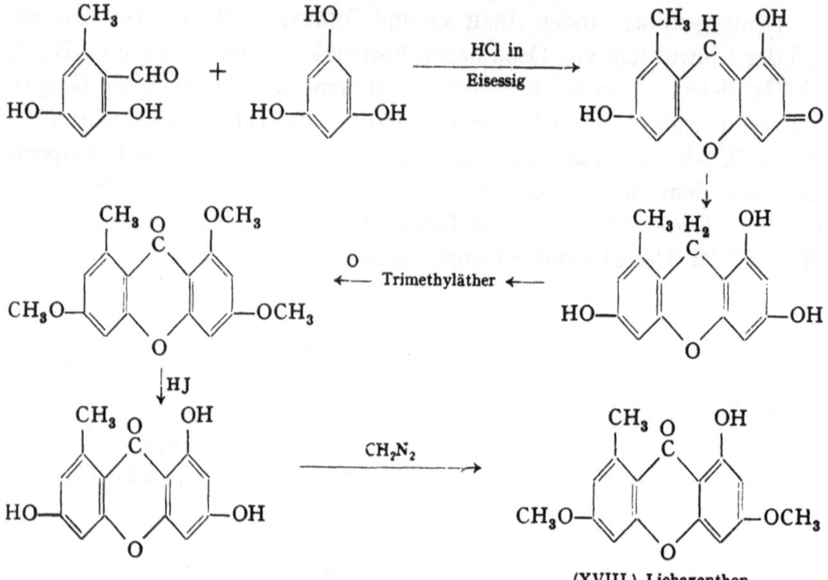

CH₃ ... —CHO + ... HO ... HCl in Eisessig →

HO— ... —OH ... HO— ... —OH

CH₃ H OH ... C ... HO— ... =O ... O

CH₃ O OCH₃ ... C ... CH₃O— ... —OCH₃ ... O

O ... ← Trimethyläther ←

CH₃ H₂ OH ... C ... HO— ... —OH ... O

↓ HJ

CH₃ O OH ... C ... HO— ... —OH ... O

CH₂N₂ ——→

CH₃ O OH ... C ... CH₃O— ... —OCH₃ ... O

(XVIII.) Lichexanthon.

Schema 5. Synthese des Lichexanthons.

4. Depside.

Depside der Orcin-Gruppe.

Lecanorsäure-Typus: Erythrin. Das Erythrin wurde zuerst von HEEREN in *Roccella fuciformis* ACH. entdeckt und gehört zu den am frühesten bekannten Flechtenstoffen. Trotz den mehrfach ausgeführten Untersuchungen von einer Anzahl von Forschern (*45, 51, 53, 63, 86, 97, 101*) blieb seine Struktur bis vor kurzem unaufgeklärt. Da das Erythrin beim Erwärmen mit Alkohol Orsellinsäureester und Pikroerythrin (Orsellinsäure-erythritester) liefert, nahm HESSE an (*45*), daß es ein Lecanorsäure-erythritester sei, was auch durch neuere Arbeiten bestätigt wurde. HESSES Behauptung, daß das Erythrin durch Alkali in die sogen. Erythrinsäure umgelagert werde, beruht jedoch auf einem Irrtum. Die von ZELLNER (*108*) aufgestellte Strukturformel, wonach zwei Orsellin-säure-Reste sich getrennt an Erythrit verestert vorfinden, stützt sich auf keine experimentell begründete Tatsache. Neuerdings erhielt SAKURAI (*80*) durch Einwirkung von Diazomethan einen Trimethyl-äther, der ein Triacetat liefert. Beim Kochen mit alkoholischem Kali wird Erythrin-trimethyläther in Dimethyläther-orsellinsäure, Isoevernin-säure und *i*-Erythrit gespalten, wodurch das Vorliegen des Lecanorsäure-Restes im Erythrin-Molekül bewiesen wurde. Anderseits hat SAKURAI Trimethyläther-lecanorsäurechlorid auf Monoaceton-erythrit einwirken lassen und das Kupplungsprodukt (XIX) entacetoniert, wobei Trimethyl-äther-erythrin (XX) (Schmp. 110°) entstand.

(XIX.)

(XX.) Erythrin-trimethyläther.

Daß das Monoaceton-erythrit eine 1,2-acetonierte Verbindung ist, wurde dadurch bewiesen, daß es sich durch Bleitetraacetat in Formaldehyd und Glycerinaldehyd spalten ließ. Ferner erhielt SAKURAI (80) beim Kochen von Erythrin mit Methanol ein Pikroerythrin vom Schmp. 136,5°, während das Pikroerythrin nach SCHUNCK (85, 86) bei 158° schmolz.

Fast zu gleicher Zeit haben RAO und SESHADRI (75) eine umfassende Untersuchung über die chemischen Bestandteile von *Roccella Montagnei* BEL. ausgeführt. Was die Konstitution des Erythrins betrifft, kamen sie zum gleichen Resultat wie SAKURAI. Dabei isolierten sie neben Erythrin auch Pikroerythrin (von ihnen „Montagnetol" genannt) und bemerkten, daß es das erste einfache Phenolcarbonsäure-Derivat in der Flechte sei. Bald darauf haben RAO und SESHADRI (76, 77) die überraschende Beobachtung gemacht, daß es zwei Arten „Montagnetol" (Schmp. 136° und 158°) gibt, wobei das erstere optisch aktiv und das letztere inaktiv ist. Das Erythrin selbst wurde als optisch aktiv erkannt. Neuerdings hat SAKURAI* den Befund der indischen Forscher bestätigt (Tabelle 1).

Tabelle 1. Drehungsvermögen des Erythrins und des Pikroerythrins.

	Nach SESHADRI und RAO	Nach SAKURAI
Erythrin	$[\alpha]_D = + 8{,}0°$	$[\alpha]_D^{29} = + 10{,}63°$
Erythrin-trimethyl-äther	—	$[\alpha]_D = + 4{,}57°$
Pikroerythrin	$[\alpha]_D = + 16{,}0°$ (in Wasser) $[\alpha]_D = + 12{,}6°$ (in Aceton)	$[\alpha]_D^{23} = + 15{,}76°$ (in Wasser)

* Privatmitteilung von Y. SAKURAI.

Olivetorsäure-Typus: Olivetonid. Um die Konstitution des Olivetonids synthetisch zu beweisen, haben ASAHINA und NOGAMI (*13, 14*) als Ausgangsmaterial den von JERDAN (*56*) dargestellten Orcin-tricarbonsäure-diäthylester (XXI) benutzt, welcher durch Erhitzen mit Kalilauge unter Abspaltung einer kernständigen Carboxyl-Gruppe in Dioxyhomophthalsäure-monoäthylester (XXII) übergeht. Dann wird der Dimethyläther der letzteren über das Chlorid mit Acetessigester gekuppelt, entacetyliert und *n*-butyliert. Das so erhaltene Produkt (XXIII) liefert bei der Ketonspaltung die Olivetonsäure (XXIV), welche durch Kochen mit Ameisensäure in Olivetonid-dimethyläther (XXV) übergeführt wird:

(XXI.) Orcin-tricarbonsäure-diäthylester.

(XXII.) Dioxyhomo-phthalsäure-monoäthylester.

(XXIII.)

(XXIV.) Olivetonsäure.

(XXV.) Olivetonid-dimethyläther.

Gyrophorsäure-Typus: Hiascinsäure. Die Flechte *Cetraria hiascens* TH. FR. enthält neben Gyrophorsäure ein bisher unbekanntes Tridepsid, die Hiascinsäure, welche von ASAHINA und KUSAKA (*11*) in reinem Zustand isoliert und strukturell aufgeklärt wurde. Der Zusammensetzung ($C_{24}H_{20}O_{11}$) nach ist sie eine Oxy-gyrophorsäure und färbt sich mit Chlorkalk rot. Abweichend von der Gyrophorsäure gibt Hyascinsäure auch mit Kaliumhydroxyd eine Rotfärbung; mit Barythydrat wird sie zuerst grün, dann rot. Durch Methanolyse wird die Hiascinsäure in 5-Oxy-orsellinsäure-methylester, Orsellinsäure-methylester und Orsellinsäure gespalten. Obwohl sie ein Pentaacetyl-Derivat gibt, so wird doch, selbst unter energischen Bedingungen, nur Tetramethyläther-methylester erhalten. Beim Verseifen mit kalter konzentrierter Schwefelsäure wird der Tetramethyläther-hiascinsäuremethylester in 2,4-Dimethyläther-5-oxy-orsellinsäure, Isoeverninsäure und Isoeverninsäure-methylester gespalten. Diese Umwandlung läßt sich mit der Formel (XXVI) für Hiascinsäure gut in Einklang bringen.

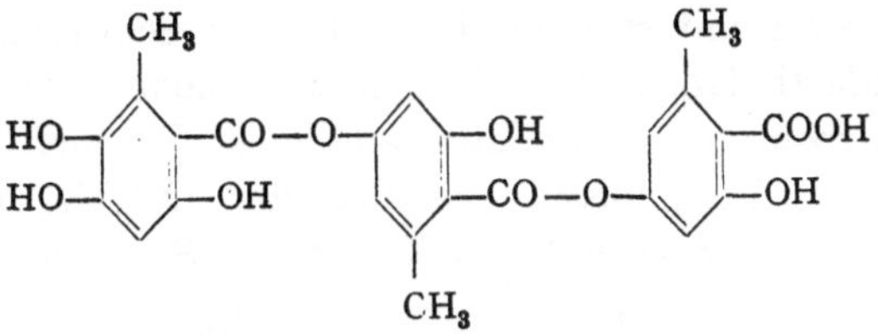

(XXVI.) Hiascinsäure.

Fast zu gleicher Zeit hat KUSAKA (*62*) Pentaacetyl-hiascinsäure und Pentacarbäthoxy-hiascinsäure synthetisch dargestellt und die soeben angegebene Struktur weiter gesichert.

Depside der β-Orcin-Gruppe.

Nichtexistenz der Isosquamatsäure. Seinerzeit beschrieben ASAHINA und YANAGITA (*22*) den Inhaltsstoff der *Cladonia Boryi* als ein Isomeres der Squamatsäure, das sich von der letzteren bloß durch die Stellung des Äther-methyls in der Komponente *S** (also 2- statt 4-Stellung) unterscheiden soll. Weiter haben sie Orcin-dicarbonsäure-diäthylester zunächst partiell carbäthoxyliert, dann methyliert und schließlich entcarbäthoxyliert. Da das so erhaltene Produkt mit der *S*-Komponente* der sogen. Isosquamatsäure übereinstimmte, glaubten die genannten Autoren, die gesuchte 2-Methyläther-orcin-dicarbonsäure (XXVII, *R* = H) synthetisiert zu haben. Später haben indessen ASAHINA und SIMOSATO (*20*) gezeigt, daß die angebliche 2-Methyläther-orcin-dicarbonsäure tatsächlich ein 4-Methyläther ist und infolgedessen kein Unterschied zwischen Iso-squamatsäure und Squamatsäure besteht. Zur Darstellung der 2-Methyläther-orcin-dicarbonsäure bedienten sich ASAHINA und SIMOSATO eines partiell benzylierten Hämatommsäure-methylesters. Der 4-Benzyläther (XXVIII) wurde weiter methyliert und dann zur Dicarbonsäure oxydiert. Beim Entbenzylieren entstand daraus 2-Methyläther-orcindicarbonsäure.

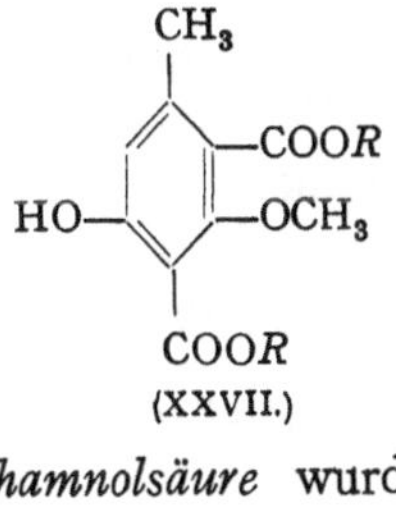

(XXVII.) 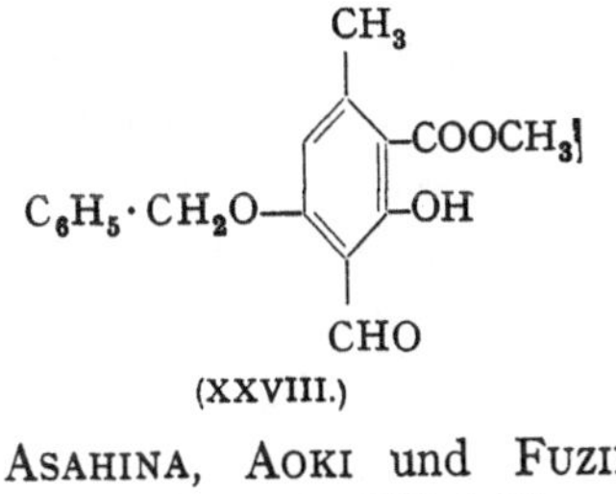　　　　　　　　　　(XXVIII.)

Hypothamnolsäure wurde von ASAHINA, AOKI und FUZIKAWA (*7*) in *Cladonia pseudostellata* ASAHINA aufgefunden**. Wird Hypothamnol-

---

　　* Als „Komponente *S*" oder „*S*-Komponente" wird die Säure-Komponente des Depsids bezeichnet, die in unserer Formulierung meist links steht.

　　** Seinerzeit haben die genannten Autoren die Flechte mit *Cladonia uncialis* var. *obtusata* identifiziert. Vgl. J. Japan. Bot. **18**, 620 (1942).

säure-dimethylester mit eiskalter konzentrierter Schwefelsäure verseift, so werden 4-Methyläther-orcin-dicarbonsäure-monomethylester(3) und Oxy-β-orcin-carbonsäure-methylester erhalten. Bei der katalytischen Hydrierung geht Thamnolsäure-dimethylester (XXIX, $R = CH_3$) in Hypothamnolsäure-dimethylester (XXX, $R = CH_3$) über.

$$CH_3 \quad\quad CH_3$$
$$CH_3O-\!\!\!\bigcirc\!\!-CO-O-\!\!\!\bigcirc\!\!-COOR$$
$$-OH \quad HO-\quad -OH$$
$$COOR \quad\quad CHO$$
(XXIX.)

$$CH_3 \quad\quad CH_3$$
$$CH_3O-\!\!\!\bigcirc\!\!-CO-O-\!\!\!\bigcirc\!\!-COOR$$
$$-OH \quad HO-\quad -OH$$
$$COOR \quad\quad CH_3$$
(XXX.)

Die Synthese des Hypothamnolsäure-dimethylesters wurde durch Kupplung von 4-Methyläther-2-carbäthoxy-orcin-dicarbonsäure-methyl-ester(3)-chlorid(1) mit Oxy-orcin-carbonsäure-methylester in Pyridin ausgeführt.

Barbatolsäure. Für diese Säure zogen Schöpf, Heuck und Duntze (*82*) als wahrscheinlichste die Formel (XXXI) in Erwägung; sie schlossen jedoch (XXXII) nicht aus.

$$CH_3 \quad\quad CHO$$
$$HO-\!\!\!\bigcirc\!\!-CO-O-\!\!\!\bigcirc\!\!-OH$$
$$-OH \quad\quad -COOH$$
$$CHO \quad\quad CH_2OH$$
(XXXI.)

$$CH_3 \quad\quad COOH$$
$$HO-\!\!\!\bigcirc\!\!-CO-O-CH_2-\!\!\!\bigcirc\!\!-OH$$
$$-OH \quad\quad -CHO$$
$$CHO \quad\quad OH$$
(XXXII.)

Das Symbol (XXXI) steht mit der Beobachtung nicht in Einklang, daß Barbatolsäure beim Erhitzen mit Alkohol zu Barbatolin decarboxy-liert wird, wobei kein Phthalid gebildet wird, während in Gegenwart von benachbarten Carbinol- und Carboxyl-Gruppen dies der Fall sein müßte.

Jüngst hat Suominen (*102*) aus *Alectoria impexa* f. *fuscidula* Barbatol-säure isoliert. Er hat den Methylester mittels Zink und Jodwasserstoff zu Hypobarbatolsäure-methylester ($C_{19}H_{20}O_8$) reduziert und den letzteren erschöpfend methyliert, wobei ein Gemisch von Tri- und Tetramethyl-äther erhalten wurde. Beim Verseifen lieferte nun der Trimethyläther

$$CH_3$$
$$CH_3O-\!\!\!\bigcirc\!\!-COOH$$
$$-OH$$
$$CH_3$$
(XXXIII.) Rhizoninsäure.

$$CH_2-O$$
$$CH_3O-\!\!\!\bigcirc\!\!-CO$$
$$-OCH_3$$
$$CH_3$$
(XXXIV.) 3,5-Dimethoxy-4-methyl-phthalid.

des Hypobarbatolsäure-methylesters die Rhizoninsäure (XXXIII) und 3,5-Dimethoxy-4-methyl-phthalid (XXXIV), woraus die obige Formel (XXXII) für Barbatolsäure folgt.

Früher hatte HESSE (*47, 48, 50*) aus *Alectoria impexa* neben Bryopogonsäure eine „Alectorsäure" (Schmp. 181°) isoliert, welche beim Erhitzen mit Wasser unter Verlust von Kohlensäure in „Alectorinsäure" (Schmp. 220°) übergeht. Daraus ist zu schließen, daß das „Alectorsäure"-Präparat eine etwas verunreinigte Barbatolsäure und die „Alectorinsäure" reines Barbatolin war.

5. Depsidone.

Grundskelett des Depsidons. Als Ausgangsmaterial für die Darstellung von unsubstituiertem Depsidon kommt vor allem 2-Oxydiphenyläther-2'-carbonsäure (XXXV) in Betracht. Wie TOMITA, INUBUSHI und KUSUTA (*103*) gezeigt haben, wird die Carbonsäure (XXXV) bei 16stündigem Stehenlassen mit Acetanhydrid unter Zusatz von Pyridin nur acetyliert und dann, nach 45tägigem Stehen, in 4-Acetoxy-xanthon (XXXVI) (Schmp. 142—143°) übergeführt. Demgegenüber liefert die Carbonsäure durch Einwirkung von Phosphorpentoxyd in Chloroform das Depsidon $C_{13}H_8O_3$ (XXXVII), welches unscharf zwischen 58—82° schmilzt. Daneben entsteht in geringer Menge ein dimeres Produkt (XXXVIII) des Depsidons. Beim Verseifen wird Depsidon in 2-Oxydiphenyläther-2'-carbonsäure rückverwandelt.

(XXXV.) 2-Oxydiphenyläther-2'-carbonsäure.

(XXXVI.) 4-Acetoxy-xanthon.

(XXXVII.) Depsidon.

(XXXVIII.) Dimeres des Depsidons.

Depsidone der Orcin-Gruppe.

Variolarsäure. Die in der älteren Literatur angegebenen Parellsäure (*85*), Ochrolechiasäure (*49*) und Variolarsäure (*113*) stellen ein und dieselbe Verbindung dar. Unter dem letzteren Namen wurde die Säure von Murphy, Keane und Nolan (64) näher untersucht und strukturell aufgeklärt.

Die Variolarsäure $C_{16}H_{10}O_7$ (XXXIX) ist ein Depsidon mit einer Phthalid-Bindung. In der Kalischmelze wird sie zu Orcin und α-Resorcylsäure gespalten.

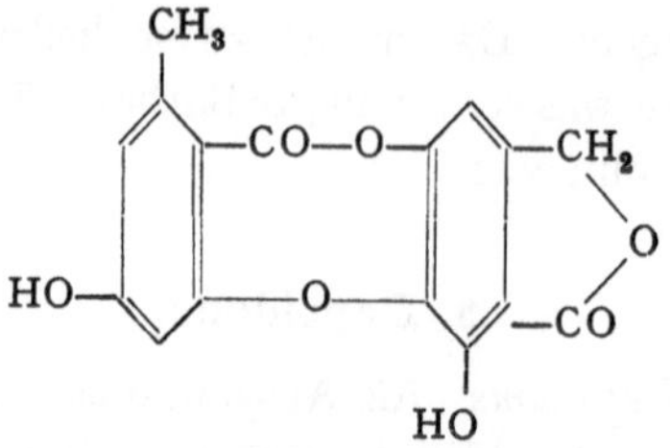

(XXXIX.) Variolarsäure.

Psoromsäure. Früher haben Asahina und Hayashi (8, 9) auf Grund der thermischen Zersetzung der Psoromsäure, bei der Atranol und 3-Oxy-4-methyl-5-methoxy-phthalsäure-anhydrid entstehen, sowie der Bildung des Desoxy-hyposalazinol-trimethyläthers (XLIa) durch elektrolytische Reduktion der Dimethyläther-hypoparellinsäure (XLI), die Konstitution (XL) für Psoromsäure aufgestellt:

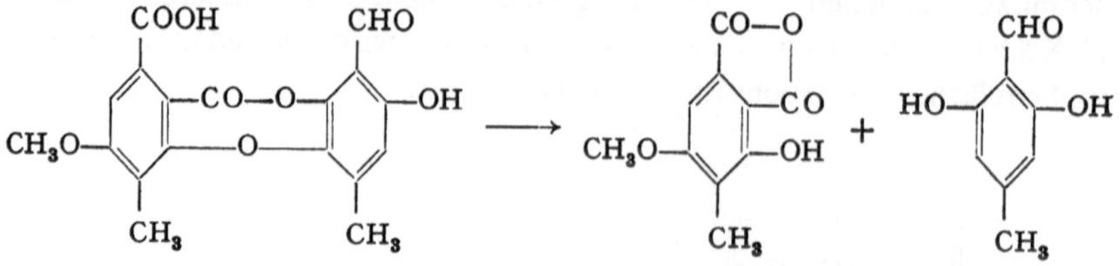

(XL.)

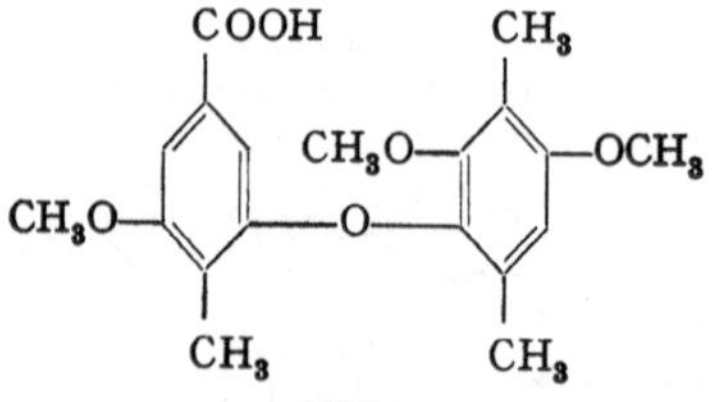

(XLI.)

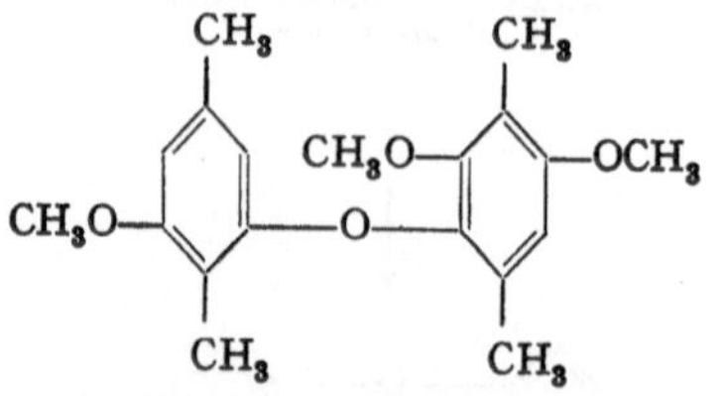

(XLIa.) Desoxy-hyposalazinol-trimethyläther.

Da aber die Dimethyläther-hypoparellinsäure durch wasserentziehende Mittel leicht in ein Xanthon-Derivat übergeht, so müßte das Carboxyl der Hypoparellinsäure zur Depsid-Bindung ortho-ständig sein. Nach Shibata (89) liefert jedoch die Dimethyläther-hypoparellinsäure (XLI) beim Decarboxylieren nicht (XLIII), sondern (XLII), woraus zu schließen ist, daß die Dimethyläther-hypoparellinsäure die Struktur (XLIV) besitzt und folglich der Psoromsäure die Formel (XLV) zukommt:

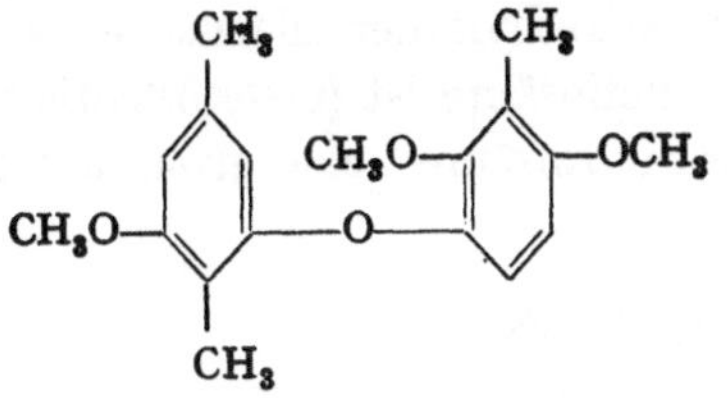

(XLII.)

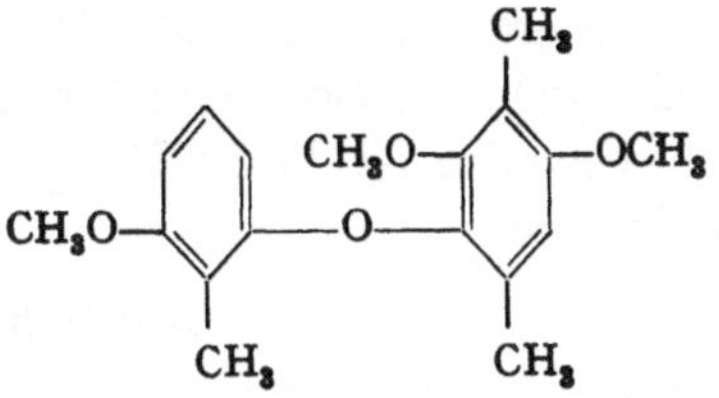

(XLIII.)

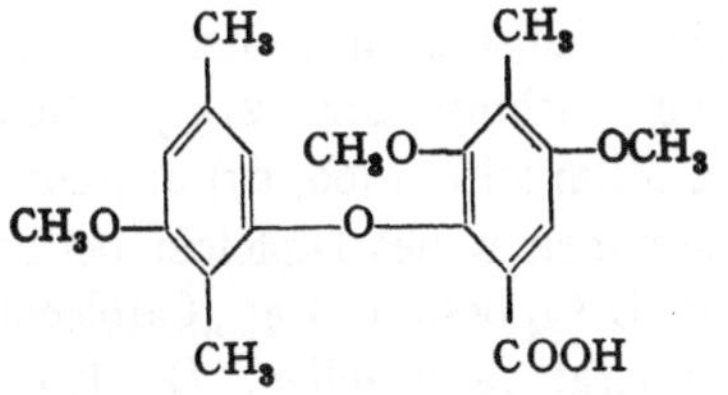

(XLIV.) Dimethyläther-hypoparellinsäure.

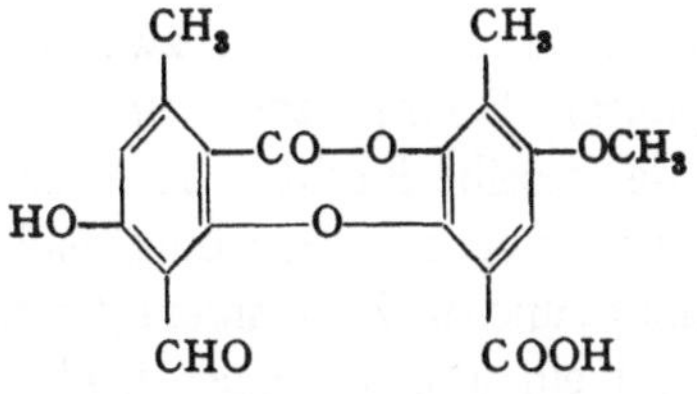

(XLV.) Psoromsäure.

Mit dem neuen Formelausdruck für die Psoromsäure läßt sich indessen der Mechanismus der thermischen Zersetzung schwer erklären. Asahina und Shibata (18) nehmen daher an, daß dabei die Carboxyl-Gruppe der Depsid-Bindung eine ortho-Verschiebung erleidet (im Sinne der Xanthon-Bildung aus Salol-Derivaten) und sich dann unter Bildung des Phtal-säure-Derivates zersetzt (vgl. die Formeln). Die frühere Annahme, daß sich dabei als Zwischenprodukt ein Xanthon-Derivat bildet, ließ sich experimentell nicht stützen.

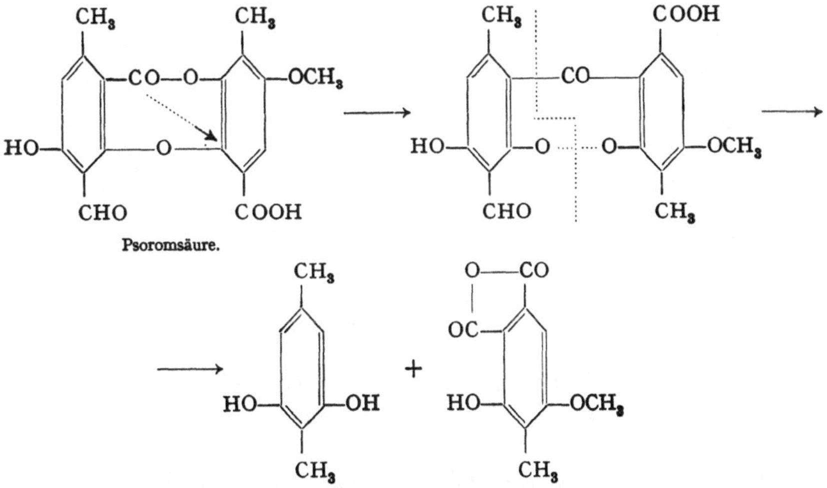

Psoromsäure.

α- und β-Methyläther-salazinsäure. Früher glaubten Asahina und Tukamoto (21), die α- und β-Methyläther-salazinsäuren aufgefunden zu haben, von denen die „α-Säure" ein in Nadeln kristallisierendes Kalium-salz lieferte. Nach der späteren mikrochemischen Untersuchung von

Asahina ergab es sich, daß die sogen. α-Methyläther-salazinsäure wohl ein Gemisch von Norstictinsäure und Stictinsäure ist (unveröffentlicht). Wahrscheinlich ist die „β-Methyläther-salazinsäure" eine etwas unreine Stictinsäure.

Chlorhaltige Depsidone.

Diploicin. Vor einiger Zeit hatte Zopf (*115*) die Flechte *Buellia canescens* (Dicks.) de Not. (unter dem Namen *Diploicia canescens* Dicks.) untersucht und daraus zwei kristalline Substanzen isoliert, nämlich Diploicin (Schmp. 225°) und Catolechin (Schmp. 214—215°). Jüngst wurde dieselbe Flechte von Nolan und Mitarbeitern (*66, 67*) ausführlich studiert, mit dem Ergebnis, daß vollkommen reines Diploicin bei 232° schmilzt und die Zusammensetzung $C_{16}H_{10}O_5Cl_4$ besitzt. Das „Catolechin" scheint ein weniger reines Diploicin-Präparat darzustellen. Die Konstitution des Diploicins wurde von Spillane, Keane und Nolan (*96*) und namentlich von Nolan, Algar, MacCann, Manahan und Nolan (*67*) im Sinne der Formel (XLVI) aufgeklärt.

(XLVI.) Diploicin.　　(XLIX.)

(XLVII.)　　(XLVIII.)　　(L.) Chlor-dioxytoluchinon.

Hierbei haben die genannten Forscher neue Arbeitsmethoden eingeführt, die auch bei künftigen Untersuchungen der Depsidone gute Dienste leisten dürften.

Durch Erwärmen mit Aluminiumchlorid in Benzol wird die Diphenyläther-Bindung des Diploicins unter Bildung des zugehörigen Depsids (XLVII) gesprengt. Dies wird auch bewerkstelligt durch Reduktion des perchlorierten Diploicins, da das so behandelte Produkt beim Verseifen mit methanolischem Kali die Säure-Komponente (XLVIII) liefert.

Löst man nun das Methanolysat des Diploïcins (XLIX) in methanolischem Kali auf und leitet in diese Lösung Sauerstoff ein, so erhält man Chlor-dioxytoluchinon (*L*), welches wohl aus dem rechtsstehenden Kern entstanden ist. Bei der Feststellung der Lage der Substituenten in beiden Ringen wurde 2,6-Dichlorchinon-chlorimid mit Vorteil angewandt. Dieses Reagens gibt nach GIBBS (*42*) in einer Natriumborat-Pufferlösung (pH 9,2) mit aromatischen Oxyverbindungen, deren *p*-Stellungen unsubstituiert sind, eine blaue Farbreaktion. Nach DAVIDSON, KEANE und NOLAN (*38*) wird durch dieses Reagens die Erkennung nicht nur des gegen die Hydroxyl-Gruppe para-ständigen Wasserstoffs, sondern auch des freien Carboxyls sowie des Chlor-atoms erzielt.

Gangaleoidin. NOLAN und Mitarbeiter (*43*) haben *Lecanora gangaleoides* NYL. chemisch untersucht und daraus, neben Chloratranorin, ein chlorhaltiges Depsidon, Gangaleoidin $C_{18}H_{14}O_7Cl_2$ (Schmp. 213—214°), isoliert. Nach DAVIDSON, KEANE und NOLAN [(*38*), cf. (*68*)] kommt dem letzteren die Formel (LI) zu, welche durch die folgenden Umwandlungen und Abbau-Versuche geschlossen wurde (Schema 6).

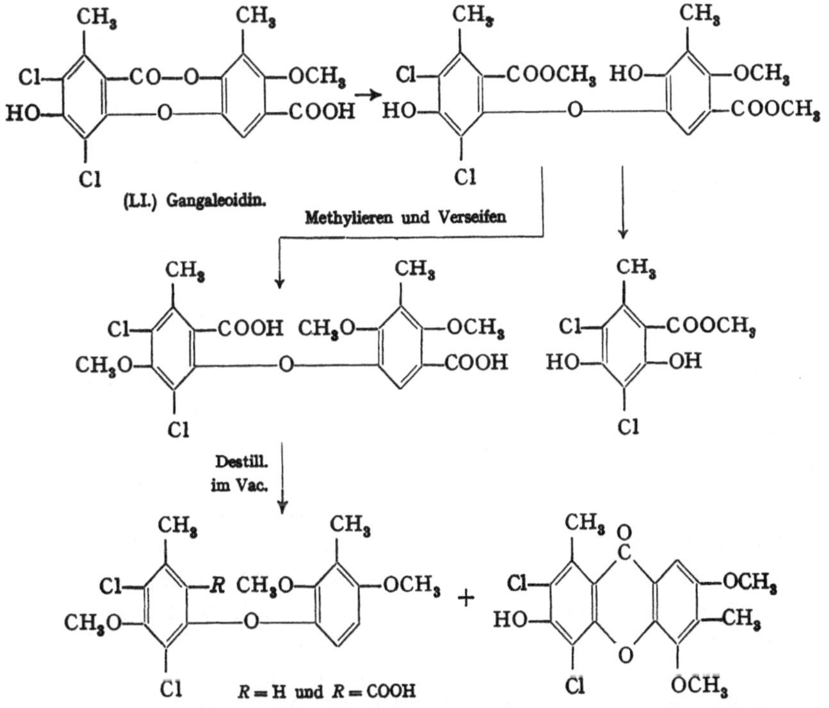

Schema 6. Umwandlungen und Abbau des Gangaleoidins.

Pannarin. ASAHINA (*3*) hat gezeigt, daß einige japanische *Pannaria*-Arten eine Substanz enthalten, die sich mit *p*-Phenylendiamin orangerot färbt und mikrochemisch durch seine eigenartige Kristallbildung charakterisiert wird. Später hat YOSIOKA (*107*) aus *Pannaria lanuginosa* KÖRB., *P. fulvescens* NYL. sowie aus *P. lurida* NYL. ein chlorhaltiges Depsidon $C_{18}H_{15}O_6Cl$ isoliert, das er Pannarin nannte. Durch eine Reihe von Umwandlungen ließ YOSIOKA dem Pannarin die Struktur (LII) zukommen (Schema 7).

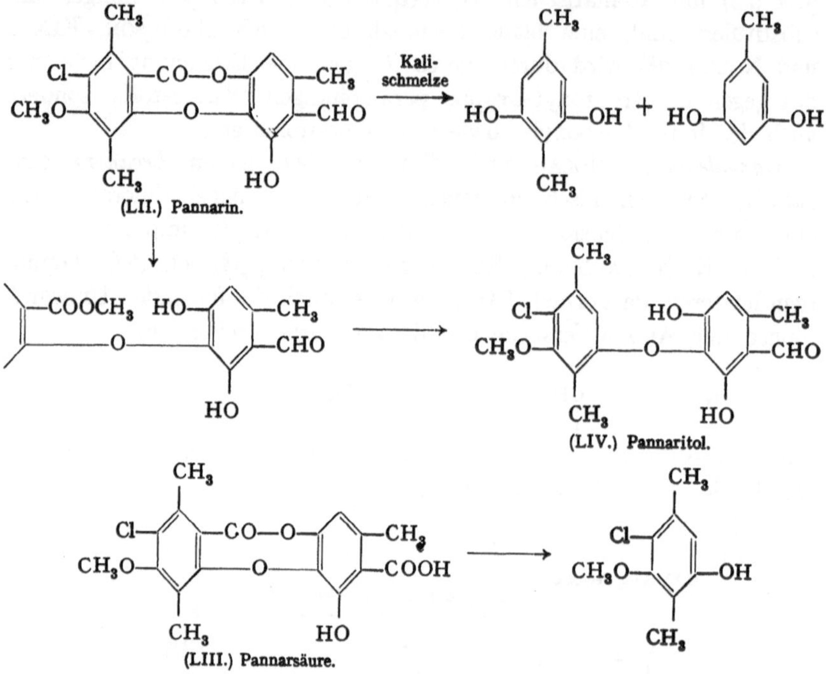

Schema 7. Umwandlungen und Abbau des Pannarins.

Pannarin ist ein Aldehyd und läßt sich in ein Anil überführen. Bei der Kalischmelze wird Pannarin in β-Orcin und Orcin gespalten. Hierbei wird das Chloratom in der *S*-Komponente nicht durch Hydroxyl, sondern durch Wasserstoff ersetzt. Dieselbe Erscheinung trafen auch NOLAN und Mitarbeiter (*38*) bei der Kalischmelze des Diploicins an. Die dem Pannarin zugehörige Pannarsäure (LIII) liefert bei der trockenen Destillation Chlor-orcin-monomethyläther. Die Feststellung der Lage der Seitenkette am rechtsstehenden Kern beruht auf den folgenden beiden Reaktionen: a) durch Einwirkung von Anilin-hydrojodid spaltet das Methyläther-pannarin-anil leicht eine Methyl-Gruppe ab unter Regenerierung von Pannarin (*27*); b) Pannaritol (LIV), das Decarboxylierungs-Produkt des Pannarin-methoxylats, liefert beim DAKINschen Abbau (Umwandlung der CHO-Gruppe in OH) eine Substanz, die alle dem Oxyhydrochinon eigenen Reaktionen zeigt.

Die Stellung der Seitenkette und der Substituenten am rechtsstehenden Ring entspricht demnach derjenigen in der Variolarsäure.

6. Anthrachinon- und Phenanthrenchinon-Derivate.

Magnesiumacetat als Reagens für Polyoxyanthrachinone. SHIBATA (90) fand, daß viele Oxyanthrachinone in alkoholischer Lösung mit Magnesium-acetat eine schöne Farbreaktion geben, welche je nach der gegenseitigen Stellung der Hydroxyle verschiedene Nuancen zeigt. Diejenigen Oxy-anthrachinone, die in der α-Stellung mindestens eine Hydroxyl-Gruppe besitzen, geben eine Rotfärbung, die durch den Eintritt weiterer meta-ständiger Hydroxyle mehr orange wird und kein scharfes Absorptions-band zeigt. Durch Eintritt von ortho- oder para-ständigen Hydroxyl-Gruppen färbt sich die magnesiumacetat-haltige Lösung violett bis blau-violett und gibt dann ein deutliches Absorptionsband. Es werden z. B. die folgenden Verbindungen durch Magnesiumacetat gefärbt:

(Mit Mg-acetat rot.)   (Orangerot.)   (Orangerot.)

(Violettrot.)   (Violettrot.)   (Blauviolett.)

Oxyanthrachinone mit ortho- oder para-ständigen Hydroxyl-Gruppen wurden bisher in Flechten nicht nachgewiesen.

Rhodocladonsäure. Nach KOLLER und HAMBURG (60) kommt dieser Verbindung die Struktur (LV) zu. Da die Rhodocladonsäure durch Magnesiumacetat in Alkohol eine orangerote Färbung gibt, hat SHIBATA (90) angenommen, daß sich in ihrem Molekül je zwei Hydroxyle in meta-Stellung befinden (LVI), wobei aus Analogie mit anderen, in Flechten vorkommenden Polyoxyanthrachinonen die 1,3,6,8-Stellungen den 1,3,5,7-Stellungen vorgezogen wurden. Diese Annahme gewinnt dadurch an Wahrscheinlichkeit, daß das von KUSAKA (61) synthetisch gewonnene 1,3,5,7-Tetraoxy-2,6-dimethylanthrachinon (LVII) in Bi-

carbonat unlöslich ist, während die Rhodocladonsäure sich darin leicht
auflöst. Ein Vergleich der ultravioletten Spektren hat ebenfalls ergeben, daß die Hydroxyle in der Rhodocladonsäure bzw. im 1,3,6,8-
Tetraoxy-anthrachinon in der gleichen Art verteilt sind.

(LV.) Rhodocladonsäure.   (LVI.)

(LVII.)

Thelephorsäure. Asahina und Shibata (*19*) stellten fest, daß der
schwarzviolette Farbstoff der Rhizinen von *Lobaria pulmonaria* Hoffm.
sowie von *L. retigera* Trev. mit dem von Kögl und Mitarbeitern strukturell
aufgeklärten Pilz-Farbstoff, Thelephorsäure (LVIII) identisch ist:

(LVIII.) Thelephorsäure.

Betreffend weitere Angaben über Thelephorsäure und ähnliche Naturstoffe
vgl. Hoffmann-Ostenhof (*54*).

7. Stickstoffhaltige Substanzen.

Pikroroccellin. Vor etwa 70 Jahren wurde diese Verbindung von
Stenhouse und Groves (*98*) neben Erythrin in *Roccella fuciformis* Ach.
entdeckt und stellt fast den einzigen stickstoffhaltigen Flechtenstoff
dar. In neuerer Zeit haben Foster und Saville (*40*) die richtige Zusammensetzung des Pikroroccellins als $C_{20}H_{22}O_4N_2$ (statt $C_{27}H_{29}O_5N_3$
nach Stenhouse und Groves) ermittelt und die Substanz als ein Diketo-

piperazin-Derivat (LIX) erkannt. Die Umwandlungen desselben, die zum Konstitutionsbeweis führten, werden durch Schema 8 veranschaulicht.

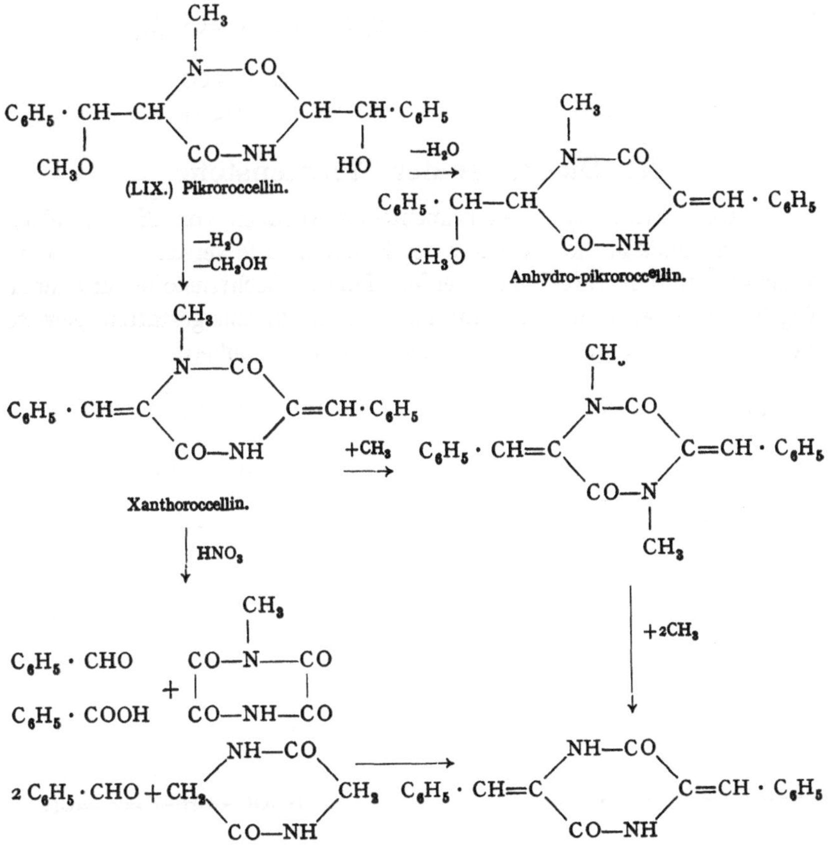

Schema 8. Die Konstitution des Pikroroccellins.

Zopf (*109, 112*) hatte aus *Lecanora epanora* Ach. Epanorin und weiters aus *Rhizocarpon geographicum* die Rhizocarpsäure isoliert. Die chemische Charakterisierung der beiden Substanzen war damals mangelhaft; von der Rhizocarpsäure war lediglich bekannt, daß sie ein Pulvinsäure-Derivat ist. Nach neueren Untersuchungen von Nolan und Mitarbeitern (*58*) sind sowohl Epanorin als auch Rhizocarpsäure stickstoffhaltig und es kommen ihnen die Bruttoformeln $C_{21}H_{21}O_5N$ bzw. $C_{26}H_{23}O_6N$ zu.

Jüngst haben Frank, Clark und Coker (*41 a*) gezeigt, daß die Rhizocarpsäure ein Säureamid ist, entstanden aus Pulvinsäure und *L*-Phenylalanin-methylester, während dem die analogen Komponenten des Epanorins Pulvinsäure und *L*-Leucin-methylester sind:

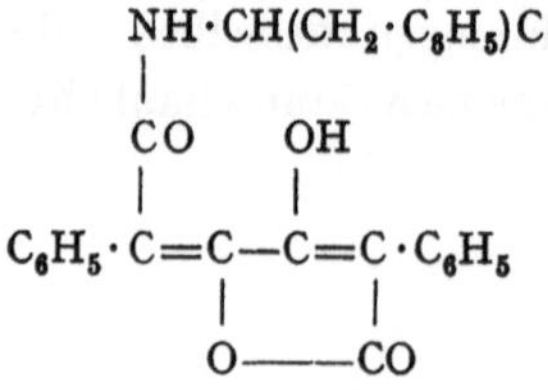

$$NH \cdot CH(CH_2 \cdot C_6H_5)COOCH_3$$
$$|$$
$$CO \qquad OH$$
$$| \qquad |$$
$$C_6H_5 \cdot C = C - C = C \cdot C_6H_5$$
$$| \qquad |$$
$$O \text{——} CO$$

Rhizocarpsäure.

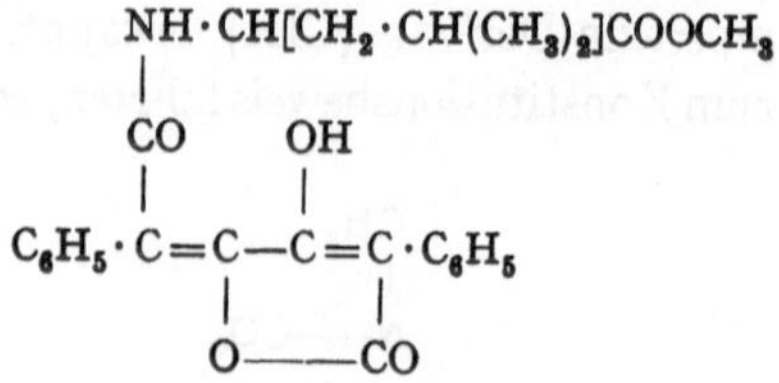

$$NH \cdot CH[CH_2 \cdot CH(CH_3)_2]COOCH_3$$
$$|$$
$$CO \qquad OH$$
$$| \qquad |$$
$$C_6H_5 \cdot C = C - C = C \cdot C_6H_5$$
$$| \qquad |$$
$$O \text{——} CO$$

Epanorin.

III. Biogenese der Flechtenstoffe.

Es liegen bisher keine experimentellen Angaben vor, die uns einen tieferen Einblick in die Biogenese der Flechtenstoffe verschaffen würden. Anderseits muß hervorgehoben werden, daß die Flechtenstoffe auffallende Regelmäßigkeiten in ihrem Bauprinzip zeigen, die uns gestatten, gewisse theoretische Ansichten über ihre Biosynthese zu äußern.

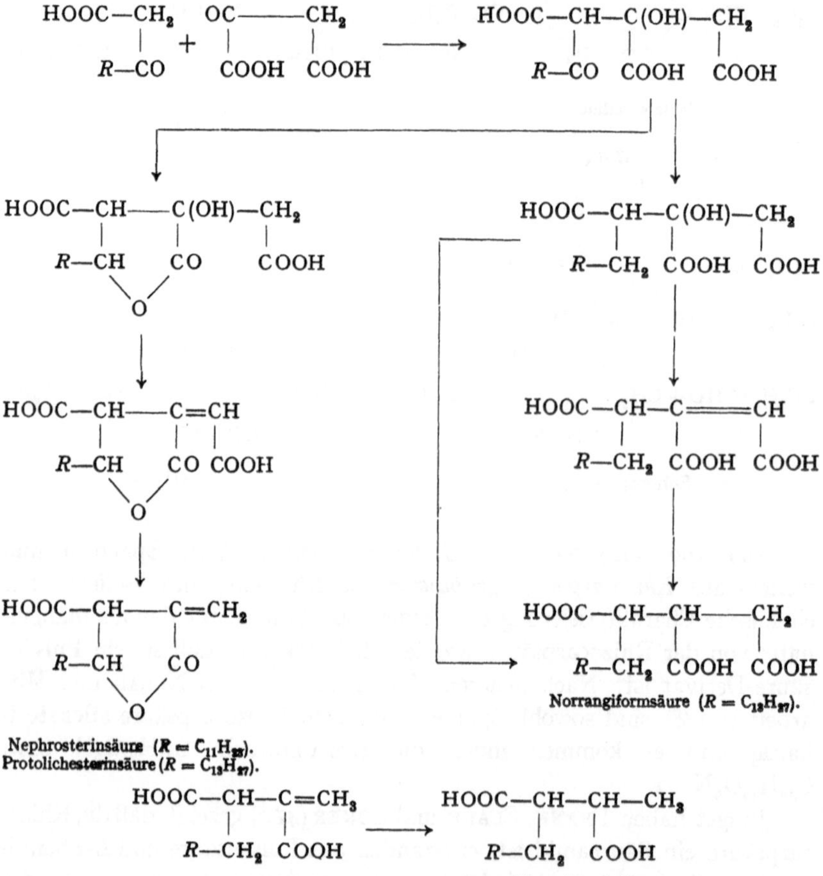

Schema 9. Biosynthese einiger aliphatischer Flechtensäuren.

ASANO ist der Ansicht (Privatmitteilung), daß die Kohlenstoff-Skelette der aliphatischen Flechtensäuren, wie Caperat-, Rangiform-, Roccell-, Protolichesterin- und Nephrosterinsäure, durch Verknüpfung von Oxalessigsäure mit β-Ketonsäuren hervorgegangen seien, entsprechend dem vorstehenden Schema 9.

Nach ROBINSON (79) soll die wichtige Grundsubstanz mancher Depside, Orsellinsäure, aus zwei C_4-Einheiten aufgebaut sein:

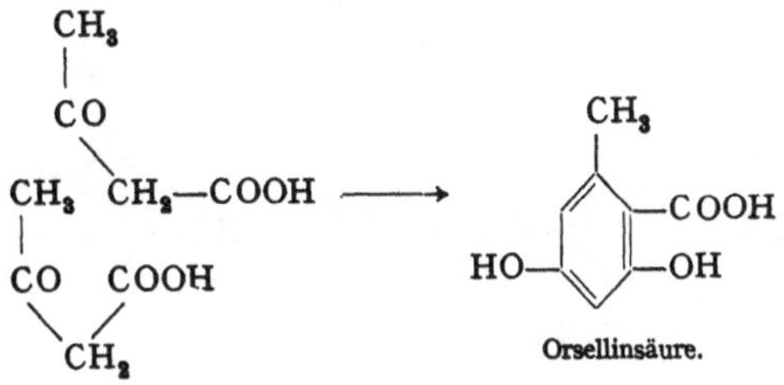

Orsellinsäure.

SESHADRI (87) nahm an, daß das Grundskelett der Depside und Depsidone dasjenige der Orsellinsäure sei (von ihm „C_8-Einheit" genannt), welches durch Kondensation von je einem Mol. Hexose und Biose wie folgt entstehen könnte:

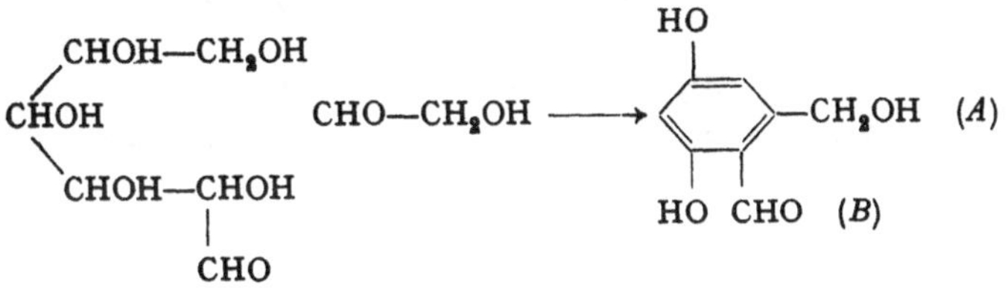

Durch eine darauffolgende Methylierung, Oxydierung und Hydrierung kann das Produkt mannigfach variiert werden. Die Verlängerung der Seitenkette soll dann durch eine weitere Aldol-Kondensation an (A) mit Biose, Tetrose und Hexose stattfinden. Der β-Orcinkern soll durch Kernmethylierung in der zwischen zwei Hydroxylen befindlichen, reaktionsfähigen 3-Stellung entstehen, welche sich in bezug auf die Carboxyl-Gruppe in meta-Stellung befindet. Als Methylierungsmittel kommt hier Formaldehyd in Betracht. Da nach CRABTREE und ROBINSON (36) bei der Kernmethylierung von Phenol-carbonyl-Verbindungen die Gegenwart eines gegen das Carbonyl para-ständigen, freien Hydroxyls erforderlich ist, so müßte bei den para-Depsiden die Kernmethylierung, zumindest im rechtsstehenden Ring, vor der Depsid-Bindung in der Pflanze stattgefunden haben.

Beim Übergang eines Depsids in das entsprechende Depsidon wird in der Regel das Wasserstoffatom in der 5-Stellung des rechtsstehenden Ringes zusammen mit demjenigen des ortho-Hydroxyls (im linksstehenden Ring) wegoxydiert. Dabei wird nach SESHADRI die Aktivität des sonst sehr reaktionsfähigen 3-Wasserstoffs durch die Depsid-Bindung des benachbarten Hydroxyls herabgesetzt, während der Wasserstoff in der 5-Stellung durch den Einfluß der para-ständigen Hydroxyl-Gruppe eine größere Aktivität erlangt. So könnten Physodsäure aus Olivetorsäure, α-Collatolsäure aus Mikrophyllinsäure und Protocetrarsäure aus Nor-atranorin entstanden sein (vgl. die Formeln). Im letzteren Falle wurde nicht nur ein Brückensauerstoff gebildet, sondern hat auch die Oxydation von Substituenten stattgefunden.

Olivetorsäure. → Physodsäure.

Microphyllinsäure. Nor-atranorin.

α-Collatolsäure. Protocetrarsäure.

Die Struktur der Variolarsäure bildet nach MURPHY, KEANE und NOLAN (64) eine Ausnahme und paßt nicht in das angeführte Schema der Depsidon-Bildung. SESHADRI (87) ist indessen der Meinung, daß man das obige Schema der Depsidon-Bildung nicht abändern sollte, bevor die Konstitution der Variolarsäure einwandfrei feststehen wird. Ich möchte in diesem Zusammenhang darauf aufmerksam machen, daß nach YOSIOKA (107) Pannarin ähnlich wie Variolarsäure konstituiert ist. Es wäre denkbar, daß im Verlaufe der Biosynthese der Variolarsäure sowie des Pannarins die Diphenyläther-Bindung früher als die Depsid-Bindung entsteht.

Bei den meta-Depsiden erfolgt keine Depsidon-Bildung trotz des Vorhandenseins von ortho-ständigen Hydroxylen in günstigen Stellen (2 und 2' oder 4 und 2'). SESHADRI führt dies darauf zurück, daß die Oxydation in der 3-Stellung der Orcin-Gruppe und in 5-Stellung der β-Orcin-Gruppe der Depsid-Bindung vorangegangen war, so daß beständige meta-Depside entstanden:

Sekikasäure.

Homosekikasäure.

Thamnolsäure.

SCHÖPF und ROSS (83, 84) nehmen an, daß die Usninsäure aus 2 Molen Methyl-phloracetophenon durch Abspaltung von 1 Mol Wasser und von 2 Wasserstoffatomen in der Pflanze aufgebaut werden könnte:

Usninsäure.

Die Flechtenstoffe der Diphenylenoxyd-Gruppe könnten nach demselben Aufbauprinzip entstehen:

Didymsäure.

Nach Foster und Saville (40) läßt sich die Biosynthese des Pikro-roccellins so auffassen, daß sich β-Oxy-phenylalanin zunächst zum Diketopiperazin-Derivat kondensiert und sodann partiell methyliert wird.

Die Frage, welche der beiden Komponenten einer Flechte, Pilz oder Alge, die sogen. Flechtenstoffe zu erzeugen imstande ist, blieb bis in neuester Zeit unbeantwortet. Nun haben aber Castle und Kubsch (35) die beiden Komponenten von Cladonia cristatella Tuck. getrennt kultiviert und in der Kultur des Pilzteils Usnin- sowie Didymsäure (die spezifischen Bestandteile der Cl. cristatella) mikrochemisch nachgewiesen. Auch wurde die Bildung des roten Farbstoffs der Früchte, Rhodocladonsäure, an gewissen Stellen des Myceliums beobachtet.

IV. Antibiotische Wirkung von Flechtenstoffen.

Burkholder, Evans, McVeigh und Thornton (33, 34) zeigten, daß gewisse Inhaltsstoffe mancher Flechten gegen die meisten gram-positiven Bakterien wachstumshemmend wirken. Unter den spezifischen Flechtenbestandteilen zeigte die Usninsäure deutlich diese Hemmungs-wirkung.

Asano und Mitarbeiter (29) prüften die antibiotische Wirkung von Vertretern der Protolichesterin-Gruppe auf Staphylococcus citreus, mit dem Ergebnis, daß die Dihydro-protolichesterinsäure sowie Lichesterin-säure weitaus wirksamer sind als Protolichesterinsäure.

Nach Stoll, Renz und Brack (100) wirkt die Usninsäure sowohl auf gram-positive als auch auf gram-negative Bakterien wachstums-hemmend, namentlich auf menschliche Tuberkelbazillen, sogar in Verdünnungen von $1:1000000$ bis $1:2000000$. Über die Wirkungs-weise s. Johnson, Feldott und Lardy (57).

Shibata, Ukita, Miura und Tamura (94, 95) haben die Versuche von Stoll nachgeprüft und fanden, daß die Usninsäure (D, L oder rac.) auf gram-positive Bakterien stärker als auf gram-negative Bakterien wachstumshemmend wirkt. Besonders hemmt die Usninsäure das Mycobacterium avium in Verdünnungen von $1:160000$ bis $1:320000$. Diese antibiotische Aktivität der Usninsäure ist bedingt durch die Gegen-wart einer α,β-ungesättigten Carbonyl-Gruppe im rechtsstehenden Kern und zwei phenolischer Hydroxyle im linksstehenden, da die Hemmungskraft der Diacetyl-, Dihydro- und Diacetyldihydro-usnin-säuren stark herabgesetzt ist. In Finnland ist man so weit gegangen, tuberkulösen Kranken Usninsäure innerlich zu verabreichen (73).

Noch wirksamer ist nach Shibata und Miura (93) die Decarbo-nordidymsäure auf Staphylococcus aureus und Mycobacterium avium in vitro, in einer Verdünnung von $1:640000$.

V. Mikrochemischer Nachweis der Flechtenstoffe.

Seit der Einführung der mikrochemischen Methode zur Erkennung der Inhaltsstoffe in Flechtenfragmenten durch ASAHINA wurde dieses Verfahren von Taxonomen wiederholt nachgeprüft. Am eingehendsten beschäftigte sich EVANS (39) mit der Methode und behandelte 19 Flechtenstoffe, die in nordamerikanischen Cladonien vorkommen.

Literaturverzeichnis.

1. AOKI, H.: Constitution of Norrangiformic Acid. J. pharmac. Soc. Japan 66 A, 52 (1946).

2. ASAHINA, Y.: Flechtenstoffe. Fortschr. Chem. organ. Naturstoffe 2, 27 (1939).

3. — Pannarin, ein Bestandteil einiger Pannaria-Arten. J. Japan. Bot. 16, 403 (1940).

4. — Nachweis der Perlatolinsäure in *Cladonia impexa* und verwandten Arten. J. Japan. Bot. 16, 185 (1940).

5. ASAHINA, Y. u. H. AKAGI: Untersuchungen über Flechtenstoffe. LXXXVIII. Mitt. Über die Zeoringruppe. I. Ber. dtsch. chem. Ges. 71, 980 (1938).

6. ASAHINA, Y. u. H. AOKI: Über Oxy-diphenylenoxyd-Derivate. J. pharmac. Soc. Japan 64, 41 (1944).

7. ASAHINA, Y., H. AOKI u. F. FUZIKAWA: Untersuchungen über Flechtenstoffe. XCVI. Mitt. Über ein neues Depsid „Hypothamnolsäure". Ber. dtsch. chem. Ges. 74, 824 (1941).

8. ASAHINA, Y. u. H. HAYASHI: Untersuchungen über Flechtenstoffe. XXVI. Mitt. Über Psoromsäure. Ber. dtsch. chem. Ges. 66, 1023 (1933).

9. — — Untersuchungen über Flechtenstoffe. LXXVIII. Mitt. Über Psoromsäure. II. Ber. dtsch. chem. Ges. 70, 810 (1937).

10. ASAHINA, Y. u. K. KIN: Thermische Zersetzung der Dihydrousninsäure. Proc. Imp. Acad. (Tokyo) 20, 371 (1944).

11. ASAHINA, Y. u. T. KUSAKA: Untersuchungen über Flechtenstoffe. XCVII. Mitt. Über ein neues Tridepsid „Hiascinsäure". Bull. chem. Soc. Japan 17, 152 (1942).

12. ASAHINA, Y. u. H. NOGAMI: Untersuchungen über Flechtenstoffe. XCVIII. Mitt. Über Lichexanthon, ein neues Stoffwechselprodukt der Flechte. Bull. chem. Soc. Japan 17, 202 (1942).

13. — — Konstitution der Dioxy-homophthalsäure und der -terephthalsäure, abgeleitet vom Orcin-tricarbonsäure-triäthylester. Proc. Imp. Acad. (Tokyo) 16, 119 (1940).

14. — — Untersuchungen über Flechtenstoffe. XCIX. Mitt. Synthese der Dimethyläther-olivetonsäure und des Dimethyläther-olivetonids. Bull. chem. Soc. Japan 17, 221 (1942).

15. ASAHINA, Y. u. K. OKAZAKI: Über die Konstitution des Decarbusnols. Proc. Imp. Acad. (Tokyo) 19, 303 (1943).

16. — — Untersuchungen über Flechtenstoffe. CI. Mitt. V. Ozonspaltung der Diacetyl-usninsäure und die Konstitution der Usninsäure. J. pharmac. Soc. Japan 63, 626 (1943).

17. ASAHINA, Y. u. T. SASAKI: Untersuchungen über Flechtenstoffe. C. Mitt. Über Rangiformsäure. Bull. chem. Soc. Japan 17, 495 (1942).

18. ASAHINA, Y. u. S. SHIBATA: Untersuchungen über Flechtenstoffe. XCII. Mitt. Über Psoromsäure. III. Ber. dtsch. chem. Ges. 72, 1399 (1939).

19. — — Untersuchungen über Flechtenstoffe. XCIV. Mitt. Über das Vorkommen der Thelephorsäure in den Flechten. Ber. dtsch. chem. Ges. 72, 1531 (1939).

20. Asahina, Y. u. Z. Simosato: Untersuchungen über Flechtenstoffe. XC. Mitt. Über Monomethyläther-orcin-dicarbonsäuren und die Nicht-Existenz der sog. Iso-squamatsäure. Ber. dtsch. chem. Ges. 71, 2561 (1938).

21. Asahina, Y. u. T. Tukamoto: Untersuchungen über Flechtenstoffe. XLII. Mitt. Bestandteile einiger Usnea-Arten unter besonderer Berücksichtigung der Verbindungen der Salazinsäure-Gruppe. II. Ber. dtsch. chem. Ges. 67, 963 (1934).

22. Asahina, Y. u. M. Yanagita: Untersuchungen über Flechtenstoffe. XVIII. Mitt. Über Iso-Squamatsäure, ein neues Depsid aus Cladonia Boryi Tuck. Ber. dtsch. chem. Ges. 66, 393 (1933).

23. — — Untersuchungen über Flechtenstoffe. LXXXII. Mitt. Über die Usninsäure. III. Ber. dtsch. chem. Ges. 70, 1500 (1937).

24. — — Untersuchungen über Flechtenstoffe. LXXXVII. Mitt. Über die Usninsäure. IV. Ber. dtsch. chem. Ges. 70, 2462 (1937).

25. — — Untersuchungen über Flechtenstoffe. LXXXIX. Mitt. Über die Usninsäure. V. Ber. dtsch. chem. Ges. 71, 2260 (1938).

26. — — Untersuchungen über Flechtenstoffe. XCI. Mitt. Über die Usninsäure. VI. Ber. dtsch. chem. Ges. 72, 1140 (1939).

27. Asahina, Y. u. I. Yosioka: Über die Methyl-Abspaltung eines o-Methoxybenzaldehyd-anils. Ber. dtsch. chem. Ges. 69, 1367 (1936).

28. — — Untersuchungen über Flechtenstoffe. XCV. Mitt. Über die Zeorin-Gruppe. II. Ber. dtsch. chem. Ges. 73, 742 (1940).

29. Asano, M., Y. Kameda, J. Komai and I. Ishino: Antibiotic Activity of Fatty Acids Upon Staphylococcus citreus. I. Igaku-sôran 1, 41 (1945).

30. Birch, H. F., D. G. Flynn and A. Robertson: Usnic Acid. IV. The Synthesis of 4:6-Dimethoxy-3:5-dimethylcoumarone-2-acetic Acid. J. chem. Soc. (London) 1936, 1834.

31. Breaden, T. W., J. Keane and T. J. Nolan: The Chemical Constituents of Lichens Found in Ireland. Cladonia impexa Harm. Sci. Proc. Roy. Dublin Soc. (N. S.) 23, 6 (1942).

32. Breen, J., J. Keane and T. J. Nolan: The Chemical Constituents of Lichens Found in Ireland. Pertusaria concreta Nyl. form Westringii Nyl. Sci. Proc. Roy. Dublin Soc. (N. S.) 21, 587 (1937).

33. Burkholder, P. R. and A. W. Evans: Further Studies on the Antibiotic Activity of Lichens. Bull. Torrey bot. Club 72, 157 (1945).

34. Burkholder, P. R., A. W. Evans, I. McVeigh and H. K. Thornton: Antibiotic Activity of Lichens. Proc. nat. Acad. Sci. (U. S. A.) 30, 250 (1944).

35. Castle, H. and F. Kubsch: The Production of Usnic, Didymic and Rhodocladonic Acids by the Fungal Component of the Lichen Cladonia cristatella. Arch. Biochemistry 23, 158 (1949).

36. Crabtree, H. G. and R. Robinson: A Synthesis of iso-Brazilein and Certain Related Anhydropyranol Salts. I. J. chem. Soc. (London), Trans. 1918, 859.

37. Curd, F. H. and A. Robertson: Usnic Acid. V. J. chem. Soc. (London) 1937, 894.

38. Davidson, V. E., J. Keane and T. J. Nolan: The Chemical Constituents of Lichens Found in Ireland. Lecanora gangaleoides. III. The Constitution of Gangaleoidin. Sci. Proc. Roy. Dublin Soc. (N. S.) 23, 143 (1943).

39. Evans, A. W.: Asahina's Microchemical Studies on the Cladoniae. Bull. Torrey bot. Club 70, 139 (1943).

40. Foster, M. O. and W. B. Saville: Constitution of Picroroccellin, a Diketopiperazine Derivative from Roccella fuciformis. J. chem. Soc. (London), Trans. 121, 816 (1922).

41. FOSTER, R. T., A. ROBERTSON and T. V. HEALY: Usnic Acid. VII. Analogues of Usnolic Acid. J. chem. Soc. (London) **1939**, 1594.

41a. FRANK, R. L., S. M. COHEN and J. N. COKER: The Structures and Syntheses of Rhizocarpic Acid and Epanorin. J. Amer. chem. Soc. **72**, 4454 (1950).

42. GIBBS, H. D.: Phenol Tests. III. The Indophenol Test. J. biol. Chemistry **72**, 649 (1927).

43. HARDIMAN, J., J. KEANE and T. J. NOLAN: The Chemical Constituents of Lichens Found in Ireland. *Lecanora gangaleoides.* I. Sci. Proc. Roy. Dublin Soc. (N. S.) **21**, 141 (1935).

44. HESSE, O.: Über einige Flechtenstoffe. Liebigs Ann. Chem. **284**, 157 (1895).

45. — Beitrag zur Kenntnis der Flechten und ihrer charakteristischen Bestandteile. I. Mitt. J. prakt. Chem. (2) **57**, 232 (1898).

46. — Beitrag zur Kenntnis der Flechten und ihrer charakteristischen Bestandteile. III. Mitt. J. prakt. Chem. (2) **58**, 465 (1898).

47. — Beitrag zur Kenntnis der Flechten und ihrer charakteristischen Bestandteile. V. Mitt. J. prakt. Chem. (2) **62**, 430 (1900).

48. — Beitrag zur Kenntnis der Flechten und ihrer charakteristischen Bestandteile. VI. Mitt. J. prakt. Chem. (2) **63**, 522 (1900).

49. — Beitrag zur Kenntnis der Flechten und ihrer charakteristischen Bestandteile. VII. Mitt. J. prakt. Chem. (2) **65**, 537 (1902).

50. — Beitrag zur Kenntnis der Flechten und ihrer charakteristischen Bestandteile. VIII. Mitt. J. prakt. Chem. (2) **68**, 1 (1903).

51. — Beitrag zur Kenntnis der Flechten und ihrer charakteristischen Bestandteile. X. Mitt. J. prakt. Chem. (2) **73**, 113 (1906).

52. — Beitrag zur Kenntnis der Flechten und ihrer charakteristischen Bestandteile. XI. Mitt. J. prakt. Chem. (2) **76**, 1 (1907).

53. — Über die wichtigsten Orseilleflechten und ihre Chromogene. Liebigs Ann. Chem. **139**, 22 (1866).

54. HOFFMANN-OSTENHOF, O.: Vorkommen und biochemisches Verhalten der Chinone. Fortschr. Chem. organ. Naturstoffe **6**, 154 (1950).

55. JEGER, O.: Über die Konstitution der Triterpene. Fortschr. Chem. organ. Naturstoffe **7**, 1 (1950).

56. JERDAN, D. S.: The Condensation of Ethylic Acetone-dicarboxylate and Constitution of Triethylic Orcinol-tricarboxylate. J. chem. Soc. (London), Trans. **75**, 808 (1899).

57. JOHNSON, R. B., G. FELDOTT and H. A. LARDY: The Mode of Action of the Antibiotic, Usnic Acid. Arch. Biochemistry **28**, 317 (1950).

58. JONES, M. P., J. KEANE and T. J. NOLAN: Lichen Substances Containing Nitrogen. Nature (London) **154**, 580 (1944).

59. KENNEDY, G., J. BREEN, J. KEANE and T. J. NOLAN: The Chemical Constituents of Lichens Found in Ireland. *Lecanora sordida* TH. FR. Sci. Proc. Roy. Dublin Soc. (N. S.) **21**, 557 (1937).

60. KOLLER, G. u. H. HAMBURG: Über die Rhodocladonsäure. Monatsh. Chem. **68**, 202 (1936).

61. KUSAKA, T.: Über die Synthese des Rubiadins. J. pharmac. Soc. Japan **55**, 682 (1935).

62. — Synthese der Hiascinsäure-Derivate. J. pharmac. Soc. Japan **61**, 353 (1941).

63. LAMPARTER, H.: Chemische Untersuchungen einiger Flechtenstoffe. Liebigs Ann. Chem. **134**, 243 (1865).

64. MURPHY, D., J. KEANE and T. J. NOLAN: The Chemical Constituents of Lichens Found in Ireland. *Lecanora parella* ARCH. The Constitution of Variolaric Acid. Sci. Proc. Roy. Dublin Soc. (N. S.) **23**, 71 (1943).

65. Nogami, H.: Über Didymsäure. J. pharmac. Soc. Japan 64, 47 (1944).
66. Nolan, T. J.: The Chemical Constituents of Lichens Found in Ireland. *Buellia canescens*. I. Sci. Proc. Roy. Dublin Soc. (N. S.) 21, 67 (1934).
67. Nolan, T. J., J. Algar, E. P. McCann, W. A. Manahan and N. Nolan: The Chemical Constituents of Lichens Found in Ireland. *Buellia canescens*. III. The Constitution of Diploicin. Sci. Proc. Roy. Dublin Soc. (N. S.) 24, 319 (1948).
68. Nolan, T. J. and J. Keane: The Chemical Constituents of Lichens Found in Ireland. *Lecanora gangaleoides*. II. Sci. Proc. Roy. Dublin Soc. (N. S.) 22, 199 (1940).
69. Nolan, T. J., J. Keane and V. E. Davidson: The Chemical Constituents of the Lichen *Parmelia latissima* Fée. Sci. Proc. Roy. Dublin Soc. (N. S.) 22, 237 (1940).
70. Paternò, E. (nach Schiffs abgekürztem Bericht). Ber. dtsch. chem. Ges. 9, 345 (1876).
71. — Über Atranorsäure und Rangiformsäure. Gazz. chim. ital. 12, 256 (1882).
72. — Ricerche sull'acido usnico e sopra altre sostanze estratte dei licheni. Gazz. chim. ital. 12, 231 (1882).
73. Pätiälä, R., J. Pätiälä, S. Siitola and P. Heilala: The Effects of *l*-Usnic Acid in vivo (on Tuberculosis). Suomen Kemistilehti 21 A, 217 (1948) [Chem. Abstr. 43, 9267 (1948)].
74. Rao, V. S. and T. R. Seshadri: Chemical Investigation of Indian Lichens. I. Chemical Components of *Roccella Montagnei*. Proc. Indian Acad. Sci., Sect. A 12, 466 (1940).
75. — — Chemical Investigation of Indian Lichens. III. Isolation of Montagnetol, a New Phenolic Compound from *Roccella Montagnei*. Proc. Indian Acad. Sci., Sect. A 13, 199 (1941).
76. — — Chemical Investigation of Indian Lichens. IV. Constitution of Montagnetol. Proc. Indian Acad. Sci., Sect. A 15, 18 (1942).
77. — — Chemical Investigation of Indian Lichens. V. Occurrence of Active Montagnetol in *Roccella Montagnei*. Proc. Indian Acad. Sci., Sect. A 15, 429 (1942).
78. — — Chemical Investigation of Indian Lichens. VI. Constitution of Erythrin. Proc. Indian Acad. Sci., Sect. A 16, 23 (1942).
79. Robinson, R.: The Molecular Architecture of Some Plant Products. IX° Congr. intern. quím. pura y aplicada, Madrid (1934).
80. Sakurai, Y.: Über die Konstitution des Erythrins. J. pharmac. Soc. Japan 61, 108 (1941).
81. Schmidt, O.: Die inneren Energie-Verhältnisse organischer Substanzen. VI. Mitt. Die Spaltung in der Hexaphenyläthan- und Zucker-Reihe auf Grund der Doppelbindungs-Regel. Ber. dtsch. chem. Ges. 68, 759 (1935).
82. Schöpf, Cl., K. Heuck u. R. Duntze: Die Konstitution der Barbatolsäure. Liebigs Ann. Chem. 491, 220 (1931).
83. Schöpf, Cl. u. F. Ross: Die Konstitution der Usninsäure. Naturwiss. 26, 772 (1938).
84. — — Die Konstitution der Usninsäure. II. Liebigs Ann. Chem. 546, 1 (1941).
85. Schunck, E.: Über die Bestandteile der *Lecanora parella*. Liebigs Ann. Chem. 54, 257 (1845).
86. — Über die in *Roccella tinctoria* enthaltenen Stoffe. Liebigs Ann. Chem. 61, 64 (1847).
87. Seshadri, T. R.: A Theory of Biogenesis of Lichen Depsides and Depsidones. Proc. Indian Acad. Sci., Sect. A 20, 1 (1944).

88. SESHADRI, T. R. and S. S. SUBRAMANIAN: Chemical Investigation of Indian Lichens. VIII. Some Lichens Growing on Sandal Trees: *Ramalina Tayloriana* and *Roccella Montagnei.* Proc. Indian Acad. Sci., Sect. A **30**, 15 (1949).

89. SHIBATA, S.: Synthese des 3,6,3'-Trimethyl-5,2',4'-trimethoxy-diphenyläthers. J. pharmac. Soc. Japan **59**, 323 (1939).

90. — Über die Rhodocladonsäure. J. pharmac. Soc. Japan **61**, 103 (1941).

91. — Über Didymsäure, einen neuen Typus der Flechtenstoffe. Acta phytochim. (Tokyo) **14**, 9 (1944).

92. — Über Strepsilin, einen Flechtenstoff vom Diphenylenoxyd-Typus. Acta phytochim. (Tokyo) **14**, 177 (1944).

93. SHIBATA, S. and Y. MIURA: Comparative Studies of Antibacterial Effects of Lichen Substances. Japan. Med. J. **2**, 22, 518 (1949).

94. SHIBATA, S., T. UKITA, Y. MIURA and T. TAMURA: Relation Between Antibacterial Properties and Chemical Constitution of Usnic Acid and its Derivatives. J. Penicillin (Japan) **1**, 588 (1948).

95. SHIBATA, S., T. UKITA, T. TAMURA and Y. MIURA: Relation Between Chemical Constitution and Antibacterial Effect of Usnic Acid and its Derivatives. Japan. Med. J. **1**, 152 (1948).

96. SPILLANE, P. A., J. KEANE and T. J. NOLAN: The Chemical Constituents of Lichens Found in Ireland. *Buellia canescens.* II. Sci. Proc. Roy. Dublin Soc. (N. S.) **21**, 333 (1936).

97. STENHOUSE, J.: Über die näheren Bestandteile einiger Flechten. Liebigs Ann. Chem. **68**, 55 (1848).

98. STENHOUSE, J. u. C. E. GROVES: Über Picroroccellin. Liebigs Ann. Chem. **185**, 14 (1877).

99. — — Note on Usnic Acid and Some Products of its Decomposition. J. chem. Soc. (London), Trans. **39**, 234 (1881).

100. STOLL, A., J. RENZ and A. BRACK: Antibiotics from Lichens. Experientia **3**, 111 (1947).

101. STRECKER, A.: Bemerkungen zu vorstehender Abhandlung [cf. ref. (97)]. Liebigs Ann. Chem. **68**, 108 (1848).

102. SUOMINEN, E. E.: Die Konstitution der Barbatolsäure (vorl. Mitt.). Suomen Kemistilehti **12** B, 26 (1939) [Chem. Zbl. **1940** I, 385].

103. TOMITA, M., Y. INUBUSHI u. F. KUSUTA: Synthese des Depsidon-Skeletts. J. pharmac. Soc. Japan **64**, 173 (1944).

104. WIDMAN, O.: Zur Kenntnis der Usninsäure. Liebigs Ann. Chem. **310**, 230 (1900).

105. — Zur Kenntnis der Usninsäure. Liebigs Ann. Chem. **324**, 139 (1902).

106. YANAGITA, M.: Synthese einiger Zersetzungsprodukte der Dihydrousninsäure. Ber. dtsch. chem. Ges. **71**, 2269 (1938).

107. YOSIOKA, I.: Konstitution des Pannarins. J. pharmac. Soc. Japan **61**, 332 (1941).

108. ZELLNER, J.: Die Struktur des Erythrins. Monatsh. Chem. **35**, 1021 (1914).

109. ZOPF, W.: Zur Kenntnis der Flechtenstoffe. I. Liebigs Ann. Chem. **284**, 107 (1895).

110. — Zur Kenntnis der Flechtenstoffe. II. Mitt. Über Atranorsäure und ihre Begleitstoffe. Liebigs Ann. Chem. **288**, 38 (1895).

111. — Zur Kenntnis der Flechtenstoffe. V. Mitt. Liebigs Ann. Chem. **300**, 322 (1898).

112. — Zur Kenntnis der Flechtenstoffe. VII. Mitt. Liebigs Ann. Chem. **313**, 317 (1900).

113. Zopf, W.: Zur Kenntnis der Flechtenstoffe. IX. Mitt. Liebigs Ann. Chem. **321,** 37 (1902).

114. — Zur Kenntnis der Flechtenstoffe. XI. Mitt. Liebigs Ann. Chem. **327,** 317 (1903).

115. — Zur Kenntnis der Flechtenstoffe. XII. Mitt. Liebigs Ann. Chem. **336,** 46 (1904).

116. — Die Flechtenstoffe. Jena. 1907.

(Eingelaufen am 14. November 1950.)

Lupinen-Alkaloide
und verwandte Verbindungen.

Von F. GALINOVSKY, Wien.

Inhaltsübersicht.

I. Einleitung.

Man kennt heute etwa hundert zumeist in Nordamerika sowie in den
Mittelmeerländern und Europa heimische Lupinenarten. Die Kultur der
Lupinen, die seit jeher nicht nur als Zierpflanzen eine Rolle spielten,
sondern auch eine gewisse wirtschaftliche Bedeutung besaßen, reicht weit
in das ägyptische und griechische Altertum zurück. Nach Mitteleuropa
kamen die Lupinen erst im Mittelalter und wurden hier im großen land-
wirtschaftlich zur Bodenverbesserung und als Futterpflanzen angebaut.
Auch für die menschliche Ernährung fanden Lupinensamen zeitweise
Verwendung. Infolge der Fähigkeit der Lupinen zur symbiotischen
Stickstoff-Fixierung enthalten ihre Samen neben Fetten und Kohlenhydraten auch reichlich Eiweiß, welches für Nahrungszwecke von Bedeutung sein könnte, wenn nicht die bitter-schmeckenden und bei reich-

lichem Genuß auch gefährlichen Alkaloide, die in den Samen vorkommen, diese Verwendung ausschließen, bzw. einschränken würden. Die Bedeutung der Lupinen als Futterpflanzen, vor allem aber für die Bodenverbesserung (Gründüngung), für die sie wegen ihrer Schnellwüchsigkeit besonders geeignet erscheinen, war deshalb immer erheblich größer.

Die Stickstoff-Fixierung, zu der die Lupinen befähigt sind, beruht darauf, daß der Stickstoff der Luft durch die mit ihnen in Symbiose lebenden Knöllchenbakterien, die in den gallenähnlichen Anschwellungen an ihren Wurzeln vorkommen, assimiliert und weiters in Aminosäuren und Eiweiß übergeführt wird. Ein Hektar (2,471 acres) Lupinen kann im Laufe einer Vegetationsperiode so über 200 kg Stickstoff aus der Luft binden (*34*). Die Lupinen können daher wie die meisten anderen Leguminosen auch auf stickstoff-armen Böden gedeihen. Sie bereichern dann nach ihrem Absterben den Boden und machen ihn für die nachfolgenden Nicht-leguminosen fruchtbarer.

Man hat sich auch in den letzten Jahrzehnten bemüht, die Lupinensamen durch Entfernung der Alkaloide aus dem Lupinenmehl mittels Extraktion und anderer Verfahren (*46*) zu entbittern und so für Nahrungszwecke brauchbar zu machen. Auch wurden in Zeiten des Rohstoffmangels die Samen zur Ölgewinnung und das Kraut für Spinnzwecke verwendet; schließlich konnten durch Auslese alkaloid-arme Pflanzen, Süßlupinen, gezüchtet werden (*91*). Trotz der erzielten Fortschritte tritt aber die Bedeutung der Lupinen für Nahrungs- und industrielle Zwecke gegenüber der anderer Leguminosen, z. B. der Sojabohnen, zurück.

Die Lupinen-Alkaloide kommen in allen Teilen der Pflanzen — nicht nur im Samen — in nach Jahreszeit und Standort wechselnden Mengen vor. Ihr pharmakologischer und medizinischer Wert ist nicht bedeutend. Gerade das Spartein, welches nur eine geringe Toxizität aufweist, wurde zeitweise als Herzmittel empfohlen.

II. Lupinen-Alkaloide bekannter Konstitution.

1. Bemerkungen zur Konstitutionsaufklärung.

Über die Konstitution der wichtigsten Lupinen-Alkaloide *Lupinin*, *Spartein* und *Lupanin*, die an sich schon sehr lange bekannt sind, war man trotz der Bedeutung der Lupinen noch vor 25 Jahren im ungewissen. Das erklärt sich durch die schon oben erwähnte, mehr negative Rolle, die sie spielen, und ihre geringere pharmakologische Bedeutung. Auch der chemischen Bearbeitung setzten die Lupinen-Alkaloide durch die besondere bicyclische Bindungsart der Stickstoffatome in meist hydrierten Ringen ziemliche Schwierigkeiten entgegen.

Erst im Jahre 1928 gelang Karrer und Mitarbeitern (*51*) ein entscheidender Eingriff in das relativ am einfachsten gebaute Lupinin (I);

in der Folge wurden auch die übrigen wichtigen Lupinen-Alkaloide aufgeklärt. Das wichtigste Alkaloid des zweiten Typs der Lupinen-
Alkaloide, welcher zwei Stickstoffatome besitzt, das Spartein, bot besondere Schwierigkeiten. Die richtige Aneinanderreihung der Ringe
konnte erst auf indirektem Wege durch die Reduktion des Anagyrins,
das zuerst in *Anagyris foetida*, in jüngster Zeit auch in Lupinen aufgefunden wurde, zum Spartein (II) festgestellt werden [ING (50)]. Die
Art der Ringverknüpfung beim Anagyrin wurde hauptsächlich durch die
Ergebnisse des HOFMANNschen Abbaus (50), der analog dem des Cytisins
[SPÄTH und GALINOVSKY (94)] verläuft, bewiesen. Die letzten Fragen
der Konstitution des Sparteins klärten dann vor allem synthetische Versuche, die zum *D,L*-Oxospartein führten [CLEMO, MORGAN und RAPER
(10)]. Damit war auch die Struktur des Lupanins (III), das einerseits
durch Hydrierung aus dem Anagyrin entsteht, andererseits sich reduktiv
in Spartein überführen läßt, gegeben. Das Grundgerüst dieser Alkaloide
ist der Norlupinanring (IV), der im Lupinin einfach und in den Alkaloiden
der Sparteingruppe mit einem zweiten kondensiert vorkommt.

(I.) Lupinin.   (II.) Spartein.   (III.) Lupanin.

Die weitere Erforschung der Lupinen-Alkaloide ist im Gange. In
letzter Zeit wurden sowohl synthetisch wie auf dem Wege der Konstitutionsaufklärung neue Fortschritte erzielt. Davon soll in den betreffenden
Kapiteln die Rede sein. Allgemein kann gesagt werden, daß aber bisher
außer den zwei Typen des Lupinins und des Sparteins kein Lupinen-
Alkaloid eines anderen Typs sicher festgestellt werden konnte.

2. Lupinin.

Lupinin, $C_{10}H_{19}ON$, kommt neben Spartein in *Lupinus luteus*,
L. niger und *L. palmeri* vor (22), außerdem wurde es in *Anabasis aphylla*,
einer Chenopodiacee, aufgefunden. Es kristallisiert (F. 69 bis 70°) und
ist optisch aktiv (linksdrehend). Durch ältere Arbeiten [Zusammen-
fassung dieser siehe (45)] war gezeigt worden, daß es ein bicyclisch gebundenes Stickstoffatom enthält, während das Sauerstoffatom in einer
primären Hydroxylgruppe vorliegt. Bei der Oxydation entsteht die
Lupininsäure, welche sich zum Norlupinan (IV) decarboxylieren läßt. Bei
vorsichtiger Oxydation konnte auch der Lupininaldehyd isoliert werden
(121). Beim Ersatz der Hydroxylgruppe durch Wasserstoff, der über das

Chlor-lupinan oder das Anhydro-lupinin durchgeführt werden kann, wurde Lupinan (V) erhalten.

(IV.) Norlupinan. (V.) Lupinan.

Karrer, Canal, Zohner und W'dmer (*51*) stellten aus Lupinin (I) durch dreimaligen Hofmannschen Abbau, wobei die ungesättigten Zwischenstufen jedesmal katalytisch hydriert wurden, einen stickstoff-freien Alkohol (VI) dar. Dieser wurde in das Bromid übergeführt, das mit Trimethylamin das quartäre Salz gab. Beim Abbau der quartären Base nach Hofmann wurde die ungesättigte Verbindung (VII) erhalten. Die Ozonisierung von (VII) lieferte ein Keton, für das die Konstitution eines *n*-Amyl-*n*-propyl-ketons (VIII) bewiesen wurde.

(I.) Lupinin. (VI.)

$$\longrightarrow CH_3 \cdot (CH_2)_4 \cdot C \cdot (CH_2)_2 \cdot CH_3 \longrightarrow CH_3 \cdot (CH_2)_4 \cdot CO \cdot (CH_2)_2 \cdot CH_3$$

(VII.) (VIII.) *n*-Amyl-*n*-propyl-keton.

Durch diesen Abbau und die folgenden Synthesen des Norlupinans (IV) wurde die Formel (I) für das Lupinin eindeutig bewiesen. Für das Norlupinan (Chinolizidin, Octahydrochinolizin, Octahydropyridocolin) gibt es heute eine Reihe von Synthesen, von denen hier die für dieses Ringsystem typischen angeführt seien [vgl. auch (*45, 6*)].

Synthesen des Norlupinans. Die ersten Synthesen des Norlupinans haben fast gleichzeitig Winterfeld und Holschneider (*110*) sowie Clemo, Ramage und Raper (*15*) durchgeführt. Die erstgenannten Autoren gingen vom Picolinsäure-äthylester aus, der mit N-Methyl-α-pyrrolidon das Kondensationsprodukt (IX) gab, das mit Salzsäure unter Decarboxylierung in die Verbindung (X) überging. Bei der kata-lytischen Hydrierung des N-Benzoylderivates von (X) wurde der Pyridinring hydriert und die Ketogruppe zur Hydroxylgruppe reduziert. Das Re-duktionsprodukt lieferte mit PBr$_5$ das Dibromid (XI). Der Ringschluß mit Natriumäthylat und die katalytische Entbromung ergaben daraus Norlupinan.

Picolinsäure-äthylester.   N-Methyl-α-pyrrolidon.   (IX.)

(X.)   (XI.)   (IV.) Norlupinan.

Die Synthese von CLEMO und Mitarbeitern (*15*) ging gleichfalls vom Picolinsäure-ester aus, der mit Bernsteinsäure-ester kondensiert, den β-(α-Pyridoyl)-propionsäure-ester (XII) gab, der nach CLEMMENSEN zur γ-(α-Pyridyl)-buttersäure (XIII) reduziert wurde. Der Ester wurde nach BOUVEAULT und BLANC zum Piperidyl-butanol (XIV) reduziert und dieses

(XII.) β-(α-Pyridoyl)-propionsäure-ester.

(XIII.)   (XIV.)   (XV.)
γ-(α-Pyridyl)-buttersäure.   Piperidyl-butanol.

(XVI.) α-Norlupinon.   (IV.) Norlupinan.

in das Bromid (XV) übergeführt, das mit Lauge leicht zum Norlupinan (IV) ringgeschlossen wurde. Die Synthese läßt sich noch vereinfachen, wenn man (XIII) zuerst katalytisch hydriert und die gebildete γ-(α-Piperidyl)-buttersäure durch Erhitzen in das α-Norlupinon (XVI) überführt, das elektrolytisch oder besser durch katalytische Hydrierung mit PtO_2 nach Adams (unter Normalbedingungen) in das Norlupinan übergeht (42).

Zwei weitere Synthesen von Clemo und Mitarbeitern (11, 9) gehen ebenfalls von Pyridinderivaten aus, die in Dicarbonsäureester mit bereits tertiärem Stickstoff übergeführt werden, wonach die Ringschlüsse mittels Esterkondensation nach Dieckmann erfolgen. Die entstandenen Keto-norlupinane geben bei der Clemmensen-Reduktion Norlupinan.

Eine weitere Norlupinansynthese von Prelog und Božičević (84) benützt das Prinzip der zweifachen intramolekularen Alkylierung. So wurde das Norlupinan aus 1,9-Dibrom-5-amino-nonan (XVII), das in einer sieben-stufigen Synthese erhalten wurde, dargestellt.

$$\text{Br}\cdot(CH_2)_4\text{—CH—}(CH_2)_4\cdot\text{Br} \longrightarrow \quad \text{(IV) Norlupinan.}$$
$$|$$
$$NH_2$$

(XVII.) 1,9-Dibrom-5-amino-nonan.

Schließlich haben Boekelheide und Rothchild (4, 5) in einer zur Darstellung größerer Mengen geeigneten Synthese das α-Norlupinon (XVI) bei 260° mit Kupferchromit als Katalysator bei einem Wasserstoffdruck von 250 atm zum Norlupinan (IV) reduziert. Die Synthese des α-Norlupinons ging von der Kondensation (nach Michael) des 2-Vinyl-pyridins mit Malonsäureester zu (XVIII) aus. Unter reduktiver Cyclisierung mit Raney-Nickel wurde der 4-Oxo-norlupinan-3-carbonsäureester (XIX) erhalten, der bei nachfolgender Verseifung und Decarboxylierung α-Norlupinon (XVI) ergab.

2-Vinyl-pyridin. (XVIII.)

(XIX.) 4-Oxo-norlupinan-3-carbonsäure-ester.

$\longrightarrow$ (XVI.) α-Norlupinon $\longrightarrow$ (IV.) Norlupinan.

Es ist noch zu bemerken, daß vom Norlupinan bisher nur eine einzige Verbindung bekannt geworden ist. Es liegt bis heute kein Beweis für eine der *cis-trans*-Isomerie des Dekalins analoge Stereoisomerie bei diesem Ringsystem vor. Eine zuerst als ein stereoisomeres Norlupinan angesehene Verbindung, welche bei der Reduktion von 1-Keto-norlupinan (XX) nach CLEMMENSEN erhalten worden war (*14*), ist in Wirklichkeit ein Umlagerungsprodukt, nämlich das 1-Aza-bicyclo-[0,3,5]-decan (XXI) (*85*). Bei der Reduktion von (XX) nach WOLFF-KISHNER wird dagegen das erwartete Norlupinan gebildet (*9*).

(XX.) 1-Keto-norlupinan.   (XXI.) 1-Aza-bicyclo-[0,3,5]-decan.

Diese eigentümliche Umlagerung von cyclischen Aminoketonen mit tertiärem Stickstoffatom, die nur bei der Reduktion nach CLEMMENSEN eintritt, wurde neuerdings von LEONARD und Mitarbeitern (*59, 55*) auch auf monocyclische Systeme übertragen. Das Studium des Reaktionsmechanismus ergab, daß immer der Ring, welcher die Ketogruppe enthält, verengt wird (*60, 18*). So ergab die Reduktion des 6-Keto-1-aza-bicyclo-[5,3,0]-decans (XXII) nach CLEMMENSEN wieder das Norlupinan (*61*).

(IV.) Norlupinan.

(XXII.) 6-Keto-1-aza-bicyclo-[5,3,0]-decan.

Von Interesse ist auch die Dehydrierung des α-Norlupinons (XVI) mit Palladium, die zu einer gelben kristallisierten Verbindung (XXIII) (F. 73°) und α-Propyl-pyridin führt (*96*).

(XVI.) α-Norlupinon.   (XXIII.)

Die Verbindung (XXIII) enthält zweifach das in manchen Eigenschaften benzoide Ringsystem des α-Pyridons und stellt den Grundkörper einer Reihe beim Abbau oder bei der Synthese von verwandten Substanzen erhaltener Verbindungen vor.

Synthese des β-Lupinans. Die eine racemische Form des Lupinans (V) haben WINTERFELD und HOLSCHNEIDER (*111*) synthetisiert. Sie setzten α-Pyridyl-methyl-keton mit γ-Brompropyläther nach GRIGNARD zur Verbindung (XXIV) um. Diese wurde katalytisch hydriert und mit HJ in das Jodid (XXV) übergeführt, das mit Alkali Ringschluß zum β-Lupinan (V) gab:

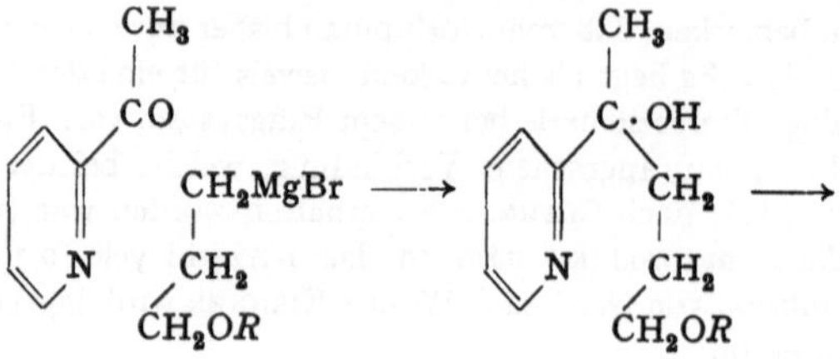

$$
\begin{array}{c}
\text{CH}_3 \\
| \\
\text{CO} \\
\end{array}
\quad \text{CH}_2\text{MgBr} \longrightarrow
\quad
\begin{array}{c}
\text{CH}_3 \\
| \\
\text{C}\cdot\text{OH} \\
\end{array}
\longrightarrow
$$

(pyridine ring with CH$_2$, CH$_2$OR chain) (XXIV.)

$$
\begin{array}{c}
\text{CH}_3 \\
| \\
\text{CH} \\
\end{array}
\longrightarrow
\quad
\begin{array}{c}
\text{CH}_3 \\
\end{array}
$$

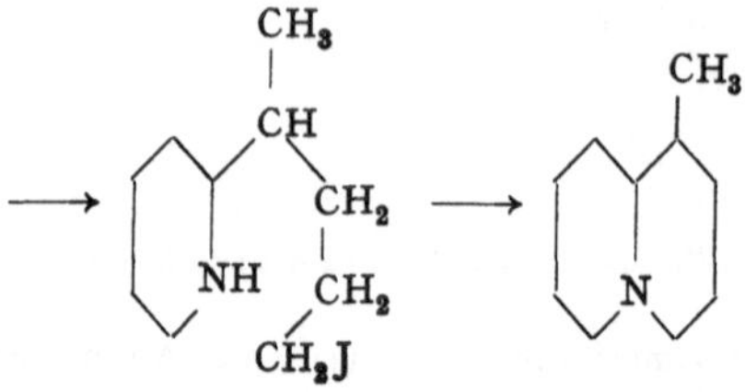

(XXV.) (V.) β-Lupinan.

Eine Übersicht über die die Stereochemie des Lupinins und seiner Derivate behandelnden Arbeiten von SCHÖPF, WINTERFELD und CLEMO siehe in (45) und (90).

Synthesen des Lupinins. Die Konstitutionsformel des Lupinins (I) wurde schließlich durch Synthesen des Alkaloids bestätigt. Die erste Synthese von CLEMO, MORGAN und RAPER (12) führte zu der natürlichen Base selbst. Sie gingen vom α-Pyridyl-essigester und γ-Phenoxy-*n*-propyl-bromid aus, hydrierten den Pyridinring des Kondensations-produktes (XXVI) und reduzierten nach BOUVEAULT und BLANC zum entsprechenden Alkohol (XXVII). Das mit HBr erhaltene Dibromid (XXVIII) gab beim Ringschluß mit PBr$_5$ ω-Brom-lupinan (XXIX). Dieses stellte das Gemisch zweier Racemverbindungen dar, deren Trennung über die Pikrolonate gelang. Die beiden, aus den Pikrolonaten in Frei-

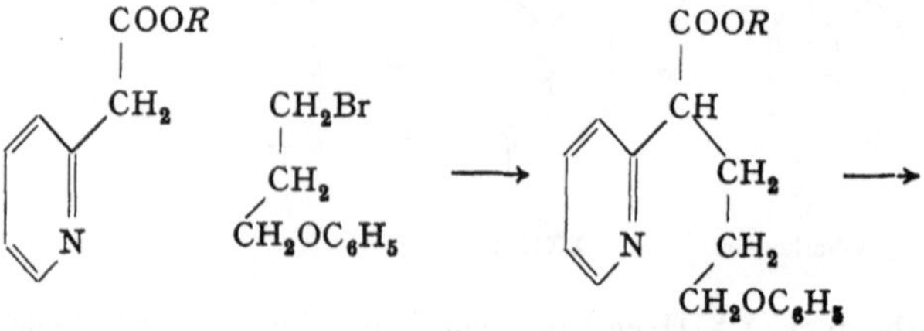

(XXVI.)

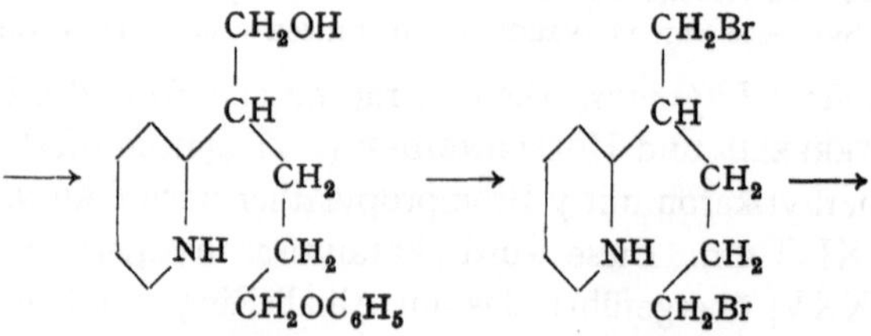

(XXVII.) (XXVIII.)

$$CH_2Br \qquad CH_2OH$$

(XXIX.) ω-Bromlupinan.   (I.) Lupinin.

heit gesetzten Brom-lupinane gaben bei der Umsetzung mit Natrium-acetat *D,L*-Lupinin und *D,L*-Isolupinin. Das *D,L*-Lupinin (F. 59°) konnte mit Weinsäure in die optischen Antipoden gespalten werden (*13*).

Die WINTERFELDsche Synthese des Lupinins (*105*) schließlich ist analog der des β-Lupinans aufgebaut, geht aber statt vom α-Pyridyl-methyl-keton vom α-Pyridyl-oxymethyl-keton, das aus Picolinsäurechlorid über das Diazoketon dargestellt wurde, aus. Es wurde nur inaktive Substanz erhalten, eine Spaltung konnte nicht durchgeführt werden.

3. Anagyrin.

Das zuerst in *Anagyris foetida*, dann in vielen anderen Papilionaceen aufgefundene Anagyrin konnte neuerdings auch in amerikanischen Lupinenarten nachgewiesen werden, so in *Lupinus laxiflorus* (*28*), *L. Macounii* (*65, 68*), *L. pusillus* (*67*) und *L. caudatus* (*23, 68*). Das Alkaloid, das die Formel $C_{15}H_{20}ON_2$ besitzt und starke Linksdrehung zeigt, wurde bisher nur amorph erhalten, gibt aber ein kristallisiertes, zur Trennung geeignetes Perchlorat. In den Nichtlupinen kommt es sehr häufig zusammen mit Cytisin (S. 265) vor, mit dem es strukturell einen α-Pyridonring, dessen Stickstoffatom ringtertiär ist, gemeinsam hat.

Mit der Konstitutionsermittlung, die, wie schon erwähnt, auch für die Frage der Struktur des Sparteins und Lupanins von besonderer Be-deutung war, hat sich besonders ING (*50*) beschäftigt.

Das Anagyrin (XXX) wurde in dreimaliger Aufeinanderfolge dem HOFMANNschen Abbau unterworfen, wobei nach jeder Stufe katalytisch hydriert wurde. Bei der ersten Stufe muß der Abbau wie beim Cytisin (siehe S. 265) besonders vorsichtig vorgenommen werden, da sonst Poly-merisation eintritt. Das Endprodukt (XXXI), das nur mehr das eine, vor dem HOFMANNschen Abbau durch seine schwache Basizität geschützte Stickstoffatom enthält, lieferte bei der Ozonisierung ein Lactam (XXXII), welches mit HCl zur Aminosäure (XXXIII) aufgespalten wurde. Diese gab bei der Oxydation mit Kaliumpermanganat α-Methyl-α'-*n*-amyl-glutarsäure (XXXIV). Mit Bariumpermanganat wurde Anagyrin zum Anagyramid (XXXV) oxydiert, das bei der Aufspaltung mit HJ unter Eliminierung der bei der Oxydation neu gebildeten CO-Gruppe leicht Kohlendioxyd abspaltet. Daraus kann aus Analogiegründen auf β-Stellung der CO-Gruppe zum Pyridonring geschlossen werden. Der Pyridonring selbst wurde sichergestellt durch die beim Spartein (S. 255) näher besprochene Hydrierung des Anagyrins (welche das Vorliegen von zwei Doppelbindungen beweist) und die Ozonisierung von (XXXI), sowie des Anagyramids (XXXV), die ganz analog verläuft und unter Verlust von C_4H_2 zu dem Dilactam (XXXVI) führt.

(XXX.) Anagyrin. (XXXI.) (XXXII.)

(XXXV.) Anagyramid. (XXXVI.) (XXXIII.)

$$CH_3 \qquad C_5H_{11}$$
$$HOOC \cdot CH \cdot CH_2 \cdot CH \cdot COOH$$

(XXXVI.) α-Methyl-α'-n-amyl-glutarsäure.

Alle diese Versuche legten das C—N-Gerüst des Alkaloids gemäß Formel (XXX) fest, mit Ausnahme des Ringes *D*, von dem auf Grund des Abbaus nicht entschieden werden konnte, ob er als Piperidin- oder α-Methylpyrrolidinring vorliegt. Erst die später beim Spartein angeführte Synthese des *D,L*-Oxosparteins von Clemo und Mitarbeitern (*10*), stellte den Piperidinring endgültig sicher.

4. Spartein.

Das Spartein, welches die Formel $C_{15}H_{26}N_2$ besitzt, kommt in der Natur in zwei optischen Antipoden vor, in linksdrehender Form ($[\alpha]_D = = -16°$, in Alkohol) neben Lupinin in *Lupinus albus* und *L. niger*. Zuerst wurde das (—)-Spartein aber im Besenginster *(Spartium scoparium)* entdeckt, aus dem es auch technisch hergestellt wird. Es kommt auch sonst in vielen Papilionaceen vor und wurde vereinzelt auch in anderen

Pflanzenfamilien aufgefunden, so in Mutterlaugen von Chelidonin aus *Chelidonium majus* (*100*), und im amorphen Aconitin (*35*), den Restbasen von *Aconitum Napellus*. Das (+)-Spartein war zuerst aus *Sophora pachycarpa* erhalten und Pachycarpin genannt worden (*80*). Später wurde es in einer Reihe anderer Papilionaceen, neuerdings auch in *Lupinus pusillus* (*67*) nachgewiesen.

Das Spartein, welches aus Pflanzenextrakten durch Wasserdampfdestillation leicht abgetrennt werden kann, stellt eine an der Luft instabile Flüssigkeit vor. Es ist eine zweisäurige Base und gibt gut kristallisierte Salze, von denen das Sulfat medizinisch verwendet wird und das Dipikrat (F. 208°) und das Monoperchlorat (F. 173°) zur Charakterisierung dienen. Weiters sind zwei Monojodmethylate des Sparteins bekannt. Zum Nachweis des Alkaloids ist auch ein bei der vorsichtigen Oxydation mit Kaliumferricyanid erhältliches Oxydationsprodukt (F. 87°) gut geeignet; es wurde zuerst Oxyspartein genannt, ist aber in Wirklichkeit ein Oxospartein, da es eine Lactamgruppe enthält. (Um Verwechslungen zu vermeiden, wird es hier immer als „Oxospartein" bezeichnet.)

Das Spartein ist das repräsentative Alkaloid der Gruppe der Lupinen-Alkaloide, die das gleiche C—N-Gerüst besitzen. Die anderen Alkaloide von diesem Typ unterscheiden sich vom Spartein durch einen Gehalt an Sauerstoff. Das Spartein hat auch von allen Lupinen-Alkaloiden das größte pharmakologische Interesse gefunden.

Die Konstitution des Sparteins. Die Konstitutionsaufklärung der Base erwies sich als ziemlich schwierig. Vor allem wegen des Mangels an funktionellen Gruppen führten die Bemühungen zahlreicher Forscher, die Struktur des Alkaloids durch direkten Abbau zu ermitteln, lange Zeit zu keinem günstigen Ergebnis. Das war der Fall bei Oxydationsversuchen und Abbauversuchen nach HOFMANN. Immerhin wurden von KARRER und Mitarbeitern (*51*) 1928 Strukturformeln des Alkaloids diskutiert, die dem vermutlichen Zusammenhang zwischen Lupinin und Spartein dadurch Rechnung trugen, daß sie das Norlupinangerüst enthielten. Einen experimentellen Beitrag dazu gaben WINTERFELD und Mitarbeiter (*113, 109*), die unter den Zersetzungsprodukten beim Erhitzen des dem Spartein nahe verwandten Lupanins mit Jodwasserstoff das β-Lupinan auffanden. Bei den von diesen Autoren aufgestellten Formeln blieb vor allem die Art der Verknüpfung des Norlupinanringes mit den zwei anderen im Spartein befindlichen Ringen noch unsicher.

Einen entscheidenden Fortschritt in dieser Richtung brachte die früher beschriebene Konstitutionsaufklärung des Anagyrins (XXX) (*50*). Dieses wurde von ING durch katalytische Hydrierung in Tetrahydroanagyrin, das sich als (—)-Lupanin erwies, und durch elektrolytische Reduktion in (+)-Spartein übergeführt. Da die Art der Ringverknüpfung beim Anagyrin sichergestellt war, erschien auch die des Sparteins und

Lupanins gegeben. Zweifelhaft blieb nur, ob Ring D ein Piperidin- oder
α-Methyl-pyrrolidin-Ring war. Clemo, Morgan und Raper (*10*) konnten
aber das D,L-Oxospartein gemäß Formel (XXXVII) synthetisch erhalten.
Winterfeld und Schirm (*118*) schlossen sich auf Grund von Abbau-
und Oxydationsversuchen am Phenyl-dehydro-spartein der Sechsring-
Formel (II) an.

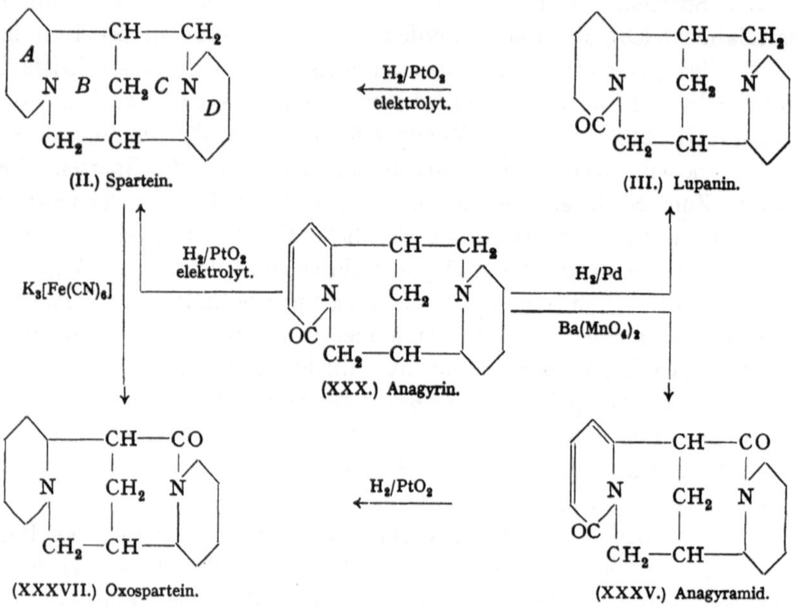

(II.) Spartein. (III.) Lupanin.

(XXX.) Anagyrin.

$K_3[Fe(CN)_6]$

H_2/PtO_2 elektrolyt.

H_2/Pd

$Ba(MnO_4)_2$

H_2/PtO_2

(XXXVII.) Oxospartein. (XXXV.) Anagyramid.

Diese wichtigen Versuche erfuhren eine weitere Stützung durch
Reduktionen des Anagyrins und Anagyramids mit Wasserstoff und
Platinoxyd nach Adams als Katalysator (*43*). Dabei wird bei Normal-
bedingungen die CO-Gruppe des Lactams zur CH_2-Gruppe reduziert, aus
Anagyrin (XXX) und Lupanin (III) wurde Spartein (II) erhalten, während
das Anagyramid (XXXV) in (+)-Oxospartein (XXXVII) überging. Damit
war auch die Struktur des Oxosparteins bewiesen, das die Oxogruppe im
Ring C enthält, der dem Ring B sterisch nicht gleichwertig ist.

Synthesen des Sparteins. Die erste Synthese des D,L-Oxosparteins
gelang, wie erwähnt, Clemo und Mitarbeitern 1936 (*10*). Zwei Moleküle
α-Pyridyl-essigsäure-äthylester [der neuerdings in einem auch zur Herstellung größerer Mengen geeigneten Verfahren aus α-Picolyl-lithium und
CO_2 mit nachfolgender Veresterung (*44, 120*) gewonnen werden kann],
wurden mit 1 Molekül Orthoameisensäure-ester bei Gegenwart von Essig-
säureanhydrid glatt zum 1-Carbäthoxy-4-oxo-3-(α-pyridyl)-pyridocolin
(XXXVIII) kondensiert. Es erfolgt zuerst Reaktion mit den Methylen-
gruppen des α-Pyridyl-essigesters und dann Ringschluß zwischen dem

Iminostickstoff des einen tautomer reagierenden Pyridinkernes und der einen Estergruppe. Dieses gut kristallisierende, gelb gefärbte Kondensationsprodukt, das später als Ausgangsmaterial für weitere Sparteinsynthesen diente, wurde katalytisch zur Verbindung (XXXIX) hydriert und diese darauf nach BOUVEAULT und BLANC zum Alkohol (XL) reduziert; (XL) gab mittels PBr_5 das Bromid (XLI), welches mit K_2CO_3 im Rohr zum D,L-Oxospartein (XXXVII) ringgeschlossen wurde. Die Ausbeute an Oxospartein war gering und die Weiterreduktion zu Spartein gelang vorerst nicht.

(XXXVIII.) (XXXIX.) (XL.) (XLI.) (XXXVII.) Oxospartein.

Die Totalsynthese des Sparteins selbst wurde erst in jüngster Zeit von mehreren Arbeitskreisen gleichzeitig durchgeführt. GALINOVSKY und KAINZ (36) stellten aus dem Hydrierungsprodukt (XXXIX) der CLEMOschen Verbindung durch Erhitzen Dioxospartein (XLII) her, das bei der katalytischen Hydrierung mit PtO_2 nach ADAMS glatt zum Oxospartein (XXXVII) reduziert wurde. Die Ausbeute an reinem D,L-Oxospartein ist bei dieser Reaktionsfolge sehr gut. Das D,L-Oxospartein läßt sich über die Bitartrate leicht in die Antipoden spalten (37). (+)- und (—)-Oxospartein wurden schließlich elektrolytisch in H_2SO_4 an Bleikathoden glatt zum (+)- und (—)-Spartein reduziert (37, 41), womit eine mit guter Ausbeute verlaufende Totalsynthese von (+)- und (—)-Spartein gegeben war. Das Dioxospartein (XLII) kann elektrolytisch auch direkt zum D,L-Spartein (II) reduziert werden. CLEMO, RAPER und SHORT (16) gelang schließlich auch die Reduktion des Oxosparteins zum Spartein mit Lithium-aluminiumhydrid.

(XLII.) Dioxospartein.

Leonard und Beyler (*56, 57, 58*) gelang eine Synthese des Sparteins und eines Isosparteins, indem sie das Clemosche Kondensationsprodukt (XXXVIII) einer Reduktion bei einem Wasserstoffdruck von 250 bis 350 atm und 250° mit Kupferchromit als Katalysator unterwarfen. Es entstand unter gleichzeitiger Reduktion und Cyclisierung in 30%iger Ausbeute ein Isomerengemisch, das aus *D,L*-Spartein neben wenig *D,L*-Isospartein bestand. Das gleiche Ergebnis wurde bei der in analoger Weise vorgenommenen Reduktion des 2,4-Di-(α-pyridyl)-glutarsäure-esters (XLIII), der aus α-Pyridyl-essigester mit Formaldehyd oder Methylenjodid dargestellt werden kann, erhalten. Zu der ziemlich schwierigen Trennung der beiden Isomeren dienten die fraktionierte Destillation und die chromatographische Analyse sowie die fraktionierte Kristallisation der Perchlorate. Das *D,L*-Spartein wurde über die β-Camphersulfonate in die optischen Antipoden gespalten. Es wurde weiters wahrscheinlich gemacht, daß das Isospartein identisch ist mit dem α-Isospartein, das Winterfeld und Rauch (*116*) durch Hydrierung des α-Didehydro-sparteins erhielten.

(XLIII.) 2,4-Di-(α-pyridyl)-glutarsäure-ester.

Šorm und Keil (*92, 93*) schließlich stellten gleichfalls aus (XLIII) und (XXXVIII) durch Hydrierung und Destillation das Dioxospartein (XLII) her, das chromatographisch in zwei kristallisierte Isomere getrennt wurde. Die eine Verbindung (F. 171°) gab bei der elektrolytischen Reduktion ein Isospartein, das wahrscheinlich mit dem Isospartein von Leonard und Beyler identisch ist, die andere (F. 135°) bei der gleichen Reduktion ein Substanzgemisch, aus dem in Form des Pikrates das *D,L*-Spartein abgetrennt werden konnte. Eine weitere Synthese des Sparteins siehe S. 270.

Stereochemie des Sparteins. Über die stereochemischen Verhältnisse des Sparteins und des Isosparteins kann beim heutigen Stand der Kenntnisse folgendes gesagt werden: Das Spartein besitzt vier asymmetrische Kohlenstoffatome, $C_{(6)}$, $C_{(7)}$, $C_{(14)}$ und $C_{(15)}$[*]. Durch den besonderen Bau dieses Ringsystems und infolge des Umstandes, daß die Methylenbrücke zwischen $C_{(7)}$ und $C_{(15)}$ wie in ähnlichen Fällen nur in *cis*-Stellung angeordnet sein kann, wird die Zahl der Stereoisomeren aber eingeschränkt.

---

[*] Die Bezifferung des Ringsystems (siehe Formel XLIIIa) erfolgt hier nach Couch (*24*).

Es ergeben sich so grundsätzlich drei Möglichkeiten (58): Es können erstens die H-Atome von $C_{(6)}$ und $C_{(14)}$ in cis-Stellung (XLIIIa) oder zweitens in trans-Stellung (XLIIIb) zur Methylenbrücke stehen. Drittens schließlich kann das eine H-Atom cis-, das andere trans-ständig angeordnet sein (XLIIIc, perspektiv gesehen XLIIId). Theoretisch wären ja zwei solche Formen möglich, die aber identisch sind, weil die beiden durch die Methylenbrücke getrennten Hälften des Ringsystems strukturell gleich sind.

Alle drei Stereoisomeren besitzen weder Symmetrieebene noch Symmetriezentrum, sind also in optische Antipoden spaltbar; (XLIIIa) und (XLIIIb) besitzen aber eine durch die Brücken-Methylengruppe gehende Symmetrieachse. In den letzteren beiden Formen sind die Ringe A und D einerseits, B und C andererseits sterisch gleichwertig. Das ist bei (XLIIIc) nicht der Fall.

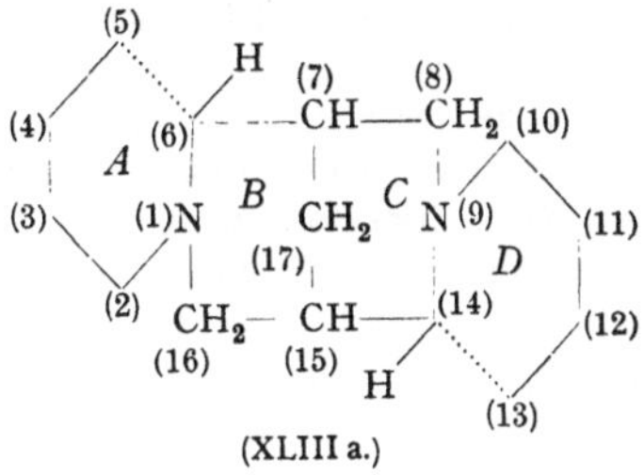

(XLIII a.) 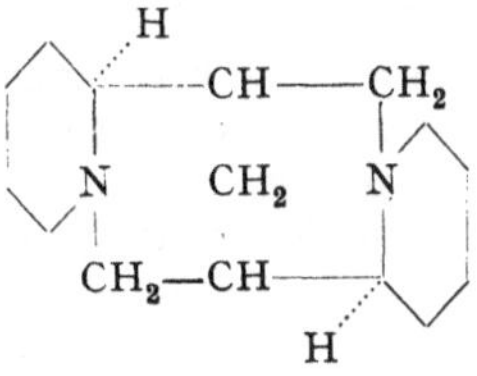   (XLIII b.)

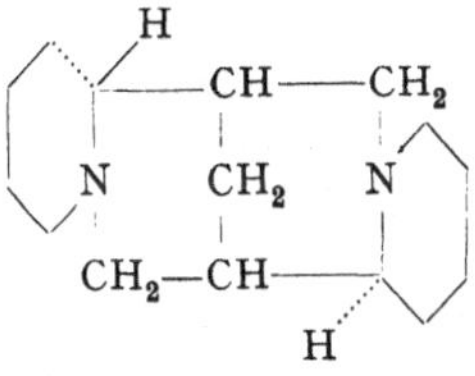

(XLIII c.) 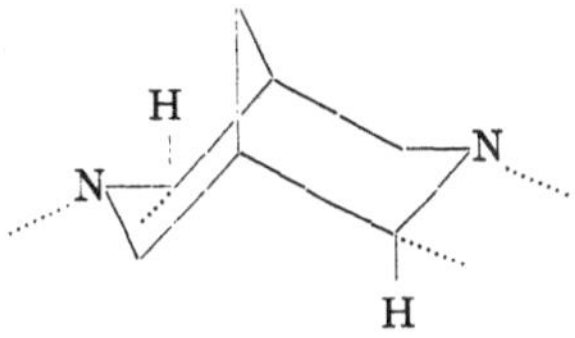   (XLIII d.)

Das Aphyllin, eine Base aus *Anabasis aphylla*, hat nun die gleiche Molekularformel wie Oxospartein und es wird ihr auch die gleiche Strukturformel (XXXVII) zugeteilt (vgl. S. 256). Da sich das Aphyllin leicht in Spartein überführen läßt, muß es also die gleiche Konfiguration wie dieses besitzen. Da es aber vom Oxospartein, das ja ebenfalls ein Sparteinderivat vorstellt, verschieden ist, müssen Aphyllin und Oxospartein sich im räumlichen Bau unterscheiden. Das ist der Fall, wenn ihre CO-Gruppen, die in den Innenringen B und C des Ringsystems liegen, konfigurativ verschieden angeordnet sind. Aphyllin und Oxospartein unterscheiden sich auch sehr stark bei der katalytischen Hydrierung mit PtO_2. Aphyllin geht dabei leicht in Spartein über, während die CO-Gruppe des Oxosparteins nicht angegriffen wird (43). Damit stimmt überein, daß im Dioxospartein (XLII), das zwei strukturell völlig gleichwertige CO-

Gruppen in den Ringen *C* und *D* enthält, bei der katalytischen Hydrierung
(*36*) nur die eine reduziert wird, wobei Oxospartein entsteht. Aus diesen
Ergebnissen folgt, daß die beiden Ringe *B* und *C* des Sparteins sterisch
nicht gleichwertig sein können (*43*). Da diese Bedingung nur von Formel
(XLIIIc) erfüllt wird, ist die Annahme gerechtfertigt, daß dem Spartein
die Konfiguration (XLIIIc) zukommt. Dem Isospartein LEONARDS
kommt dann eine der beiden Konfigurationen (XLIIIa) oder (XLIIIb) zu.

Einige weitere Derivate des Sparteins und einige Abbauversuche, die für die
Konstitutionsermittlung nicht von wesentlicher Bedeutung waren, seien hier nur
kurz besprochen [siehe (*45, 6*)]. Das Spartein bildet mit Wasserstoffsuperoxyd nur
ein Mono-aminoxyd ($C_{15}H_{26}ON_2 \cdot H_2O$) (*82*), kein Dioxyspartein oder Di-N-oxyd,
wie man früher angenommen hatte. Durch einige Oxydationsmittel wird eine
partielle Dehydrierung des Sparteins erzielt. Man erhält mit Natriumhypobromit
(*119*), Chromsäure (*104*) und Mercuriacetat (*116*) verschiedene Dehydro-sparteine,
deren Struktur noch nicht völlig sichergestellt erscheint. Die Dehydrierung mit
Mercuriacetat geht noch weiter und führt zu α- und β-Didehydro-spartein. Das
erstere gibt bei der Hydrierung das α-Isospartein (siehe S. 258). Die Ergebnisse der
von WINTERFELD und RÖNSBERG (*117*) am α-Didehydro-spartein durchgeführten
Oxydationsversuche bestätigten die Sechsring-Struktur des Ringes *D* im Spartein.

Beim HOFMANNschen Abbau des Sparteins, das zwei isomere Monojodmethylate
gibt, beteiligen sich von der zweiten Stufe an beide Stickstoffatome, so daß Gemische komplizierter Abbauprodukte auftreten, deren Bearbeitung wenig Erfolg
gezeitigt hat. Durch konsequent bis zur Eliminierung der beiden Stickstoffatome
durchgeführten HOFMANNschen Abbau mit jeweiliger Hydrierung der ungesättigten
Zwischenstufen, konnte ein Kohlenwasserstoff $C_{15}H_{32}$ erhalten werden (*52*), der ein
6,8-Dimethyl-tridecan vorstellen muß (*87*). Wenn man vom Oxospartein ausgeht,
wird nur ein Stickstoffatom eliminiert, da das laktamartig gebundene der erschöpfenden Methylierung entzogen ist. Es wurde so ein Hexahydro-hemisparteylen
(*98*) gewonnen, das die Struktur eines Methyl-amyl-α-norlupinons (XLIV) besitzen
muß.

(XLIV.) Methyl-amyl-α-norlupinon.

5. Lupanin.

Das Lupanin, $C_{15}H_{24}ON_2$, wurde in der (+)-Form als Hauptalkaloid in
Lupinus angustifolius, dann in *L. albus* und *L. laxus* (*26*) aufgefunden. Die
(—)-Form wurde neuerdings in *L. Macounii* (*65, 68*) und *L. pusillus.* (*67*)
nachgewiesen. Das *D,L*-Lupanin kommt in *L. termis* (*8*) und neben der
rechtsdrehenden Form in *L. albus* vor.

Für (+)-Lupanin ist ein Schmelzpunkt von 44° angegeben, das *D,L*-
Lupanin schmilzt bei 98°. Das Lupanin ist eine einsäurige Base, gibt

aber neben dem Mono- auch ein Dihydrochlorid. Die optisch aktiven Formen können als Bitartrat [(+)-Lupanin-D-bitartrat: F. 194°], als Pikrat (F. 180°) oder als Perchlorat (F. 213°) charakterisiert werden.

Die Konstitution des Alkaloids (III) ist eindeutig durch die Hydrierung des Anagyrins (XXX) zum (—)-Lupanin (50) und die Reduktion des Lupanins zum Spartein gegeben (8, 17, 43). Konfigurativ entspricht das (+)-Lupanin dem (—)-Spartein. Das (—)-Lupanin war zuerst durch Spaltung von D,L-Lupanin in die optischen Antipoden erhalten worden (17). Alle sonstigen Ergebnisse und Befunde sind mit dieser Konstitution (III) vereinbar. So läßt sich das Lupanin zu einer Aminosäure aufspalten (47). Bemerkenswert ist, daß das Lupanin wie ein Keton mit GRIGNARDscher Lösung reagiert. Diese Reaktion wurde besonders von WINTERFELD und Mitarbeitern studiert. Die Reaktion des (+)-Lupanins mit Phenylmagnesiumbromid (106) führt intermediär zu einem Carbinol, das sofort Wasser abspaltet und in das Phenyl-dehydro-spartein (XLV) übergeht. Dieses setzt sich in wäßriger Lösung durch Wasseraufnahme zum ringoffenen Aminoketon um, das als Benzoylprodukt der Oxydation unterworfen wurde (118). Die Ergebnisse der Oxydation standen mit der Strukturformel (III) des Lupanins durchaus im Einklang.

(XLV.) Phenyl-dehydro-spartein.   (III.) Lupanin.   (XLVI.)

Bei der Umsetzung von Lupanin mit Äthylmagnesiumbromid gelang es auch, die Carbinol-Zwischenstufe zu fassen (XLVI) (115).

6. Oxylupanin.

Das Oxylupanin wurde neben (+)-Lupanin in *Lupinus perennis* und *L. angustifolius*, in neuester Zeit noch in *L. hilarianus* (32) aufgefunden. Die Base, welche die Formel $C_{15}H_{24}O_2N_2$ hat, ist rechtsdrehend. Sie kristallisiert wasserfrei (F. 173 bis 174°) oder mit zwei Molekülen Kristallwasser (F. 76 bis 77°).

Über die Funktion der beiden Sauerstoffatome war durch ältere Arbeiten Klarheit erzielt worden. Das eine liegt als Hydroxylgruppe vor (3), während das zweite einer Lactamgruppe angehört, was durch die Reduktion mit HJ und Phosphor zu (+)-Lupanin bewiesen wurde (2). GALINOVSKY und PÖHM (64) zeigten durch Oxydation des Oxylupanins (XLVII) zu einem Keton (nach OPPENAUER), daß die Hydroxylgruppe sekundärer Natur ist. Bei der Wasserabspaltung mit nachfolgender

Hydrierung erhielten sie optisch reines (+)-Lupanin (III). Bei der katalytischen Hydrierung mit PtO_2 nach Adams wurde die Lactamgruppe
reduziert und ein Oxyspartein (XLVIII) erhalten, aus dem durch Wasserabspaltung und Hydrierung wieder (—)-Spartein (II) gewonnen wurde.

Die Stellung der Hydroxylgruppe wurde folgendermaßen ermittelt
[Galinovsky, Pöhm und Riedl (40)]: Bei der erschöpfenden Oxydation
mit Chromsäure geben das Spartein und die verwandten Basen Bernsteinsäure und Aminosäuren (53). Es wurden nun das Oxylupanin und die
aus ihm erhaltenen Basen Oxyspartein, Lupanin und Spartein unter ver-
gleichbaren Bedingungen mit CrO_3 erschöpfend oxydiert. Der Vergleich
der Mengen der aus diesen vier Basen entstandenen Bernsteinsäure, die
stets quantitativ erfaßt wurde, ließ schon den Schluß zu, daß sich die
Hydroxylgruppe im Oxylupanin und Oxyspartein im Ring D befindet.
Die bei der Oxydation gebildeten Aminosäuren wurden durch Papier-
chromatographie nach Consden, Gordon und Martin (19) getrennt und
identifiziert.

Aus Spartein wurde in Übereinstimmung mit Karrer und Widmer
(53) als Hauptprodukt γ-Aminobuttersäure, aus Lupanin γ-Aminobuttersäure und Glycin, aus Oxylupanin β-Alanin und Glycin (dagegen keine
Spur von γ-Aminobuttersäure) und schließlich aus Oxyspartein γ-Aminobuttersäure, β-Alanin und Glycin erhalten. Das völlige Fehlen der
γ-Aminobuttersäure unter den Oxydationsprodukten des Oxylupanins
und das Auftreten von β-Alanin als Hauptprodukt der Oxydation lassen
für das Oxylupanin selbst die Formel (XLVII), für das Oxyspartein die
Formel (XLVIII) annehmen. Die Bindung der OH-Gruppe an $C_{(11)}$*
kommt wegen des Auftretens von β-Alanin nicht in Betracht; bei der

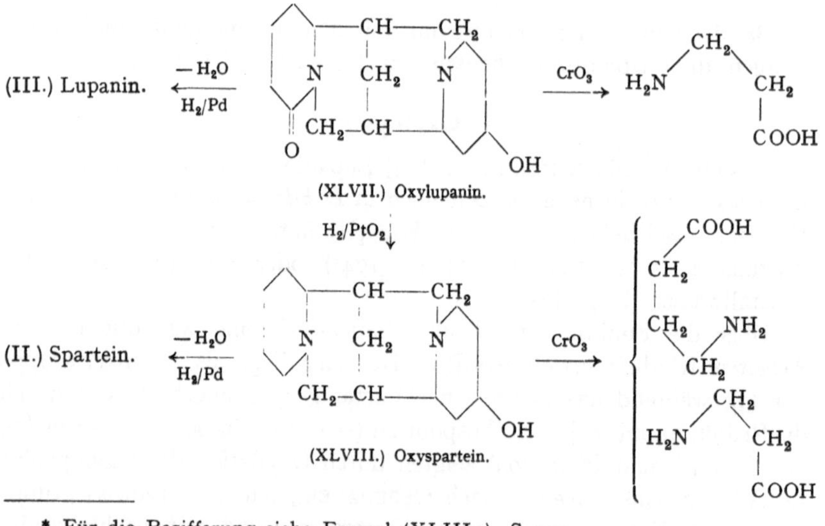

(III.) Lupanin. $\xrightarrow[H_2/Pd]{-H_2O}$

(XLVII.) Oxylupanin.

H_2/PtO_2

(II.) Spartein. $\xrightarrow[H_2/Pd]{-H_2O}$

(XLVIII.) Oxyspartein.

---

* Für die Bezifferung siehe Formel (XLIIIa), S. 259.

Bindung an $C_{(13)}$ wäre auch γ-Aminobuttersäure als Oxydationsprodukt zu erwarten.

Die Formeln (XLVII) und (XLVIII) für das Oxylupanin und Oxyspartein werden weiters dadurch gestützt, daß das Anhydro-oxylupanin und das Anhydro-oxyspartein die gleichen Aminosäuren wie Oxylupanin und Oxyspartein, aus denen sie durch Wasserabspaltung entstanden waren, gaben, daß aber eine deutliche Verschiebung im Mengenverhältnis von β-Alanin und Glycin zugunsten des Glycins eingetreten war. Daß beim Oxylupanin und Oxyspartein größere Mengen β-Alanin entstehen, erklärt sich daraus, daß die Oxydation hier vorwiegend zwischen $C_{(12)}$ und $C_{(13)}$ angreift. Bei der Wasserabspaltung aus Oxylupanin (XLVII) und Oxyspartein (XLVIII) bilden sich erwartungsgemäß in annähernd gleichen Mengen die beiden ungesättigten Basen mit den Doppelbindungen zwischen $C_{(11)}$ und $C_{(12)}$ bzw. $C_{(12)}$ und $C_{(13)}$, die bei der Oxydation β-Alanin und Glycin in annähernd äquivalenten Mengen ergeben.

III. Lupinen-Alkaloide unbekannter Konstitution.

Die folgende Tabelle 1 gibt eine Übersicht über die Lupinen-Alkaloide von unbekannter oder unsicherer Konstitution.

Tabelle 1. Lupinen-Alkaloide von unbekannter oder unsicherer Konstitution.

Alkaloid	Formel	Vorkommen	F.	$[\alpha]_D$
Tetralupin	$C_{10}H_{19}ON$	· *Lupinus palmeri (22)*	$81-83°$	$+\ \ 4,6°$
Pusillin	$C_{15}H_{28}N_2$	*L. pusillus (67)*	Perchlorat: $219°$	$-\ 15,3°$ (Alk.)
Hexalupin.....	$C_{15}H_{20}ON_2$	*L. corymbosus (21)*	$197-198°$	$+ 126,1°$ (Alk.)
Alkaloid *P* 1 ..	$C_{15}H_{22}ON_2$	*L. Macounii (65)*	$126°$	—
Octalupin	$C_{15}H_{22}O_2N_2$	*L. sericeus (27)*	$167,5-169,5°$	$+\ 52,3°$ (Alk.)
Nonalupin.....	$C_{15}H_{24}ON_2$	*L. sericeus (29)* *L. Andersonii (30)*	$235°$ Hydrat: $92°$	$21,3$ (Alk.)
Isomeres des Oxylupanins.	$C_{15}H_{24}O_2N_2$	*L. laxus (26)*	$176-177°$	$+ 133,2°$ (Wasser)
Trilupin.......	$C_{15}H_{24}O_3N_2$	*L. laxus (26)* *L. barbiger (25)*	$256-257°$	$+\ 63,8°$ (Wasser)
Dilupin	$C_{16}H_{26}O_2N_2$	*L. barbiger (25)*	(Öl)	$+\ 65,6°$ (Wasser)
Pentalupin	$C_{16}H_{30}ON_2$?	*L. palmeri (22)*	(Öl)	$-\ \ 3,2°$ (Alk.)
Spathulatin ...	$C_{33}H_{64}O_5N_4$	*L. spathulatus (20)* *L. sericeus (29)*	$234°$	$-\ \ 2,3°$ (Alk.)

Das Tetralupin ist nach der Formel isomer mit Lupinin. Die anderen Alkaloide entsprechen dagegen formelgemäß dem Typ des Sparteins mit Ausnahme des Spathulatins, das die für die Lupinen-Alkaloide ungewöhnliche Formel $C_{33}H_{64}O_5N_4$ besitzt. Beim Erhitzen mit HCl geht dieses aber in eine Base von der Formel $C_{15}H_{24}N_2$ über, die also wieder dem Spartein-Typ entspricht.

Für das Tetralupin wurde eine Identität mit Allo-lupinin, das nach WINTERFELD (*108*) möglicherweise im rohen Lupinin enthalten ist, in Erwägung gezogen. Das Allo-lupinin (XLIX) konnte auch synthetisch dargestellt werden (*112*). Das Allo-lupinan, in welchem die Hydroxylgruppe durch Wasserstoff ersetzt ist, wurde gleichfalls synthetisiert (*63*). Das Trilupin wurde von COUCH (*25*) als Diaminoxyd

CH$_2$OH

(XLIX.) Allo-lupinin.

des Lupanins angesehen und soll auch bei der Einwirkung von Calciumperoxyd auf Lupanin entstehen. Es ist an sich aber unwahrscheinlich, daß dieses Alkaloid, das ein sehr schwach basisches, da lactamartig gebundenes N-Atom enthält, ein Diaminoxyd gebe, zumal das Spartein mit zwei basischen N-Atomen nur ein Mono-N-oxyd liefert. Es konnte auch neuerdings kein derartiges Diaminoxyd des Lupanins erhalten werden (*38*).

Das Dilupin soll nach COUCH (*25*) das Aminoxyd eines C-Methyl-lupanins und das Octalupin ein 2,16-Dioxospartein sein (*27*). Bei der elektrolytischen Reduktion liefert letzteres (—)-Spartein und (+)-Lupanin.

IV. Verwandte Alkaloide anderer Papilionaceen oder Pflanzenfamilien.

Von den für die Lupinen typischen Alkaloiden von bekannter Konstitution, Lupinin, Spartein, Lupanin, Oxylupanin und Anagyrin sind das Spartein, Lupanin und Anagyrin auch in anderen Papilionaceen sehr oft aufgefunden worden, besonders häufig das Spartein. Das Lupinin ist interessanterweise bisher sonst nicht in Papilionaceen beobachtet worden, dagegen kommt es in *Anabasis aphylla*, einer Cheniopodacee, vor. Auch das Spartein kommt noch in anderen Pflanzenfamilien vor (S. 255).

Die Zahl der bisher aufgefundenen Papilionaceen-Alkaloide, die nicht in Lupinen enthalten, aber den Lupinen-Alkaloiden strukturell verwandt sind, ist kaum größer als die Zahl der Lupinen-Alkaloide selbst. Das verbreitetste davon ist das Cytisin, das auch chemisch am besten erforscht ist. Die Konstitutionsermittlung dieses Alkaloids hat, wie schon erwähnt, auch die Bearbeitung der übrigen verwandten Alkaloide sehr gefördert. Strukturell stellt es einen neuen Typ vor, der durch Kondensation eines Norlupinanringes mit einem Piperidinring entstanden gedacht werden kann. Es ist stark giftig; medizinisch besitzt es keine Bedeutung.

1. Cytisin.

Das Cytisin, welches im Goldregen und in zahlreichen anderen *Cytisus*-sowie *Sophora*-Arten aufgefunden wurde, hat die Formel $C_{11}H_{14}ON_2$. Es kristallisiert gut (F. 153°) und ist optisch aktiv ($[\alpha]_D = -120°$). Ältere Arbeiten ergaben (45, 6), daß von den beiden Stickstoffatomen eines tertiärer Natur ist, während das andere sekundär vorliegt und sich methylieren läßt (Methyl-cytisin, F. 136 bis 137°). Es wurde weiters gefunden, daß das Sauerstoffatom in einem Pyridonring enthalten ist. Typisch für diesen Ring ist die VAN DE MOERsche Reaktion, eine Rotfärbung mit Ferrichlorid, die in gleicher Weise auch beim Anagyrin auftritt. Da bei einigen Reaktionen des Cytisins, so beim Erhitzen mit Jodwasserstoff und Phosphor sowie bei einem Dehydrierungsversuch mit Pd-Asbest, Chinolinderivate erhalten wurden, stellte man Strukturformeln zur Diskussion, die einen Chinolinring als Grundgerüst enthielten. Beim HOFMANNschen Abbau, dem bereits frühzeitig ein besonderes Interesse zugewendet wurde, trat schon, wie sich später erwies, bei der ersten Stufe ein bimolekulares Produkt auf, so daß auch durch diese, für die Konstitutionsermittlung von Alkaloiden so wichtige Abbaumethode vorerst keine Möglichkeit zur Aufklärung der Struktur des Cytisins gegeben war.

SPÄTH und GALINOVSKY (94) gelang es nun, den HOFMANNschen Abbau des Cytisins (L) durch besonders schonende Zersetzung der aus dem Methyl-cytisin-jodmethylat dargestellten quartären Base völlig normal durchzuführen. Die ungesättigte Base wurde sofort katalytisch hydriert und so der Dimerisation entzogen. Das Dihydro-*des*-N-dimethyl-cytisin gab beim weiteren Abbau mit nachfolgender Hydrierung der Doppelbindung, unter Verlust des im Cytisin ursprünglich als Iminogruppe vorliegenden Stickstoffatoms, das Tetrahydro-hemi-cytisylen (LI). Dieses gab bei der Weiterhydrierung die Octahydroverbindung (LII), die bei der Oxydation Glutarsäure lieferte. Bei der Ozonisierung von (LI) konnte ein Lactam (LIII) erhalten werden, das zur Aminosäure aufgespalten wurde, die bei der Oxydation mit Kaliumpermanganat die α,α'-Dimethyl-glutarsäure ergab.

Aus diesen Versuchen folgte für das Tetrahydro-hemi-cytisylen die Strukturformel (LI). Sie wurde weiters durch die Synthese der durch Pd-Dehydrierung aus (LI) gewonnenen Verbindung (LIV) — daneben wurde 3,5-Dimethyl-2-*n*-propyl-pyridin (LV) gebildet — bestätigt (96). Die Synthese des Octahydro-hemi-cytisylens (LII) wurde, ausgehend von 3,5-Dimethyl-picolinsäureester, analog der des α-Norlupinons (S. 250) durchgeführt (97). Die Dehydrierung der synthetischen Verbindung mit Palladium ergab wieder die beiden Dehydrierungsprodukte (LIV) und (LV), die aus dem End-

(L.) Cytisin. (LI.) Tetrahydro-hemi-cytisylen. (LIII.)

(LV.) 3,5-Dimethyl-2-*n*-propylpyridin. (LIV.) (LII.) Octahydro-hemi-cytisylen. Glutarsäre.

produkt des HOFMANNschen Abbaus, dem Tetrahydro-hemi-cytisylen, erhalten worden waren. Durch einen zweiten HOFMANNschen Abbau konnte weiters, unter Aufspaltung aller anderen Ringe, der Ring des Cytisins, der die Iminogruppe enthält, aus dem Molekül herausgeschält werden (95). Es wurde dazu zuerst der Pyridonring elektrolytisch redu-ziert, das erhaltene Tetrahydro-desoxy-cytisin acetyliert und die Acetyl-

(LVI.) N-Acetyl-tetrahydro-desoxy-cytisin.

(LVII.) (LVIII.) 3-Methyl-5-*n*-amyl-pyridin. (LIX.) β-Methyl-nicotinsäure.

(LX.) (LXI.) (LXII.)

verbindung (LVI) nach HOFMANN abgebaut. Durch dreimaligen Abbau, begleitet von der katalytischen Hydrierung der Zwischenstufen, wurde das bicyclische N-Atom eliminiert. Dann wurde aus dem resultierenden Acetylprodukt (LVII) Acetyl abgespalten und die entstandene Base dehydriert. Bei der Oxydation des Dehydrierungsproduktes, das 3-Methyl-5-*n*-amyl-pyridin (LVIII) vorstellte, wurde β-Methylnicotinsäure (LIX) erhalten.

Nach diesen beiden Abbauergebnissen kamen für das Cytisin nur mehr die Strukturformeln (L) und (LX) in Betracht. Gegen die Formel (LX) sprach der Umstand, daß das Cytisin selbst bei der Oxydation keine Methylbernsteinsäure gibt und für Formel (L) die Tatsache, daß das Methyl-cytisin bei der Oxydation mit Bariumpermanganat [ING (49)] zwei strukturisomere Lactame, (LXI) und (LXII), liefert. Die aus dem Lactam (LXI) bei der Aufspaltung gebildete Aminosäure gibt als N-Benzolsulfonylderivat beim Erhitzen leicht CO_2 ab, während die Aminosäure aus (LXII) dieses Verhalten nicht zeigt. Damit erscheint die Strukturformel (L) des Cytisins völlig gesichert. Das Entstehen der bei den vorhin erwähnten Reaktionen des Cytisins auftretenden Chinolinderivate ist auf Umlagerungen zurückzuführen. Auf die Möglichkeit solcher Umlagerungsreaktionen hatte übrigens schon früher ING (48) hingewiesen.

Eine Synthese des Cytisins selbst oder eines Derivates, welches noch das vollständige C—N-Gerüst enthält, wurde bisher nicht durchgeführt.

Auch das N-Methyl-cytisin wurde häufig in Papilionaceen, meist als Begleiter des Cytisins, gelegentlich auch ohne Cytisin aufgefunden. Zuerst wurde es interessanterweise aber in der Berberidacee *Caulophyllum thalictroides (83)* entdeckt.

2. Aphyllin und Aphyllidin.

ORECHOFF und MENSCHIKOFF (74) haben aus der hochsiedenden Alkaloidfraktion von *Anabasis aphylla*, einer in Mittelasien verbreiteten giftigen Pflanze, zwei Alkaloide, das Aphyllin und das Aphyllidin, gewonnen, die konstitutionell zu den Basen der Sparteingruppe gehören. Das kristallisierte Aphyllidin (F. 112 bis 113°), das rechtsdrehend ist, geht bei der Hydrierung unter Aufnahme von zwei H-Atomen in das Aphyllin über (Pikrolonat: F. 233 bis 234°). Bei der elektrolytischen Reduktion (75), auch bei der katalytischen Hydrierung mit PtO_2 (43) bekommt man aus beiden Alkaloiden (+)-Spartein. Für das eine Stickstoffatom läßt sich eine lactamartige Bindung nachweisen. Die Lactamgruppe kann mit Salzsäure leicht zur Aminosäure aufgespalten werden, bei der Veresterung der Aphyllinsäure erhält man gut kristallisierte Monohydrate des Methyl- bzw. Äthylesters (99). Zur Trennung der beiden Alkaloide hat sich die Chromatographie als sehr brauchbar erwiesen, wobei bei Verwendung von Aluminiumoxyd als Adsorbens das ungesättigte Aphyllidin zuerst eluiert wird.

Orechoff (73) hat für das Aphyllin und Aphyllidin die Strukturformeln (LXIII) und (LXIV) vorgeschlagen:

(LXIII.)   (LXIV.)

Nach diesen Formeln befindet sich die CO-Gruppe beider Verbindungen in einem Innenring. Da aber Aphyllin (LXIII) und Oxospartein (XXXVII), das bei der Oxydation des Sparteins erhalten wird und das Carbonyl im Ring C enthält, verschieden sind, muß die CO-Gruppe des Aphyllins im Ring B liegen, sofern die beiden Ringe B und C sterisch ungleichwertig sind. Es ist, wie auf S. 259 ausgeführt wurde, ein Raummodell möglich, das dieser Bedingung entspricht. Allerdings ist zu bemerken, daß auch eine Verbindung mit der Lactamgruppe im Ring D vom Lupanin (III, S. 247), das die CO-Gruppe im Ring A enthält, verschieden sein müßte.

Aphyllin unterscheidet sich in seinen Reaktionen erheblich vom Oxospartein und Lupanin. Eine weitgehende Analogie besteht nur im Verlauf des Hofmannschen Abbaus beim Oxospartein und beim Aphyllin, bzw. Aphyllidin. Von letzterem führt er zu einem Hemi-aphylliden (76). Ganz verschieden verhalten sich Aphyllin und Oxospartein dagegen bei der Aufspaltung und Hydrierung. Oxospartein wird mit Salzsäure im Gegensatz zum Aphyllin schwer aufgespalten und kann katalytisch unter Normalbedingungen nicht zum Spartein reduziert werden. Mit Lupanin hat das Aphyllin die leichtere Reduzierbarkeit gemeinsam, dagegen läßt sich Lupanin nicht nach Hofmann abbauen.

Ein strenger Beweis für die Lage der Lactamgruppe im Aphyllin ist jedenfalls noch nicht erbracht worden. Ebenso verhält es sich mit der Lage der Doppelbindung im Aphyllidin. Diese addiert Brom; aus dem Additionsprodukt spaltet sich leicht HBr ab, wobei eine stabile Monobromverbindung erhalten wird. Bei der Hydrierung des Aphyllidins bekommt man reines Aphyllin. Oxydationsversuche zur Ermittlung der Lage der Doppelbindung sind noch nicht durchgeführt worden.

3. Weitere verwandte Alkaloide (fraglicher Konstitution).

In den Papilionaceen wurde noch eine Reihe weiterer Alkaloide aufgefunden, die, was die Molekularformel und manche Eigenschaften anbelangt, Ähnlichkeit mit den Lupinen-Alkaloiden aufweisen. Ob sie mit ihnen verwandt sind, ist nicht sicher zu behaupten, solange ihre Konstitution ungewiß ist. Bei manchen Basen, die gemeinsam mit in Lupinen nachgewiesenen Alkaloiden vorkommen, kann eine solche Verwandtschaft wohl vermutet werden.

Das einzige Alkaloid, das möglicherweise dem Lupinin-Typ entspricht, ist das mit Lupinin isomere Virgilidin aus *Virgilia capensis* (*103*). Andere Alkaloide besitzen die gleichen oder ähnlichen Formeln wie die Spartein-Alkaloide von Lactam-Charakter. Zu ihnen gehören das Sophocarpin ($C_{15}H_{24}ON_2$) aus *Sophora pachycarpa* (*78*, *81*), *S. alopecuroides* (*79*), *Ammothamnus Lehmannii* (*54*), das Sophoridin ($C_{15}H_{26}ON_2$) aus *Sophora alopecuroides* (*72*), das Monspessulanin ($C_{15}H_{22}ON_2$) aus *Cytisus monspessulanus* (*102*), das Sophoramin ($C_{15}H_{20}ON_2$) aus *Sophora alopecuroides* (*72*) und das Thermopsin ($C_{15}H_{20}ON_2$) aus *Thermopsis lanceolata* (*77*) sowie *Th. rhombifolia* (*64*). Die letzten drei Basen sind ungesättigter Natur, wie nach ihrer Formel zu erwarten ist.

Weiters wurden noch einige ähnliche Alkaloide mit zwei Sauerstoffatomen aufgefunden, das Rhombifolin ($C_{15}H_{20}O_2N_2$) aus *Thermopsis rhombifolia* (*64*) sowie das Baptifolin mit der gleichen Formel aus *Baptisia perfoliata* (*69*) und *B. minor* (*66*). Diese Basen scheinen zum Teil ein lactamartig gebundenes Sauerstoffatom zu enthalten, es ist aber fraglich, ob sie demselben Struktur-Typus angehören wie das Spartein. So ist für das Matrin, das in *Sophora flavescens* und in anderen Sophora-arten aufgefunden wurde und mit Lupanin isomer ist, von KONDO und Mitarbeitern (*45*, *6*) eine Strukturformel (LXV) wahrscheinlich gemacht worden, die bei mancher Ähnlichkeit mit der des Lupanins (Vorliegen von zwei kondensierten Norlupinanringen, Lactamgruppe) aber eine andere Verknüpfung der Ringe aufweist, als sie den Alkaloiden der Sparteingruppe zugrunde liegt. Eine weitere Base vom gleichen Typ wurde im Oxymatrin gefunden, das das Matrin-N-oxyd sein dürfte.

(LXV.)

In *Retama sphaerocarpa* wurde neben Spartein eine Base von der Formel $C_{15}H_{26}ON_2$, das Retamin (*86*), nachgewiesen, das eine tertiäre Hydroxylgruppe enthält und vielleicht die Struktur eines 6-Oxyparteins besitzt.

Von Interesse sind noch die Nebenalkaloide des Sparteins in *Spartium scoparium*, von denen einige schon lange bekannt sind, so das Genistein ($C_{16}H_{28}N_2$) (*114*) und das Sarothamnin, für das früher die Formel $C_{15}H_{24}N_2$, neuerdings aber eine Formel $C_{30}H_{50}N_4$ (*31*), die also dem bimolekularen Typus eines Disparteins entspricht, aufgestellt wurde. Über die genaue Konstitution dieses Alkaloids sowie über die des Genisteins, das ein C-Atom mehr als das Spartein enthält, ist noch nichts bekannt. Eine Reihe weiterer Nebenalkaloide des Sparteins wurde noch von WINTERFELD und NITZSCHE (*114*) dargestellt. Sie sind mit Spartein isomer, und es wurde von den genannten Autoren teils Struktur-, teils Stereoisomerie angenommen, doch steht eine weitere Untersuchung dieser Basen noch aus.

V. Versuche einer zell-möglichen Synthese der Lupinen-Alkaloide.

Die Versuche einer zell-möglichen Synthese, einer Synthese unter „physiologischen" Bedingungen, wie sie in einigen Alkaloidgruppen hauptsächlich von R. ROBINSON und CL. SCHÖPF mit großem Erfolg zuerst theoretisch formuliert und in vielen Fällen auch experimentell

verwirklicht werden konnte, stehen bei den Lupinen-Alkaloiden erst im
Anfangsstadium. Immerhin sind gerade in letzter Zeit auch hier Erfolge
erzielt worden. Schöpf (88) hat bezüglich der Frage der zell-möglichen
Synthese des Lupinins (I) zuerst angenommen, daß es in der Pflanze aus
δ-Amino-valeraldehyd, das aus Lysin entstehen kann, und Glutaraldehyd
durch Aldehydammoniak-Bildung, Aldolkondensation und Reduktion des
entstandenen Lupininaldehyds (LXVI) gebildet wird.

(LXVI.) Lupininaldehyd.   (I.) Lupinin.

Da Versuche, ausgehend vom Δ^1-Piperidein, das die durch innere
Wasserabspaltung entstandene Schiffsche Base des δ-Amino-valer-
aldehyds vorstellt, und Glutaraldehyd zu (LXVI) zu gelangen, ergebnislos
blieben, wurde neuerdings auch der Iminoaldehyd (LXVII) als Zwischenstufe der zell-möglichen Lupininsynthese in Betracht gezogen (89).

In letzter Zeit wurde aber eine Sparteinsynthese durchgeführt, die
unter zell-möglichen Bedingungen abläuft. Anet, Hughes und Ritchie
(1) gelang es, δ-Amino-valeraldehyd mit Acetondicarbonsäure zuerst bei
pH 13 und dann mit Formaldehyd bei pH 7, bei Zimmertemperatur und in
gepufferter Lösung zum Ketospartein (LXVIII) zu kondensieren. Das
17-Ketospartein, das in 15%iger Ausbeute erhalten wurde, ging bei der
Reduktion nach Clemmensen in Spartein über.

(LXVII.)

(LXVIII.) 17-Ketospartein.

Über den Alkaloid- und Stickstoffwechsel in den Lupinen und zur
Frage der Bildung der Lupinen-Alkaloide in den Lupinen selbst liegen

noch wenige Arbeiten vor. Bei der Keimung von *Lupinus luteus* wurde durch quantitative Bestimmung der Alkaloide festgestellt [eine Übersicht der Alkaloidbestimmungs-Methoden in Lupinensamen siehe (7)] (*101*), daß auch schon gebildete Alkaloide wieder in den Stoffwechsel einbezogen werden können. Neuerdings wurde gezeigt (*71*), daß in isolierten Lupinenwurzeln auf Nährboden mit 2% Saccharose und Stickstoff in Form von Calciumnitrit Lupanin gebildet wird. Eine Mitwirkung des Sprosses ist für diese Synthese nicht erforderlich.

VI. Pharmakologie der Lupinen-Alkaloide.

Wie schon erwähnt, hat von den Lupinen-Alkaloiden nur das Spartein eine gewisse therapeutische Verwendung gefunden. Von den verwandten Alkaloiden besitzt keines medizinisches, wohl aber das Cytisin toxikologisches Interesse. Cytisin macht die Gewächse, in denen es vorkommt, je nach der Menge, in der es auftritt, zu mehr oder minder starken Giftpflanzen. Die Wirkungen des Cytisins sind nikotin-ähnlich, zu tödlichen Vergiftungen kommt es aber selten, da das Alkaloid ähnlich wie Nicotin schnell Übelkeit und Erbrechen hervorruft. Die Grundwirkungen des Cytisins sind bei höheren Dosen Lähmung des zentralen wie peripheren Nervensystems, Gefäßverengung, damit Blutdrucksteigerung, Erregung der Drüsentätigkeit und Verstärkung der Darmbewegungen.

Manche nikotin-ähnliche Wirkungen liegen auch beim Spartein vor, nur ist die Giftwirkung wie bei den anderen Lupinen-Alkaloiden weit geringer. Mengen von 0,05 bis 0,2 g Sparteinsulfat können beim Menschen mehrmals am Tage, auch subkutan, gegeben werden (*70*). Die relative Harmlosigkeit und ein gewisser Einfluß auf die Regulierung der Herztätigkeit hat dazu geführt, daß das Spartein als Herzmittel empfohlen wurde und in manchen Ländern, so in Frankreich und in Nordamerika, therapeutische Verwendung gefunden hat, obwohl die Meinungen über die Wirkungsweise des Alkaloids und die arzneiliche Verwendbarkeit voneinander oft stark abweichen. Das Spartein besitzt eine pulsverlangsamende Wirkung, die atropin-resistent ist, und reguliert den Kreislauf durch Verlangsamung der Reizleitung und Behebung des Vorhof- und Kammerflimmerns. Es wurde daher bei manchen Erkrankungen und bei Irregularitäten des Herzens als Beruhigungsmittel empfohlen, ist aber kein Digitalis-Ersatzmittel. [Eine neuere Studie der Sparteinwirkung siehe (*33*).] Interessant ist auch die Wirkung des Sparteins auf den Uterus (*62*). Es wurde daher auch als Wehenmittel in Betracht gezogen.

Als Nebenwirkung des Alkaloids ist vor allem der lähmende Einfluß auf das Nervensystem zu nennen, der die volle Ausnützung der Herzwirkung beim Menschen verhindert. Voraussetzung für eine erfolgreiche Anwendung beim Menschen wäre die Auffindung von Derivaten des

Sparteins, bei denen die Nebenwirkungen wesentlich abgeschwächt sind. Einige Versuche dazu wurden wohl unternommen (*107, 119*), doch steht ein systematisches Studium der Zusammenhänge zwischen Konstitution und Wirkung der Lupinen-Alkaloide und ihrer Derivate im Hinblick auf dieses Ziel noch aus.

Literaturverzeichnis.

1. Anet, E., G. K. Hughes and E. Ritchie: A Synthesis of Sparteine. Nature (London) **165**, 35 (1950).

2. Beckel, A.: Über das Oxylupanin. Arch. Pharmaz. Ber. dtsch. pharmaz. Ges. **248**, 451 (1910).

3. Bergh, F.: Über die Alkaloide der perennierenden Lupine. Arch. Pharmaz. Ber. dtsch. pharmaz. Ges. **242**, 416 (1904).

4. Boekelheide, V. and S. Rothchild: Quinolizidine. J. Amer. chem. Soc. **69**, 3149 (1947).

5. — — Curariform Activity and Chemical Structure. V. Syntheses in the Quinolizidine Series. J. Amer. chem. Soc. **71**, 879 (1949).

6. Boit, H.-G.: Fortschritte der Alkaloidchemie seit 1933. Berlin: Akademie-Verlag. 1950.

7. Brahm, C. u. G. Andresen: Die Bestimmung des Alkaloidgehaltes in Lupinen-samen nach verschiedenen Methoden. Angew. Chem. **39**, 1348 (1926).

8. Clemo, G. R. and G. C. Leitch: The Lupin Alkaloids. I. J. chem. Soc. (London) **1928**, 1811.

9. Clemo, G. R., T. P. Metcalfe and R. Raper: The Lupin Alkaloids. XI. The Octahydropyridocoline-nor-Lupinane Relationship. J. chem. Soc. (London) **1936**, 1429.

10. Clemo, G. R., W. Morgan and R. Raper: The Lupin Alkaloids. X. The Synthesis of *d,l*-Oxysparteine. J. chem. Soc. (London) **1936**, 1025.

11. — — — The Lupin Alkaloids. VIII. J. chem. Soc. (London) **1935**, 1743.

12. — — — The Lupin Alkaloids. XII. The Synthesis of *d,l*-Lupinine and *d,l*-iso-Lupinine. J. chem. Soc. (London) **1937**, 965.

13. — — — The Lupin Alkaloids. XIII. The Resolution of *d,l*-Lupinine. J. chem. Soc. (London) **1938**, 1574.

14. Clemo, G. R. and G. R. Ramage: The Lupin Alkaloids. IV. The Synthesis of Octahydropyridocoline. J. chem. Soc. (London) **1931**, 437.

15. Clemo, G. R., G. R. Ramage and R. Raper: The Lupin Alkaloids. VI. J. chem. Soc. (London) **1932**, 2959.

16. Clemo, G. R., R. Raper and W. S. Short: The Lupin Alkaloids. XIV. J. chem. Soc. (London) **1949**, 663.

17. Clemo, G. R., R. Raper and Ch. Tenniswood: The Lupin Alkaloids. III. J. chem. Soc. (London) **1931**, 429.

18. Clemo, G. R., R. Raper and H. J. Vipond: The Clemmensen Reduction of Certain α-Amino-ketones and its Bearing on the Reduction of 1-Keto-octa-hydropyridocoline. J. chem. Soc. (London) **1949**, 2095.

19. Consden, R., A. H. Gordon and A. J. P. Martin: Qualitative Analysis of Proteins: A Partition Chromatographic Method Using Paper. Biochemic. J. **38**, 224 (1944).

20. Couch, J. F.: A New Lupine Alkaloid, Spathulatine, Isolated from *Lupinus spathulatus* (Rydb.). J. Amer. chem. Soc. **46**, 2507 (1924).

21. — Lupine Studies. VI. The Alkaloids of *Lupinus corymbosus*, Heller. Part 1. J. Amer. chem. Soc. **56**, 155 (1934).

22. COUCH, J. F., VIII. The Alkaloids of *Lupinus palmeri*, S. WATS. J. Amer. chem. Soc. 56, 2434 (1934).

23. — IX. Monolupine, a New Alkaloid from *Lupinus caudatus* KELLOG. J. Amer. chem. Soc. 58, 686 (1936).

24. — The Numbering of the Sparteine Molecule and its Derivatives. J. Amer. chem. Soc. 58, 688 (1936).

25. — XI. The Alkaloids of *Lupinus barbiger* S. WATS. J. Amer. chem. Soc. 58, 1296 (1936).

26. — XII. The Alkaloids of *Lupinus laxus* RYDB. J. Amer. chem. Soc. 59, 1469 (1937).

27. — XIII. Octalupine, a New Alkaloid from *Lupinus sericeus* var. *flexuosus* C. P. SMITH. J. Amer. chem. Soc. 61, 1523 (1939).

28. — XIV. The Isolation of Anagyrine from *Lupinus laxiflorus* var. *silvicola* C. P. SMITH. J. Amer. chem. Soc. 61, 3327 (1939).

29. — XV. The Alkaloids of *Lupinus sericeus* PURSH. J. Amer. chem. Soc. 62, 554 (1940).

30. — XVI. The Isolation of Nonalupine from *Lupinus Andersonii* WATS. J. Amer. chem. Soc. 62, 986 (1940).

31. DELABY, R., R. BARONNET et J. GUY: La sarothamnine, alcaloide fixe du genêt à balai *(Sarothamnus scoparius)*. Bull. Soc. chim. France [5] 16, 152 (1949).

32. DEULOFEU, V. and A. GATTI: Studies on Argentine Plants. VI. Hydroxylupanine in *Lupinus hilarïanus*. J. org. Chemistry 10, 179 (1945).

33. DONGEN, R. VAN, J. A. BOS and J. GUY: Sparteine. Arch. int. Pharmacodynamic Thérap. 76, 163 (1948); [Chem. Abstr. 42, 6453 (1948)].

34. FITTING, H., W. SCHUMACHER, R. HARDER u. F. FIRBAS: Lehrbuch der Botanik für Hochschulen, 22. Aufl., S. 207. Jena: Verlag G. Fischer. 1944.

35. FREUDENBERG, W. and E. F. ROGERS: New Alkaloids in *Aconitum Napellus*. J. Amer. chem. Soc. 59, 2572 (1937).

36. GALINOVSKY, F. u. G. KAINZ: Über die Synthese des *d,l*-Oxysparteins. Monatsh. Chem. 77, 137 (1947).

37. — — Synthese des *d*-Sparteins und *l*-Sparteins. Monatsh. Chem. 80, 112 (1949).

38. — — Über das Monohydrochlorid und Monoaminoxyd des Lupanins. Monatsh. Chem. [im Druck].

39. GALINOVSKY, F. u. M. PÖHM: Zur Kenntnis des Oxylupanins. Monatsh. Chem. 80, 864 (1949).

40. GALINOVSKY, F., M. PÖHM u. K. RIEDL: Über die Konstitution des Oxylupanins. Die Stellung der Hydroxylgruppe. Monatsh. Chem. 81, 77 (1950).

41. GALINOVSKY, F. u. H. SCHMID: Die Reduktion des Oxysparteins zum Spartein. Monatsh. Chem. 79, 322 (1948).

42. GALINOVSKY, F. u. E. STERN: Zur Kenntnis des α-Norlupinons: Reduktion zum Norlupinan. Ber. dtsch. chem. Ges. 76, 1034 (1943).

43. — — Über die katalytische Reduktion einiger Alkaloide der Sparteingruppe, die einen Lactam- oder α-Pyridonring enthalten. Ber. dtsch. chem. Ges. 77, 132 (1944).

44. GRUBER, W. u. K. SCHLÖGL: Lithiumreaktionen an α-Picolinen. Monatsh. Chem. 81, 473 (1950).

45. HENRY, TH. A.: The Plant Alkaloids. 4th. ed., London: J. & A. Churchill Ltd. 1949.

46. HENTRICH, W. u. F. HOERMANN-GUTTENBERG: Ein neues Verfahren zur Lupinenentbitterung. Chemiker-Ztg. 61, 621 (1935).

47. Hoffmann, E., F. W. Holschneider u. K. Winterfeld: Über die Spaltung des Lupanin-Piperidonringes mittels rauchender Salzsäure. Arch. Pharmaz. Ber. dtsch. pharmaz. Ges. **275**, 65 (1937).
48. Ing, H. R.: Cytisine. Part I. J. chem. Soc. (London) **1931**, 2195.
49. — Cytisine. Part II. J. chem. Soc. (London) **1932**, 2778.
50. — The Alkaloids of *Anagyris foetida* and their Relation to the Lupin-Alkaloids. J. chem. Soc. (London) **1933**, 504.
51. Karrer, P., F. Canal, K. Zohner u. R. Widmer: Über Lupinin. Helv. chim. Acta **11**, 1062 (1928).
52. Karrer, P., B. Shibata, A. Wettstein u. L. Jacubowicz: Über Spartein. Helv. chim. Acta **13**, 1292 (1930).
53. Karrer, P. u. A. Widmer: Über erschöpfende Chromsäureoxydation hydrierter, cyclischer Basen. Helv. chim. Acta **9**, 886 (1926).
54. Lasurjewski, G. W. u. A. S. Ssadykow: Chemische Untersuchung von *Ammothamnus Lehmanii* Bge. I. Bull. Univ. Asie Centrale **22**, 171; [Chem. Zbl. **1940** II, 1308].
55. Leonard, N. J. and E. Barthel, Jr.: Rearrangement of α-Amino-ketones During Clemmensen Reduktion. III. Contraction of a Seven-membered Ring in the Monocyclic Series. J. Amer. chem. Soc. **71**, 3098 (1949).
56. Leonard, N. J. and R. E. Beyler: The Total Synthesis of Sparteine. J. Amer. chem. Soc. **70**, 2298 (1948).
57. — — Resolution of Sparteine. J. Amer. chem. Soc. **71**, 757 (1949).
58. — — The Synthesis of Sparteine and Isosparteine. J. Amer. chem. Soc. **72**, 1316 (1950).
59. Leonard, N. J. and W. V. Ruyle: Rearrangement of α-Amino-ketones During Clemmensen Reduction. II. Contraction of a Six-membered Ring in the Monocyclic Series. J. Amer. chem. Soc. **71**, 3094 (1949).
60. Leonard, N. J. and W. C. Wildman: I. Bicyclic Compounds Containing a Bridgehead Nitrogen. J. Amer. chem. Soc. **71**, 3089 (1949).
61. — — IV. Contraction of a Seven-membered Ring in the Bicyclic Series. J. Amer. chem. Soc. **71**, 3100 (1949).
62. Ligon, W., Jr.: Die Wirkung von Lupinen-Alkaloiden auf die Motilität des isolierten Kaninchenuterus. J. Pharmacol. exp. Therapeut. **73**, 151 (1941); [Chem. Zbl. **1943** II, 1560].
63. Lukeš, R. and F. Šorm: Syntheses in the Allolupinane Series. Collect. Czechoslov. Chem. Communs. **12**, 356 (1947); [Chem. Abstr. **42**, 7780 (1948)].
64. Manske, R. H. F. and L. Marion: The Alkaloids of *Thermopsis rhombifolia* (Nutt.) Richards. Canad. J. Res., B **21**, 144 (1943).
65. Marion, L.: The Papilionaceous Alkaloids. I. *Lupinus Macounii*, Rydb. J. Amer. chem. Soc. **68**, 759 (1946).
66. Marion, L. and W. F. Cockburn: V. *Baptisia minor*, Lehm. J. Amer. chem. Soc. **70**, 3472 (1948).
67. Marion, L. and S. W. Fenton: VI. *Lupinus pusillus* Pursh. J. org. Chemistry **13**, 780 (1948).
68. Marion, L. and J. Quellet: III. Identity of Rhombinine and Monolupine with Anagyrine. J. Amer. chem. Soc. **70**, 3076 (1948).
69. Marion, L. and F. Turcotte: IV. *Baptisia perfoliata* (L.) R. Br. J. Amer. chem. Soc. **70**, 3253 (1948).
70. Meyer, H. H.: Experimentelle Pharmakologie, 9. Aufl., S. 379. Berlin u. Wien: Urban & Schwarzenberg. 1936.
71. Mothes, K. u. D. Kretschmer: Über die Alkaloidsynthese in isolierten Lupinenwurzeln. Naturwiss. **33**, 26 (1946).

72. ORECHOFF, A.: Über die Alkaloide des Krautes von *Sophora alopecuroides*. (III. Mitt. über Sophora-Alkaloide.) Ber. dtsch. chem. Ges. **66**, 948 (1933).

73. — Über die Alkaloide von *Anabasis aphylla*. XIV. Über die Struktur von Aphyllidin und Aphyllin. J. Gen. Chem. (USSR.) **7**, 2048 (1937); [Chem. Zbl. **1938** I, 2365].

74. ORECHOFF, A. u. G. MENSCHIKOFF: Über die Alkaloide von *Anabasis aphylla*. III. Mitt.: Über die hochsiedenden Basen. Ber. dtsch. chem. Ges. **65**, 234 (1932).

75. ORECHOFF, A. u. S. NORKINA: Über die Alkaloide von *Anabasis aphylla*. X. Mitt.: Über die Reduktion des Aphyllidins. Ber. dtsch. chem. Ges. **67**, 1845 (1934).

76. — — Über die Alkaloide von *Anabasis aphylla*. XI. Mitt.: Über den HOF-MANNschen Abbau des Aphyllidins. Ber. dtsch. chem. Ges. **67**, 1974 (1934).

77. ORECHOFF, A., S. NORKINA u. H. GUREWITSCH: Über Sophora-Alkaloide. II. Mitt.: Über die Alkaloide von *Thermopsis lanceolata*. Ber. dtsch. chem. Ges. **66**, 625 (1933).

78. ORECHOFF, A. u. N. PROSKURNINA: Über Sophora-Alkaloide. IV. Mitt.: Über die Alkaloide der Samen von *Sophora pachycarpa*. Ber. dtsch. chem. Ges. **67**, 77 (1934).

79. ORECHOFF, A., N. PROSKURNINA u. R. KONOWALOWA: Über Sophora-Alkaloide. VIII. Mitt.: Alkaloide aus den Samen und aus dem Kraute von *Sophora alopecuroides*. Ber. dtsch. chem. Ges. **68**, 431 (1935).

80. ORECHOFF, A., M. RABINOWITCH u. R. KONOWALOWA: Über Sophora-Alkaloide. I. Mitt.: Über die Alkaloide des Krautes von *Sophora pachycarpa*. Ber. dtsch. chem. Ges. **66**, 621 (1933).

81. — — — Über Sophora-Alkaloide. VI. Mitt.: Über die hochsiedenden Basen des Krautes von *Sophora pachycarpa*. Beitrag zur Kenntnis des Sophoridins und Sophocarpins. Ber. dtsch. chem. Ges. **67**, 1850 (1934).

82. POLONOVSKI, MAX et MICHEL POLONOVSKI: Sur les aminoxydes des alcaloïdes. XII. Spartéine-N-oxyde. Bull. Soc. chim. France [5] **3**, 891 (1936).

83. POWER, F. B. and A. H. SALWAY: The Constituents of the Rhizome and Roots of *Caulophyllum thalictroides*. J. chem. Soc. (London) **103**, 191 (1913).

84. PRELOG, V. u. K. BOŽIČEVIĆ: Über eine neue Synthese von Norlupinan (1-Aza-bicyclo-[0,4,4]-decan). Ber. dtsch. chem. Ges. **72**, 1103 (1939).

85. PRELOG, V. u. R. SEIWERTH: Über die Konstitution des sogenannten Nor-lupinans B. Ber. dtsch. chem. Ges. **72**, 1638 (1939).

86. RIBAS, I., A. SANCHEZ and E. PRIMO: The Alkaloids of *Retama sphaerocarpa* and the Constitution of Retamine. Anales fis. y quím. (Madrid) **42**, 516 (1946); [Chem. Abstr. **41**, 4894 (1947)].

87. SCHIRM, M. u. H. BESENDORF: Über die Konstitution des dem Sparteinmolekül zugrunde liegenden Kohlenwasserstoffes $C_{15}H_{32}$. Arch. Pharmaz. Ber. dtsch. pharmaz. Ges. **280**, 64 (1942).

88. SCHÖPF, CL.: Zur Frage der Biogenese der Naturstoffe: Alkaloidsynthesen unter physiologischen Bedingungen. Congr. int. Quím. pura apl. **9** V, 189; [Chem. Zbl. **1936** II, 3302].

89. — Neuere Synthesen unter physiologischen Bedingungen. Angew. Chem. **61**, 31 (1949) [Vortragsreferat].

90. SEKA, R.: Alkaloide, Ergänzung 1927—1932. In: ABDERHALDEN, Biologische Arbeitsmethoden. Berlin u. Wien: Urban & Schwarzenberg.

91. SENGBUSCH, R. v.: Süßlupinen und Öllupinen. Die Entstehungsgeschichte einiger neuer Kulturpflanzen. Landw. Jahrb. **91**, 719 (1942); [Chem. Zbl. **1942** II, 1841].

92. Šorm, F. and B. Keil: The Synthesis of an Isomer of Dioxosparteine. Collect. Czechoslov. Chem. Communs. 12, 655 (1947); [Chem. Abstr. 42, 5920 (1948)].

93. — — The Synthesis of Sparteine and Isosparteine. Collect. Czechoslov. Chem. Communs. 13, 544 (1948); [Chem. Abstr. 43, 3828 (1949)].

94. Späth, E. u. F. Galinovsky: Über die Konstitution des Cytisins. Ber. dtsch. chem. Ges. 65, 1526 (1932).

95. — — Die Konstitution des Cytisins. Ber. dtsch. chem. Ges. 66, 1338 (1933).

96. — — Über die Dehydrierung des Cytisins und einiger Abbauprodukte dieser Base. Ber. dtsch. chem. Ges. 69, 761 (1936).

97. — — Synthese des 2,4-Dimethyl-8-oxo-*ps*-chinolizins-(8), eines Abbauproduktes des Cytisins. Ber. dtsch. chem. Ges. 71, 721 (1938).

98. — — Zur Kenntnis des Sparteins. Der Hofmannsche Abbau des Oxysparteins. Ber. dtsch. chem. Ges. 71, 1282 (1938).

99. Späth, E., F. Galinovsky u. M. Mayer: Über die hochsiedenden Basen von *Anabasis aphylla* L. Ber. dtsch. chem. Ges. 75, 805 (1942).

100. Späth, E. u. F. Kuffner: Über das Vorkommen von Spartein in *Chelidonium majus*. Ber. dtsch. chem. Ges. 64, 1127 (1931).

101. Wallebroek, J. C. P.: Alkaloid- und Stickstoffwechsel bei der Keimung von *Lupinus luteus*. Rec. Trav. bot. neerl. 37, 78 (1940); [Chem. Zbl. 1940 II, 2485].

102. White, E. P.: Alkaloids of the Leguminosae. XIV. Alkaloids of *Cytisus canariensis*, *Cytisus stenopetalus*, and Allied Species. New Zealand J. Sci. Technol., B 27, 335 (1946); [Chem. Abstr. 40, 5750 (1946)].

103. — XVI. Alkaloids of *Genista aethnensis* DC. New Zealand J. Sci. Technol., B 27, 474 (1946); [Chem. Abstr. 41, 6574 (1947)].

104. Willstätter, R. u. W. Marx: Über die Oxydation von Spartein. Ber. dtsch. chem. Ges. 38, 1772 (1905).

105. Winterfeld, K. u. H. v. Cosel: Synthese des razem. Lupinins. Arch. Pharmaz. Ber. dtsch. pharmaz. Ges. 278, 70 (1940).

106. Winterfeld, K. u. E. Hoffmann: Über das Verhalten des Lupanins bei der Grignardierung. Arch. Pharmaz. Ber. dtsch. pharmaz. Ges. 275, 5 (1937).

107. — — Zur Kenntnis des Anisoyl-Sparteins. Arch. Pharmaz. Ber. dtsch. pharmaz. Ges. 275, 526 (1937).

108. Winterfeld, K. u. F. W. Holschneider: Über die Konstitution des Lupinins (I. Mitt.). Ber. dtsch. chem. Ges. 64, 137 (1931).

109. — — Über die Einwirkung von rauchender Jodwasserstoffsäure auf Lupanin. Ber. dtsch. chem. Ges. 64, 2415 (1931).

110. — — Über die Alkaloide der Lupinen. VII. Die Synthese des Norlupinans. Liebigs Ann. Chem. 499, 109 (1932).

111. — — Die Synthese des β-Lupinans. Ber. dtsch. chem. Ges. 66, 1751 (1933).

112. — — Synthese des Allolupinins. Arch. Pharmaz. Ber. dtsch. pharmaz. Ges. 277, 221 (1939).

113. Winterfeld, K. u. A. Kneuer: Zur Kenntnis des Lupanins. Ber. dtsch. chem. Ges. 64, 150 (1931).

114. Winterfeld, K. u. F. Nitzsche: Über die Nebenalkaloide von *Spartium scoparium*. Arch. Pharmaz. Ber. dtsch. pharmaz. Ges. 278, 393 (1941).

115. Winterfeld, K. u. P. Petkow: Die Umsetzung von *d,l*-Lupanin mit Grignard-Lösungen. Chem. Ber. 82, 156 (1949).

116. Winterfeld, K. u. C. Rauch: Zur Kenntnis des Sparteins. Arch. Pharmaz. Ber. dtsch. pharmaz. Ges. 272, 273 (1934).

117. Winterfeld, K. u. H. Rönsberg: Der oxydative Abbau des α-Didehydrosparteins. Arch. Pharmaz. Ber. dtsch. pharmaz. Ges. 274, 48 (1936).

118. WINTERFELD, K. u. M. SCHIRM: Der oxydative Abbau des Phenyldehydro-sparteins. Ein Beitrag zur Konstitutionsaufklärung des Sparteins bzw. Lu-panins. Arch. Pharmaz. Ber. dtsch. pharmaz. Ges. **275**, 630 (1937).

119. WOLFFENSTEIN, R. u. J. REITMANN: Zur Kenntnis des Sparteins. Biochem. Z. **186**, 269 (1927).

120. WOODWARD, R. B. and E. C. KORNFELD: Ethyl-2-Pyridylacetate. Org. Syn-theses **29**, 44 (1949).

121. ZABOEV, S. A.: Lupinine and its Oxidation. J. Gen. Chem. (USSR.) **18**, 194 (1948); [Chem. Abstr. **42**, 7299 (1948)].

(Eingelaufen am 16. Februar 1951.)

Brechwurzel-Alkaloide.

Von **M. Pailer**, Wien.

Inhaltsübersicht.

I. Einleitung.

Zur Geschichte. Die *Ipecacuanha* wird erstmalig um das Jahr 1570 von einem damals in Brasilien weilenden portugiesischen Mönch erwähnt, wonach die Pflanze von den Eingeborenen gegen verschiedene Blutungen verwendet wurde. Ungefähr ein Jahrhundert später wird dann von einer Verwendung der Droge gegen Ruhr in Frankreich berichtet. Infolge falscher Dosierungen ereigneten sich aber tödliche Vergiftungen, so daß man von dieser Verwendung vorerst wieder abkam. Einige Jahre später (1680) verstand es der Medizinstudent Jan Adrian Helvetius, der durch den Pariser Kaufmann Grenier in den Besitz einer größeren Menge der „Ruhrwurzel" gekommen war, diese durch öffentliche Anschläge in Paris einerseits berühmt zu machen, anderseits die Natur seines Wundermittels geheimzuhalten. Es gelang ihm dann die Heilung verschiedener adeliger Persönlichkeiten, so daß schließlich Ludwig XIV. auf ihn aufmerksam wurde, und ihm zuerst im Jahre 1688 das alleinige Verkaufsrecht erteilte und schließlich für die Preisgabe seines Geheimnisses 1000 Louisdor und besondere Stellungen bot.

Dadurch war jedenfalls diese wichtige Droge bekannt geworden und war nach und nach auch in verschiedenen anderen europäischen Ländern anzutreffen. Man betrachtete sie anfangs als Wundermittel, das man gegen alle möglichen menschlichen Leiden anwandte. Ihre wesentlichste, auch heute noch geschätzte Anwendung findet sie als Expektorans und besonders zur Bekämpfung der Amöbenruhr (Amöbendysenterie). In der Therapie dieser Krankheit stieg die Bedeutung der *Ipecacuanha* sukzessive mit der Ausdehnung des europäischen Kolonialbesitzes. Durch die Arbeiten verschiedener europäischer Kliniken sowie verschiedener Kolonial- und Militärärzte gelang es, die fürchterliche Seuche, wie sie die Amöbendysenterie mit ihren Komplikationen, besonders den Leberabszessen in den Tropen darstellte, entscheidend zu bekämpfen und die Dysenteriefälle und Sterblichkeit zu Beginn des 20. Jahrhunderts auf ein Minimum herabzusetzen.

Verfälschungen. Mit zunehmender Bedeutung erlitt die Wurzel, vorerst wahrscheinlich unabsichtlich, später aber mit Absicht, Verfälschungen verschiedenster Art. Zu Ende des 17. Jahrhunderts kannte man fünf verschiedene Sorten der Brechwurzel, von denen drei, nämlich die aus Rio de Janeiro über Porto und Lissabon importierte schwärzliche, die über Peru und Cadiz kommende gelbe und die von Holländern importierte Droge, wohl Varianten der echten Wurzel sind. Die anderen beiden, die „*Ipecacuanha blanca*" von *Piso* (Wurzel von Richardsonia) sowie die „*Ipecacuanha Griscendré-glicyrrhizé*" (Pomet, Lemery), die Wurzel von *Psychotria emetica* Multis sind dagegen als Surrogate zu betrachten. Mit der Zeit nahm die Zahl dieser Drogen natürlich zu, worüber sich genauere Angaben bei Tschirch (57) sowie bei Staub (55) finden.

Staub gibt eine Zusammenfassung der älteren, vor allem auch chemischen Literatur über die Ipecacuanha-Alkaloide bis zum Jahre 1927, so daß diese hier nur in Form einer kurzen Übersicht berücksichtigt werden soll.

Auch die gepulverte Droge wurde ausgiebig verfälscht, indem verschiedene andere Produkte beigemahlen wurden. Früher setzte man zur Verstärkung der Brechwirkung Brechweinstein zu. Heute finden sich als Beimengungen Stamm- und Holzteile der Ipecacuanha, Pulver von Richardsonia-Wurzeln, Süßholz, entölte Birkenrinde, verschiedene Stärkearten, Dextrin, Pflanzengallen, gemahlene bittere Mandeln, gemahlene Oliven sowie Olivenpreßrückstände.

Handelssorten. Alle im Laufe der Zeit als emetin- bzw. cephaelin-haltig gefundenen Brechwurzeln gehören zum Tribus *Psychotriae* der Familie *Rubiaceae.* Die heute unter der Bezeichnung „Brechwurzel" verwendeten Drogen sind im allgemeinen die gemahlenen Wurzeln von *Cephaelis ipecacuanha* (BROT) RICHARD, die im Handel meist als brasilianische oder Rio Ipecacuanha bezeichnet werden, bzw. von *Cephaelis acuminata* KARSTEN, die unter den Namen Cartagena, Nicaragua oder Panama Ipecacuanha bekannt sind.

Die brasilianische Ipecacuanha enthält bis zu 2,5% Gesamtalkaloide, wovon ungefähr die Hälfte Emetin ist. Wegen des höheren Emetingehaltes ist die Rio Ipecacuanha für die Emetindarstellung die weit begehrtere, während die Cartagena Ipecacuanha wegen ihres verhältnismäßig großen Cephaelin-Gehaltes weniger gefragt ist.

Die günstigste Zeit für das Sammeln der Wurzeln und Rhizome liegt zwischen Jänner und März, wobei drei- bis vierjährige Pflanzen die höchste Alkaloidausbeute liefern.

Die Rio Ipecacuanha wächst in den feuchten Wäldern von Brasilien und Bolivien. Sie stellt eine ungefähr 40 cm hohe, immergrüne, strauchartige Pflanze dar, deren dünne, unverästelte Nebenwurzeln, die durch Verdickungen ein perlschnur-ähnliches Aussehen haben, die offizinelle Droge bilden. Die Blüten sind klein und weiß, die Früchte rote Beeren. Ähnliches Aussehen zeigt die in den Wäldern von Columbien vorkommende Nicaragua Ipecacuanha. Eine Unterscheidung der beiden Arten ist durch verschiedene Eigenschaften möglich. Die Rio Ipecacuanha ist mehr rotbraun, während die andere graubraun ist. Weiters unterscheiden sie sich im histologischen Bild durch die verschiedene Größe der Stärkekörner, in den Zellen der Holzteile und schließlich durch das verschiedene Verhältnis von Emetin und Cephaelin in ihrem Alkaloidgehalt.

II. Zusammenhänge der einzelnen Brechwurzel-Alkaloide.

Obwohl das Emetin, das Hauptalkaloid der Brechwurzel, eines der frühzeitig entdeckten Alkaloide ist, wurde es in chemischer Hinsicht vorerst eigentlich wenig bearbeitet. Ein Hauptgrund dafür mag wohl darin liegen, daß man lange Zeit keine reinen Produkte, sondern Gemische des Emetins mit seinen Nebenalkaloiden in Händen hatte. Das Emetin wurde 1817 von PELLETIER und MAGENDIE erstmalig isoliert, rein erhalten wurde es erst 1879 von PODWYSSOTZKI, noch wahrscheinlicher aber erst 1893 durch PAUL und COWNLEY (*43, 44*), die daneben ein zweites Alkaloid feststellten, das Cephaelin genannt wurde. Einer dritten Base, die PAUL und COWNLEY beschrieben, wurde später der

Name Psychotrin gegeben. Zwei weitere Nebenbasen, die HESSE (22) 1914
erhalten haben will, wurden von PYMAN nicht als einheitliche Stoffe
anerkannt und auch seither von keinem Bearbeiter dieses Gebietes mehr
genannt. Nach HESSES Angaben sollten diese beiden Alkaloide neben
dem Emetin den nicht-phenolischen Anteil der Brechwurzel-Alkaloide
ausmachen. Nach PYMAN (46) enthält dieser Anteil zwei weitere Basen,
die er Methylpsychotrin und Emetamin nannte. Bei einer genaueren
Betrachtung der Bruttoformel der Alkaloide dieser Gruppe findet man,
daß hier eine nahe Verwandtschaft bestehen muß:

> Psychotrin $C_{28}H_{36}O_4N_2$.
> O-Methylpsychotrin $C_{29}H_{38}O_4N_2$.
> Emetin $C_{29}H_{40}O_4N_2$.
> Cephaelin $C_{28}H_{38}O_4N_2$.
> Emetamin $C_{29}H_{36}O_4N_2$.

Tatsächlich lassen sich alle diese Basen in unmittelbare Beziehung
zum Emetin setzen, wodurch die Konstitutionsermittlung der ganzen
Gruppe letzten Endes auf die diesbezügliche Untersuchung des Emetins
hinausläuft (Schema 1).

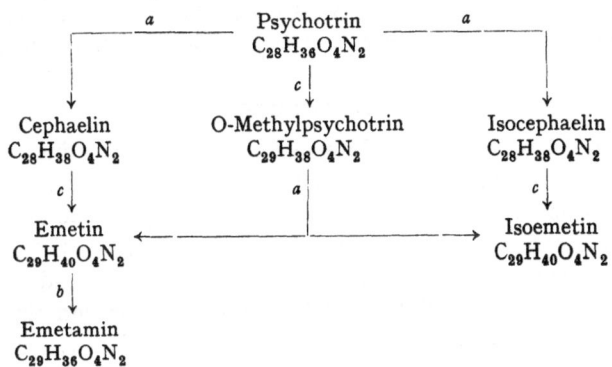

Schema 1. Zusammenhänge zwischen den Brechwurzel-Alkaloiden. (*a* = Hydrierung; *b* = Dehydrierung;
c = Methylierung.)

Die genaueren Zusammenhänge, bzw. die Reaktionen, durch die die
Alkaloide ineinander übergeführt werden können, sollen bei den einzelnen
Basen näher besprochen werden.

Bei der Entmethylierung des Emetins und Cephaelins entsteht das-
selbe nor-Produkt, Emetolin, $C_{25}H_{32}O_4N_2$. Von den zwei Stickstoffatomen ist in allen Brechwurzel-Alkaloiden eines sekundär und das
zweite tertiär, mit Ausnahme des Emetamins, in welchem beide Stick-
stoffe tertiär sind.

III. Emetin.

Das Emetin besitzt die Formel $C_{29}H_{40}O_4N_2$, welche von CARR und
PYMAN (12) sowie von KARRER (23, 24) als endgültig und richtig er-

mittelt wurde, nachdem vorher verschiedene andere Formulierungen diskutiert worden waren. Dieses Ergebnis wurde auf Grund der Elementaranalysen und Molekulargewichtsbestimmungen verschiedener Salze und Derivate sowie auf Grund des genetischen Zusammenhanges mit den verschiedenen Nebenalkaloiden erreicht. Die Elementaranalyse der freien Base führte im allgemeinen zu keinen brauchbaren Werten, weil die Entfernung der letzten Wasserspuren äußerst schwierig ist.

Die Emetinbase ist ein weißes, amorphes Pulver vom Schmelzpunkt 74°; $[\alpha]_D = -25,8$ bis $-32,7°$ (in 50% Alkohol, $c = 1,8$ bis 4,1) oder $[\alpha]_D = -49,8°$, in Chloroform. Das Alkaloid ist leicht löslich in Äthanol, Äther oder Chloroform, weniger leicht in Benzol und Petroläther und sehr wenig in Wasser.

Das Hydrochlorid, $B \cdot 2 HCl \cdot 7 H_2O$ $(B = Base)$ (27, 43) scheidet sich aus Wasser in feinen Nadeln, aus kalten gesättigten Lösungen in Form von Prismen aus; $[\alpha]_D = +11$ bis 21° (Wasser, $c = 1$ bis 8); $[\alpha]_D = +53°$ (Chloroform). Als Schmelzpunkt geben CARR und PYMAN für ein bei 100° getrocknetes Salz (nach Sintern bei 253°) 265° (korr., Zers.) an. Die Schmelzpunkte der Ipecacuanha-Alkaloide und ihrer Derivate haben im allgemeinen sehr streuende Werte. Meist ist zuerst eine Sinterung und dann Schmelzen, im allgemeinen unter Zersetzung zu beobachten. Emetinhydrobromid: $B \cdot 2 HBr \cdot 4 H_2O$; Nadeln aus Wasser, kristallisiert aus Alkohol in farblosen Nadeln, Schmelzp. 235°.

Nitrat: $B \cdot 2 HNO_3 \cdot 3 H_2O$ aus Alkohol oder Wasser, sintert bei 188° und schmilzt bei 245°. Sulfat: $B \cdot H_2SO_4 \cdot 7 H_2O$, weiße Nadeln, Schmelzp. 205 bis 245°.

1. Konstitution.

Die älteren chemischen Untersuchungen des Emetins, an denen sich vor allem PYMAN und Mitarbeiter (8, 12, 46, 47), KELLER (27, 28, 29, 30), HESSE (22), WINDAUS und HERMANNS (58), KARRER (23, 24), SPÄTH und LEITHE (51), STAUB (55), KUNZ-KRAUSE (32, 33, 34, 35) sowie PAUL und COWNLEY (43, 44) beteiligten, führten zu folgenden Erkenntnissen: Von den beiden N-Atomen ist eines sekundär und eines tertiär, was durch Benzoylierung, wobei ein Monobenzoylderivat erhalten wurde, bewiesen werden konnte. Dieses Monobenzoylderivat ist eine einsäurige Base. Die vier Sauerstoff-Atome liegen in Form von vier Methoxylgruppen vor. KARRER führte einen HOFMANNschen Abbau des Alkaloids durch. Durch Einwirkung von Jodmethyl erhielt er ein Methylemetindijodmethylat. Beim Abbau wurde daraus eine amorphe Methinbase erhalten. Bei nochmaligem HOFMANNschem Abbau wurde ein N-Atom in Form von Trimethylamin abgespalten. Damit hat KARRER bewiesen, daß ein N-Atom nur einem Ring, das zweite dagegen zwei Ringen gleichzeitig angehört. Beim EMDE-Abbau, der von SPÄTH und LEITHE durchgeführt wurde, konnte in den ersten beiden Stufen Übereinstimmung mit dem HOFMANNschen Abbau (KARRER) beobachtet werden. Bei der dritten Stufe wurde auch das zweite Stickstoff-Atom teilweise eliminiert und das Endprodukt bestand zu 25% aus einem stickstoff-freien Neutralprodukt und zu 75% aus Basen.

Durch *Oxydation* von Emetin mit Permanganat erhielten WINDAUS und HERMANNS *m*-Hemipinsäure (I). CARR und PYMAN isolierten neben der *m*-Hemipinsäure bei der Oxydation in Aceton 6,7-Dimethoxyisochinolin-1-carbonsäure (II). Dies war der erste Beweis für das Vorhandensein von wenigstens einem Isochinolinring. SPÄTH und LEITHE erhielten bei der gelinden Oxydation von Emetin in schwach alkalischer Lösung Corydaldin (III). Sie konnten auch zeigen, daß der Komplex, aus dem das Corydaldin gebildet wurde, schon in Form eines Tetrahydroisochinolins mit einem sekundären N-Atom vorhanden war, denn die Oxydation des N-Benzoylemetins führte zur Verbindung (IV).

SPÄTH und LEITHE (*51*) nahmen ferner an, daß die 6,7-Dimethoxyisochinolin-1-carbonsäure, die CARR und PYMAN erhalten hatten, von einem anderen Tetrahydroisochinolin-Komplex stammen dürfte, da derartige Dehydrierungen von Tetrahydroisochinolin-Derivaten durch Permanganat vorher nicht beobachtet worden waren. Daß ein zweiter Isochinolinring vorhanden sein mußte, zeigte ihnen ferner die Ausbeute an *m*-

(I.) *m*-Hemipinsäure.

(II.) 6,7-Dimethoxy-isochinolin-1-carbonsäure.

(III.) Corydaldin.

(IV.)

(V.) Emetin, nach STAUB.

(VI.) Emetin, nach BRINDLEY und PYMAN.

(VII.) Emetin, nach SPÄTH und LEITHE.

Hemipinsäure bei der Oxydation des Emetins. Es wurden nämlich bei dieser Oxydation 96% der für einen Dimethoxyisochinolinring berechneten Menge *m*-Hemipinsäure erhalten, während bei der Oxydation ähnlicher bekannter Verbindungen nur 30—40% festgestellt werden konnten. Cephaelin lieferte unter den gleichen Bedingungen ebenfalls *m*-Hemipinsäure, allerdings nur in halber Ausbeute wie Emetin. Weiters gab das am Sauerstoff äthylierte Cephaelin (O-Äthylcephaelin) bei der gelinden Oxydation mit Permanganat neben Corydaldin eine Verbindung, die zwar als solche nicht identifiziert wurde, aber wohl die entsprechende 6-Äthyl-7-methylverbindung sein mußte, da sie bei der Oxydation den der *m*-Hemipinsäure analogen Äthyl-methyläther lieferte.

Die bisher aufgezählten Erkenntnisse führten im Jahre 1927 zur Aufstellung der drei provisorischen Formeln (V), (VI) und (VII), die natürlich noch reichlich hypothetisch waren.

In neuer Zeit konnten AHL und REICHSTEIN (*1*) durch den HOFMANN-schen Abbau des N-Acetylemetins weiters bekräftigen, daß das sekundäre Stickstoff-Atom einem Dimethoxyisochinolinring angehört. Sie isolierten nämlich bei der Oxydation des, nach dreistufigem Abbau gewonnenen N-Acetylderivates, in welchem das ursprünglich tertiäre N-Atom entfernt war, 4,5-Dimethoxyphthalonimid.

Eine Aufklärung des in der Formel von SPÄTH und LEITHE (*51*) zwischen den beiden Isochinolinringen gelegenen Molekülteiles und damit die endgültige Konstitutionsermittlung wurde durch Arbeiten von SPÄTH und PAILER (*53*), PAILER (*38, 39*) sowie PAILER und PORSCHINSKI (*41*) erbracht.

Zuerst wurde das Emetin auf zwei verschiedenen Wegen nach HOF-MANN abgebaut, wodurch die ersten und wesentlichen Aufschlüsse erhalten wurden. Emetinbase gab in der üblichen Weise durch Behandeln mit Jodmethyl und Natriummethylat das Methylemetin-dijodmethylat (VIII), welches dann mit Silberoxyd in die entsprechende diquartäre Base übergeführt wurde. Diese erlitt nach Verdampfen des Lösungsmittels leicht Aufspaltung beider Ringe zur Verbindung (IX). Die gewonnene Dimethinbase wurde an beiden Doppelbindungen hydriert und nun eine neuerliche Behandlung mit Methyljodid, und zwar durch 6-stündiges Kochen in Methanol vorgenommen. Diese Umsetzung führte zu einem Jodmethylat (Schmelzp. 232—234°) (X), dessen nähere Untersuchung zeigte, daß es sich hierbei um ein ungesättigtes Monojodmethylat handelt, das durch Abspaltung eines N-Atoms entstanden war. Es tritt also Eliminierung des einen Stickstoffes vor der zweiten Stufe des HOF-MANNschen Abbaues ein — eine Tatsache, die über die von KARRER bei dem vorher durchgeführten Abbau gemachten Beobachtungen hinausgeht. Bei der Untersuchung der Mutterlauge nach der Abtrennung dieses Jodmethylats konnte der abgespaltene Stickstoff in Form des Trimethyl-

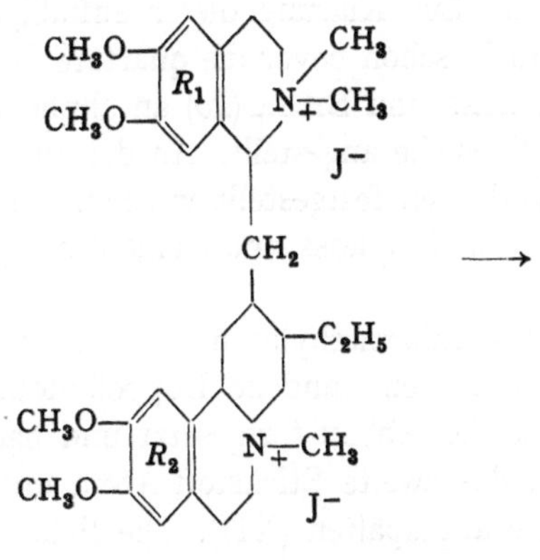

(VIII.) Methylemetin-dijodmethylat.

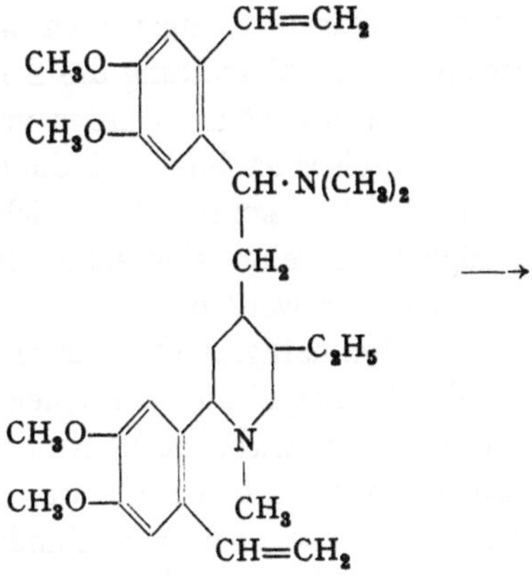

(IX.)

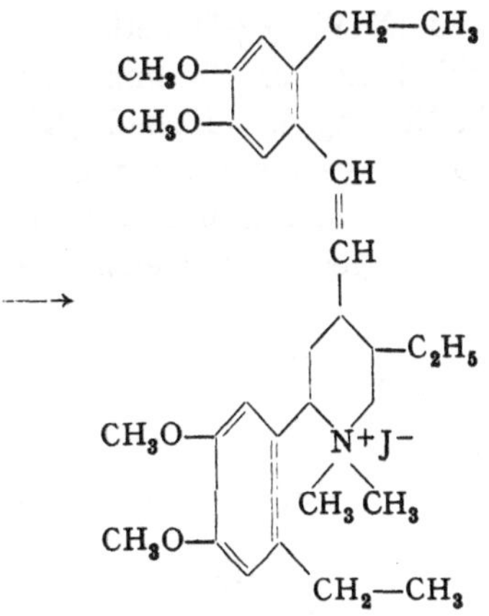

(X.)

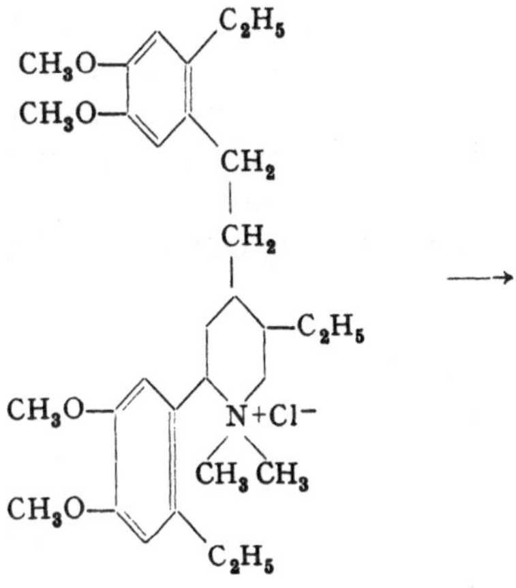

(XI.)

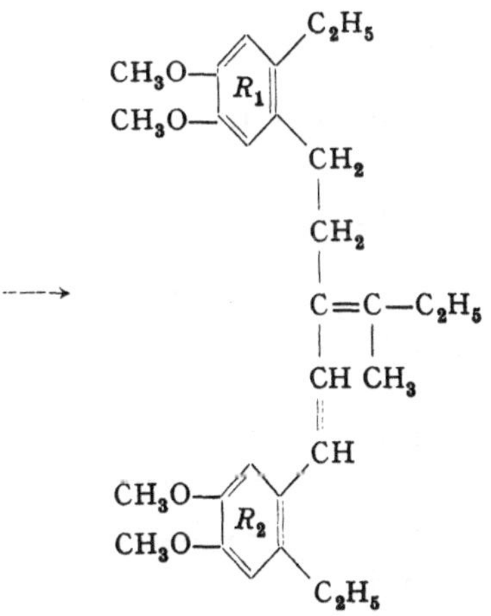

(XII.)

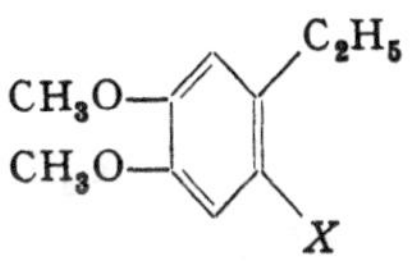

(XIII.) $X = -COOH$ (Äthylveratrumsäure).

(XIV.) $X = -CHO$ (Äthylveratrum-aldehyd).

(XV.) $X = -OH$ (3,4-Dimethoxy-6-äthyl-phenol).

(XVI.) $X = -CH_2-CH_2 \cdot COOH$ (3,4-Dimethoxy-6-äthyl-hydrozimtsäure).

amin-jodhydrats nachgewiesen werden. Zur Klärung dieser auffälligen Reaktion, der Abspaltung des Stickstoffs, schon bevor die quartäre Base in Freiheit gesetzt ist, wurden von Pailer und Bilek (40) an ähnlichen Tetrahydroisochinolin-Verbindungen Versuche angestellt. An den untersuchten Fällen konnte das gleiche Verhalten festgestellt werden. Diesbezügliche Arbeiten sind später auch von Norcross und Openshaw (37) veröffentlicht worden.

Das Jodmethylat (X) wurde in das Chlormethylat übergeführt und die durch Abspaltung des einen N-Atoms entstandene Doppelbindung katalytisch hydriert (XI). Nun wurde der Abbau fortgesetzt und nach zwei weiteren Abbaustufen war auch das zweite Stickstoff-Atom unter Ausbildung von zwei Doppelbindungen abgespalten (XII). Die Behandlung dieser N-freien Verbindung mit Ozon in Chloräthyl lieferte nach der üblichen Verkochung des gebildeten Ozonides die Spaltstücke: Äthylveratrumsäure (XIII), Äthylveratrumaldehyd (XIV), 3,4-Dimethoxy-6-äthylphenol (XV) und eine Carbonylverbindung der Zusammensetzung $C_{18}H_{26}O_3$. Außergewöhnlich war hier die Bildung eines Phenols (XV) neben dem erwarteten Aldehyd (XIV). Dem Austausch einer CHO-Gruppe, die primär bei der Ozonisierung des zweifach ungesättigten N-freien Abbauproduktes (XII) zu erwarten war, gegen ein phenolisches

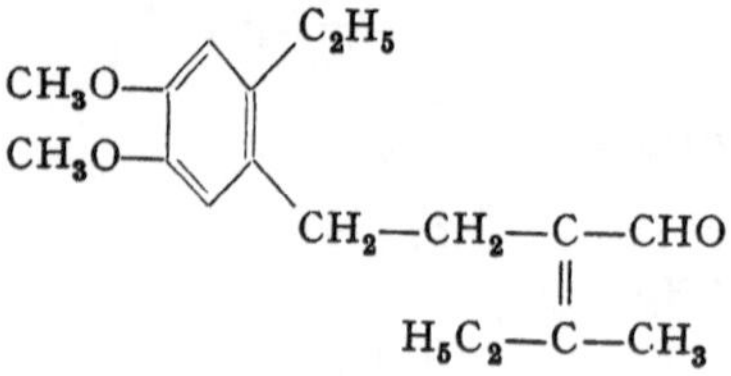

(XVII.)

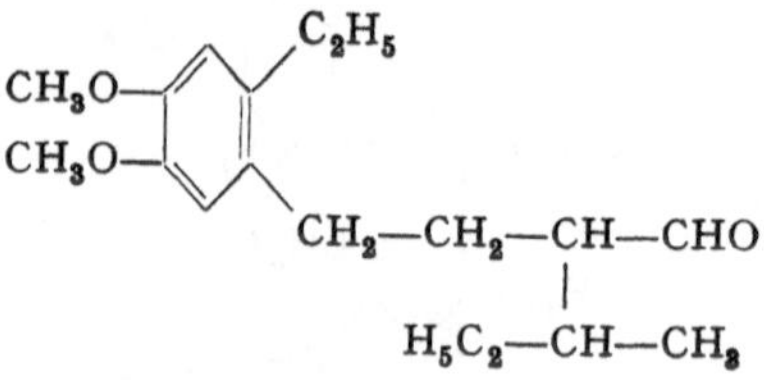

(XVII a.)

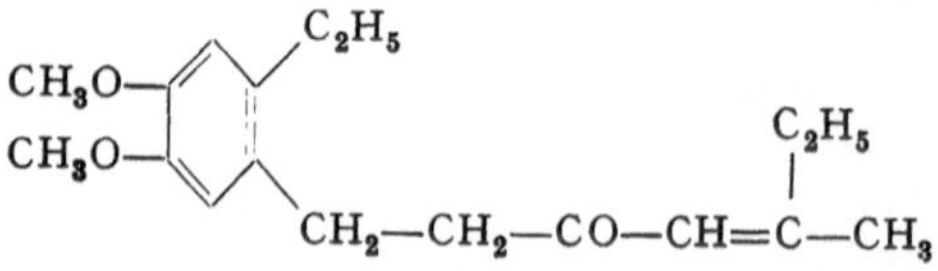

(XVIII.)

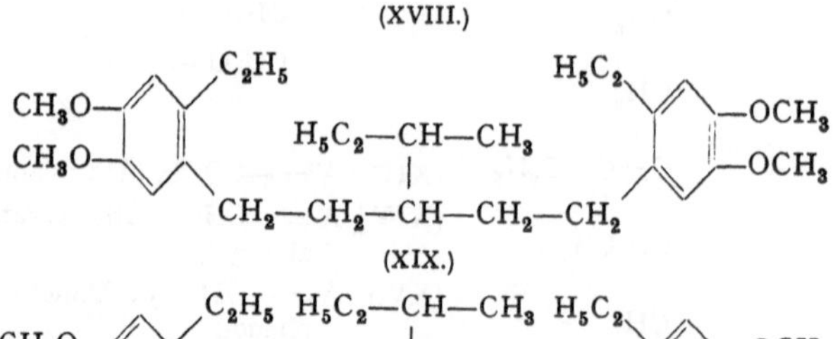

(XIX.)

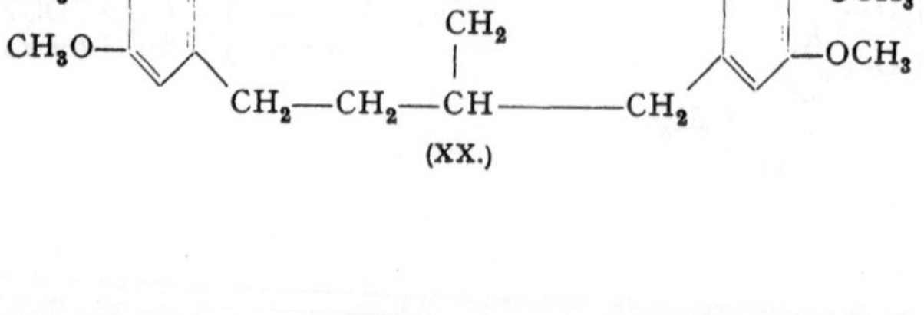

(XX.)

OH sind SPÄTH und PAILER nachgegangen und haben darüber in einer kurzen Arbeit berichtet (*52*). In einer weiteren Untersuchung (*54*) konnten sie zeigen, daß diese Reaktion allgemeineren Charakter besitzt und daß besonders bei Phenoläthern neben den zu erwartenden Aldehyden in wechselnder Ausbeute Phenole auftreten.

Die Carbonylverbindung $C_{18}H_{26}O_3$ gab ein bei 156 bis 157° schmelzendes Semicarbazon. Über dieses ließ sich die Abbauverbindung gut reinigen und durch Kochen mit einer konzentrierten Oxalsäurelösung wieder zurückgewinnen. Durch katalytische Hydrierung wurde in dieser Verbindung 1 Doppelbindung festgestellt. Die so erhaltene hydrierte Carbonylverbindung (XVII a) lieferte ebenfalls ein Semicarbazon, das den Schmelzp. 156 bis 158° hatte. Dieses war natürlich, wie vorauszusehen, mit der ursprünglichen Verbindung vom Schmelzp. 156 bis 157° nicht identisch. Es zeigte im Gemisch mit dieser starke Depression.

Bei der Oxydation der ungesättigten, öligen Carbonylverbindung mit Kaliumpermanganat. in Aceton erhielten SPÄTH und PAILER eine Säure von der Formel $C_{13}H_{18}O_4$ (oder $C_{11}H_{12}O_2 + 2\ OCH_3$) vom Schmelzp. 64—65°, die sich als 3,4-Dimethoxy-6-äthyl-hydrozimtsäure (XVI) charakterisieren ließ. Weiters wurde die ungesättigte Carbonylverbindung mit Ozon behandelt, wobei Methyläthylketon nachgewiesen werden konnte.

Unter Berücksichtigung dieser beiden Abbauprodukte, nämlich der 3,4-Dimethoxyhydrozimtsäure und des Methyläthylketons ergaben sich zwei mögliche Formeln für die Verbindung $C_{18}H_{26}O_3$ (XVII, XVIII) und daraus weiters zwei Formeln (XIX, XX) für die stickstoff-freie Abbauverbindung. PAILER (*38*) führte einen zweiten HOFMANNschen Abbau durch, der nunmehr so geleitet wurde, daß im Gegensatz zum vorher beschriebenen, schließlich der Ring R_2 erhalten blieb und der Ring R_1 abgespalten wurde [vgl. (VIII) bzw. (XXI)—(XXIII)].

N-Acetylemetin (XXI) wurde unter vollständiger Eliminierung des tertiären N-Atoms nach HOFMANN abgebaut. Dann wurde die erste entstehende Doppelbindung allein, die beiden weiteren zusammen aushydriert und nunmehr der Benzoylrest abgespalten (XXII). Unter Entfernung des zweiten Stickstoffes wurde diese Verbindung nun neuerlich abgebaut. Das so gewonnene stickstoff-freie Abbauprodukt (XXIII) lieferte bei der Ozonspaltung nun ebenfalls eine Carbonylverbindung der Zusammensetzung $C_{18}H_{26}O_3$, die sich mit der anfangs beschriebenen Verbindung (XVII a) (Semicarbazon: Schmelzp. 156—158°), die durch Hydrierung des Abbaualdehyds erhalten worden war, identisch erwies. Durch die Identität dieser beiden Spaltstücke war der symmetrische Bau der N-freien Abbauverbindung entsprechend der Formel (XIX) bewiesen.

Der Abbaualdehyd $C_{18}H_{26}O_3$ (XVII a) wurde außerdem von PAILER (*39*) wie folgt synthetisch erhalten. Aus dem β-sek.-Butyl-acrylsäureäthylester (XXIV) wurde durch Anlagerung von Natriummalonester der

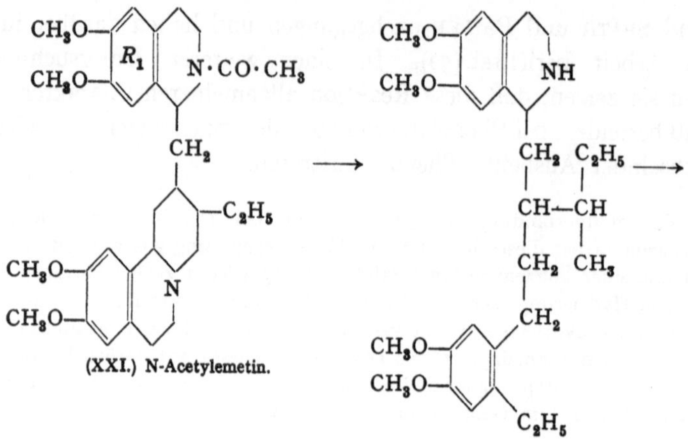

(XXI.) N-Acetylemetin.

(XXII.)

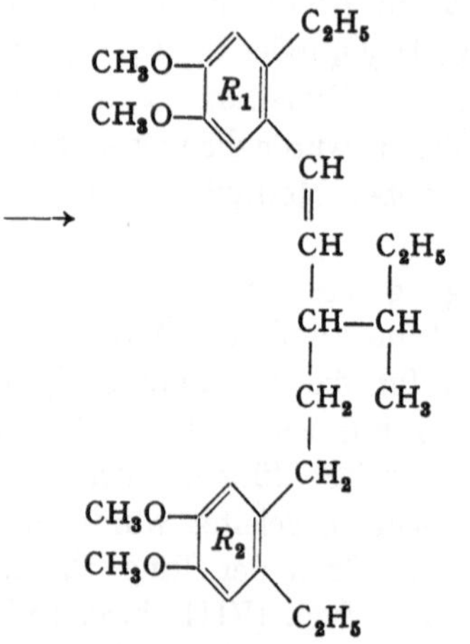

(XXIII.)

β-sek.-Butyl-α-carbäthoxyglutarsäure-äthylester (XXV) dargestellt und diese Verbindung zur β-sek.-Butylglutarsäure (XXVI) verseift bzw. decarboxyliert. Aus dem Anhydrid dieser Säure und 3,4-Dimethoxy-äthylbenzol wurde die Ketosäure (XXVII) dargestellt. Die Ketogruppe wurde reduziert und aus der so erhaltenen Säure mit Homoveratrylamin in der üblichen Weise das entsprechende Isochinolinderivat (XXVIII) dargestellt. Eine Verbindung derselben Konstitution wurde auch beim stufenweisen Abbau des Emetins erhalten (XXII). Die sterischen Ver-hältnisse dieser beiden Substanzen sind ungeklärt. Die synthetische Verbindung wurde nun ebenso wie beim Naturstoff nach HOFMANN ab-gebaut, wobei schließlich die ungesättigte N-freie Verbindung (XXIII) erhalten wurde. Diese gab bei der Behandlung mit Ozon die bei der Spaltung des Naturstoff-Abbauproduktes erhaltenen Verbindungen, vor allem aber den erwarteten Aldehyd (XVIIa) von der Zusammensetzung $C_{18}H_{28}O_3$, mit dem bei 156—158° schmelzenden Semicarbazon.

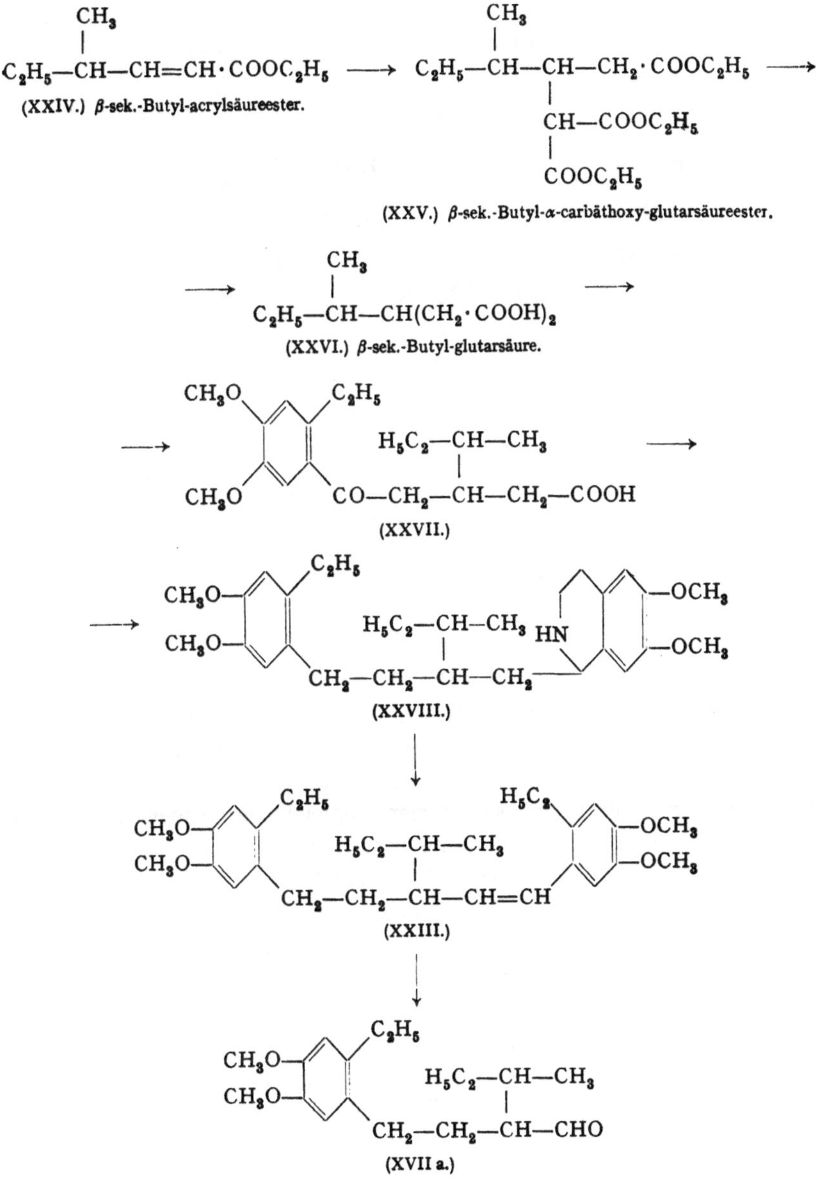

$$CH_3$$
$$|$$
$$C_2H_5-CH-CH=CH \cdot COOC_2H_5 \longrightarrow$$

(XXIV.) β-sek.-Butyl-acrylsäureester.

$$CH_3$$
$$|$$
$$C_2H_5-CH-CH-CH_2 \cdot COOC_2H_5 \longrightarrow$$
$$|$$
$$CH-COOC_2H_5$$
$$|$$
$$COOC_2H_5$$

(XXV.) β-sek.-Butyl-α-carbäthoxy-glutarsäureester.

$$CH_3$$
$$|$$
$$\longrightarrow \quad C_2H_5-CH-CH(CH_2 \cdot COOH)_2 \longrightarrow$$

(XXVI.) β-sek.-Butyl-glutarsäure.

(XXVII.)

(XXVIII.)

(XXIII.)

(XVII a.)

Damit ergab sich für die N-freie Abbauverbindung eindeutig die Formel (XIX) und daraus ließen sich weiter nur noch drei Möglichkeiten für die Formel des Emetins ableiten, nämlich (XXX), (XXXI) und (XXXII).

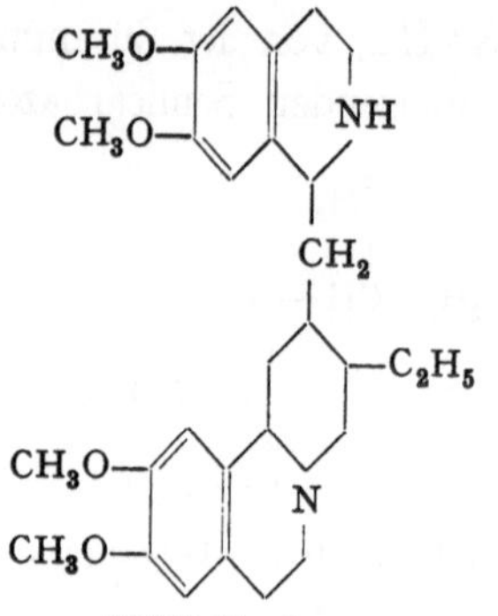

CH₃O— ... NH ... CH₂ ... —C₂H₅ ... CH₃O— ... N

(XXX.) Emetin.

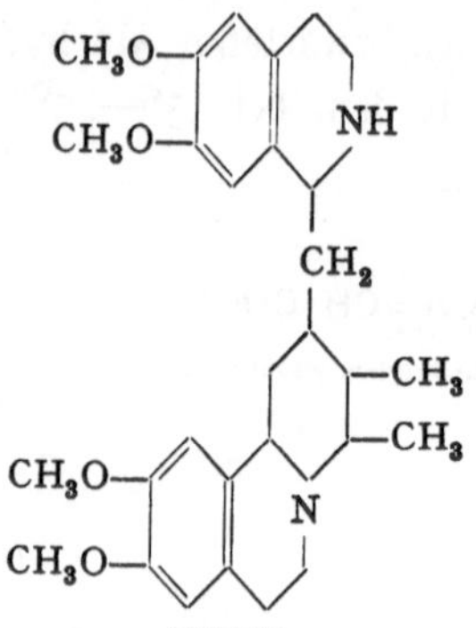

CH₃O— ... NH ... CH₂ ... —CH₃ ... —CH₃ ... CH₃O— ... N

(XXXI.)

CH₃O— ... NH ... CH₂ ... CH ... CH₂ ... C ... CH₃ ... C₂H₅ ... N

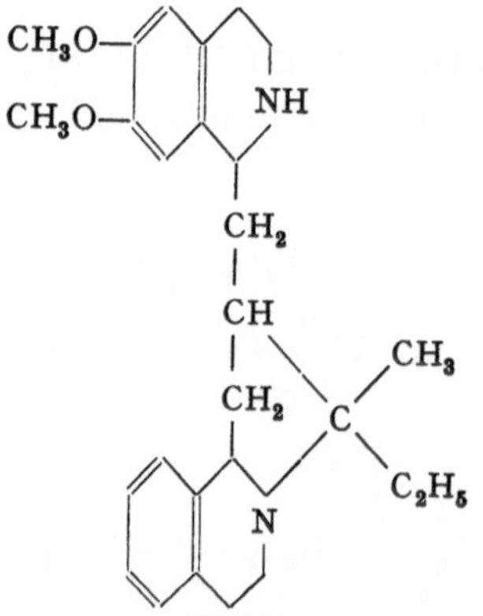

(XXXII.)

Um zwischen diesen drei Formeln zu unterscheiden, haben Pailer und Porschinski (*41*), ausgehend vom Methylemetin-dijodbenzylat, einen neuerlichen Hofmannschen Abbau durchgeführt. Bei der katalytischen Hydrierung der Dimethinbase wurden zugleich die beiden Doppelbindungen hydriert und eine Benzylgruppe als Toluol abgespalten. Auf diese Weise wurde die Verbindung (XXXIII) erhalten, welche nunmehr durch Erhitzen mit Palladiummohr auf 300—310° dehydriert wurde. Dabei konnte β-Collidin-(β-Äthyl-γ-methylpyridin) (XXXIV) erhalten werden. Auch bei der gleichen Dehydrierung der Verbindung (XXXV), die durch Abbau des N-Benzoylemetins erhalten wurde, ließ sich die Bildung von β-Collidin nachweisen. Diese Ergebnisse berechtigten Pailer und Porschinski, für Emetin die Formel (XXX) anzugeben. Sie wiesen auch auf die Möglichkeit einer Ringerweiterung bei der hohen Dehydrierungstemperatur, die zur Bildung des β-Collidins führte, hin. Sie betonten aber, daß auf Grund von Untersuchungen von Ehrenstein, welcher verschiedene N- und C-methylierte Pyrrolidine durch zwei- bis dreistündiges Leiten über Palladiumasbest bei 310—330° glatt und ohne Ringerweiterung zu den Pyrrolen dehydrierte, ein solcher Vorgang äußerst unwahrscheinlich sei.

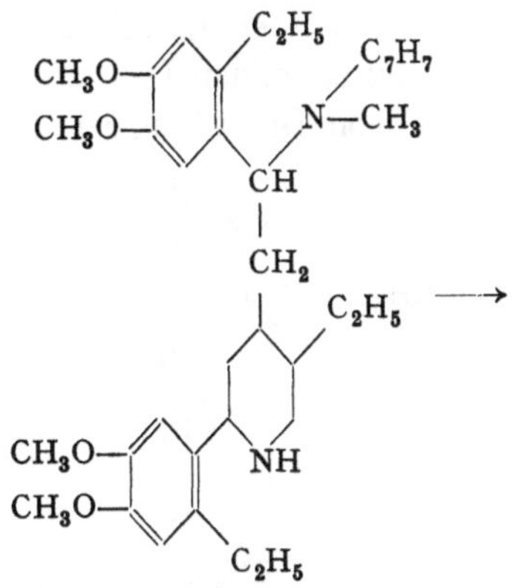

CH_3O- CH_3O- C_2H_5 C_7H_7 $N-CH_3$ CH CH_2 C_2H_5 $\longrightarrow$ CH_3O- CH_3O- NH C_2H_5

(XXXIII.)

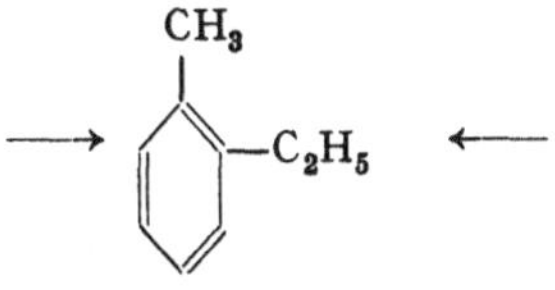

CH_3O- CH_3O- $N \cdot CO \cdot C_6H_5$ CH_2 C_2H_5 CH_3O- NH CH_3O- C_2H_5

(XXXV.)

CH_3 $-C_2H_5$

$\longrightarrow$ $\longleftarrow$

(XXXIV.)
β-Collidin.

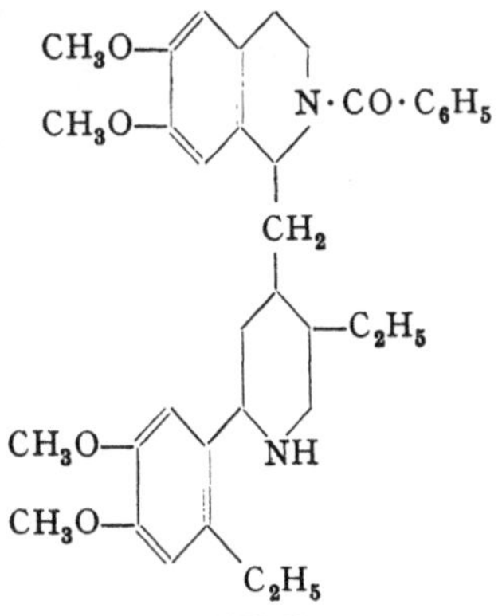

Eine schöne Bestätigung erhielt die Formel (XXX) durch verschiedene, nach den oben referierten Arbeiten erschienene, unabhängig durchgeführte Untersuchungen von OPENSHAW und Mitarbeitern. So konnten BATTERSBY, OPENSHAW und WOOD (2, 5a) bei der Ozonspaltung eines N-freien Produktes, das durch einen HOFMANNschen Abbau des Emetins erhalten wurde, Formaldehyd (33%, als Dimedonderivat) und ein nicht näher untersuchtes Keton erhalten. Dieses Ergebnis läßt sich mit der vorher beschriebenen Formel gut vereinbaren. Der N-freien Abbauverbindung, die bei der Ozonbehandlung Formaldehyd liefert, entspricht dann Formel (XXXVI). BATTERSBY und OPENSHAW (4, 4a) haben weiters ein Zwischenprodukt des HOFMANNschen Abbaues (XXXVII), einerseits durch Erhitzen mit Palladium auf 260—270° und anderseits durch Erhitzen mit Silberacetat auf 180°, in das kristallisierte Pyridinderivat (XXXVIII) übergeführt. Diese Verbindung gab bei der Oxydation mit konz. Salpetersäure 5-Äthylpyridin-2,4-dicarbonsäure, welche bei weiterer Oxydation mit Permanganat Berberonsäure (Pyridin-2,4,5-tricarbonsäure) lieferte. Das letztere Ergebnis deckt sich mit dem von

PAILER und PORSCHINSKI erhaltenen (β-Collidin), wobei besonders der Erhalt dieses Ringskeletts unter milden Dehydrierungsbedingungen, wie im Falle des Silberacetats, von Bedeutung ist.

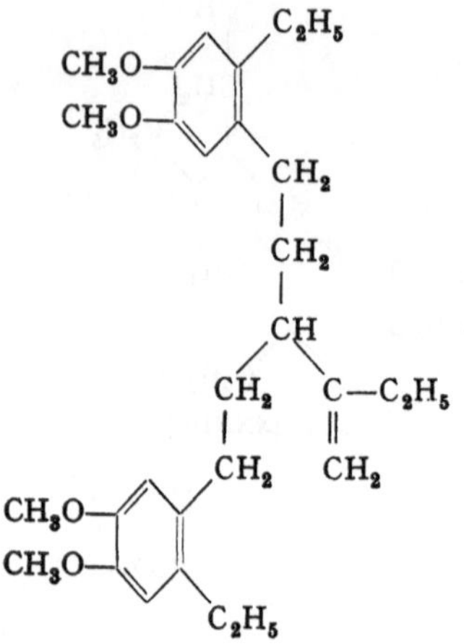

(XXXVI.)

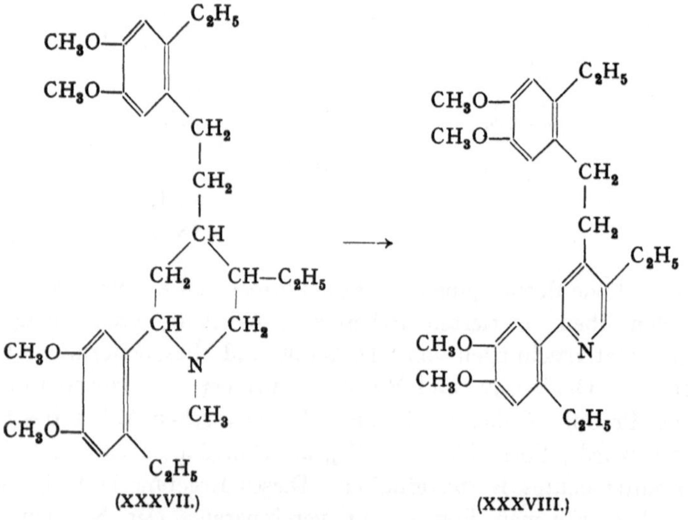

(XXXVII.)

(XXXVIII.)

2. Rubremetin (Dehydroemetin).

Bei der Einwirkung von milden sauren Oxydationsmitteln (Ferrichlorid, Brom in Chloroform, Jod in Alkohol) entsteht aus Emetin, Isoemetin oder O-Methylpsychotrin unter Dehydrierung das rote Rubremetin oder Dehydroemetin in Form seiner quartären Halogensalze ($C_{29}H_{33}O_4N_2X$, X = Halogen) (12, 23). Bei der Bildung des Rubremetins aus Emetin (Isoemetin) muß das O-Methylpsychotrin als Zwischenprodukt angenommen werden, da es einerseits bei Behandlung mit einem Unterschuß von Jod aus Emetin erhalten werden konnte, anderseits durch Brom zum Rubremetinbromid oxydiert wurde (47).

Eine weitere Methode, die in 45%iger Ausbeute zum Rubremetin führte und bei welcher Mercuriacetat als Oxydationsmittel angewandt wurde, beschrieben BATTERSBY und OPENSHAW (2, 3, 5a). Als Nebenprodukt dieser Dehydrierung, bzw. bei der Oxydation des Emetins mit der zwei Sauerstoffen entsprechenden Menge Mercuriacetat, erhielten die genannten Autoren eine Verbindung $C_{28}H_{36}O_4N_2$, ein Tetradehydroemetin mit zwei zueinander einerseits und zu einem Benzolring anderseits konjugierten Doppelbindungen.

Bei der Bildung von Rubremetinchlorid aus Emetin-hydrochlorid werden acht Wasserstoff-Atome eliminiert, ein Stickstoff-Atom wird quartär und das zweite verliert seinen basischen Charakter. Weder N-Methylemetin noch Emetamin führen bei der Oxydation zu Rubremetin-Salzen (8).

Das durch Dehydrierung von Emetamin mit Brom erhaltene Dehydroemetin unterscheidet sich wesentlich von dem aus Emetin oder O-Methylpsychotrin erhaltenen Produkt, wie bereits PYMAN (9, 47) fand, und KARRER, EUGSTER und RÜTTNER (25) durch die Verschiedenheit der Absorptionsspektren bestätigen konnten.

Die Konstitution des Rubremetins ist noch umstritten. Die alten hypothetischen Formeln von PYMAN sowie von STAUB konnten von KARRER, EUGSTER und RÜTTNER (25) im Zusammenhang mit verschiedenen Reduktionsversuchen des Rubremetins abgelehnt werden und sind natürlich auch mit der Aufstellung der richtigen Emetin-Formel hinfällig. Bei der Hydrierung von Dehydroemetinbromid bildet sich nämlich unter Aufnahme von zwei Molen Wasserstoff eine Verbindung, die von den Schweizer Autoren als Tetrahydro-dehydroemetin bezeichnet wird. Diese Verbindung ist gegen weitere Hydrierungen beständig, obwohl sie entsprechend den Formeln von STAUB und PYMAN noch eine hydrierbare Doppelbindung besitzen müßte. KARRER, EUGSTER und

(XXX.) Emetin. → (XXXIX.) Dehydroemetin-Salz. → (XL.) Tetrahydro-dehydroemetin.

RÜTTNER formulieren daher den Übergang des Emetins in das Dehydroemetin und dessen Reduktion zum Tetrahydro-dehydro-emetin entsprechend den Formelbildern (XXXIX) und (XL). (Das Mittelstück der Formeln wurde vom Verfasser entsprechend der nunmehr feststehenden Konstitutionsformel des Emetins ergänzt).

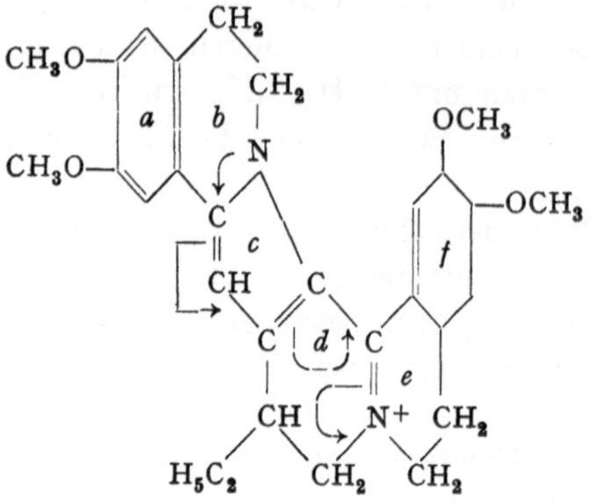

(XLI.)

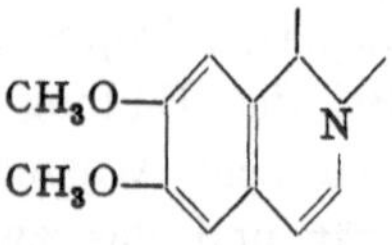

(XLII.)

BATTERSBY, OPENSHAW und WOOD (*2, 3, 5a*) finden diese Theorie von KARRER und Mitarbeitern nicht annehmbar, weil sie verschiedenen Tatsachen nicht gerecht wird. Sie erklärt nicht a) das Fehlen der Basizität des nicht quartären N-Atoms; b) die tieforangerote Farbe der Salze; c) die Nicht-identität des Rubremetins und des unter gleichen Bedingungen erhaltenen Produktes aus Emetamin und d) die katalytische Hydrierung zu einem Dihydroderivat, welches durch Luft rasch wieder re-oxydiert wird. Außerdem wird Emetin durch Mercuriacetat in Rubremetin übergeführt; dagegen gelingt eine Dehydrierung von 1-*n*-Butyl-3,4-dihydroisochinolin unter den gleichen Bedingungen nicht. BATTERSBY, OPENSHAW und WOOD nahmen daher für das Rubremetin die Formel (XLI) an. Die gelbe Farbe dieses Dehydrierungsproduktes soll demnach ihre Ursache darin haben, daß die Verbindung ein Resonanzhybrid darstellt, in dem ähnlich, wie in den Cyaninfarbstoffen, die positive Ladung zwischen den beiden N-Atomen wechseln kann.

Diese Formulierung wird jedoch von KARRER und RÜTTNER (*26*) abgelehnt. Die Verbindung (XLI) enthält nämlich zwei ortho-Dihydro-pyridinringe (*d* und *e*), von denen man, entsprechend der allgemeinen Empfindlichkeit der *o*-Dihydropyridinderivate gegen Oxydationsmittel, nach KARRER und RÜTTNER kaum annehmen kann, daß sie beim Kochen mit überschüssigem Brom oder Jod (also unter den Bedingungen der Rubremetin-Darstellung) beständig sein könnten. Auch die Bildung eines *o*-Dihydro-dehydroemetins [Teilformel (XLII)], erhalten aus Dehydroemetinbromid durch Reduktion mit LiAlH₄, schließt nach KARRER und RÜTTNER die Formel (XLI) für Dehydroemetin aus.

Die Totalsynthese des *D,L*-Rubremetin-hydrobromids wurde von BATTERSBY und OPENSHAW (5) beschrieben. Es handelt sich hierbei um eine vollständige, konstitutionsbeweisende Synthese des Emetingerüstes entsprechend der Formel (XXX), ohne Berücksichtigung der sterischen Verhältnisse. Der Vergleich mit dem Naturprodukt wurde daher auf dem Umweg über das Rubremetin durchgeführt, bei welchem asymmetrische C-Atome wegfallen. Verbindung (XLIII), die aus Homoveratrylamin und Carbäthoxy-acetylchlorid in der üblichen Weise erhalten worden war, wurde mit α-Formylbuttersäureester kondensiert und ergab bei der Hydrierung Verbindung (XLIV). Durch DIECKMANNschen Ringschluß wurde das Keton (XLV) erhalten, das durch Kondensation mit Cyanessigester, nachherige Verseifung, Decarboxylierung, Veresterung, zur Verbindung (XLVI) führte. Dieser Ester wurde nun durch Erhitzen mit Homoveratrylamin (180°) in das entsprechende Amid übergeführt, welches mit Phosphoroxychlorid zur Dihydro-isochinolin-Verbindung (XLVII) ringgeschlossen wurde.

3. Technische Gewinnung des Emetins (20).

Bei der Darstellung des Emetins im großen sind verschiedene Vorsichts-maßnahmen geboten, weil Emetin, Cephaelin und auch die gepulverte Droge sehr unangenehme Entzündungen der Augen, der Nase und der Haut unter den Fingernägeln hervorrufen.

Die gepulverte Brechwurzel wird in geeigneten Extraktoren mit Methanol, 50%igem Methanol oder 70%igem denaturiertem Alkohol extrahiert. Die Auszüge werden zuerst unter gewöhnlichem Druck und schließlich im Vakuum eingedampft, wobei ein dunkler, dicker Sirup hinterbleibt. Dieser wird dann noch warm mit der gleichen Menge warmen Wassers vermischt, gerührt, bis alles gelöst ist und dann in ein geeignetes Gefäß gebracht, das sich entweder im Freien oder in einem gut ventilierten Raum befindet und mit einem guten Rührer versehen ist. Diese Lösung wird nun unter Zusatz eines größeren Überschusses von konzentriertem Ammoniak stark alkalisch gemacht und mit einem, mit Wasser nicht mischbaren Lösungsmittel extrahiert. Besonders empfohlen werden Isopropyläther und mit Wasser nicht mischbare Ketone, z. Beisp. Methyl-isobutylketon. Zur vollständigen Entfernung der Alkaloide aus dem Rohauszug ist eine mehrmalige Extraktion, womöglich bei leichter Erwärmung nötig. Nach Abtrennung der so gewonnenen Alkaloidlösung wird diese zweimal mit 10—15%igem Kaliumhydroxyd ausge-schüttelt, um das Cephaelin und andere phenolische Basen abzutrennen. Der zweite Extrakt wird aufgehoben und bei der nächsten Aufarbeitung für die erste Extraktion verwendet. Die alkalischen Lösungen werden gesammelt; daraus wird das Cephaelin gewonnen und dieses wird dann in Emetin übergeführt.

Die nun cephaelin-freie Lösung wird noch mit Wasser gewaschen und dann mit einem geringen Überschuß von verdünnter, 15—20%iger Schwefelsäure extrahiert. Im allgemeinen genügt eine Extraktion, um alles Emetin zu entfernen. Das Lösungs-mittel, das nun frei von Alkaloiden ist, kann wieder für eine neue Extraktion der ammoniakalischen Lösung verwendet werden.

Die schwefelsaure Lösung soll für die weitere Abtrennung des Emetins nicht zu stark sauer sein und wird nötigenfalls durch vorsichtiges Zufügen von Ammoniak soweit neutralisiert, bis sie gegen Kongopapier gerade noch sauer reagiert. Dann scheidet man das Emetin in Form seines Hydrobromids oder Hydrojodids ab, und zwar durch Zufügen von NH_4Br bzw. NH_4J u. dgl. Nach längerem Stehen werden die Niederschläge abgetrennt und mit wenig Eiswasser gewaschen, bis sie farblos sind. Die Filtrate und Waschflüssigkeiten werden gesammelt, mit Ammoniak alkalisch gemacht, die in Freiheit gesetzte Base extrahiert und dem Aufarbeitungs-prozeß neuerdings zugeführt.

Das rohe Emetin-hydrobromid wird in heißem Wasser gelöst, die Lösung mit Tierkohle behandelt, filtriert und 24 Stunden der Kristallisation überlassen. Das Kristallisat wird dann abgesaugt und gewaschen. Die vereinigten Filtrate und Waschflüssigkeiten werden aufgearbeitet und einem neuen Ansatz zugesetzt.

Das so erhaltene Hydrobromid (Hydrojodid) wird nun in das Hydrochlorid übergeführt. Zu diesem Zweck wird es in wenig Wasser suspendiert, mit viel Äthyläther überschichtet und dann die Base durch Zugabe von überschüssigem Ammoniak in Freiheit gesetzt. Die Ätherausschüttelung wird wiederholt, die Ätherauszüge werden vereinigt und mit einer verdünnten NaOH-Lösung und dann mit Wasser gewaschen. Die Ätherlösung wird nun mit Salzsäure (verdünnt 2 : 1) ausgeschüttelt, und zwar wird die Säurezugabe vorsichtig vorgenommen, um einen Überschuß zu vermeiden. Zum Schluß soll der Extrakt gerade kongosauer sein. Nach längerem Stehen im Dunkeln wird das ausgeschiedene Emetin-hydrochlorid abgesaugt und im Dunkeln an der Luft bei Raumtemperatur getrocknet.

Um die Erzeugung von Emetin besonders wirtschaftlich zu gestalten, ist es notwendig, auch das Cephaelin in Emetin überzuführen. Zu diesem Zweck wird das Cephaelin, das vorher durch Ausschüttelung mit KOH abgetrennt wurde, vorerst gereinigt. Die Lösung wird schwach angesäuert und das Cephaelin dann durch Zusatz von Ammoniak ausgefällt und mit einem geeigneten Lösungsmittel aufgenommen (Äthylenchlorid, Isopropyläther, Methyl-Isobutylketon). Aus dieser Lösung wird das Cephaelin wieder mit verdünnter Salzsäure oder Schwefelsäure extrahiert. Die Lösung wird heiß mit Tierkohle behandelt, filtriert und mehrere Tage der Kristallisation überlassen.

Die O-Methylierung des Cephaelins kann nach mehreren Methoden durchgeführt werden, von denen einige in Patenten festgelegt sind (60). Als Methylierungsmittel wurden verwendet: Dimethylsulfat, Na-methylsulfat, Trimethylphenylammoniumhydroxyd, Nitrosomethylharnstoff und Nitrosomethylurethan (Diazomethan).

IV. Nebenalkaloide der Brechwurzel.

1. Cephaelin.

Cephaelin $C_{28}H_{38}O_4N_2$ ist, wie bereits mehrfach betont wurde, das hauptsächlichste Nebenalkaloid des Emetins.

Es wird am besten über das Hydrochlorid oder Hydrobromid gereinigt und bildet farblose Nadeln vom Schmelzp. $115-116°$ oder $120-130°$ (bei $100°$ getrocknet). $[\alpha]_D = -43,4°$ (in Chloroform). Die Base ist unlöslich in Wasser, schwer löslich in Äther und leicht in Chloroform, Alkohol und Alkalilösungen. Hydrochlorid, $B \cdot 2\ HCl \cdot 7\ H_2O$, Schmelzp. $245-270°$, $[\alpha]_D = +25°$ bis $29,5°$ (Wasser; $c = 1,7$ bis $6,7$). Aus stark sauren Lösungen scheidet sich die Verbindung $B \cdot 5\ HCl$ vom Schmelzp. $84-86°$ aus. Hydrobromid, $B \cdot 2\ HBr \cdot 7\ H_2O$, farblose Prismen, Schmelzp. $266-293°$. Das Sulfat, Hydrojodid und Nitrat sind amorph.

Das Cephaelin enthält drei Methoxylgruppen und eine freie phenolische Hydroxylgruppe und läßt sich, wie bereits im vorhergehenden Kapitel eingehender beschrieben wurde, durch einfache O-Methylierung in Emetin überführen. Bei der Hydrierung von Psychotrin (S. 299), das sich durch einen Mindergehalt von 2 H vom Cephaelin unterscheidet, entstehen Cephaelin und Isocephaelin. BRINDLEY und PYMAN (8) haben auf Grund theoretischer Überlegungen für das Cephaelin und für das Psychotrin die freie OH-Gruppe am $C_{(6)}$ des Ringes A ihrer ziemlich spekulativen Emetinformel (VI, S. 283) angenommen.

Zur eindeutigen Klärung dieser Frage haben PAILER und PORSCHINSKI (42) O-Äthylcephaelin zur Verbindung (XLVIII) abgebaut. Dieser Abbau erfolgte in analoger Weise zu dem von SPÄTH und PAILER (51) beim Emetin durchgeführten. Bei der Ozonspaltung der Verbindung

(XLVIII) wurde 2-Äthyl-4-äthoxy-5-methoxybenzaldehyd (XLIX) isoliert, so daß dem Cephaelin die Formel (L) zukommt.

(XLIX.) 2-Äthyl-4-äthoxy-5-
methoxy-benzaldehyd.

(XLVIII.)

(L.) Cephaelin.

2. Psychotrin und O-Methylpsychotrin.

Psychotrin $C_{28}H_{36}O_4N_2$ kristallisiert am besten aus wäßrigem Aceton oder
Alkohol. Nach dem Trocknen bei 100° sintert es bei 120° und schmilzt bei 138°.
Hydrojodid, $B \cdot 2$ HJ, gelbe Nadeln, Schmelzp. 200—220°; Nitrat, $B \cdot 2$ $HNO_3 \cdot H_2O$
schmilzt (bei 100° getrocknet) bei 184—187°. Sulfat, $B \cdot H_2SO_4 \cdot 3$ H_2O, Schmelzp.
214—217° (bei 100° getrocknet), $[\alpha]_D = +39,2°$ (Wasser). Das saure Oxalat,
$B \cdot 2$ $H_2C_2O_4 \cdot 4,5$ H_2O, farblose Nadeln, die bei 130° weich werden und bei 145°
schmelzen (Zers.).

Psychotrin kann aus O-Methylpsychotrin durch partielle Entmethylierung mit HCl bei 170° erhalten werden (8). Pyman hat gezeigt, daß
bei der Methylierung des Psychotrins mit Natriumamylat und Methyl-
sulfat O-Methylpsychotrin entsteht. Diese Verbindung wurde aber
auch von Pyman (46, vgl. 1, 10) aus der nicht-phenolischen Alkaloidfraktion bei der Aufarbeitung des Extraktes der Gesamtalkaloide isoliert.

Das O-Methylpsychotrin, $C_{29}H_{38}O_4N_2$, schmilzt (aus Äther kristallisiert) bei
123—124°; $[\alpha]_D = + 43,2°$ (Alkohol). Verdünnte wäßrige Lösungen fluoreszieren.
Das Sulfat, $B \cdot H_2SO_4 \cdot 7 H_2O$, schmilzt bei 247° (Zers.), nach vorheriger Trocknung
bei 160—170°; $[\alpha]_D = + 54,1°$ (Wasser). Hydrobromid, $B \cdot 2 HBr \cdot 4 H_2O$, schwach
gelbe Nadeln, Schmelzp. 190—200° (im Vakuum getrocknet); $[\alpha]_D = + 48°$
(Wasser). Saures Oxalat, $B \cdot 2 H_2C_2O_4 \cdot 3,5 H_2O$, Schmelzp. 150—162° (Zers.),
$[\alpha]_D = + 45,9°$ (Wasser). Pikrat, langsames Schmelzen zwischen 142—175°.

Die freie Hydroxyl-Gruppe des Psychotrins muß ebenso wie beim
Cephaelin am $C_{(6)}$ haften, da sich Psychotrin durch Reduktion (unter
Aufnahme von 1 Mol Wasserstoff) zu einem Gemisch von Cephaelin
und Isocephaelin reduzieren läßt (*12*). Aus diesem entsteht durch
Methylierung, wie bereits erwähnt wurde, Emetin bzw. Isoemetin (*47*).
Die beiden letzteren entstehen auch bei der Reduktion von O-Methyl-
psychotrin (*46, 47*). Bei dieser Operation wird eine dritte Base $C_{28}H_{38}O_3N_2$
(Schmelzp. 126—128°), $[\alpha]_D = — 66,2°$ (in Chloroform) erhalten, welche
durch Verlust einer Methoxyl-Gruppe entstanden ist und als ein De-
methoxy-emetin oder Demethoxy-isoemetin angesehen werden kann.

KARRER, EUGSTER und RÜTTNER (*25*) haben das O-Methylpsychotrin
zum Unterschied von PYMAN (*46, 47*), welcher zur Reduktion Natrium
und Alkohol verwendete, katalytisch reduziert und dabei, unter Verwendung von Platin und Wasserstoff in neutraler Lösung, Isoemetin
und in alkalischer Lösung Emetin erhalten. Bei Behandlung mit RANEY-
Nickel entstand ebenfalls Emetin. Aus der gleichzeitigen Bildung von
Emetin und Isoemetin wird geschlossen, daß die Kohlenstoffbindung im
Methylpsychotrin (Psychotrin) so angeordnet ist, daß durch den Reduk-
tionsvorgang ein asymmetrisches C gebildet wird. BRINDLEY und
PYMAN (*8*) haben, ohne irgendeinen Beweis, die Doppelbindung zwischen
$C_{(18)}$ und $C_{(17)}$ angenommen. Daß die Doppelbindung tatsächlich diese
Lage besitzt, konnten KARRER und Mitarbeiter durch Einwirkung von
Phthalpersäure auf N-Benzoyl-O-methylpsychotrin beweisen. Bei dieser
Reaktion erhielten sie Benzoyl-corydaldin (LI), das durch die oxydative

(LI.) Benzoyl-corydaldin.

(LII.) $R = — OH$ (Psychotrin).
(LIII.) $R = — OCH_3$ (O-Methylpsychotrin).

Spaltung der semizyklischen Doppelbindung zwischen $C_{(17)}$ und $C_{(18)}$ erhalten wurde. Etwas N-Benzoyl-corydaldin konnten sie auch bei einem Abbau des N-Benzoyl-O-methylpsychotrins mit Ozon gewinnen. Dem Psychotrin bzw. dem O-Methylpsychotrin kommt daher die Formel (LII) bzw. (LIII) zu.

3. Emetamin.

$C_{29}H_{36}O_4N_2$, Schmelzp. (aus Essigester kristall.) 153—154°; aus Äther (mit $^1/_2 C_2H_5 \cdot O \cdot C_2H_5$), Schmelzp. 138—139° oder 142—143°; $[\alpha]_D = + 13,6°$ (in Alkohol). Hydrobromid, $B \cdot 2 HBr \cdot 7 H_2O$, Schmelzp. 210—225°, $[\alpha]_D = - 24,3°$ bis — 22,0° (Wasser; $c = 8,04$ bis 4,15). Saures Oxalat, $B \cdot 2 H_2C_2O_4 \cdot 3 H_2O$, Schmelzp. 172° (Zers.); $[\alpha]_D = - 6,1°$ (Wasser). Hydrochlorid, $B \cdot 2 HCl \cdot 8,5 H_2O$, Schmelzp. 77—80° oder 218—223° (trocken), $[\alpha]_D = - 17,5°$ (lufttrockenes Salz in Wasser). Nitrat, $B \cdot 2 HNO_3 \cdot 2 H_2O$, Schmelzp. 165—166°. Das Pikrat erweicht bei 147° und schmilzt bei 173°.

Die Verbindung enthält vier Methoxyl-Gruppen und beide N-Atome sind tertiär. Bei der Reduktion des Emetamins mit Natrium und Alkohol erhielt Pyman (46) Isoemetin. Anderseits gewannen Ahl und Reich-stein (1) durch Dehydrierung des Emetins mit Palladium-Tierkohle bei 190—200° Emetamin und daneben 1-Methyl-6,7-dimethoxyiso-chinolin. Entsprechend diesen Ergebnissen wird für Emetamin die Formel (LIV) vorgeschlagen.

(LIV.) Emetamin.

V. Nachweisreaktionen der Ipecacuanha-Alkaloide.

Für die einzelnen Brechwurzel-Alkaloide gibt es eine große Zahl von Fällungs- und Farbreaktionen. Sie werden besonders zur Kontrolle der Drogen auf Verfälschungen und zum Nachweis von Emetin-Aus-scheidungen bei amöbenruhrkranken Menschen verwendet. Eine ein-gehendere Besprechung der älteren Methoden sowie eine tabellarische Zusammenfassung der einzelnen Reaktionen findet sich bei Staub (55). Die wesentlichsten Nachweise seien im folgenden kurz erwähnt.

Emetin gibt mit konz. Schwefelsäure und Molybdänsäure eine gelbliche Grünfärbung, das Emetin-hydrochlorid liefert unter denselben Bedingungen ein reines Grün. Vanadinpentoxyd und Schwefelsäure erzeugen ebenfalls eine Grünfärbung. Konz. Salpetersäure gibt eine Orangefärbung.

Cephaelin gibt mit Schwefelsäure-Molybdänsäure ein mattes Blaugrün, wenn eine Spur HCl anwesend ist. Mit *p*-Nitrobenzoldiazoniumsalzen gibt die Base einen Farbstoff, der sich in wäßriger Natronlauge mit purpurner Farbe löst. FRÖHDES Reagens und HCl bringen ein helles Blau hervor.

Die beiden Alkaloide Emetin und Cephaelin geben mit Goldchlorid, Platinchlorid, Phosphorwolframsäure, MAYERS Reagens, Kaliumwismutjodid und Jodkalium Fällungen.

Psychotrin zeigt mit Sulfomolybdänsäure eine Grünfärbung und kuppelt ähnlich wie Cephaelin mit *p*-Nitrobenzoldiazonium-Salzen zu einem Farbstoff, der sich in Alkali mit purpurroter Farbe löst. Auch mit FRÖHDES Reagens und HCl entsteht eine Grünfärbung. Ferrichlorid erzeugt eine Blaufärbung oder einen dunklen Niederschlag. *O-Methylpsychotrin* und *Emetamin* geben mit FRÖHDES Reagens ebenfalls Grünfärbungen.

Von ROSENTHALER (*50*) wird eine mikrochemische Methode zur Identifizierung der Ipecacuanha-Droge beschrieben. Über die quantitative Bestimmung der Alkaloide gibt es eine umfangreiche Literatur, auf die im besonderen verwiesen sei (*6, 7, 11, 15, 15a, 16, 18, 36*).

VI. Biogenetische Betrachtungen zur Konstitution des Emetins.

ROBINSON (*49*) versuchte die Biogenese des Emetins zu erklären, angeregt durch Betrachtungen von WOODWARD (*59*) über die mögliche Bildung des Strychnins in der Pflanze. Durch Kondensation von zwei Molekülen Dioxyphenylalanin und einem Molekül Formaldehyd kann man sich, wie dies bei anderen derartigen Betrachtungen allgemein angenommen wird, die Verbindung (LV) entstanden denken (ein Dioxyphenylalanin geht dabei in den entsprechenden Dioxyphenylacetaldehyd über; das zweite spaltet Kohlendioxyd ab). Nimmt man nun eine Ringspaltung zwischen den beiden OH-Gruppen am X und Y des Ringes B analog WOODWARD an, wobei X sich zu einer Formyl-Gruppe verwandelt, während die Kette an Y vollkommen reduziert wird (LVI), so ergibt sich eine neue Kondensationsmöglichkeit mit einem weiteren Dioxyphenylalanin-Molekül. Unter CO_2-Abspaltung in derselben Reaktionsstufe entsteht so eine Verbindung, die bei vollkommener O-Methylierung der Konstitution des Emetins entspricht.

(LV.)   (LVI.)

VII. Therapeutische Verwendung des Emetins.

Emetin wurde früher wegen seiner Brechwirkung vielfach als Emeticum verwendet, kommt aber heute als solches kaum mehr in Frage, da man weit besser und schneller wirkende Präparate zur Hand hat.

Als Expectorans ist es ebenfalls medizinisch in Verwendung, und zwar besonders in Form von Extrakten der Gesamtalkaloide. Hervorragende Bedeutung hat aber das Emetin zur Bekämpfung der Amöbenruhr (Amöbendysenterie) erlangt. Es ist dies eine in den Tropen sehr verbreitete Krankheit, deren Erreger die im Darm schmarotzende, den Protozoen zugehörige *Entamoeba histolytica (E. dysenterie)* ist. Die in Europa heimische Ruhr wird dagegen durch einen Bacillus *(Bact. dysenterie)* hervorgerufen.

Emetin wirkt in vitro schon in sehr beachtlichen Verdünnungen (1 : 5 000 000) entwicklungshemmend auf Ruhramöben, indem es deren Teilung verhindert. Ein wesentlicher Nachteil dieser Wirkungsweise liegt darin, daß auch im Organismus nur die vegetativen Formen, nicht aber die Dauerformen, die Amöbencysten beeinflußt werden. Die Anwendung des Emetins erfolgt im allgemeinen als Hydrochlorid in Form von Injektionen oder als schwer lösliches Wismutsalz peroral. Unangenehm und hindernd für seine Anwendung sind seine brechenerregende Wirkung und seine kumulativen Eigenschaften. Mehrere Einzeldosierungen des Emetins verstärken sich nämlich additiv, so daß wiederholte Gaben von sonst recht gut verträglichen Mengen zu schweren Schädigungen führen können. Seine unerwünschte Nebenwirkung als Capillargift sei ebenfalls erwähnt, welche zu unangenehmen Störungen im Magen und Darm Veranlassung geben kann.

Ebenso wie das Emetin ist auch das dazu stereoisomere Isoemetin, sowie das N-Methylemetin und das Psychotrin in vitro gegen *Entamoeba histolytica* wirksam. Diese Verbindungen haben sich aber bei der klinischen Erprobung als vollkommen unwirksam erwiesen.

Cephaelin ist doppelt so giftig wie Emetin. Man hat nun diese entgiftende Wirkung der O-Alkylierung auch an höheren Äthern (z. B.

Amyläther u. dgl.) untersucht und dabei festgestellt, daß mit zunehmender Länge der Kette die Giftigkeit abnimmt, daß mit dem Absinken der Toxizität aber auch die Wirksamkeit gegen Amöben sich verringert.

Die Emetinbehandlung führt im allgemeinen aus den oben erwähnten Gründen selten zur vollständigen Heilung. Man zieht daher heute vielfach Yatren, Vioform und Rivanol vor. Diese Verbindungen wirken zum Unterschied von Emetin gewissermaßen als Desinfektionsmittel. Sie töten die Amöben und auch ihre Dauerformen, die Cysten, ab. Neben diesen Mitteln werden noch die Arsenpräparate Carbason und Spirocid angewendet, welche sich auch bei Amöbenhepatitis bewährt haben sollen. Ferner wurden Wismutsubnitrat, Hexyl-, Heptyl- und Octylresorcin angewendet.

Emetin-hydrochlorid hat weiters zur Behandlung der Bilharziose Anwendung gefunden (TSAMIS, 1913). Es ist dies eine durch einen kleinen Wurm hervorgerufene, äußerst unangenehme Seuche, welche in Ägypten beheimatet ist, sich aber mit der Zeit über ganz Afrika ausgebreitet hat.

VIII. Synthesen von Substanzen mit Emetin-Wirksamkeit.

Die Synthese von Verbindungen mit Emetin-Wirksamkeit wurde vielfach versucht, wobei man die neuen Verbindungen teilweise von der alten hypothetischen Emetinformel ableitete, teils aber auch Verbindungen darstellte, deren Konstitution keinerlei Zusammenhänge mit dieser erkennen lassen.

So haben PYMAN und Mitarbeiter (*13, 13a, 14, 48*) Substanzen synthetisiert, deren Grundkörper den Formeln (LVII), (LVIII) und (LIX) entsprechen.

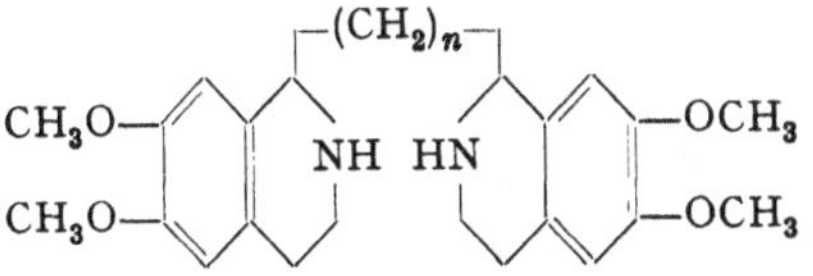

(LVII.)

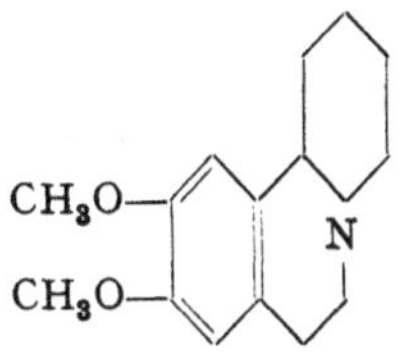

(LVIII.)

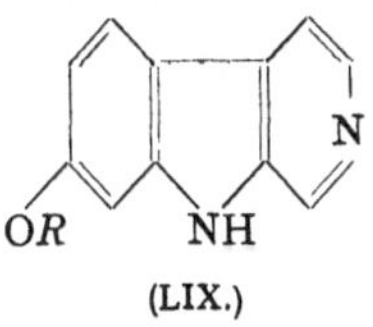

(LIX.)

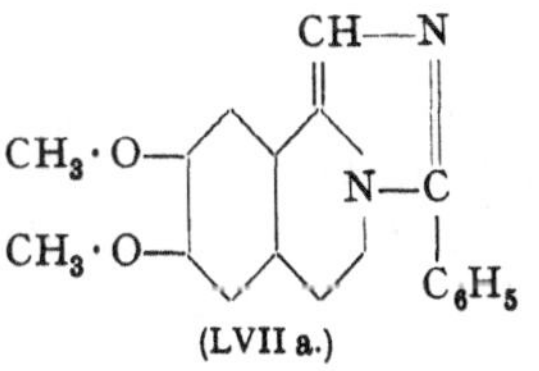

(LVII a.)

Die Verbindungen des Typs (LVII) zeigten in vitro nur $^1/_{100}$ und die beste Verbindung des Typs (LVIII) besaß $^1/_{20}$ der Emetin-Wirksamkeit.

Die Verbindungen vom Typ (LIX) zeigten in vitro ebenfalls Aktivität. Der Rest R wurde vom Äthyl- bis zum n-Dodecyl variiert und hierbei beim n-Nonylderivat die beste Wirksamkeit erzielt. Eine besondere Wirksamkeits-Steigerung erfuhren diese Verbindungen durch das Einführen einer Dialkylaminoalkyl-Seitenkette. Das O-λ-di-n-butylamino-undecylharmol [$R = (C_4H_9)_2N(CH_2)_{11}$—] zeigte die stärkste Wirksamkeit. Diese Aktivitätssteigerung wurde auch beobachtet, als man den Harmolring durch andere basische Ringe oder auch durch einfache, substituierte Amino-Gruppen ersetzte. Besonders das ω,ω'-Tetra-n-amyl-diamino-n-decan [$R,R' > N$—$(CH_2)_{10}$—$N < R,R'$] zeigte sich in vitro unter Einhaltung eines günstigen p$_H$ (6,2 bis 6,3) wirksamer als Emetin. Die klinische Erprobung ergab aber dann, daß diese Substanz für die praktische Anwendung nicht in Frage kommt.

Eine größere Anzahl von Untersuchungen in dieser Richtung wurden von Goodwin und Sharp (*19*, *19b*) mit einer Reihe von Mitarbeitern veröffentlicht. Weitere diesbezügliche Literatur findet sich bei Henry (*21*) referiert.

Andere Arbeiten, die hauptsächlich die Synthese von Verbindungen, die sich von der alten Formel des Emetins ableiten, zum Ziel hatten und schließlich auch zur Synthese dieser Verbindung selbst führten, wurden von Sugasawa und seinen Mitarbeitern beschrieben (*31*, *56*). In den FIAT-Berichten über „Naturforschung und Medizin in Deutschland" (1939 bis 1946) Band 43 (Chemotherapie), berichtet Schönhöfer über die Synthese von organischen Basen, die sich an *Entamoeba histolytica* chemotherapeutisch wirksam erwiesen. Bei Untersuchungen, die eigentlich einem anderen Gebiet galten, fand Kikuth, das das 2-Diäthylamino-äthylamino-diphenylenoxyd (LX) im Vitro-versuch bei der *Entamoeba*

(LX.) 2-Diäthylamino-äthylamino-diphenylenoxyd.

histolytica Wirksamkeit zeigte. Gemeinsam mit Pützer (*45*) wurde daher eine systematische Durchforschung dieser Körperklasse in Angriff genommen. Verbindung (LX) wurde nach den verschiedenen Richtungen geändert und abgewandelt. So zeigte es sich, daß das 2-Bis-(diäthylamino-äthyl)-amino-diphenylenoxyd eine etwa zehnmal größere Wirkung auf *Entamoeba histolytica* hatte als das monosubstituierte Produkt (LX). Die Verbindung war damit in das Bereich der praktischen Anwendung gerückt und erhielt den Prüfungsnamen „Gavan". Verschiedene andere Verbindungen, die man durch Variation des Gavans erhielt, zeigten alle einen Wirkungsabfall. Man hatte vor allem auch versucht, durch Einführung entsprechender Reste der Verbindung die ungefähre Größe des Emetinmoleküls zu geben, weil die Höhe des Molekulargewichtes neben vielen anderen, noch unbekannten Faktoren für eine bestimmte chemotherapeutische Wirkung wesentlich ist. Alle diese Verbindungen zeigten aber keine besondere Wirksamkeit.

Es wurden nun noch Verbindungen anderer Ringsysteme untersucht, z. B. (LXI, LXII, LXIII), die aber alle eine geringere Wirkung zeigten.

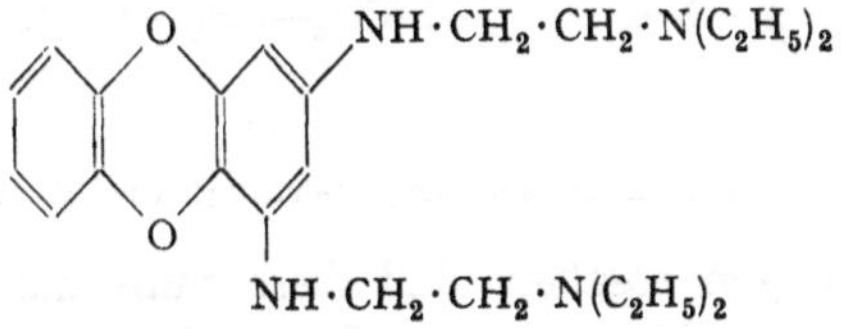

$$NH \cdot CH_2 \cdot CH_2 \cdot N(C_2H_5)_2$$

$$NH \cdot CH_2 \cdot CH_2 \cdot N(C_2H_5)_2$$

(LXI.)

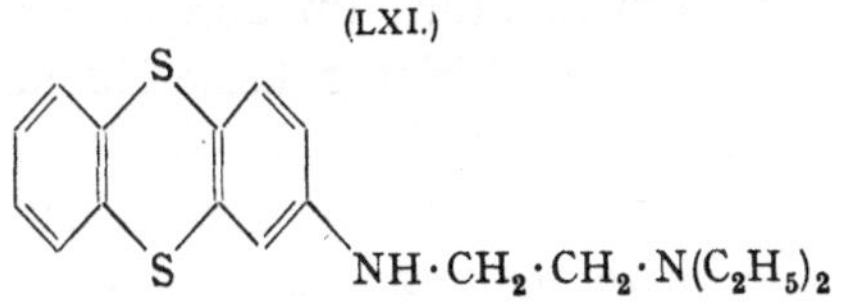

(LXII.)

$$(CH_3)_2N—CH_2—CH_2 \qquad CH_2—CH_2—N(CH_3)_2$$

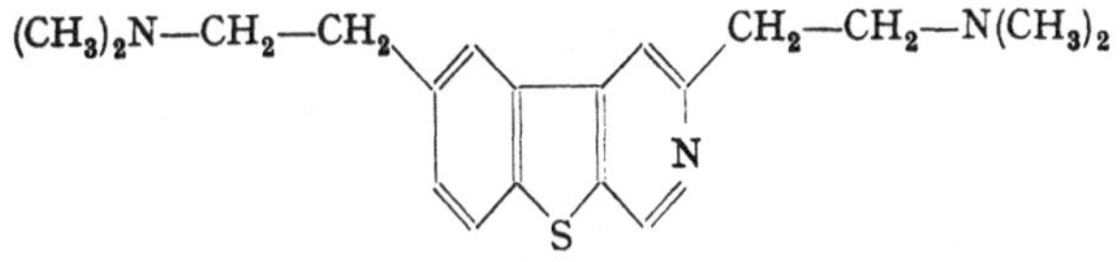

(LXIII.)

Im Zusammenhang mit dem Bau des Emetins versuchte man weiters verschiedene Tetrahydroisochinoline und Tetrahydrochinoline zum Aufbau wirksamer Substanzen heranzuziehen. Die einfachen N-Diäthylaminoäthyl-tetrahydroisochinoline und Chinoline waren praktisch wirkungslos. Man griff nun auf Verbindungen der Plasmochin-Reihe zurück und stellte Substanzen z. B. von der Art des 8-Diäthylamino-6-phenoxychinolins her. Diese Stoffe entsprachen in der Wirkung dem 2-Diäthylamino-äthylamino-diphenylenoxyd. Verbindungen, bei welchen der Heteroring hydriert war (LXIV), zeigten keine Wirkungssteigerung. Von diesen Verbindungen leitete man nun wieder einfachere Verbindungen ab (LXV).

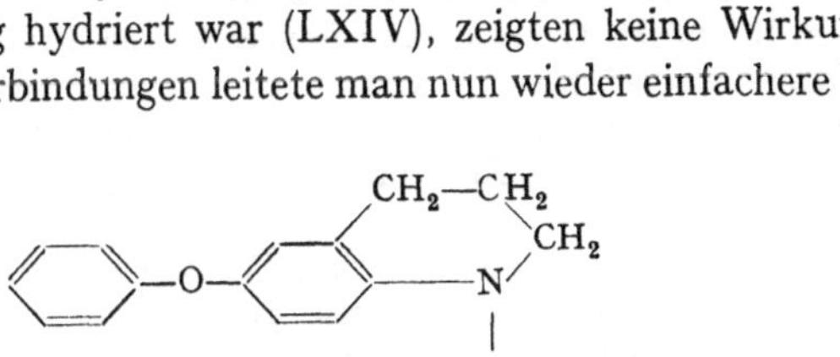

$$CH_2—CH_2$$
$$CH_2$$
$$—N$$
$$CH_2—CH_2—N(C_2H_5)_2$$

(LXIV.)

$$—O—\qquad —NH \cdot CH_2 \cdot CH_2 \cdot N(C_2H_5)_2$$

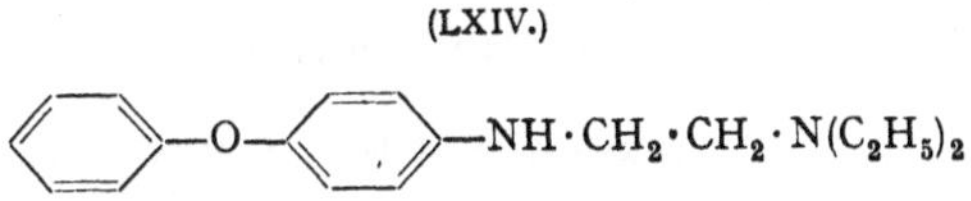

(LXV.)

Verschiedene Untersuchungen, die über den Einfluß der Substituenten bei solchen Substanzen auf deren Amöbenwirksamkeit durchgeführt wurden, ergaben schließlich den 4-Bis-(diäthylaminoäthyl)-amino-3′,5′-dimethylphenyläther (LXVI) als praktisch am besten brauchbares Heilmittel. Es erhielt für die klinische Untersuchung den Prüfnamen „Gavano".

(LXVI.) 4-Bis-(diäthylaminoäthyl)-amino-3′,5′-dimethylphenyläther („Gavano").

Genauere Prüfungen ergaben, daß diese Substanz halb so wirksam und halb so giftig wie Emetin war. Im übrigen verhielt sie sich wie „Gavan" (keine brecherregende Wirkung, keine Kumulation). Es wurde aber beim Gavano ebenso wie beim Gavan bei Injektionen eine besondere Schmerzerzeugung festgestellt. Bei der Verabreichung in Tabletten, in Form schwerlöslicher Salze, stellte sich als Nachteil heraus, daß Gavano eine längere Anlaufzeit zur Wirkung braucht, während das Emetin in weit kürzerer Zeit zu heilen vermag. Aus den vorerwähnten Gründen wurden das Gavan und das Gavano nicht in den Arzneischatz aufgenommen.

Schließlich zeigte 5-Chlor-8-oxychinolin, wie auch dessen 7-Alkylen-substituenten eine gute Antiamöben-Wirkung. Acylierungen und Alkylierungen am Sauerstoff schwächen die Wirkung ab, bzw. lassen sie verschwinden. Im Gegensatz zum Emetin und den Verbindungen der Gavano-Reihe war hier der Angriffspunkt nicht auf die Teilungsfähigkeit und Vermehrung der Erreger festzustellen, sondern es lag dieselbe Aktivität vor, wie sie dem Yatren und Rivanol eigen ist.

Literaturverzeichnis.

1. Ahl, A. u. T. Reichstein: Partieller Hofmannscher Abbau des Emetins sowie seine Dehydrierung zu Emetamin. Helv. chim. Acta 27, 366 (1944).

2. Battersby, A. R. and H. T. Openshaw: Studies on the Structure of Emetine. I. The Hofmann Degradation. J. chem. Soc. (London) 1949, 59.

3. — — Studies on the Structure of Emetine. II. Oxydation with Mercuric Acetate, and the Properties of Rubremetinium Salts. J. chem. Soc. (London) 1949, 67.

4. — — Studies on the Structure of Emetine. IV. Elucidation of the Structure of Emetine. J. chem. Soc. (London) 1949, 3207.

4a. — — Further Evidence Regarding the Structure of Emetine. Experientia 5, 398 (1949).

5. — — The Totalsynthesis of *dl*-Rubremetinium Bromide. Experientia 6, 387 (1950).

5a. Battersby, A. R., H. T. Openshaw and H. C. S. Wood: The Constitution of Emetine. Experientia 5, 114 (1949).

6. BAUER, K. H. u. K. HEBER: Über *Infusum radicis Ipecacuanhae*. Pharmaz. Zentralhalle Deutschland **71**, 513 (1930); [Chem. Zbl. **1930** II, 2416].

7. BLISS, A. R.: Bericht über Ipecacuanha-Alkaloide. J. Assoc. off. agric. Chemists **9**, 301 (1926); [Chem. Zbl. **1926** II, 2619].

8. BRINDLEY, W. H. and F. L. PYMAN: The Alkaloids of Ipecacuanha. IV. J. chem. Soc. (London) **1927**, 1067.

9. — — The Alkaloids of Ipecacuanha. J. chem. Soc. (London) **1927**, 1076.

10. BUCK, J. S.: Some Substituted Tetrahydroisoquinoline Hydrochlorides. J. Amer. chem. Soc. **56**, 1769 (1934).

11. BÜCHI, J.: Alkaloidbestimmung von *Radix Ipecacuanhae*. Pharmaz. Ztg. **80**, 1121 (1935); **81**, 806 (1936).

12. CARR, F. H. and F. L. PYMAN: The Alkaloids of Ipecacuanha. J. chem. Soc. (London) **105**, 1591 (1914).

13. CHILD, R. and F. L. PYMAN: Bases Containing Two *iso*-Quinoline Rings. J. chem. Soc. (London) **1929**, 2010.

13a. — — 1-ω-Halogenalkyl-*iso*-quinolines and Their Derivatives. J. chem. Soc. (London) **1931**, 36.

14. COULTHARD, C. E., H. H. L. LEVENE and F. L. PYMAN: The Chemotherapy of Derivatives of Harmine and Harmaline. I.—II. Biochemic. J. **27**, 727 (1933); **28**, 264 (1934).

15. DÁVID, L.: Über die Bestimmung des Alkaloidgehaltes der Ipekakuanhawurzel. Pharmaz. Ztg. **80**, 1121 (1935).

15a. — Anweisungen zur zweckmäßigen Herstellung des Dekoktums bzw. Infusums der Ipekakuanhawurzel, sowie des Wurzelpulvers mit gleichbleibender Wirkungskraft. Pharmaz. Ztg. **81**, 806 (1936).

16. DIETERLE, H.: Gehaltsbestimmungen alkaloidhaltiger Drogen des deutschen Arzneibuches **5**, ausgeführt mit möglichst kleinen Drogenmengen. Arch. Pharmaz. u. Ber. dtsch. pharmaz. Ges. **261**, 77 (1923).

17. EHRENSTEIN, M.: Über die katalytische Dehydrierung cyklischer Basen. I. Ber. dtsch. chem. Ges. **64**, 1137 (1931).

18. FIGDOR, W.: Application of Micro Methods to Control Work in Pharmaceutical Manufacturing. Amer. J. Pharmac. **98**, 157 (1926).

19. GOODSON, J. A., L. G. GOODWIN, J. H. GORVIN, M. D. GROSS, K. S. KIRBY, J. A. LOCK, R. A. NEAL, T. M. SHARP and W. SOLOMON: Chemotherapy of Amebiasis. II. Amines Derived Formally from Emetine. Brit. J. Pharmac. **3**, 49 (1948) [Chem. Abstr. **43**, 3378 (1949)].

19a. — — — — — — — — Chemotherapy of Amebiasis. III. Variants of Bis(diamylamino) decane. Brit. J. Pharmac. **3**, 62 (1948); [Chem. Abstr. **43**, 3379 (1949)].

19b. GOODWIN, L. G., C. A. HOARE and T. M. SHARP: The Chemotherapy of Amebiasis. I. Introduction and Methods of Biological Assay. Brit. J. Pharmac. **3**, 44 (1948); [Chem. Abstr. **43**, 3066 (1949)].

20. HAMERSHAG, F. E.: The Technology and Chemistry of Alkaloids. New York: van NOSTRAND. 1950.

21. HENRY, Th. A.: The Plant Alkaloids. 4th ed. Philadelphia: P. Blakiston's Son & Co. 1949.

22. HESSE, O.: Beitrag zur Kenntnis der Alkaloide der echten Brechwurzel. Liebigs Ann. Chem. **405**, 1 (1914).

23. KARRER, P.: Über die Brechwurzel-Alkaloide. Ber. dtsch. chem. Ges. **49**, 2057 (1916).

24. — Über die Brechwurzel-Alkaloide. II. Ber. dtsch. chem. Ges. **50**, 582 (1917).

25. Karrer, P., C. H. Eugster u. O. Rüttner: Zur Kenntnis des Emetins und seiner Nebenalkaloide. Helv. chim. Acta **31**, 1219 (1948).

26. Karrer, P. u. O. Rüttner: Zur Kenntnis des Dehydroemetins. Helv. chim. Acta **33**, 291 (1950).

27. Keller, O.: Untersuchungen über die Alkaloide der Brechwurzel *Uragoga Ipecacuanha*. I. Arch. Pharmaz. u. Ber. dtsch. pharmaz. Ges. **249**, 512 (1911).

28. — Untersuchungen über die Alkaloide der Brechwurzel *Uragoga Ipecacuanha*. II. Arch. Pharmaz. u. Ber. dtsch. pharmaz. Ges. **251**, 701 (1913).

29. — Untersuchungen über die Alkaloide der Brechwurzel *Uragoga Ipecacuanha*. III. Arch. Pharmaz. u. Ber. dtsch. pharmaz. Ges. **255**, 75 (1917)'

30. Keller, O. u. X. Bernhard: Untersuchungen über die Alkaloide der Brechwurzel *Uragoga Jpecacuanha*. IV. Arch. Pharmaz. u. Ber. dtsch. pharmaz. Ges. **263**, 401 (1925).

31. Kobayashi, K.: Synthesis of Racemic Emetamine (Pyman). J. pharmac. Soc. Japan **69**, 91 (1949); [Chem. Abstr. **44**, 1514 (1950)].

32. Kunz, H.: Beiträge zur Kenntnis des Emetins. Arch. Pharmaz. u. Ber. dtsch. pharmaz. Ges. **225**, 461 (1887).

33. Kunz-Krause, H.: Beiträge zur Kenntnis des Emetins. Arch. Pharmaz. u. Ber. dtsch. pharmaz. Ges. **232**, 466 (1894).

34. — Beiträge zur Kenntnis des Emetins. Schweiz. Wschr. chem. Pharmaz. **32**, 961 (1894).

35. — Beiträge zur Kenntnis des Emetins. Arch. Gen. (3) **34**, 290 (1895).

36. Léger, E.: Über die Titrierung des Ipecacuanhaextrakts. J. Pharmac. Chim. (8) **6**, 501 (1927); Dansk Tidsskr. Farmac. **2**, 145 (1928); **3**, 250 (1929).

37. Norcross, G. and H. T. Openshaw: Studies on the Structure of Emetine. III. Stability of Some α-Arylalkyltrimethyl-ammonium-iodides. J. chem. Soc. (London) **1949**, 1174.

38. Pailer, M.: Zur Konstitution des Emetins. II. Mh. Chem. **79**, 128 (1948).

39. — Zur Konstitution des Emetins. III. Synthese einer Abbauverbindung. Monatsh. Chem. **79**, 331 (1948).

40. Pailer, M. u. L. Bilek: Studien über den Hofmannschen Abbau quartärer Ammoniumverbindungen. Monatsh. Chem. **79**, 135 (1948)

41. Pailer, M. u. K. Porschinski: Über Brechwurzelalkaloide: Die Konstitution des Emetins. Monatsh. Chem. **80**, 94 (1949).

42. — — Über Brechwurzelalkaloide: Die Konstitution des Cephaelins. Monatsh. Chem. **80**, 101 (1949).

43. Paul, B. H. and A. J. Cownley: Chemistry of the Ipecacuanha. Pharmac. J. (3) **25**, 111, 373 (1894).

44. — — Chemistry of the Ipecacuanha. Pharmac. J. (3) **25**, 690 (1895).

45. Pützer, B. u. M. Schönhöfer: D. R. P. 550327.

46. Pyman, F. L.: The Alkaloids of Ipecacuanha. II. J. chem. Soc. (London) **111**, 419 (1917).

47. — The Alkaloids of Ipecacuanha. III. J. chem. Soc. (London) **113**, 222 (1918).

48. — Researches in Chemotherapy. Chem. and Ind. **15**, 789 (1937).

49. Robinson, R.: Structure and Biogenesis of Emetine. Nature (London) **162**, 524 (1948).

50. Rosenthaler, R.: Beiträge zur angewandten Drogenkunde: Chemische Charakterisierung von Drogen. Schweiz. Apoth.-Ztg. **62**, 152 (1924).

51. Späth, E. u. W. Leithe: Über Brechwurzelalkaloide. I. Zur Konstitution des Emetins und des Cephaelins. Ber. dtsch. chem. Ges. **60**, 688 (1927).

52. Späth, E. u. M. Pailer: Über die Bildung von 1-Oxy-3,4-dimethoxy-6-äthyl-
benzol bei der Ozonisation des 3,4-Dimethoxy-6-äthyl-zimtsäure-methylesters.
Ber. dtsch. chem. Ges. **73**, 238 (1940).

53. — — Zur Konstitution des Emetins. Monatsh. Chem. **79**, 128 (1948).

54. Späth, E., M. Pailer u. G. Gergely: Über die Bildung von Phenolen bei
der Ozonisierung von Benzolderivaten. Ber. dtsch. chem. Ges. **73**, 795 (1940).

55. Staub, H.: Ein Jahrhundert chemischer Forschung über Ipecacuanha-Alkaloide.
Dissert. Zürich, 1927.

56. Sugasawa, S. and K. Kobayashi: Synthesis of Emetine and its Allied
Compounds. X. An Improved Synthetic Method for 4′,5′-Dimethoxy-6-carb-
ethoxy-8-methyl-3,4,5,6,7,8-hexahydrobenzo-(1′,2′,1,2,)-quinolizine. J. pharmac.
Soc. Japan **69**, 85 (1949); [Chem. Abstr. **44**, 1514 (1950)].

57. Tschirch, A.: Handbuch der Pharmakognosie, Bd. III, S. 683. Leipzig:
Bernhard Tauchnitz. 1923. — Realenzyklopädie der gesamten Pharmazie,
2. Aufl., Bd. VII, S. 113. Wien u. Berlin: Urban & Schwarzenberg. 1906. —
Pharmaz. Ztg. **32**, 115 (1887).

58. Windaus, A. u. L. Hermanns: Untersuchungen über Emetin. I. Ber. dtsch.
chem. Ges. **47**, 1470 (1914); vgl. Hermanns, L.: Dissert. Freiburg i. Br., 1915.

59. Woodward, R. B.: Biogenesis of the Strychnosalkaloids. Nature (London)
162, 155 (1948).

60. D. R. P. 267219, 301498, 298678, 99090. — Engl. P. 11717—719, 14677,
17483, 103881, 104652, 105722. — U. S. P. 1219575.

(Eingelaufen am 7. März 1951.)

X-Ray Diffraction Studies of Crystalline Amino Acids and Peptides.

By **Robert B. Corey**, Pasadena, California.

With 25 Figures.

Contents.

Introduction.

Within the last ten years workers in the physical, biological, and medical sciences have become increasingly aware of the intimate relationship which exists between the chemistry of living organisms and the structure of protein molecules. The general properties of a molecule of an individual protein depend primarily upon the nature and number of the amino acid residues of which it is composed and upon the sequence in which they are arranged along the polypeptide chains. If this information alone were available for any protein it would go far toward explaining its chemical behavior and many of its biological properties; it would serve as a basis for the understanding and control of its chemical reactions, for carrying out proposed alterations of its composition, and, hence, for its possible adaptation to new and useful purposes. But this chemical

information alone would not be sufficient to provide adequate understanding of those remarkably specific properties which are especially
significant in determining the behavior of proteins in biological systems.
These properties are directly dependent upon the spatial relationships
between the carbon, nitrogen, oxygen, hydrogen, and other atoms of
which the proteins are composed, and upon the nature and the direction
of the forces which act between the atoms. These spatial relationships
and interatomic forces are largely responsible for the ability of certain
proteins to crystallize, for the architecture of protein fibers, for the
specificity of enzymes, and for immunological reactions.

For the study of the spatial relationships and geometrical fine structure
of protein molecules the methods of X-ray diffraction have been found
to be a most useful tool. Early X-ray diffraction studies of proteins,
and other natural products, are discussed in the excellent review by
KRATKY and MARK (16) appearing in an earlier volume of this series.
At the time that review was written, many X-ray investigations had been
made of both fibrous and crystalline proteins; in the intervening twelve
years they have been greatly multiplied and extended, especially through
the use of new and powerful techniques, both in experimental method
and in theoretical interpretation. Although they have provided a basis
for interesting and stimulating speculations regarding the structure of
fibrous proteins, such as keratin and the protein of muscle, and have
thrown considerable light on the gross architecture and swelling of
protein crystals, these investigations have yielded no information concerning the relative positions of individual atoms in proteins nor the
precise dimensions and geometry of the polypeptide chain.

What information we have concerning these fundamental structural
details is largely derived from complete X-ray analyses of crystals of
amino acids and peptides. These X-ray analyses, and their contribution
to our knowledge of interatomic distances in proteins, comprise the scope
of the present paper.

I. X-Ray Analysis of Crystals in General.

A comprehensive discussion of X-ray theory and techniques is quite
beyond the scope of this review; the reader is referred to general references
(B 1–B 9). During the last decade, great progress has been made in the
methods of crystal analysis, especially in the use of PATTERSON
interatomic vector diagrams for arriving at the correct trial structure
and in the use of FOURIER plots of electron density, and of the method
of least squares for the refinement of atomic parameters. The recent
application of punched card calculating machines and other aids to computation have added greatly to the power of these methods, so that the

structures of many crystals of relatively complex organic compounds, heretofore beyond the practical limits of X-ray analysis, have been determined, and simpler structures can be established with greatly increased precision. A brief discussion of the X-ray method as applied to crystals of amino acids and similar substances will be given in the following sections; in part, it is patterned after an outline published by Dunitz (*10*).

1. Outline of the X-Ray Method.

When a crystal is exposed to a beam of X-rays, its regularly spaced atoms act as a three-dimensional diffraction grating; in certain directions, constructive interference produces diffraction maxima which may be recorded on photographic film as a diffraction pattern. These diffraction maxima were found by Bragg to correspond to reflections from crystallographic planes according to the equation

$$n \lambda = 2\, d_{(hkl)} \sin \Theta$$

in which n is an integer, λ the wavelength of the X-rays, $d_{(hkl)}$ the spacing of the equivalent atomic planes parallel to the face hkl and Θ is the glancing angle of incidence which the primary X-ray beam makes with the plane hkl.

The determination of the positions of the atoms in a crystal by the analysis of its X-ray diffraction pattern usually involves the following successive steps.

First, from the directions of the diffracted beams, as determined from the positions of the corresponding spots on the photographic film, can be calculated the dimensions of the unit cell, or unit of structure, which, when translationally repeated in three dimensions, generates the crystal. From the symmetry of the X-ray pattern, and from the observation of systematic absences among the diffraction spots, the space group or distribution in the crystal of centers, planes, and axes of symmetry, can be derived. For most crystals this information can be obtained quickly from a few diffraction photographs.

The second step, and by far the more difficult one, is the determination of the positions of the atoms. This information is derived, not from the positions of the diffraction spots on the photographic film, but from their relative intensities. The intensities of the diffracted X-ray beams, or X-ray reflections, are determined from measurements of the relative intensities of the spots on the films, and from each of these intensities a quantity $\left|F_{hkl}\right|$ is calculated. This quantity is a function of the scattering powers of the atoms and of their positions in the unit cell. From assumed atomic positions the value of $\left|F_{hkl}\right|$ for each (hkl) reflection which would be obtained from a hypothetical crystal can be calculated; but, in general, it is not possible to reverse the process, i. e., to calculate the positions of the atoms directly from the values of $\left|F_{hkl}\right|$ derived from measurements of the intensities of X-ray reflections from a crystal. This arises from the fact that F_{hkl}, comprising the combined contributions of the waves scattered by all of the atoms of the unit cell, involves a resultant amplitude and phase, the former of which, only, may be determined experimentally from the intensity data. For any but the simplest structures it is therefore necessary to proceed in an indirect manner: to make as intelligent a guess as possible concerning the structure, to calculate the various values of $\left|F_{hkl}\right|$ corresponding to this hypothetical structure, to compare these calculated values with those obtained from the observed intensities, and to judge the correctness of the postulated structure by the extent to which the observed and calculated values of $\left|F_{hkl}\right|$ agree. Generally, this procedure comprises two stages: (a) the selection of a satisfactory trial structure, and (b) refinement of the trial structure during which the atomic positions undergo successive changes designed to minimize the discrepancies between observed and calculated values of $\left|F_{hkl}\right|$. We shall describe these stages in more detail, but first we should discuss one of the most generally useful techniques of crystal structure analysis—the application of FOURIER series to the calculation of the electron density distribution in a crystal.

For the most general case (a crystal having no center of symmetry) the electron density distribution throughout the unit cell of a crystal may be found by evaluating the triple FOURIER series

$$\varrho(xyz) = \frac{1}{V} \sum_h \sum_k \sum_l |F_{hkl}| \cos \left[2\pi (hx + ky + lz) - \alpha(hkl) \right]$$

in which $\varrho(xyz)$ is the electron density at a point designated by the parameters x, y, and z, V is the volume of the unit cell, $|F_{hkl}|$ is the amplitude, and $\alpha(hkl)$ the phase of the reflection (hkl). For centrosymmetric structures $\alpha(hkl)$ is either o or π and the formula reduces to

$$\varrho(xyz) = \frac{1}{V} \sum_h \sum_k \sum_l \pm |F_{hkl}| \cos 2\pi (hx + ky + lz)$$

In general, the evaluation of the three-dimensional summation is unnecessary, and, indeed, without the recent developments in computational techniques, is impracticable because of the excessive labor involved in making the calculation. It is usually sufficient to calculate one or more two-dimensional projections of electron density on a face of a unit cell, say (oo1), by use of the intensity data from a single zone, in this case from the (hko) reflections. If the projection has a center of symmetry, the formula is:

$$\varrho(xy) = \frac{1}{A} \sum_h \sum_k \pm |F_{hko}| \cos 2\pi (hx + ky)$$

in which A is the area of the face (oo1). Since, in general, the phases or signs of the reflections cannot be derived from the observed intensities,

FOURIER plots of electron density distribution are usually based on an approximate or trial structure which has been found to give rough agreement between observed and calculated values of $|F_{hkl}|$. From the trial structure approximate values of $\alpha(hkl)$ can be calculated and used with the corresponding observed values of $|F_{hkl}|$ for the calculation of the electron density distribution, in which the positions of the atoms in the cell will be indicated by the centers of the peaks. If now, these centers are chosen as new and more nearly correct positions for the atoms, new and additional values of $\alpha(hkl)$ can be calculated from them, and the computation of the electron density distribution can be repeated to give atomic positions of even greater accuracy.

The usefulness and limitations of electron density plots in crystal analysis will be discussed in a later section; as already mentioned, a trial or postulated structure is generally necessary for their calculation.

2. The Trial Structure and the Patterson Diagram.

It is sometimes possible to arrive at an approximate or trial structure by trial and error methods. A model of the molecule may be constructed in accord with chemical information and this model may then be oriented in the unit cell of the crystal in ways suggested by intermolecular packing, the optical, magnetic, or cleavage properties of the crystal, the relative intensities of certain X-ray reflections, and other considerations. The values of $|F_{hkl}|$ corresponding to the selected arrangement are then calculated, and if they are found to be in approximate agreement with those obtained from the observed intensities, the structure is assumed to be approximately correct. One then proceeds with its refinement by methods outlined in the following Part 3 (p. 316). A more direct method of arriving at a trial structure, and one which has been especially useful in determinations of the structures of amino acids and peptides, is the use of a PATTERSON function. In 1935 PATTERSON showed that the FOURIER series

$$P(uvw) = \sum_h \sum_k \sum_l |F_{hkl}|^2 \cos 2\pi \, (hu + kv + lw)$$

will have maxima at positions corresponding to interatomic vectors in the crystal. Since this expression does not involve the phase angles of the reflections, but their amplitudes, $|F_{hkl}|$, only, it can be evaluated directly from the intensity data. As in the plot of electron density distribution, the evaluation of the three-dimensional PATTERSON function is very laborious and has become practicable only with the recent development of methods for high speed calculation. Indeed, the first crystal structure determination utilizing the complete PATTERSON function was that of L-threonine published in 1950 (p. 328).

Figure 1 below is designed to illustrate the relationship between the positions of atoms in the unit cell of a crystal (a) and the positions of the maxima in the corresponding PATTERSON interatomic vector diagram (b), and electron density plot (c), respectively. If Figure 1 (a) represents the unit cell of a crystal containing three atoms A, B, and C, then the corresponding PATTERSON interatomic vector diagram would be that shown in (b). Besides the maximum at the origin, which arises from the fact that each atom is at zero distance from itself, maxima occur at positions such that a line drawn from the origin to a maximum corresponds in length and direction to an interatomic vector connecting two atoms in the crystal. Both positive and negative vectors are represented, as indicated in Figure 1; n atoms will generate n (n — 1) interatomic vector maxima.

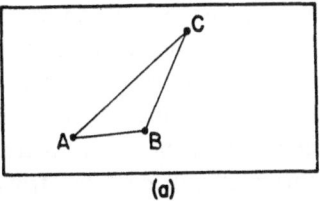

(a)

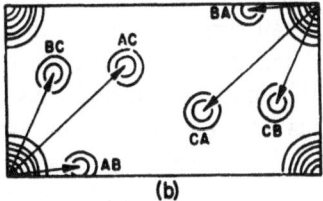

(b)

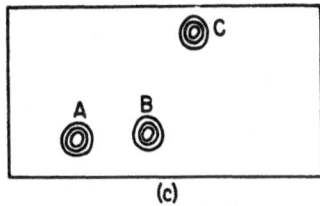

(c)

Fig. 1. A diagramatic representation of the relationships between (a) the unit cell of a hypothetical crystal containing three atoms at positions A, B, and C respectively, (b) the corresponding PATTERSON interatomic vector diagram, and (c) the corresponding FOURIER plot of electron density distribution.

The usefulness of a PATTERSON diagram depends upon its successful interpretation, that is, upon the derivation of atomic positions from the plot of interatomic vectors so as to obtain a correct approximate or trial structure. For this interpretation it is necessary that a reasonably large number of the peaks corresponding to individual interatomic vectors be resolved, a condition which becomes more unlikely with increasing numbers of atoms in the unit cell. The volume of a unit cell is roughly proportional to the number of atoms which it contains, but the number of PATTERSON peaks is proportional to n (n — 1), so that for crystals of great complexity the selection of resolved peaks and their interpretation in terms of atomic positions is virtually impossible. For example, the crystal of L-threonine contains only thirty-two atoms (exclusive of hydrogen) in the unit cell, but the calculation of the complete three-dimensional PATTERSON plot was found to be necessary in order to obtain a sufficient number of resolved and identifiable interatomic vectors to lead to a trial structure.

If a crystal contains a few atoms of much higher atomic number than the others, the interatomic vectors associated with them will be represented in the PATTERSON diagram by correspondingly higher maxima which can generally be recognized and interpreted to establish the approximate positions of the heavier atoms in the crystal. Since the scattering of X-rays by an atom is roughly proportional to its atomic number, and since the resultant scattering by an assemblage of single light atoms of equivalent scattering power will be reduced by interference until it is quite ineffective relative to the scattering of a single heavy atom, both the amplitudes and the phases of the diffracted beams will be largely determined by the heavy atoms. If the positions of the heavy atoms have been found by use of a Patterson diagram, they may be used to calculate the corresponding phase angles $\alpha(hkl)$. A plot of the electron density distribution may then be calculated incorporating these phase angles with the experimentally determined amplitudes $|F_{hkl}|$. In general, this plot will contain strong, well-resolved peaks corresponding to the heavy atoms, and in addition less pronounced and more poorly resolved detail corresponding to the distribution of the lighter atoms. From this detail it is often possible to derive approximately correct positions for the lighter atoms and thus to obtain a satisfactory trial structure which may then be refined by straightforward methods.

If isomorphous crystals are available whose structures differ only in the presence or absence of a particular atom, or in its replacement by another atom of different atomic number, and if the positions of these atoms can be ascertained by the use of Patterson diagrams or by other means, then it is often possible to determine the approximate positions of the remaining atoms directly by a procedure closely related to the heavy atom technique described above.

The methods of analysis used in the determination of the structure of a particular crystal will depend upon the nature of the crystal itself, and upon the preferences and experience of the analyst. Each investigation will present problems of its own. Certainly no hard and fast procedure can be recommended for the determination of crystal structures.

3. Refinement of the Structure.

The accuracy with which the structure of a crystal is determined depends upon many factors—the objectives of the investigation, the quantity and quality of the X-ray diffraction data, the time and effort available for the determination, and other considerations.

Interest in the structures of amino acids and peptides centers around the contribution which they will make to an understanding of the fine structure of protein molecules, namely, the interatomic distances and bond angles in the peptide chain and in specific side chains, the forces which act between molecules and between different parts of the same molecule, the directional character of these forces and the extent to which they determine the configuration and distribution of molecules in solid proteins. For information of this sort it is desirable that atomic positions be determined with maximum accuracy, that interatomic distances be known to within about 0,02 Å. and bond angles within 2° or 3°. An accuracy of this order has not generally been obtained in determinations fo the crystal structures of organic compounds.

Until recently, the methods commonly used for the estimation of intensities were lacking in precision, and even when adequate data were obtained two-dimensional plots of electron density calculated from the reflections of prism zones often failed to resolve the atoms. The latter limitation has been largely removed by modern computational methods which permit the use of all of the data for the refinement of parameters by the method of least squares, or by the calculation of three-dimensional electron density plots in which good resolution can be obtained for all atoms. Nevertheless, a prodigious amount of work is still required for the precise analysis of most crystals of the complexity of amino acids or peptides.

The method which is probably in most general use for the refinement of atomic parameters, namely, the calculation of successive FOURIER plots of electron density distribution, has already been described. In some crystals the molecules may be so oriented in the unit cell that two-dimensional projections of electron density give peaks which are so poorly resolved as to make them of little use in refining the parameters of the trial structure (see glycine and alanine, p. 322, 326). In such cases structure factor contour plots have often been helpful as a guide in determining the adjustments in the atomic parameters leading to a minimum discrepancy between observed and calculated values of $|F_{hkl}|$. These plots consist of a set of separate two-dimensional plans of the unit cell, the (001) face, for example, on each one of which is copied the values of the trigonometric component of the structure factor $|F_{hkl}|$ for a single prism zone reflection, in this case for each reflection (hko). For a given reflection, the contours of this plot in the neighborhood of an atom indicate directly the effect which any change in the atom's position will have on the value of $|F_{hkl}|$. The usefulness of these plots is practically confined to zones for which the structure factors have a center of symmetry. Their severe limitation, in contrast to the FOURIER method, rests in their inability to select the direction and extent of desirable simultaneous changes of the positions of several atoms.

A method of refinement which is equivalent to the FOURIER in its results, and which has certain advantages in some cases, is that developed by HUGHES (12). In this procedure the quantity

$$\sum_{hkl} w \left[\, \left| F_{observ.} \right| - \left| F_{calc.} \right| \,\right]^2$$

is regarded as a function of the parameters of the atoms, and is minimized by the method of least squares.

II. Studies of Crystals of Amino Acids.

The first X-ray diffraction study of amino acid crystals was published by BERNAL in 1931 (3). It contained a tabulation of the unit cell dimensions, space groups, and optical data for crystals of α-glycine, β-glycine, D-alanine, DL-alanine, D-phenylalanine, L-cystine, L-aspartic acid, L-glutamic acid, and asparagine monohydrate. A brief discussion was given of probable arrangements of the molecules; no intensity data were included and no attempt was made to determine the positions of any atoms. ALBRECHT and collaborators (2) have published data concerning the unit cells and space groups of DL-valine, DL-threonine*, DL-serine, DL-norleucine, and DL-methionine. Corresponding data for L-proline have been published by WRIGHT and COLE (26) and for DL- and D-leucin by MÖLLER (21a). Two determinations of the structure of glycine (11, 15), both based upon inadequate data, are incorrect.

Unit cell and space group data were given by BERNAL (3) for the addition compound of glycine with barium chloride, $BaCl_2 \cdot (NH_2CH_2COOH)_2$. STOSICK (24) published an incomplete study of copper DL-α-aminobutyrate, $Cu[CH_3 \cdot CH_2 \cdot CH(NH_2) \cdot COO]_2$, in which he reported the unit cell dimensions and probable space group. Intensity data from 126 (h0l) reflections served to establish x and z parameters for copper, carbon, nitrogen, and oxygen atoms, but no values for interatomic distances were obtained. Crystals of nickel glycine dihydrate, $Ni(NH_2CH_2COO)_2 \cdot 2 H_2O$, were studied by the same author (25). The unit cell dimensions and space group were determined, and intensity data from 198 prism reflections were used to obtain two-dimensional projections of electron density along each of the crystallographic axes. Unfortunately the carbon, nitrogen, and oxygen atoms were so poorly resolved in these projections that several parameters remained undetermined.

1. Glycine.

Crystals of glycine, $NH_2 \cdot CH_2 \cdot COOH$, were the first amino acid crystals to be analyzed by means of X-rays.

In an investigation published by ALBRECHT and COREY (1), estimated intensities from nearly three hundred reflections were used for the calculation of PATTERSON plots of interatomic vectors from which a satisfactory trial structure was obtained. Two-dimensional projections of electron density along the crystallographic axes fail to resolve any atom and,

---

* The space group assigned in (2) to threonine is incorrect, cf. (23).

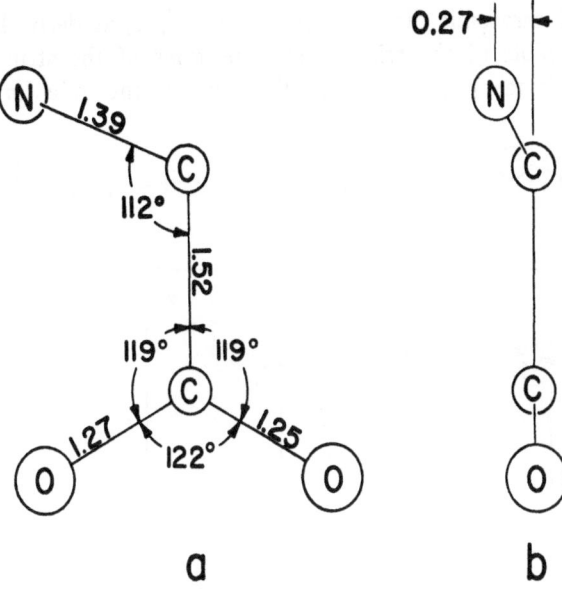

Fig. 2. The dimensions of the glycine molecule.

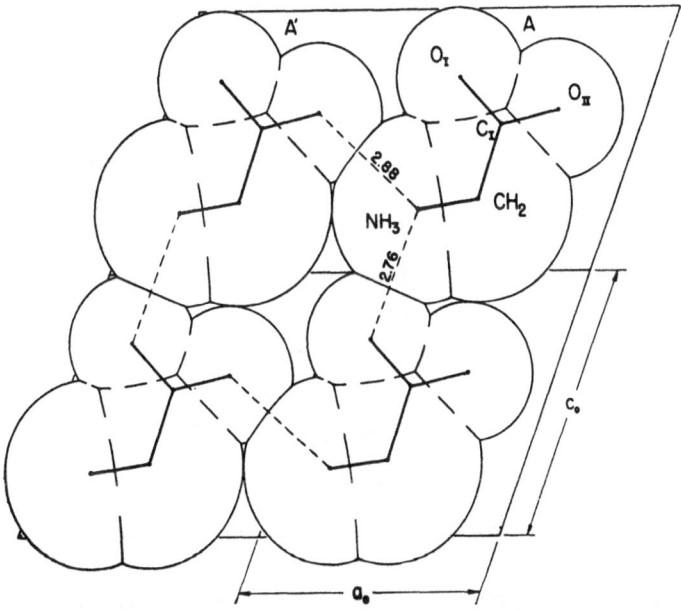

Fig. 3. A portion of a single layer of glycine molecules showing how the molecules are bound firmly together by hydrogen bonds between NH_3 groups and carboxyl oxygen atoms.

since the calculation of three-dimensional plots of electron density distribution had not yet been made feasible, the refinement of the structure was carried out by essentially trial and error methods. Reasonable values for interatomic distances and bond angles, as derived from other structures, plots of the trigonometric portions of the structure factor, and other devices were used as guides in arriving at a final structure

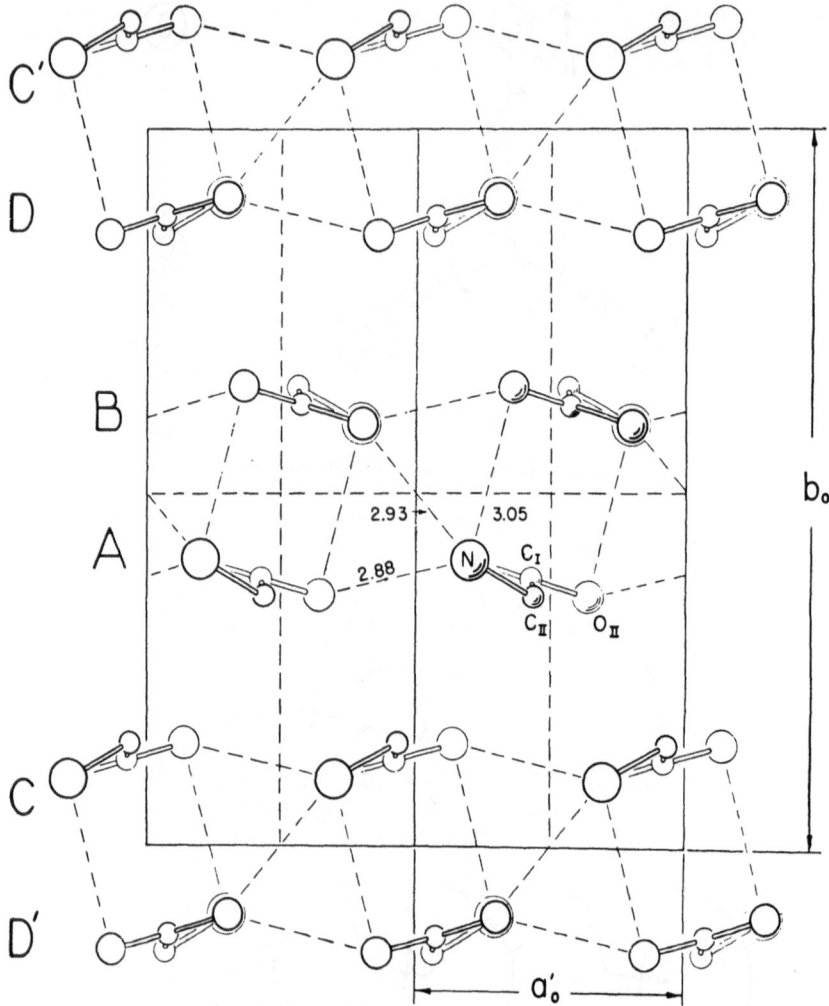

Fig. 4. The structure of the glycine viewed along the *c*-axis showing the molecular layers C'D, BA, CD'.

giving satisfactory agreement between observed and calculated values of $|F_{hkl}|$. Because of the somewhat poorer quality of the intensity data and the necessarily cruder methods used for arriving at the positions of the atoms, the results of this determination are lacking in the precision which characterizes some more recent crystal structure studies. Nevertheless, they are adequate to define the shape and orientation of the glycine molecule and to provide detailed information regarding the architecture of the crystal.

The dimensions of the glycine molecule as determined in this investigation are shown in Figure 2. All are in reasonable agreement with those to be anticipated, with the exception of the carbon-nitrogen distance, 1,39 Å. More recent and more accurate crystal structure studies of other amino acids and of acetylglycine and β-glycylglycine have all found carbon-nitrogen bond lengths close to 1,47 Å. Consequently, there seems to be good reason to believe that this short distance (1,39 Å.) reported for glycine (2) is in error. It is hoped that this question can be settled, and other details of the glycine structure clarified, by a redetermination of the crystal structure of glycine based on first class intensity data and refined by three-dimensional methods. As in the molecules of other amino acids thus far studied, the nitrogen atom of glycine lies a short distance out of the plane which contains the carboxyl group and the α-carbon atom.

In the crystal the molecules of glycine are arranged in sheets or layers. In each single layer the molecules are bound firmly to one another by hydrogen bonds between adjacent

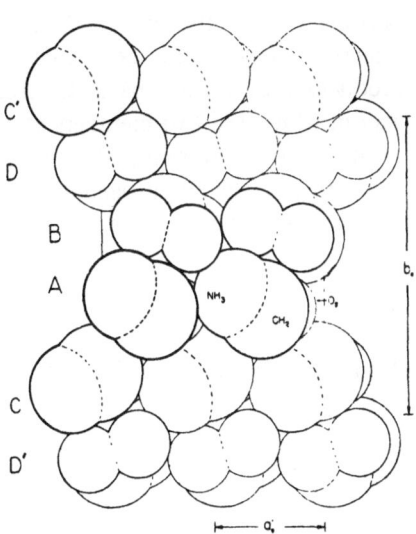

Fig. 5. A packing drawing of the molecules of Fig. 4, showing how the NH$_3$ groups are turned toward the centers of the double layers which involve carboxyl and CH$_2$ groups only.

oxygen and nitrogen atoms. A portion of a single layer is shown in Figure 3.

Figure 4 shows the way these molecular layers are arranged in the crystal. The molecules of each single layer (A) are bound closely to those of a particular adjacent layer (B) to form a sort of double layer, which in turn is separated by ordinary VAN DER WAALS contacts from adjacent double layers (C, D', and D, C'). The structure strongly suggests that the glycine molecule exists in the crystal as a dipolar ion. Within a single layer the angle between the hydrogen bonds (2,88 Å. and 2,76 Å.) is approximately the tetrahedral angle, and these bonds make virtually equal angles with the nitrogen-carbon covalent bonds. The third hydrogen atom of a positively charged NH$_3$ group probably occupies a position close to and about equidistant from two oxygen atoms on different molecules of the opposing member of the double layer.

Between double layers the contacts are made between CH$_2$ groups or between CH_2 groups and oxygen atoms—a fact which explains the excellent cleavage of glycine crystals parallel to the xz plane. The VAN DER WAALS contacts and the structure of the double layers are shown in the packing drawing (Figure 5, p. 321).

Many factors influence the arrangement which the molecules assume in the glycine crystal—the size and shape of the molecules, their ability to form intermolecular hydrogen bonds between nitrogen and oxygen atoms and the directional character of such bonds, VAN DER WAALS interactions, and electrostatic interactions between oppositely charged portions of the dipolar ions. The shape of the glycine molecule and the directional characteristics of hydrogen bond formation cooperate in the production of closely packed, strongly bonded molecular layers. The way in which the two members of a double layer are fitted to each other seems to be determined largely by their residual hydrogen bonding capacity (the third hydrogen atom of the NH_3^+ group) and by the distribution of electrical charges of opposite sign carried by the dipolar ions.

2. DL-Alanine.

The crystal structure of DL-alanine, $CH_3 \cdot CH(NH_2) \cdot COOH$, was determined by LEVY and COREY (19) in 1941. The determination was

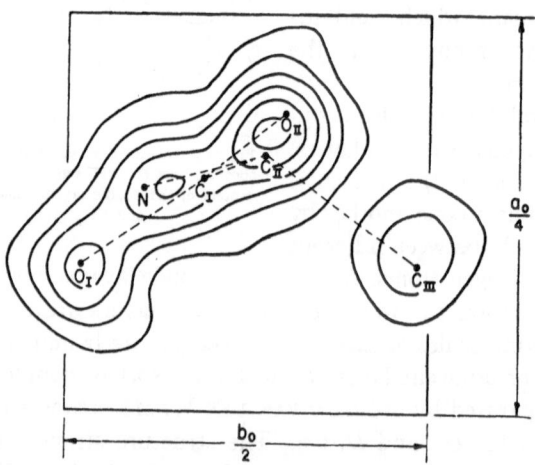

Fig. 6. Projection of the electron density distribution along the c-axis of the crystal of DL-alanine. The methyl carbon atom only is resolved.

based on good intensity data from 141 prism reflections and 238 additional general reflections (hkl). PATTERSON interatomic vector plots were useful guides in arriving at the correct trial structure. The symmetry of DL-alanine crystals is such that the calculation of two-dimensional FOURIER projections of electron density is feasible along the c-axis only, and in this direction the methyl carbon atom alone is resolved (see Figure 6). It was therefore necessary to rely essentially on trial and error methods for the refinement of the parameters. The dimensions thus obtained for the alanine molecule were in general agreement with those anticipated, but, as in glycine, the length of the carbon-nitrogen bond, normally about 1,47 Å., was exceptionally short (1,43 Å.) and quite inexplicable. A redetermination of the atomic positions based on the same original data was therefore made by DONOHUE (9) in 1949. Whereas in the original determination the intensities of prism reflections only could be utilized in placing the atoms, the revision made use of all 379 reflections for the computation of a three-dimensional FOURIER plot of electron density distribution in which all atoms were completely resolved. A composite drawing showing the electron density contours in the neighborhood of each atom is reproduced in Figure 7. The dimensions of the alanine molecule as determined from the three-dimensional plot are shown in Figure 8.

The revised dimensions, differ in two respects from those originally

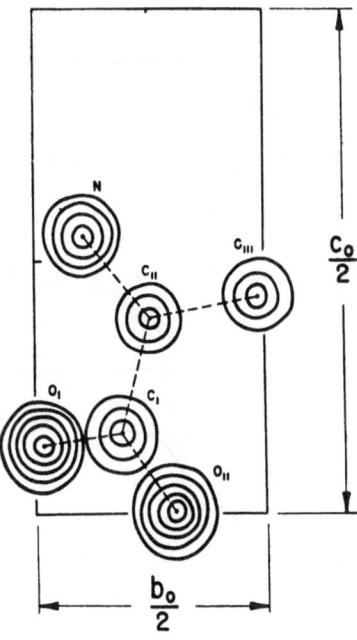

Fig. 7. A composite representation of the three-dimensional plot of electron density distribution in the crystal of DL-alanine. The contours associated with each atom are those corresponding to planes passing through the respective atomic centers. All atoms are completely resolved.

obtained by trial and error methods: "The C—N bond distance is 1,50 Å. and is no longer anomalously short; and the C—O distances in the carboxyl group differ by 0,06 Å., indicating that hydrogen bond formation caused

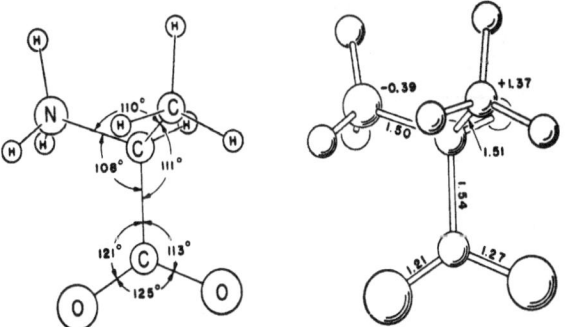

Fig. 8. Dimensions of the alanine molecule.

one of the resonating structures to be favored over the other, an effect which is substantiated by the inequality of the C—C—O bond angles.''

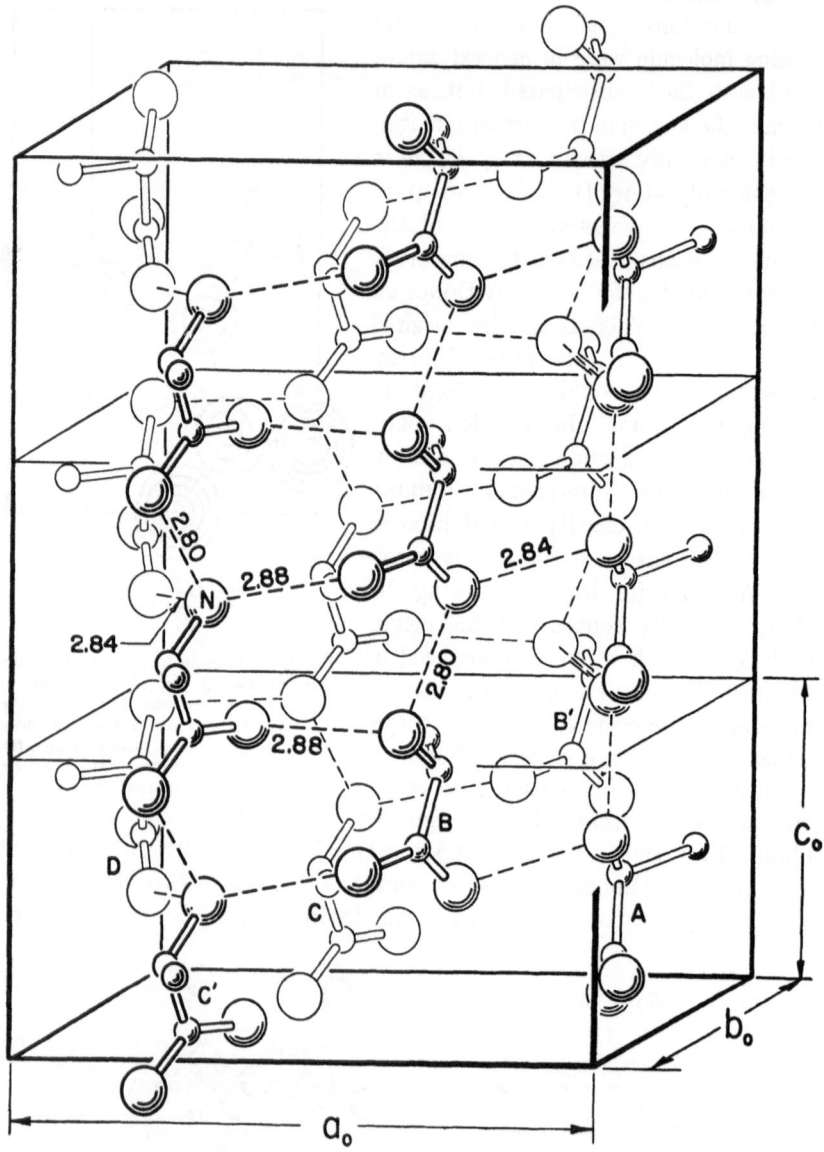

Fig. 9. The arrangement of the molecules in crystals of *DL*-alanine. Small circles represent carbon atoms, large circles oxygen and nitrogen atoms; hydrogen atoms are not shown.

As in glycine, the nitrogen atom is a slight distance (0,39 Å.) from the plane of the carboxyl group and the α-carbon atom; the methyl carbon is 1,37 Å. distant from and on the opposite side of this plane. Although the hydrogen atoms contribute too little to the X-ray scattering to permit their positions to be definitely determined, there is little doubt that in crystals of *DL*-alanine their arrangement relative to the other atoms of the molecule is close to that shown in Figure 8. Indeed, if the contribu-

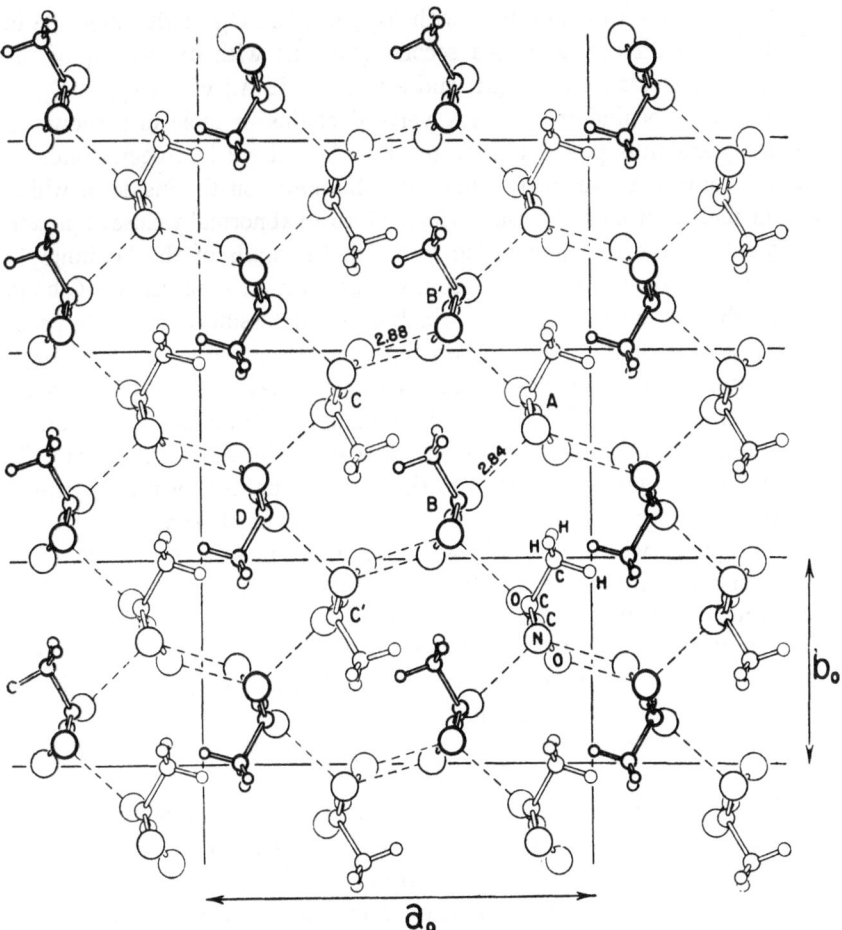

Fig. 10. The structure of *DL*-alanine viewed along the *c*-axis of the crystal.

tions of the hydrogen atoms are calculated for this arrangement, and these calculated values are included in the comparison of calculated and observed magnitudes of $|F_{hkl}|$, the agreement for many reflections is found to be significantly improved.

The arrangement of the molecules in the *DL*-alanine crystal is shown in Figure 9. Unlike crystals of glycine, which are made up of layers tightly bonded together in two directions but held together in the third direction by VAN DER WAALS forces only, crystals of *DL*-alanine comprise a three-dimensional network in which the molecules are bound firmly to each other by hydrogen bonds in all directions.

If we consider an individual nitrogen atom, say N in Figure 9, we see that three N—H$\cdots$O hydrogen bonds radiate from it, making approximately tetrahedral angles with each other and with the covalent N—C bond; one (2,80 Å.) connects it with oxygen atom O_{II} of the molecule in the unit cell directly above, a second (2,88 Å.) with oxygen O_I of the adjacent molecule on the right, and a third (2,84 Å.) with oxygen O_{II} of the molecule behind it. The six vertical chains of molecules shown in Figure 9 are thus pulled tightly together by lateral hydrogen bonds to form a continuous vertical "tube" or "chimney" on the inside of which are the methyl groups, themselves forced into an abnormally close approach (3,65 Å.) by the strong bonding forces in the walls of the "chimney". Indeed, so close are the methyl groups, that rotation about the C—C bond is sharply restricted, a condition which makes it possible to select the probable positions of the methyl hydrogen atoms shown in Figure 8 (p. 323).

Figure 10 shows a view of the structure looking down from above. The lateral hydrogen bonds and the rectangular "chimneys" containing the methyl groups are clearly seen. The six molecules A, B, B', C, C', and D in this figure correspond to the vertical chains shown in Figure 9.

It is clear that the dipolar ions in solutions of DL-alanine possess a size and shape which make it possible for them to solve the problems of crystallization in an unusually successful manner. Their dimensions conform to a nearly tetrahedral packing of negatively charged carboxyl oxygen atoms around positively charged NH_3 groups, thus making possible the formation of unusually strong uniform hydrogen bonds in all three directions throughout the crystal.

3. L-Threonine.

The determination of the crystal structure of L-threonine (threo-α-amino-β-hydroxy-n-butyric acid)*, $CH_3 \cdot CHOH \cdot CH(NH_2) \cdot COOH$, by Shoemaker, Donohue, Schomaker and Corey (23) is of special interest for several reasons: based upon excellent intensity data for over six hundred X-ray reflections, this investigation established the atomic positions in the crystal with an accuracy greater than that previously obtained for any amino acid, and perhaps for any organic compound of similar complexity; the analysis involved the development of new computational techniques and the free use of three-dimensional Fourier plots, notably the complete Patterson function which was used for the first time; it established on unambiguous physical grounds the relative configurations about the two asymmetric centers of the threonine molecule.

---

* "threo-" denotes the relative configurations about the two asymmetric centers corresponding to that in the molecule of the sugar threose.

No approximate structure for *L*-threonine could be found by trial and error methods, and two-dimensional PATTERSON plots of interatomic

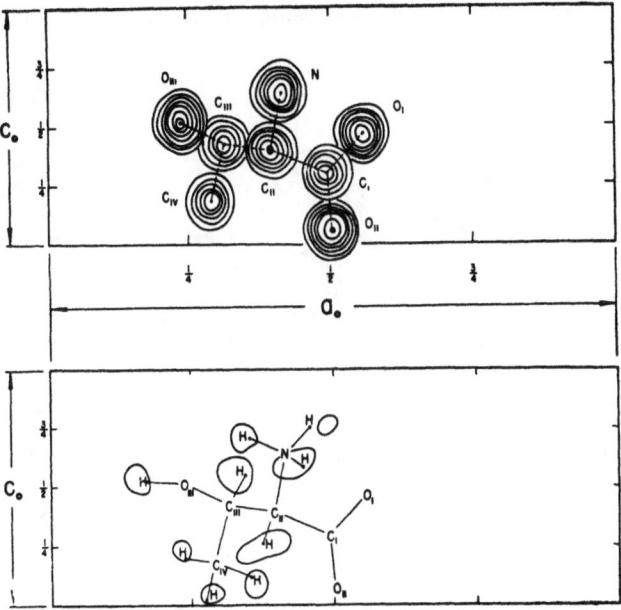

Fig. 11. Above: a composite presentation of sections of the three-dimensional electron density plot for *L*-threonine, each section being taken close to an atomic center. Below: a similar plot showing maxima corresponding to positions of the hydrogen atoms.

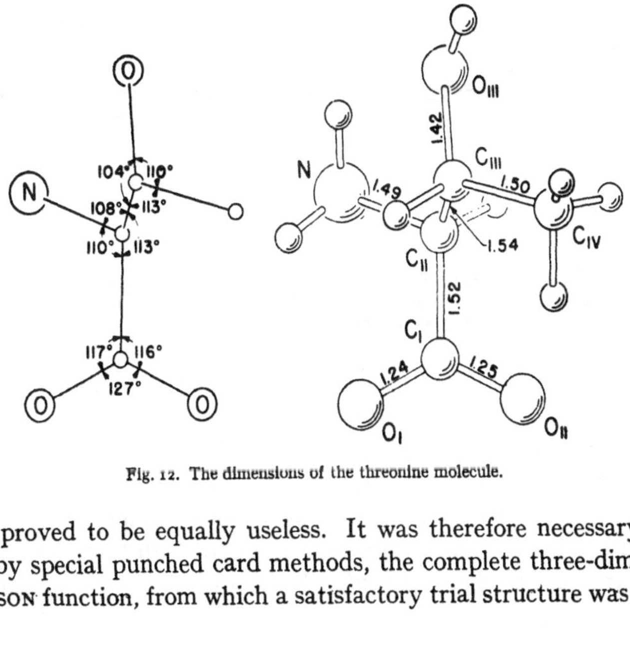

Fig. 12. The dimensions of the threonine molecule.

vectors proved to be equally useless. It was therefore necessary to calculate, by special punched card methods, the complete three-dimensional PATTERSON function, from which a satisfactory trial structure was derived.

This structure was refined by the calculation of three-dimensional Fourier plots of electron density (five, in all) and by the method of least squares. Figure 11 is a composite presentation of sections of the three-dimensional electron density plot, each section being taken close to an atomic center.

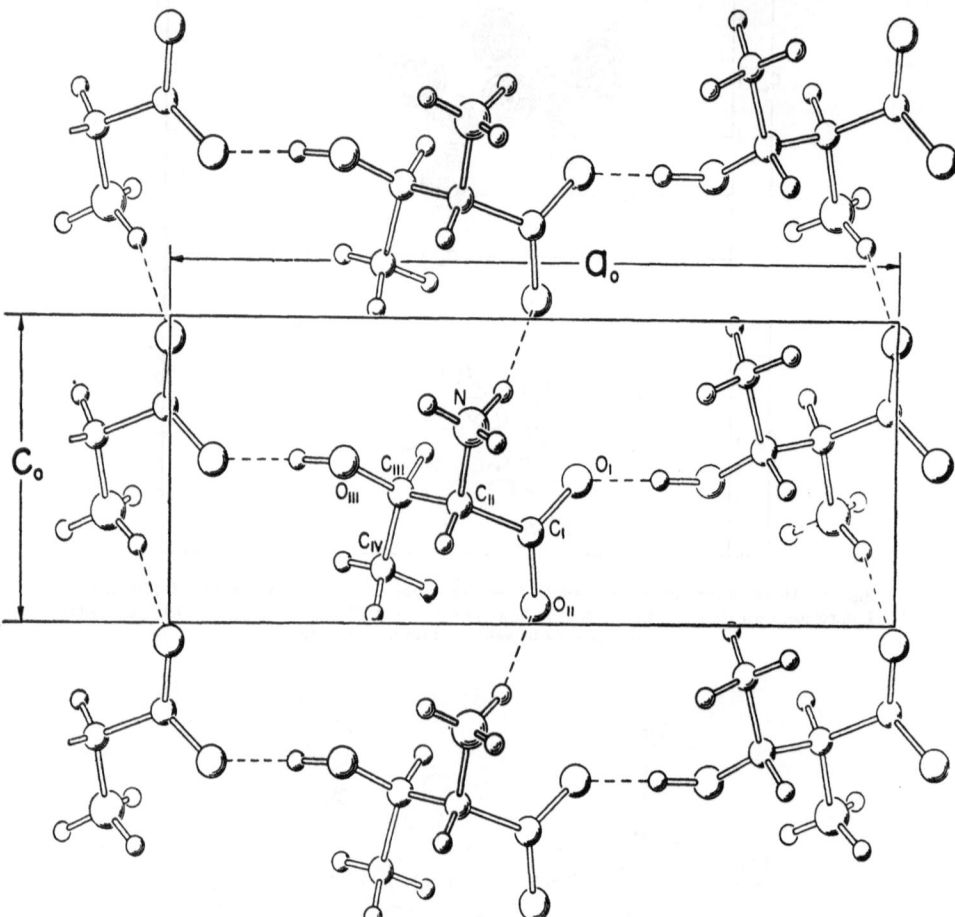

Fig. 13. Some of the molecules of the L-threonine crystal viewed along the b-axis showing hydrogen bonding of amino and hydroxyl groups with oxygen atoms. Interleaving layers of molecules which bind these molecules together in the b direction are omitted for the sake of clarity.

Small electron density peaks were also found at all positions where hydrogen atoms would be expected. These are shown in the lower half of the figure.

Figure 12 shows the dimensions of the threonine molecule. The relative stereochemical configuration around the two asymmetric carbon atoms is identical with that assigned on chemical grounds (21). In general, the bond lengths and bond angles are in accord with those found in other crystal structure studies. The αC—N distance (1,49 Å.) is in agreement with the sum of the covalent radii (B 7) and with the corresponding distance found in the revised structure of DL-alanine (1,50 Å.). The bond joining the β-carbon and the methyl-carbon atoms (1,50 Å.) is surprisingly short. It is also difficult to discover reasons for the small angle (104°) between the C—C and C—O bonds associated with the β-carbon atom.

In crystals of DL-alanine, the molecules are bound together into a continuous three-dimensional network by means of hydrogen bonds and

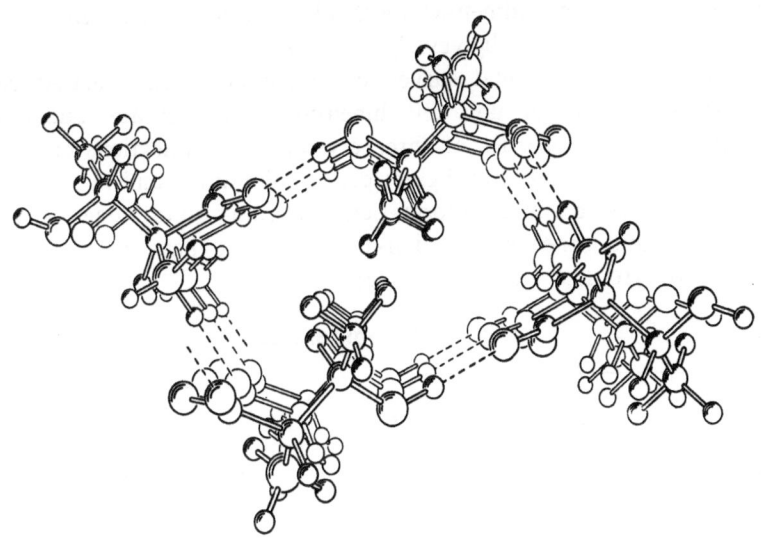

Fig. 14. A view of the crystal of L-threonine looking along the c-axis.

electrostatic forces between positively charged NH_3 groups and carboxyl oxygen atoms. In crystals of L-threonine the situation is similar; in two dimensions the structure appears to be tied together by hydrogen bonds and electrostatic forces between amino groups and carboxyl oxygen atoms, in the third dimension primarily by hydrogen bonds between hydroxyl groups and carboxyl oxygen atoms. Figure 13 shows some of the molecules of the threonine crystal as viewed along the b-axis. In the vertical direction molecules are bound by hydrogen bonds between NH_3 groups and carboxyl oxygen atoms of the molecules above; horizontally the bonding is between hydroxyl groups and carboxyl oxygen atoms. A view of the structure looking down from the top along the c-axis is shown in Figure 14. The molecular spiral forming a continuous "chimney" containing methyl groups is reminiscent of the alanine structure. Both O—H$\cdots$O and N—H$\cdots$O bonds are clearly visible.

III. Studies of Crystals of Peptides and Some Related Compounds.

Early diffraction studies of peptides were restricted to recordings of X-ray powder patterns demonstrating the crystallinity of the compound and to determinations of unit cell dimensions and space groups for a few well-characterized crystals. Four of EMIL FISCHER's preparations of peptides containing some six to eighteen amino acid residues were shown to be crystalline [LENEL (17)] and powder patterns were also obtained for the glycine polymers, di-, tri-, tetra-, penta-, hexa-, and heptaglycylglycine (18). Unit cell dimensions were obtained for D-alanylglycine and for anhydrous α- and β-diglycylglycine (18).

BERNAL in his original survey of amino acids and related compounds determined the unit cell dimensions, space groups, and cleavage and optical properties of three crystalline modifications of glycylglycine (α, β, and γ), and of diglycylglycine dihydrate, and in a separate work (22) obtained similar information concerning glutathione. The cell dimensions and space group of β-diglycylglycine have been reported recently by HUGHES and MOORE (14).

I. 2,5-Diketopiperazine.

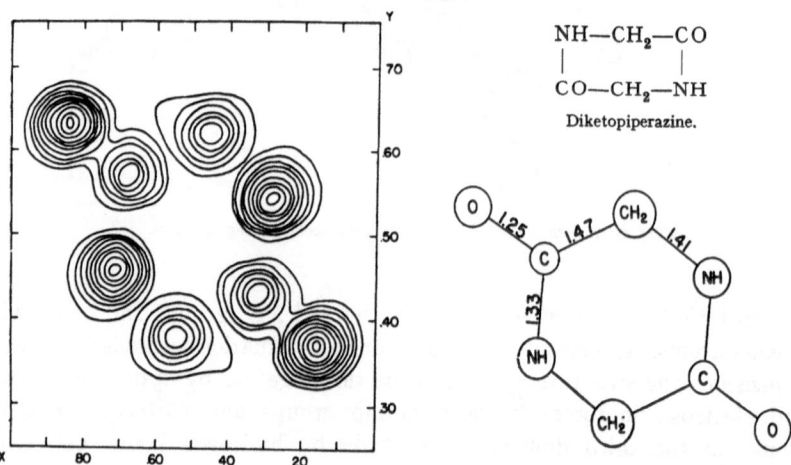

$$NH—CH_2—CO$$
$$|\qquad\qquad|$$
$$CO—CH_2—NH$$

Diketopiperazine.

Fig. 15. A projection on the c-face of the electron density distribution in a crystal of diketopiperazine.

Fig. 16. The dimensions of the diketopiperazine molecule. All bond angles are close to 120°.

The first crystalline peptide to be analyzed by X-ray diffraction was the cyclic dipeptide, diketopiperazine (glycine anhydride). The structure (5) was based on intensity data for 60 (hk0) and 22 additional (h0l) reflections. Trial and error methods established the trial structure which was refined by means of two-dimensional FOURIER plots of electron density distribution. (See Figure 15.)

The dimensions of the molecule are shown in Figure 16. It is essentially a planar hexagon, all bond angles being close to 120°. The length

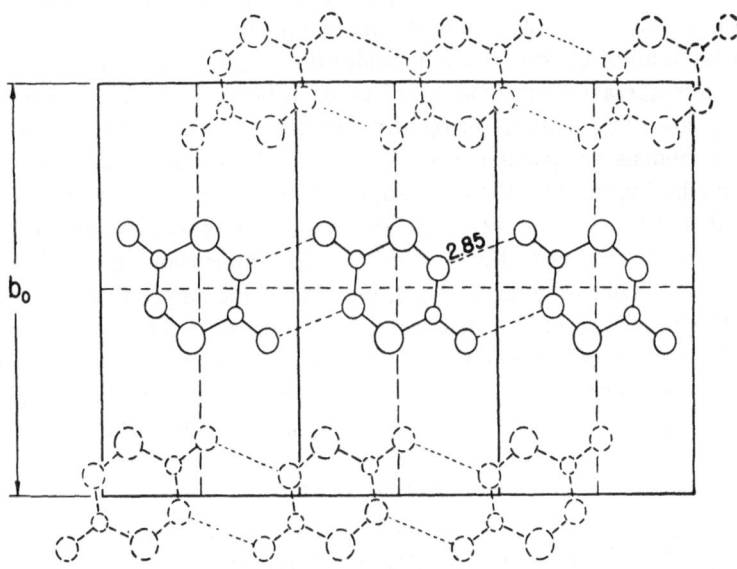

Fig. 17. The crystal structure of diketopiperazine. Hydrogen bonds (2,85 Å.) between NH groups and oxygen atoms bind the molecules together into continuous parallel chains or ribbons.

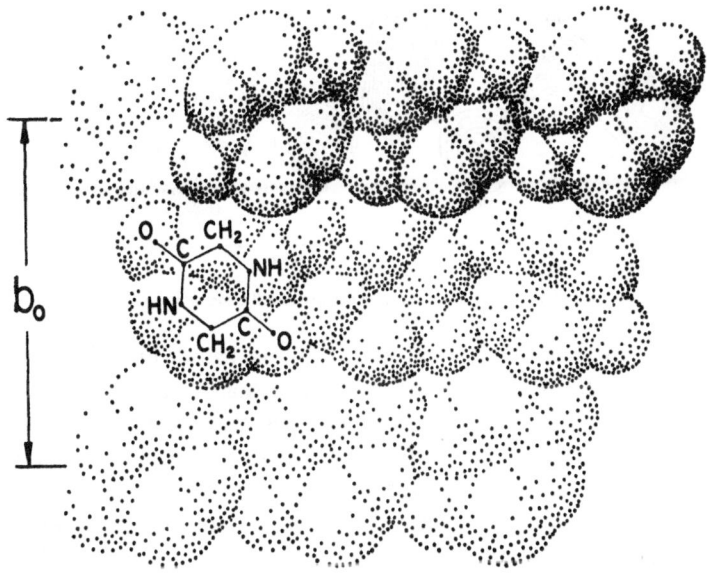

Fig. 18. A packing drawing of Fig. 17.

of the bonds from the keto carbon atom to the oxygen and nitrogen atom, 1,25 Å. and 1,33 Å., respectively, are in accord with those to be anticipated from other structures. The remaining ring distances, C—C, 1,47 Å., and C—N, 1,41 Å., are surprisingly short.

In contrast to crystals of the amino acids, crystals of diketopiperazine are remarkably simple. Indeed, this simplicity of structure is charac-teristic of all of the crystalline peptides thus far analyzed. The molecules of diketopiperazine are bound by hydrogen bonds (2,85 Å.) between NH groups and oxygen atoms into continuous flat chains or ribbons (Figure 17). These ribbons are packed together laterally into long parallel arrays as shown in Figure 18. Between adjacent chains the molecules are held together by VAN DER WAALS forces. The structure explains the observa-tion that the crystals cleave readily in two dimensions parallel to the molecular chains, but cannot be cleaved in a plane perpendicular to them.

The simplicity of the structure of this crystal is doubtless the result of the geometrical simplicity of the diketopiperazine molecule combined with the tendency of the hydrogen bonds formed between the NH groups and oxygen atoms to lie in or close to the molecular plane.

2. β-Glycylglycine.

The first linear peptide for which atomic positions have been deter-mined is β-glycylglycine*, $NH_2 \cdot CH_2 \cdot CO \cdot NH \cdot CH_2 \cdot COOH$. The structure,

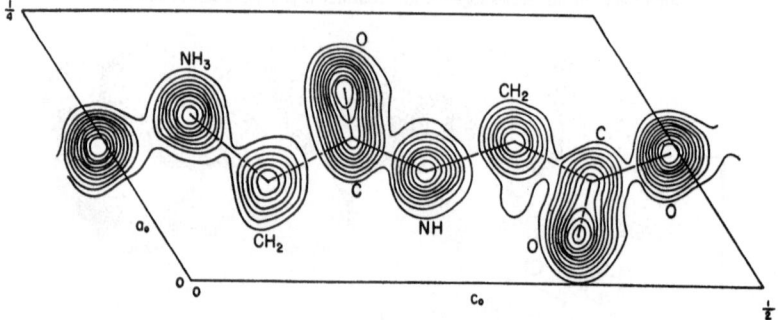

Fig. 19. A projection on the b face of the electron density distribution in a crystal of β-glycylglycine.

determined by HUGHES and MOORE (13), is based on intensity data for 158 (h0l) and 38 additional (0kl) reflections. Although, exclusive of hydrogen atoms, its molecule contains one more atom than threonine, the molecular arrangement in the crystal is so remarkably simple that a correct trial structure could be derived from a two-dimensional PATTERSON projection of interatomic vectors. Refinement of the parameters was carried out by successive FOURIER plots of electron density and by the method of least squares. Figure 19 shows a plot of electron density pro-

---

* "β-"designates one of the three crystallographic modifications mentioned on p. 330.

jected along the b-axis. The peak at the extreme left represents a carboxyl oxygen atom of an adjacent molecule.

The dimensions of the molecule of glycylglycine as determined in this investigation (*13*) are shown in Figure 20. The molecule is coplanar except for the nitrogen atom which falls 0,64 Å. out of the molecular plane. Within the limits of error of the determination, the bond lengths and bond angles appear to be in agreement with those found in other structures. The arrangement of the three hydrogen bonds about the terminal nitrogen atom indicates strongly that the molecule is in the form of a dipolar ion in the crystal. This view is also supported by other features of the structure.

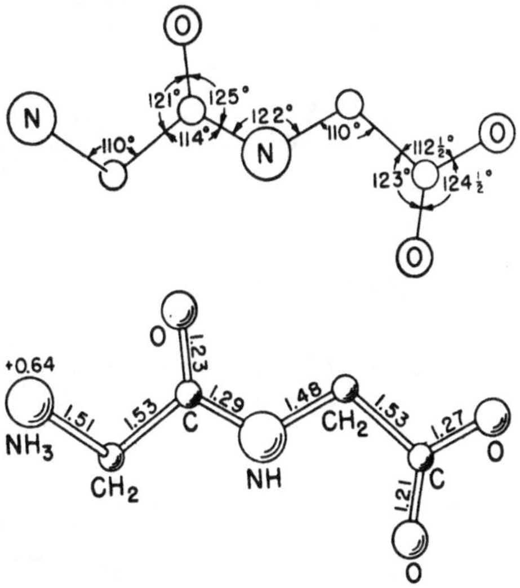

Figure 21 shows the way in which molecules of glycylglycine are arranged in the crystal. Hydrogen bonds between nitrogen and oxygen atoms hold the structure together in all three directions. Two types of bonds extend vertically, those joining terminal NH_3 groups with keto oxygen atoms (2,81 Å.), and those joining amid NH groups

Fig. 20. The dimensions of the molecule of glycylglycine.

with carboxyl oxygen atoms (3,07 Å.). Bonds between terminal NH_3 and carboxyl groups knit all molecules together in the horizontal direction.

3. N-Acetylglycine.

Although N-acetylglycine, $CH_3 \cdot CO \cdot NH \cdot CH_2 \cdot COOH$, is not a true peptide, it contains a peptide bond. A detailed determination of its structure was therefore made in order to add to the rather meager information available concerning the dimensions of this fundamental link in the polypeptide chain of proteins. The investigation by CARPENTER and DONOHUE (*4*) was based on good intensity data for about 800 observed reflections. The trial structure was derived from a two-dimensional PATTERSON projection of interatomic vectors, and refined by means of two- and three-dimensional FOURIER plots of electron density distribution. Figure 22 shows a section of the three-dimensional electron density plot taken close to the plane of the molecule.

The dimensions of the molecule of acetylglycine are shown in Figure 23. All atoms of the molecule, exclusive of hydrogen atoms, are within 0,1 Å.

of a common plane. An even closer approach to planarity exists in the two halves of the molecule: the carbon and oxygen atoms of the

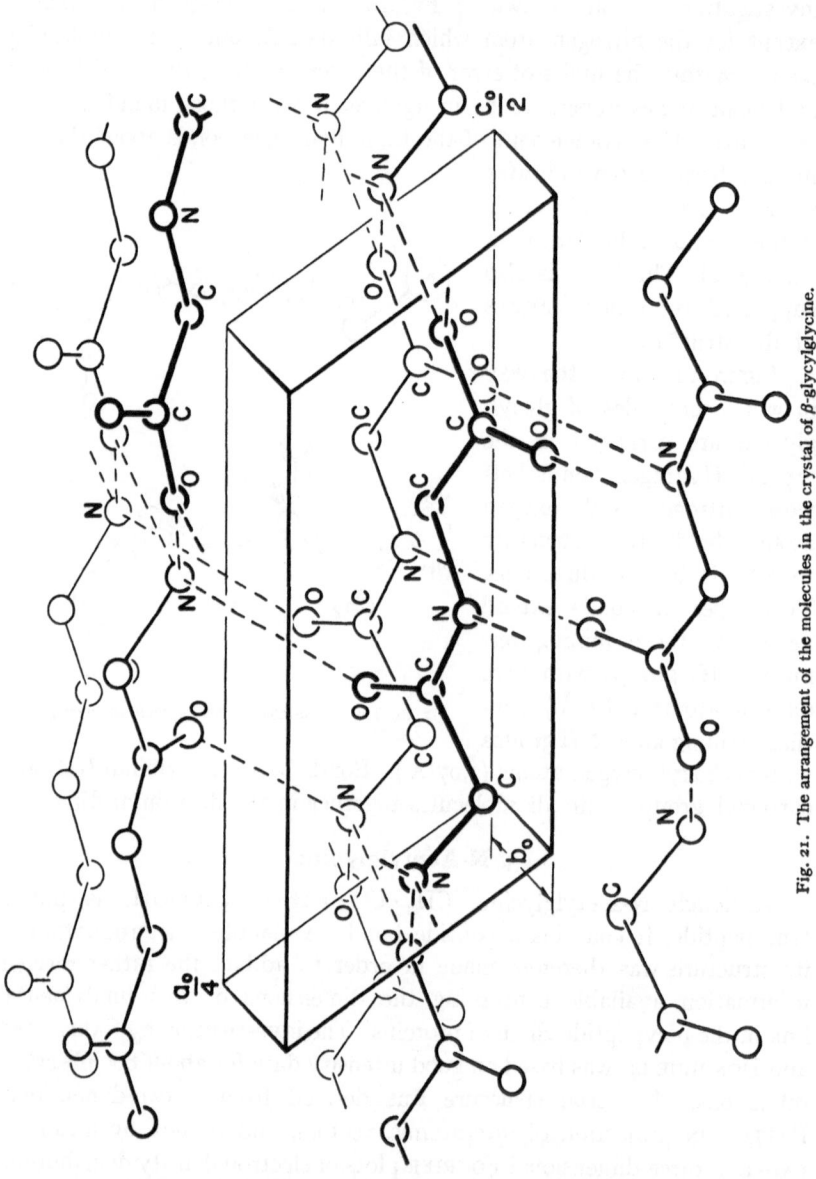

Fig. 21. The arrangement of the molecules in the crystal of β-glycylglycine.

—CH$_2$COOH half are less than 0,001 Å. from a common plane, and the carbon, oxygen, and nitrogen atoms of the CH$_3$CONH— half are less than 0,005 Å. from another common plane.

The length of the carbon-nitrogen peptide bond, 1,32 Å., is close to that found in diketopiperazine (1,33Å.), urea (20, 27), and other crystals showing resonance between the structures

$$-\underset{\underset{O}{\parallel}}{C}-NH- \quad \text{and} \quad -\underset{\underset{O^-}{\mid}}{C}=\overset{+}{N}H-$$

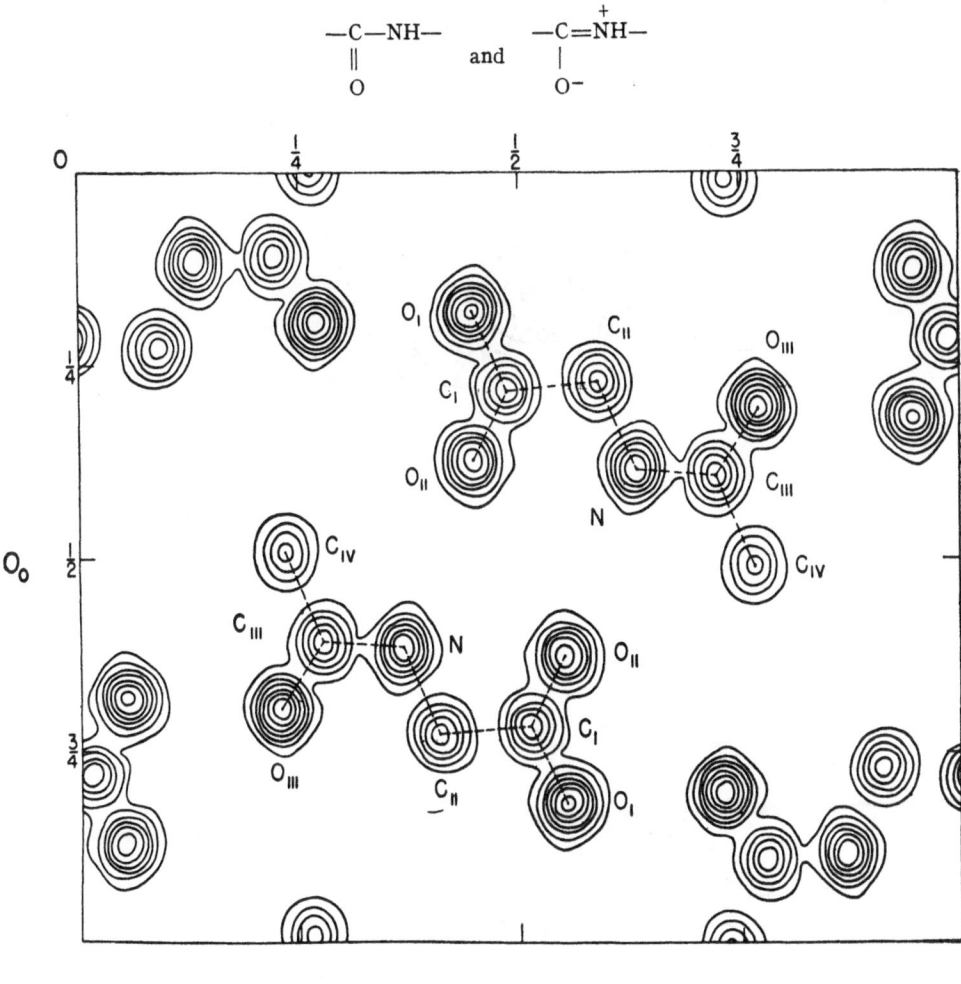

Fig. 22. A section of the three dimensional plot of electron density distribution in the crystal of N-acetylglycine.

It is interesting to compare the structure of *N*-acetylglycine with that of β-glycylglycine. A relatively large probable error is associated with the value 1,29 Å. found for the C—N peptide bond in β-glycylglycine, so that the discrepancy between this length and 1,32 Å. found in acetylglycine is probably not significant. Other differences in the structure, however, are certainly well beyond the limits of error. Since *N*-acetylglycine is not a dipolar ion, the hydrogen atom of the carboxyl group

is associated with one of the oxygen atoms; this is indicated by the considerable difference in the lengths of the C—O bonds (1,19 Å. and 1,31 Å.)
and by the distribution of the bond angles about the carboxyl carbon

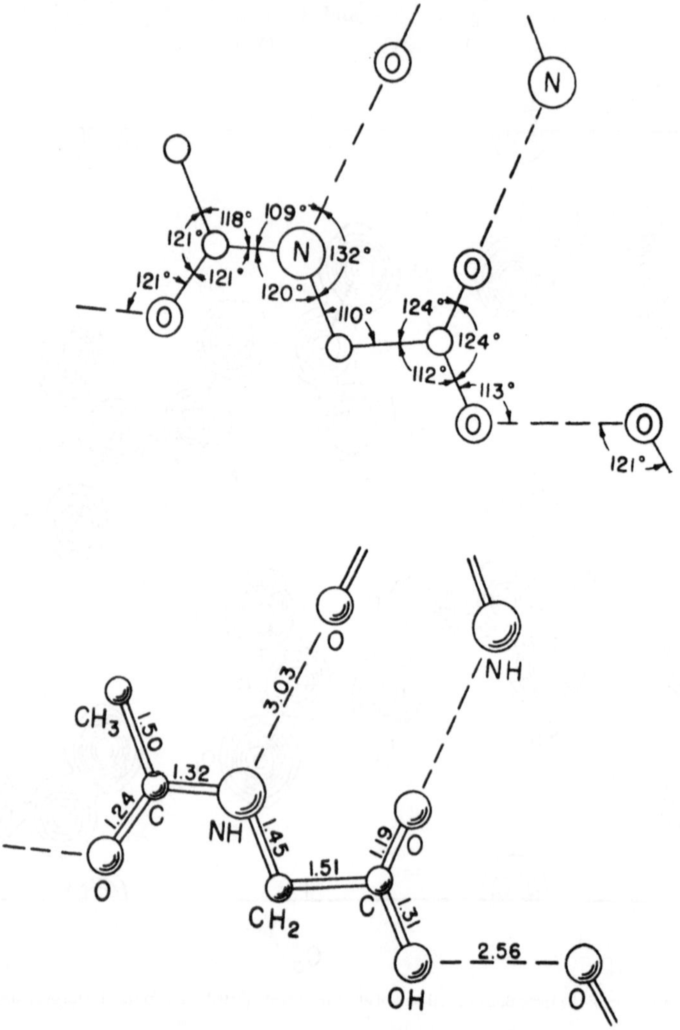

Fig. 23. The dimensions of the molecule of N-acetylglycine.

atom. The bond angles around the keto carbon atom in β-glycylglycine
are in good agreement with the high degree of double bond character
associated with the C—N peptide bond. It is not clear as to just why this
sort of difference in the bond angles adjacent to the C—O bond is not
found in acetylglycine. It is to be noted that in both crystals the arrange-

ment around the peptide N—C bond is *trans* with respect to the C—N—C—C peptide "chain".

The packing of the molecules in the crystal of acetylglycine is shown in Figure 24. Hydrogen bonds between keto oxygen atoms and hydroxyl groups bind the molecules into continuous chains which are linked together longitudinally to form two-dimensional sheets by hydrogen bonds between carboxyl oxygen atoms and —NH groups. The molecular layers appear to be held together almost entirely by VAN DER WAALS forces.

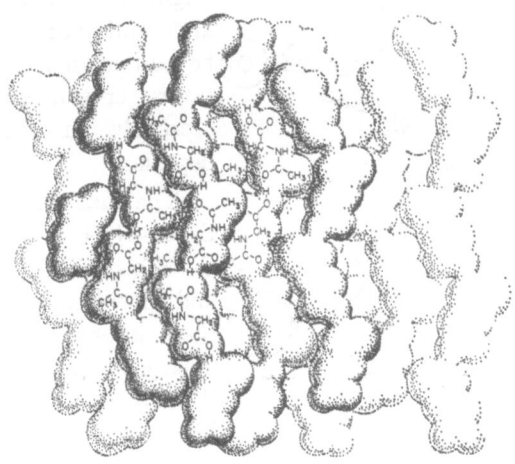

Fig. 24. A drawing illustrating the packing of the molecules in the crystal of N-acetylglycine.

IV. Inferences Concerning Protein Structure.

The crystal structure studies of amino acids and peptides provide specific information concerning the interatomic distances and bond angles occurring in these compounds. As the results of these diffraction studies accumulated, they were used (*6, 7*) from time to time as a basis for assigning probable spatial relationships to the carbon, nitrogen, and oxygen atoms constituting the polypeptide chains of proteins. Dimensions of the fully extended polypeptide chain as derived from the most recent crystal structure studies (*8*) of alanine, threonine, serine, acetylglycine, and glycylglycine, are shown in Figure 25. Around the nitrogen atom the bond angles are close to 120°. The revised length of the αC—N bond (1,47 Å.) is in complete agreement with the sum of the covalent radii. The bond between the keto carbon and the nitrogen atom is 1,32 Å. The proper choice of bond angles around the keto carbon atom is still a bit in doubt. Those shown in Figure 25 represent values intermediate between the angles found in glycylglycine and acetylglycine, with weighting slightly in favor of the latter structure which was based on more adequate data and three-dimensional refinement. In all amino acid structures recently determined the C—C bond has been found slightly but significantly shorter than the classical 1,54 Å. These findings are expressed by the distance 1,53 Å. for the C—C distance in the peptide chain.

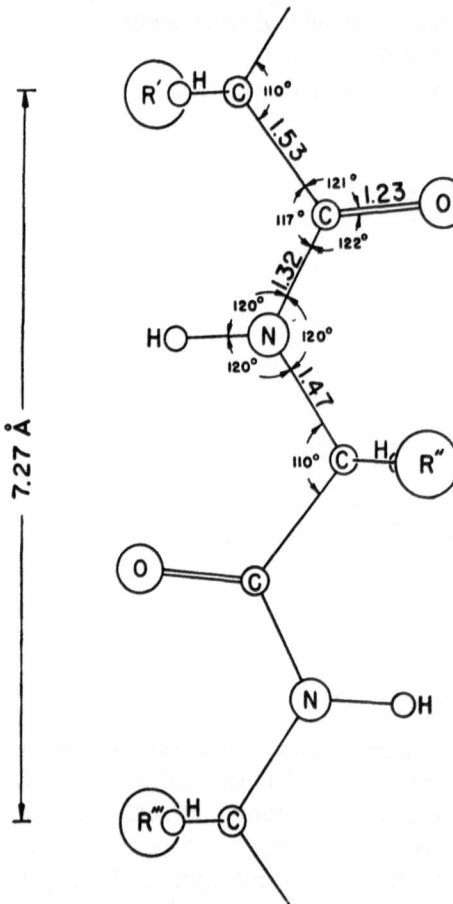

Fig. 25. Dimensions of the fully-extended polypeptide chain as derived from crystal structure studies of amino acids and peptides.

There seems to be little doubt that the geometry of the polypeptide chain in proteins differs little from that here represented. Slight revisions of some features and confirmation of others may be anticipated as a result of future crystal structure work, especially on additional linear peptides.

The *trans* configuration of the C—C—N—C chain about the C—N bond in both acetylglycine and glycylglycine suggests that there is probably a strong preference for this configuration in the polypeptide chain in proteins. The length (1,32 Å.) of the C—N bond indicates about 40 per cent double bond character. It is therefore to be expected that rotation about this bond will be severely restricted, a conclusion which is confirmed by the planarity of the peptide molecules in the crystals.

We have seen the extent to which the crystallization of amino acids and peptides is influenced by the size and shape of the molecules, the distribution of electrical charges, and the formation of hydrogen bonds between nitrogen and oxygen atoms. The same considerations may be expected to determine the configuration of protein molecules and their relative arrangements in crystals and in plant and animal tissues, perhaps also in the relations of enzyme to substrate and in the formation of antibodies.

Acknowledgement. — The drawings used to illustrate this article were prepared by Mrs. Betty Bell and Miss Lillian Casler.

References.

Books.

B 1. BRAGG, W. L.: The Crystalline State. I. General Survey. London: Bell & Sons, Ltd. 1933.

B 2. BUERGER, M. J.: X-Ray Crystallography. New York: J. Wiley & Sons, Inc. 1942.

B 3. BUNN, C. W.: Chemical Crystallography. Oxford: Clarendon Press. 1945.

B 4. EWALD, P. P.: Die Erforschung des Aufbaues der Materie mit Röntgenstrahlen. Handbuch der Physik, Bd. XXIII/2. Berlin: Springer-Verlag. 1933.

B 5. LONSDALE, K.: Crystals and X-Rays. New York: Van Nostrand Co., Inc. 1949.

B 6. MEYER, K. H. u. H. MARK: Der Aufbau der hochpolymeren organischen Naturstoffe. Leipzig: Akadem. Verlagsges. 1930.

B 7. PAULING, L.: The Nature of the Chemical Bond. 2nd. ed. Ithaca: Cornell University Press. 1940.

B 8. WYCKOFF, R. W. G.: The Structure of Crystals. New York: Chemical Catalog Co., Inc. 1931. Supplement. New York: Reinhold Publ. Corp. 1935.

B 9. — Crystal Structures. New York: Interscience Pub., Inc. 1948.

Journal Articles.

1. ALBRECHT, G. and R. B. COREY: The Crystal Structure of Glycine. J. Amer. chem. Soc. 61, 1087 (1939)

2. ALBRECHT, G., G. W. SCHNAKENBERG, M. S. DUNN and J. D. MCCULLOUGH: Quantitative Investigations of Amino Acids and Peptides. XII. Structural Characteristics of Some Amino Acids. J. physic. Chem. 47, 24 (1943).

3. BERNAL, J. D.: The Crystal Structure of the Natural Amino Acids and Related Compounds. Z. Kristallogr., Mineral., Petrogr. 78, 363 (1931).

4. CARPENTER, G. B. and J. DONOHUE: The Crystal Structure of N-Acetylglycine. J. Amer. chem. Soc. 72, 2315 (1950).

5. COREY, R. B.: The Crystal Structure of Diketopiperazine. J. Amer. chem. Soc. 60, 1598 (1938).

6. — Interatomic Distances in Proteins and Related Substances. Chem. Reviews 26, 227 (1940).

7. — X-Ray Studies of Amino Acids and Peptides. Adv. Protein Chem. 4, 385 (1948).

8. COREY, R. B. and J. DONOHUE: Interatomic Distances and Bond Angles in the Polypeptide Chain of Proteins. J. Amer. chem. Soc. 72, 2899 (1950).

9. DONOHUE, J.: The Crystal Structure of DL-Alanine. II. Revision of Parameters by Three-Dimensional Fourier Analysis. J. Amer. chem. Soc. 72, 949 (1950).

10. DUNITZ, J. D.: X-Ray Structure Analysis of Complex Organic Compounds. Research 2, 6 (1949).

11. HENGSTENBERG, J. u. F. V. LENEL: Die Struktur des Glycins $NH_2 \cdot CH_2 \cdot COOH$. Z. Kristallogr., Mineral., Petrogr. 77, 424 (1931).

12. HUGHES, E. W.: The Crystal Structure of Melamine. J. Amer. chem. Soc. 63, 1737 (1941).

13. HUGHES, E. W. and W. J. MOORE: The Crystal Structure of β-Glycylglycine. J. Amer. chem. Soc. 71, 2618 (1949).

14. — — β-Diglycylglycine. Acta Cryst. 3, 313 (1950).

15. KITAYGORODSKY, A.: The Structure of Amino Acetic Acid. Acta physicochim. URSS 5, 749 (1936).

16. KRATKY, O. u. H. MARK: Anwendung physikalischer Methoden zur Erforschung von Naturstoffen: Form und Größe dispergierter Moleküle. — Röntgenographie. Fortschr. Chem. organ. Naturstoffe 1, 255 (1938).

17. LENEL, F. V.: Untersuchung der Polypeptide E. FISCHERS mit Röntgenstrahlen. Naturwiss. 19, 19 (1931).

18. — Die Struktur der einfachen Polypeptide des Glycins. Z. Kristallogr., Mineral., Petrogr. 81, 224 (1932).

19. LEVY, H. A. and R. B. COREY: The Crystal Structure of DL-Alanine. J. Amer. chem. Soc. 63, 2095 (1941).

20. MARK, H. u. K. WEISSENBERG: Röntgenographische Bestimmung der Struktur des Harnstoffs und des Zinntetrajodids. Z. Physik 16, 1 (1923).

21. MEYER, C. E. and W. C. ROSE: The Spatial Configuration of α-Amino-β-hydroxy-n-butyric acid. J. biol. Chemistry 115, 721 (1936).

21a. MÖLLER, Chr. K.: The Structure of DL- and D-Leucine. Acta Chem. Scand. 3, 1326 (1949).

22. PIRIE, N. W. and J. D. BERNAL: Cuprous Glutathione. With a Note on the Crystallography of Glutathione. Biochemic. J. 26, 75 (1932).

23. SHOEMAKER, D. P., J. DONOHUE, V. SCHOMAKER and R. B. COREY: The Crystal Structure of L_s-Threonine. J. Amer. chem. Soc. 72, 2328 (1950).

24. STOSICK, A. J.: The X-Ray Investigation of Copper DL-α-Aminobutyrate. J. Amer. chem. Soc. 67, 362 (1945).

25. — The Crystal Structure of Nickel Glycine Dihydrate. J. Amer. chem. Soc. 67, 365 (1945).

26. WRIGHT, B. A. and P. A. COLE: Preliminary Examination of the Crystal Structure of L-Proline. Acta Cryst. 2, 129 (1949).

27. WYCKOFF, R. W. G. and R. B. COREY: Spectrometric Measurements on Hexamethylene Tetramine and Urea. Z. Kristallogr., Mineral., Petrogr. 89, 462 (1934).

(Received, November 30, 1950.)

Some Aspects of Enzyme Chromatography.

By L. ZECHMEISTER, Pasadena, and M. ROHDEWALD, Bonn.

With 6 Figures.

I. General Remarks on Enzyme Chromatography.

Introduction. The articles appearing in our volumes usually present either a survey of a field in which a rather complete picture can be given, or they discuss less explored fields in which it is hoped that a survey offered at an early stage of research may have a stimulating effect on further progress. The present article belongs to the second category.

The term *"Chromatography"* includes those processes which allow the resolution of solute mixtures by selective fixation and liberation on a solid surface or support, with the aid of a fluid streaming in a definite direction. The methods of chromatography are now so generally known that it suffices to mention a few reference works (*105, 103, 75, 95, 43, 8, 7*). Comprehensive surveys on protein and enzyme chromatography were published by Grundmann (*29*) in 1940 and by Turba (*85*) in 1948/49.

Since enzyme chemistry as a whole constitutes a special section of protein chemistry, enzyme chromatography is protein chromatography applied to a specific and biologically active subclass of high-molecular catalysts.

The chromatography of low-molecular components of enzyme systems such as co-enzymes will not be treated in the present review. Cf., for example, contributions by von Euler and Adler (*18*), Das (*14*), von Euler and Schlenk (*23*), Sumner, Krishnan and Sisler (*77*), von Euler and Fonó (*19*). However, the latter paper in which the resolution of β-glycerophosphatase into an apo-enzyme and a co-enzyme had been described, will be discussed on p. 344.

According to Willstätter (*97, 96*), Brücke in 1861 (*4*) was the first investigator who eliminated an enzyme from a solution by direct treatment with an adsorbent, although previous authors had noticed that precipitates formed in a solution may occlude the enzyme content. As is well known, extensive use of enzyme adsorption was made by Willstätter and his colleagues (*97*) in the course of which, however, no chromatographic methods were employed. One of the remarkable features of that monumental work is the characterization of various adsorbent preparations which showed highly differentiating properties when in contact with enzyme solutions.

The following three, newer examples, selected from many others, may illustrate the great variability of adsorption effects. *a)* Thyroglobulin can be resolved on aluminum hydroxide columns (precipitated on kieselguhr) into several fractions containing the same globulin but various amounts of iodine [Riviere, Gautron and Thély (*61*)]. — *b)* Celite scarcely adsorbs trypsin while non-enzymatic proteins are retained [Schormüller (*65*)]. — *c)* A separation of thymus nucleoproteids from the protein component can be carried out on alumina or floridin columns [von Euler and Hahn (*20*)].

Molecular Weight and Adsorption Affinity. The adsorption affinity of proteins, like that of lower-molecular compounds, is subject to variations within broad limits which can be used as a basis for separations.

In several instances the observation was made that the adsorbability in a given system decreases with increasing molecular weight, perhaps because the size of the solute molecules surpasses the average diameter of the active spots. Thus, it was found by MORING-CLAESSON (*53*), in frontal analysis studies on carbon, that the adsorption affinity of pure, crystalline egg albumin was weaker than that of most amino acids and peptides tested. Likewise, according to TISELIUS (*80*), casein shows less adsorbability on charcoal than do its digestion products. Furthermore, it was reported by CLAESSON and CLAESSON (*9*) that within the respective classes of nitrocellulose, polyvinyl acetate, and synthetic rubber the adsorbability decreases with lengthening of the chain. On the other hand, INGELMAN and HALLING (*33*) found that the adsorption affinity for active carbon passes through a maximum value in the series leading from glucose to the highest dextrans.

It should be noted that, in spite of some of the mentioned phenomena, even very high-molecular, protein-like substances are accessible to chromatography [TISELIUS (*83*)]. RILEY (*58*) carried out an accumulation of the chicken tumor agent *(Rous sarcoma virus)* on celite, while a "salting out adsorption" (p. 345) on filter paper pulp columns of THEILER's virus was recently reported by LEYON (*44*). RILEY et al. (*60*) have even adapted chromatography to (in part) enzymatically active particulates ranging from virus to mitochondrial and bacterial size.

Rate of Adsorption. On the practical side, one of the pitfalls in chromatography of high-molecular compounds may be slowness of the fixation process in some instances as recently studied by CLAESSON (*10, 9*). SCHWIMMER and BALLS (*67, 68*) found that a wheat starch, at 0°, adsorbed only 50% of a purified α-amylase preparation (obtained from germinated barley) from 40% methanol solution within 30 minutes, while the corresponding figure even after 90 minutes amounted to only 70%. Although delayed adsorption is by no means a general phenomenon in the class of proteinic substances, this possibility should not be disregarded in pertinent experiments.

Possible Chemical Conversions During Chromatography. It is a fundamental requirement of all chromatographic procedures that the analysis should not modify the molecular weight, structure or configuration of the solute or, to be more exact, that all components of the starting material should be recovered in unaltered state at the end of the experiment. Whether or not this requirement was fulfilled in a given instance, can be decided easily in the case of the adsorption and elution of a low-molecular compound. In contrast, with complicated high-molecular substances, elaborate physical-chemical tests will be needed in order to ascertain that no irreversible interaction between the adsorbent and the solute and no irreversible processes of another kind such as autoxidation have taken place.

The situation is considerably simpler in the chromatographic resolution or purification of proteinic substances possessing specific immunological or other properties which can be checked rapidly and with high accuracy. The same statement is valid for enzymes. Unchanged catalytic effect will justify the claim that the enzyme molecules have remained unaltered during the chromatographic treatment. On the other hand, a decrease of the enzymatic potency may, possibly, indicate changes in the structure so mild that they might have been overlooked when working with non-enzymatic proteins.

Of course, the (partial) destruction of an enzyme during an adsorption experiment should be sharply differentiated from the elimination of a component, a co-enzyme, for example, which passes into the chromatographic filtrate. The inactivation is then reversible, and a re-introduction of the component should restore full enzyme potency. Under favorable circumstances such a reversible inactivation may even differentiate between two enzyme systems, only one of which might have been affected. Thus, the activity of a glycerophosphatase preparation (*ex* mucosa of the small intestine of the calf) was inhibited by adsorption on an alumina column while the ribonuclease potency of the solution remained unaltered [VON EULER and FONÓ (*19*)].

The preservation or irreversible destruction of some delicate active groups in enzyme molecules during chromatographic procedures may depend strongly on such modifications of the adsorbent which perhaps seem a priori to be irrelevant. Thus, some tissue extracts yielded active phosphorylase preparations after a treatment with "C_β" aluminum hydroxide [prepared according to WILLSTÄTTER, KRAUT and ERBACHER (*98, 99*)] while the application of the "C_γ" modification of the same adsorbent frequently resulted in the inactivation of this enzyme [SHAPIRO and WERTHEIMER (*69*)]. The influence of accompanying solutes on similar effects can be illustrated by the observation that "C_γ" aluminum hydroxide destroyed the phosphorylase activity of some heart extracts while it did not inactivate similar enzyme solutions obtained from red and white muscle extracts. [MIRSKI and WERTHEIMER (*50*); cf. CORI, CORI and SCHMIDT (*12*).]

II. Methods.

Exchange Adsorption.

The principle of exchange adsorption was applied recently to egg white albumin (from hen's eggs) by SOBER, KEGELES and GUTTER (*73*) who made use of the cation exchanger Dowex 50. The schlieren scanning pattern showed, in accordance with the well-known results of electrophoretic experiments, the presence of three major components. No alteration in the charge distribution of the components (no marked denaturation) took place, although the ultracentrifuge indicated small changes in the initial size distribution. Similar methods are now available for the treatment of enzymes as shown for ribonuclease and lysozyme by HIRS, STEIN and MOORE (32 a).

"Salting out" Adsorption.

An important development also in enzyme chromatography was initiated by the observation of the "salting out" adsorption. In 1948 TISELIUS (81–83) discovered the enhancement of protein adsorption by the presence of certain inorganic salts such as ammonium sulfate or potassium phosphate. Some materials which otherwise do not display any marked affinity for proteins, thus become effective adsorbents at moderately high salt concentrations which amount, for example, to $1/3$ of that required for salting out the protein. Elution takes place when the adsorbent is washed with a weaker salt solution or with pure water.

It was reported recently by SHEPARD and TISELIUS (70) that the various proteins behave very differently in "salting out" chromatographic treatments. The albumin and globulin fractions of human sera were distinctly separated on silica gel plus supercel under certain conditions. For a number of other interesting observations the reader must be referred to the original paper.

Evidently, the "salting out" principle will likewise be applicable in the resolution of certain protein and enzyme mixtures.

Chromatography of Enzymes on Filter Paper.

When adsorbed and moved on filter paper under suitable conditions, many enzymes will retain their activity. This was demonstrated for urease quantitatively by means of manometric readings [FRANKLIN and QUASTEL (26)].

As a forerunner to modern methods of paper chromatography, the preparation of "chromograms" by "capillarization", according to GRÜSS (30, 31), should be mentioned. The solution is spread out on filter paper in a drum-like arrangement, and segments of the paper are treated individually for the detection of certain enzymes. For example, by means of p-phenylenediamine and hydrogen peroxide the presence of peroxidase was demonstrated; on filter paper which had been pre-treated with starch solution and dried, amylase was detected by exposure to iodine vapors which did not develop any coloration where the enzyme had cleaved the starch.

JONES and MICHAEL (34) pointed out that protein spots or zones can be located on paper strips or cellulose columns by means of a suitable acid dye (e. g., Solway Purple) which combines with intact protein while neither the adsorbent nor peptides and amino acids are stained [cf. Klotz, Triwush and Walker (37)]. On the other hand, it was reported simultaneously by Carroll (5) that while native albumin does show an affinity for some dye anions, pepsin and other proteolytic enzymes do not have such combining capacity. Hence, some dyes can be used as indicators for following the course of proteolytic cleavage. On the basis of these observations Simonart and Chow (71) propose the following procedure for the detection of proteinase spots: The paper strip carrying the enzyme is treated first with a casein solution and, half a day later, with Gentian Violet, then with water, iodine, water and alcohol. The

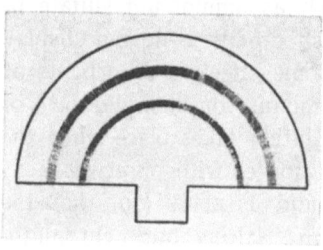

proteinase spot will be colorless or mauve, in contrast to its violet surroundings which contain unchanged casein. Analogous operations were carried out for the detection of amylase.

As is well known, the two-dimensional chromatographic technique, invented by Consden, Gordon and Martin (11) is exerting a profound influence on biochemical research.

Fig. 1. The application of fan-form filter paper for the separation of colored proteins in the presence of salt (Tiselius). [From: Ark. Kem., Mineral. Geol. 26 B, Nr. 1 (1948).]

This method has been subject to a number of modifications during the last few years. Thus, Williams and Kirby (94) propose the principle of capillary ascent, instead of descent; Yanofsky, Wasserman and Bonner (101) use a revolving drum to which the paper is attached and onto which the solution is fed by a capillary tube. A fan-shaped filter paper (Fig. 1) is successfully employed by Tiselius (81) for the separation of colored proteins in the presence of salts; the usefulness of this simple technique was demonstrated with phycoerythrin, a red protein from Ceramium.

Recently, it was recognized by several authors that great advantages could be derived from such modifications of paper chromatography which would permit purifications and resolutions on a larger scale. Thus, Hedén (31a) working with protein hydrolysates on a 5-mg. scale developed 20 spots simultaneously on the same paper sheet. It had been pointed out earlier by Datta, Dent and Harris (15, 16) that a great number of two-dimensional runs can be operated by means of the equipment represented in Fig. 2. The solution is placed near the corner of each sheet and the frame holding the sheets is placed in a dish which contains the first solvent (the whole is enclosed in a box). After the first development upward has taken place, the sheets while still in the frame, are dried, turned, then run with another edge downward in a second developer, and finally dried again in the frame.

Mitchell and Haskins (52) must be credited with the first use of filter paper chromatography on large scale. They devised for this purpose a "chromatopile" (Fig. 3) which was employed by Mitchell, Gordon and Haskins (51) for the resolution of some enzyme mixtures.

A solution of 300 mg. of a good commercial Taka diastase sample in 20 ml. of water was taken up in twenty 9-cm. sheets of Whatman No. 1 paper. After drying, these disks were incorporated at 40 disks from the top of a 500-sheet pile, and developed with a 60%-saturated ammonium sulfate solution at p_H 6,5. When, after 8 hours, the column had taken up 275 ml. of the liquid in 450 disks, alternate sections of five disks each were cut into $4 \times 0,5$-cm. strips which, for the estimation, were dropped directly into the respective substrate solutions.

The movement of the enzymes on the pile is influenced by the presence of other proteins.

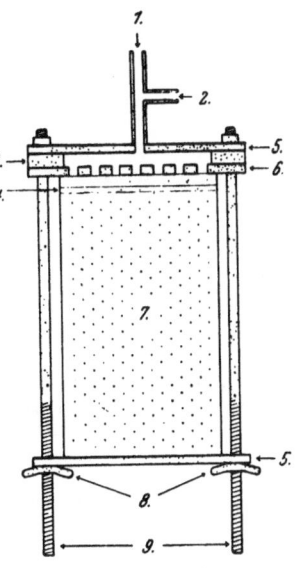

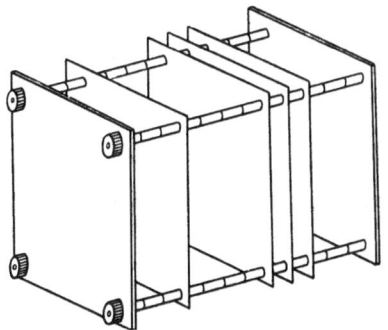

Fig. 2. Apparatus for mass production two-way paper chromatography (Datta, Dent and Harris). [From: Biochemical J. 46, XLII (1950).]

Fig. 3. Chromatopile (Mitchell and Haskins). 1. Connection for a rubber tube for filling the siphon 2. Connection for the siphon tube. 3. Rubber gasket. 4. Filter paper disks containing the sample. 5. Stainless steel plates. 6. Perforated stainless steel plate. 7. Filter paper disk pile. 8. Wing nuts. 9. Bolts at four corners of steel plates. [From: Science 110, 278 (1949).]

The effect obtained is demonstrated in Fig. 4 referring to the separation of adenosine deaminase, amylase, and phosphatase (p. 348).

For some purposes it seemed to be desirable to fill the classical chromatographic tube with filter paper disks in order to handle larger amounts of material in a closed system. It was found (104) that in such systems channeling and irregularities in the flow can be avoided if the size of the disks is adapted to the average diameter of the cylindrical tube by cutting the paper with unusually high precision (0,01–0,005 cm.), using a stainless steel die and a corresponding punch. Solutions, under suction, then flow evenly through the paper column, and the duration of the chromatographic experiment does not exceed that usually required for a powder column of similar size. The packing and removing of the filter paper disks is shown in Fig. 5 (p. 348).

Filter paper columns are now also being used for the electrophoretic resolution of protein mixtures. A bibliography of this field was given by TISELIUS (*83*) [cf. also CREMER and TISELIUS (*13*)]; and comprehensive photographs referring to normal and pathological human sera have been

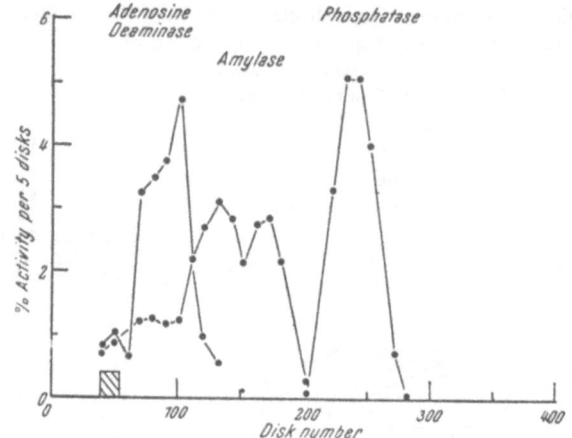

Fig. 4. Separation of adenosine deaminase, amylase and phosphatase on the chromatopile (MITCHELL, GORDON and HASKINS). (Salt concentrations in % saturation at 10°; disk 30 : 14%; disk 120 : 20%; disk 202 : 40%; and disk 280 : 50%). [From: J. biol. Chemistry *180*, 1071 (1949).]

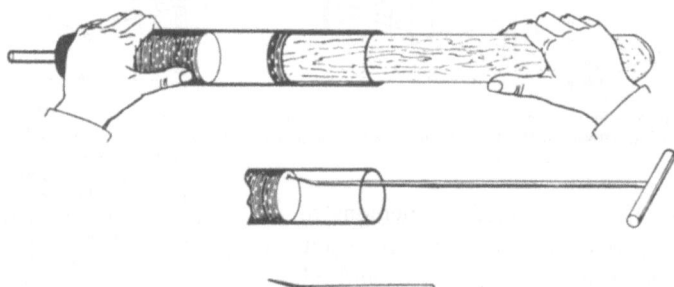

Fig. 5. Packing and removing a filter paper disk column, using a glass chromatographic tube (*104*). [From: Science *113*, 35 (1951).]

published by TURBA and ENENKEL (*86*). Recently, WALLENFELS and VON PECHMANN (*89*) succeeded in the electrophoretic separation of several enzymes on a filter paper sheet; cf. also the "retention analysis" as proposed by WIELAND and WIRTH (*93*).

Location of Enzyme Zones on the Column by Brushing with Color Reagents.

As is well known, the colorless zones of many compounds can be detected on the column by painting a streak along the whole main axis with a camel hair brush dipped into a suitable reagent; a color change is then observed where the streak crosses the zone (*106*). Since the surface of a moist enzyme adsorbate is able to display its activity when in contact

Table 1. Examples of the Application of the Brush Method in the Location of Enzyme Zones on the Alumina Column (*107, 62*).

Enzyme	First brush	Second brush	Observed coloration	Site of the coloration
Amylase	Starch	Iodine in KI	dark blue	enzyme-free zone
Saccharase	Saccharose	FEHLING's reagent, with hot air or N_2 (from nozzle)	brick red	enzyme zone
α-Glucosidase	Phenyl-α-D-glucoside	2,6-Dibromo-quinone-chlor-imide + alkali	blue	enzyme zone
β-Glucosidase	Phenyl-β-D-glucoside	2,6-Dibromo-quinone-chlor-imide + alkali	blue	enzyme zone
α-Galactosidase	Phenyl-α-D-galactoside	2,6-Dibromo-quinone-chlor-imide + alkali	blue	enzyme zone
β-Galactosidase (Lactase)	Phenyl-β-D-galactoside	2,6-Dibromo-quinone-chlor-imide + alkali	blue	enzyme zone
Phosphorylase (ex potatoes)	Glucose-1-phosphate	Iodine in KI	blue	enzyme zone
β-Glucuronidase	Phenolphthaleine-mono-β-glucuronide	NaOH	pink or red	enzyme zone
Phosphatase	Sodium-phenol-phthaleine-phosphate	NaOH	pink or red	enzyme zone
Phosphatase	Monophenyl-mono-sodium-phosphate	2,6-Dibromo-quinone-chlor-imide + alkali	blue	enzyme zone
Esterase	Ethyl butyrate	Na_2CO_3 + phe-nolphthaleine	disappearing of red color	enzyme zone
Catalase	H_2O_2	Titanium sulfate in sulfuric acid	yellow	enzyme-free zone
Catalase	H_2O_2	KI + sulfuric acid + starch	dark blue	enzyme-free zone
Mono- and poly-phenolase	p-Cresol	—	reddish-violet	enzyme zone
Mono- and poly-phenolase	Pyrogallol	—	reddish-brown	enzyme zone
Urease	Urea	NESSLER's reagent	brownish-yellow	enzyme zone
Urease	Urea	Phenolphthaleine	pink	enzyme zone*
Hydroxynitrilese	Amygdalin	Benzidine-copper acetate	blue	enzyme zone

* On a $Ca_3(PO_4)_2$ column.

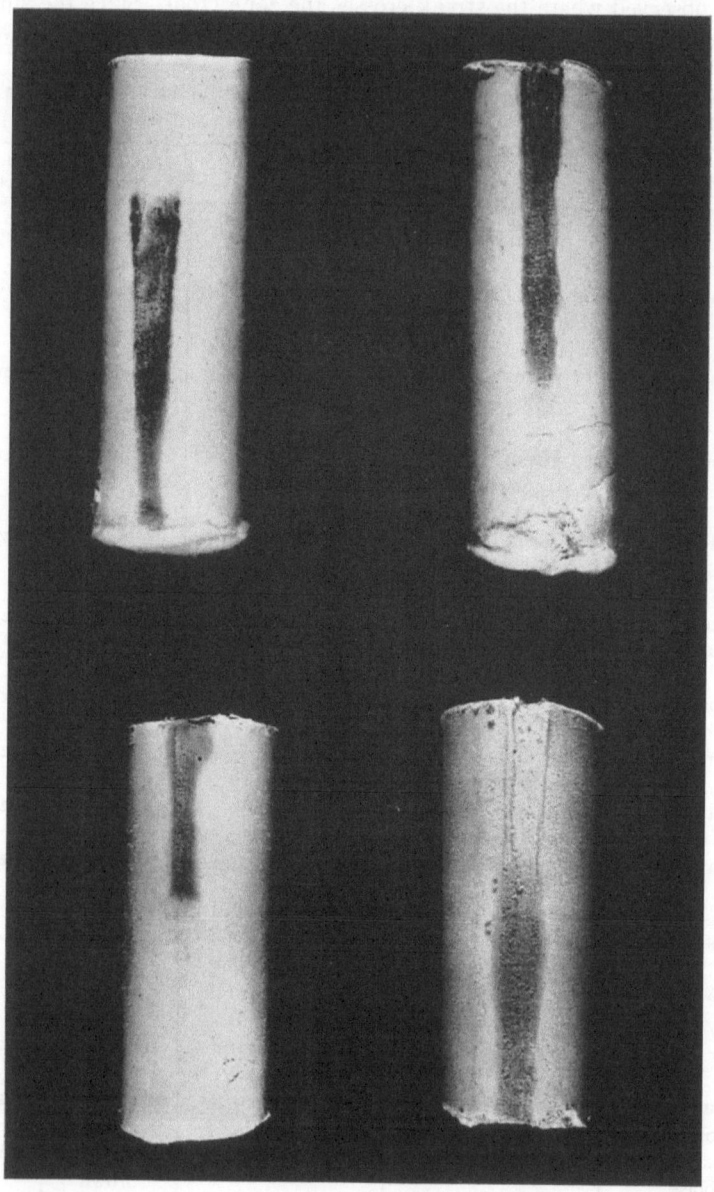

Fig. 6. Location of enzyme zones by the chromatographic brush method on alumina (*107*). Upper left: detection of amylase (starch; iodine); upper right: detection of potato phosphorylase [Cori-ester; iodine (for synthesized starch)]; lower left: detection of peroxidase (H_2O_2; guaiacol); lower right: detection of catalase (H_2O_2; titanium sulfate). [Unpublished photographs.]

with the corresponding substrate solution, the application of the brush method can be extended to enzyme analyses as follows (*107, 62*).

After the extrusion of the column, the substrate solution is painted on by a first brush, and then followed (after an incubation period of 1–60 minutes) by a reagent brush over the same streak. The latter will indicate the border line between the enzyme-containing and the enzyme-free sections of the column. In some instances only a single brush is necessary. Examples are summarized in Table 1 (p. 349), and some illustrations are given in Figure 6.

It is desirable to adapt these principles to the detection of several enzymes contained in a solution by drawing parallel streaks with different reagents on the same column. This has been achieved so far in a few instances only.

The prospect for the elaboration of more generally applicable methods seems to be rather good for the following reasons. First, a geometrical differentiation of the individual enzymes into discrete zones is not a prerequisite of a successful analysis; and, second, all the reactions available for enzyme detection in spot tests are, in principle, applicable in enzyme chromatography. As a matter of fact, the location of the saccharase or catalase zones (Table 1) was based on some reactions used previously in spot tests, e. g. on paper strips [SASTRI and SREENIVASAYA (*63*); NOSAKA (*56*); cf. FEIGL (*24*)]; for amylase, a similar principle had been used by GRÜSS (*30, 31*, see p. 345); cf. SIMONART and CHOW (*71*, see p. 346); WALLENFELS and VON PECHMANN (*89*).

In biochemical experimentation, of course, the limits of the described method should be established in each study; and, if necessary, disturbing substances should be eliminated from the solution before the application of the brush. Thus, the location of the α-glucosidase zone (*ex* crude extracts of dried brewer's yeast) became possible only after some impurities had been removed by adsorption (*62*).

The Treatment of Colored Enzymes and Colored Derivatives of Colorless Enzymes.

The chromatography of enzymes is, of course, facilitated in the relatively few instances in which the presence of color permits direct observation. Thus, as shown by photographs in SUMNER and SOMERS' book (*78*) a colored catalase zone appeared on tricalcium phosphate columns. As mentioned, colored proteins were chromatographed on paper by TISELIUS (*81*, see p. 346).

On the basis of some chromatographic observations on alumina, the possibility of the existence of two forms of hemoglobin was considered [Altschul, Sidwell and Hogness (3); cf. also Liebecq (45)].

Occasionally, a cleavage of chromoproteids has been observed in adsorption experiments, — a phenomenon which may become a useful tool in the study of certain enzyme systems. It is well known from organic-chemical practice that alumina columns may be used for a clean liberation of terpenes, steroids, etc. from their picrates, trinitrobenzene complexes, and similar derivatives. While, for example, trinitrobenzene remains adsorbed, the colorless component passes into the filtrate. A similar principle was successfully applied by Weygand and Birkofer (92) who carried out a cleavage of the "old" yellow enzyme into its protein and flavin-phosphoric acid components.

When a column was used whose top section consisted of franconite SB, and the bottom section of granosil, the enzyme solution in $N/50$-HCl yielded a colorless filtrate containing unaltered protein; upon the elimination of the acid by dialysis, this protein coupled with lactoflavin-phosphoric acid, and at the same time, the fluorescence in ultraviolet light disappeared.

The literature also contains a few examples of the conversion of proteinic substances into colored derivatives previous to or during a chromatographic treatment; similar methods might well become applicable in the field of enzymes.

Westphal, Gedigk and Meyer (91) observed that when an acid azo dye such as azorubin is adsorbed on anionotropic alumina (pretreated with HCl), a sharp pigment zone appeared. However, in the presence of certain proteins a fraction of this pigment is carried into the filtrate as a result of the competition for the dye between the protein and the adsorbent. While the albumin fraction of human sera behaved as just described, the globulin did not combine with the dye and did not color the filtrate under the applied conditions. Turba (85) also worked with azoprotein mixtures and briefly reported resolutions on neutral zinc carbonate and alumina. Some other colored protein derivatives, viz. picrates were used by Spies, Coulson, Bernton and Stevens (74) for the fractionation of allergens contained in cottonseed extracts. By development with 50% ethanol on activated alumina three clinically active zones were obtained.

Franklin and Quastel (26) have proposed a paper chromatographic technique for those proteins which combine with hemin; it is based on the detection of the hemin-protein complex by brushing with benzidine-hydroperoxide reagent.

The ascending principle suggested by Williams and Kirby (94) was employed; and aqueous salt solutions and buffers served as developers. Two-dimensional paper chromatograms were also obtained by similar techniques.

III. Chromatography as a Step in the Purification of Enzymes.

So far as we know chromatography alone will usually not lead to preparations of highest purity or to crystalline enzymes from a crude extract. However, in combination with other procedures such as precipitation, salting out, etc. it represents a valuable tool and offers, under favorable circumstances, a short cut on an otherwise lengthy road. Since no experimental details are given in the present survey, we may refer to

Table 2. Some Characteristic Examples of the Application of Adsorption Processes as a Step in the Purification of Enzymes.

Enzyme	Origin	Adsorbent	Reference
α-Amylase (cryst.)	Pancreas (human)	Amberlite IR–4 B	FISCHER, DUCKERT and BERNFELD (25)
α-Amylase (cryst.)	Saliva (human)	Amberlite IR–4 B	MEYER, FISCHER, BERNFELD and STAUB (48) MEYER, FISCHER, STAUB and BERNFELD (49)
α-Amylase (cryst.)	Germinated barley	Wheat starch + Celite 1 : 1	SCHWIMMER and BALLS (67, 68)
Pectin-esterase	Orange	Celite	MACDONNELL, JANG, JANSEN and LINEWEAVER (46)
Pectinase	Bact. aroideae	Amberlite IRC–50	TALBOYS (79)
Phosphorylase	Rabbits	Activa carbon	KRITSKII (39)*
Catalase	Horse liver	Tricalcium phosphate	AGNER (2)
Catalase (cryst.)	Horse liver	Tricalcium phosphate	SUMNER, DOUNCE and FRAMPTON (76)**
Catalase (cryst.)	Erythrocytes (cow)	Aluminum hydroxide	LASKOWSKI and SUMNER (42)
L-Aminooxydase	Snake venom	Calcium phosphate (gel)	SINGER and KEARNEYL (72)
Penicillinase	PD–2	Hyflo Super Cel	McQUARRIE, LIEBMANN, KLUENER and VENOSA (47)
Trypsin	Pancreas	Celite	SCHORMÜLLER (65)
Rennin	(commercial)	Alumina (Brockmann)	SCHÖBERL and RAMBACHER (64)
Prothrombin	Undiluted plasma	Aluminum hydroxide	MUNRO and MUNRO (54)
"Old" yellow enzyme	Yeast	Franconite SB	WEYGAND and BIRKOFER (92) (cf. p. 352)

* The enzyme passed through but the pentose content was retained by the column.

** Photographs of beef liver catalase chromatograms were published by SUMNER and SOMERS (78).

the papers listed in Table 2, p. 353. (Some of the adsorbents mentioned in that Table were not employed in column form.)

IV. The Resolution of Enzyme Mixtures by Partial Adsorption.

It is desirable to find chromatographic methods of general applicability for the separation of different enzymes present in the same solution.

Although one of the main goals of Willstätter's studies (97) was the purification and isolation of individual enzymes, numerous and valuable observations were made by his group concerning the resolution of enzyme mixtures. Thus, by treatment with alumina, pancreas lipase was quantitatively separated from several accompanying enzymes, especially amylase and trypsin. Although, as mentioned, Willstätter and his colleagues did not make use of chromatographic techniques, their papers contain the description of many separations which could be translated easily into the language of the chromatographer. This statement is also valid for the data given by some other authors.

A characteristic difference between enzyme mixture resolutions and many other chromatographic analyses is the circumstance that in the former instance the compound in which the experimenter is interested is accompanied by overwhelming quantities of foreign material, in part of proteinic nature. Hence, empiricism plays an unusually great role in such experiments; for example, it has been repeatedly observed that the length of the column is an essential factor.

Chromatography may also become a useful tool for a clearer definition of the specificity limits of some enzyme effects. We believe that if the relative activities of an enzyme solution on two different substrates are markedly altered by a chromatographic procedure, then the presence of two different enzymes should be assumed. It seems less probable that an unequal elimination of activators or inhibitors from the respective fractions is responsible for such effects.

With reference to some carbohydrases this interpretation is not shared by Helferich (32). On the basis of Weidenhagen's theoretical arguments (90) and because the sum of the effects of β-glucosidase (eluate) and α-galactosidase (filtrate) is evidently more than 100% (32, there p. 83), Helferich believes that the chromatographic separation of β-glucosidase and α-galactosidase as described below does prove the existence of two separate enzymes but he does not acknowledge β-glucosidase and cellobiase to be different, in spite of the separability of their effects. Helferich writes:

„Tritt bei den Operationen mit dem Ausgangsmaterial eine Verschiebung des Verhältnisses der beiden Fermentwirkungen ein, so kann dies verschiedene Ursachen haben. Es kann auf einer mehr oder weniger vollständigen Trennung von zwei verschiedenen Fermenten beruhen. Es kann aber auch die Schädigung eines der beiden Fermente oder beider Fermente in verschiedenem Maße bedeuten. Eine solche Schädigung kann schließlich aber auch bei einem einheitlichen Ferment (oder Fermentsystem) mit zwei verschiedenen Wirkungen die Fermentwirksamkeiten ändern und das Verhältnis der Fermentwirkungen verschieben."... „Einwirkungen und damit Veränderungen am Enzym können auch an *einem* Ferment oder Fermentsystem sich gegenüber verschiedenen Substraten verschieden stark auswirken."

This is certainly not the place for a detailed discussion of enzyme specificity problems, and we may refer to some pertinent surveys [OPPENHEIMER (*56a*); KRAUT and ROHDEWALD (*38*)]. However, a few comments seem to be justified. A satisfactory theoretical basis which would prove that some successful chromatographic separations are just pseudo-separations is not available at the present time. According to KUHN (*40, 41, 100*), the respective reaction rates in equimolecular concentrations of the enzyme-substrate compounds should be compared in order to reach a clear decision with reference to similar specificity problems. However, when KUHN studied the saccharase and raffinase content of a well-defined yeast preparation, he came to the result that those accompanying substances which can be eliminated by *purification* procedures do not influence the affinity of the enzyme to the substrate; thus, the "ZW.-Quotient" remained constant in the course of all operations. With reference to β-glucosidase and β-galactosidase it should also be mentioned that recent, careful kinetic studies, concerning the affinities of these enzymes to the substrates and to the "inhibitory" carbohydrates, indicate a difference between β-glucosidase and β-galactosidase [VEIBEL, WANGEL and ÖSTRUP (*87*)].

The following paragraphs contain typical examples of the resolution of enzyme mixtures by partial adsorption.

Amylase and Maltase. The amylase present in maltase containing extracts (obtained from *Clostridium aceto-butylicum*) was eliminated by passing the solution plus 20% acetone through starch; the amylase remained quantitatively on the adsorbent [FRENCH and KNAPP (*27*)].

Lichenase and Cellobiase. Cellobiase is adsorbed on aluminum hydroxide from acetic acid solution much more strongly than is lichenase, the enzyme which acts on lichenin [KARRER and LIER (*36*)].

Cellulase and Cellobiase. Aluminum metahydroxide or bauxite or diaspore shows a selective adsorption toward cellobiase. This oligosaccharase can thus be removed from solutions quantitatively, while most of the polysaccharase remains unadsorbed [GRASSMANN, ZECHMEISTER, TÓTH and STADLER (*28*)].

Cellobiase and Salicinase (110). Since a sample of Merck's emulsin contained by far more salicinase than cellobiase while extracts of the house (dry rot) fungus *(Merulius lacrimans)* showed the opposite relative proportion, some experiments were carried out with such a mixture of both extracts which contained these two enzymes in similar concentrations. For their separation Brockmann's alumina was found to be superior to bauxite. When the enzyme mixture (in acetate buffer p_H 4,6) had passed through the column, the filtrate showed in two experiments 88% and 101% of the initial cellobiase activity, respectively; the corresponding figures for salicinase were, 19% and 55%.

Gentiobiase, Salicinase and *Amygdalase (110)*. A partial separation of these enzymes was carried out by filtration of a house fungus extract (buffer, p_H 4,6) first through a bauxite and then through an alumina column. The final filtrate showed 78% of the initial gentiobiase and 43% of the salicinase activity but was found free of amygdalase.

β-Glucosidase, α-Galactosidase, and Chitinase (109). When a crude solution of Merck's emulsin (in acetate buffer p_H 4,7) was filtered through short bauxite columns (material from Hungary, mixed with sand), the following resolutions took place (the initial amount of each of the enzymes is designated as 100%):

Adsorption on the first column.

Adsorbate
(after elution with
0,1 *N*-ammonia):

78% β-Glucosidase
20% α-Galactosidase

Filtrate:

33% β-Glucosidase
70% α-Galactosidase
90% Chitinase

Adsorption on the second column.

Adsorbate
(after elution with
dilute ammonia):

78% α-Galactosidase

Filtrate:

0% β-Glucosidase
2,5% α-Galactosidase
80% Chitinase

Chitinase and Chitobiase (108). A commercial emulsin sample (Merck) contained the enzyme chitinase which attacked chitodextrin vigorously but practically did not hydrolyse chitobiose. A chromatographic treatment showed, however, that a second enzyme, termed chitobiase was also present which displayed the opposite specificity (to the chitinase) towards the various chain-length members of this series. Thus, a separation of a polysaccharase and an oligosaccharase was made possible. Under the conditions applied (p_H 4,7), 80—90% of the chitobiase but only 20% of the chitinase was retained by a 8—10 cm. high bauxite-sand column.

The main portion of the chitinase could then be quantitatively adsorbed on a second column. The adsorbed fractions were recovered by elution with disodium hydrogen phosphate.

A separation of the chitinase and chitobiase content of the hepatopancreas of the edible snail *(Helix pomatia)* can be carried out by a similar technique. However, many individual cases demonstrated that sometimes the chitinase and in other tests the chitobiase was preferentially adsorbed on bauxite.

In two limiting instances the chromatographic filtrate showed the following percentage of the respective initial enzyme activities:

a) 98% chitinase and 0% chitobiase;

b) 22% chitinase and 100% chitobiase.

The results of the type *b)* were obtained with extracts which were unusually rich in the biase.

Chitinase, Chitobiase, α- and β-Phenyl-N-acetyl-D-glucosaminase, and Salicinase. After a commercial emulsin solution had passed through a bauxite column, the resulting filtrate was found to be practically free of β-glucosaminase but showed about one half of the initial salicinase activity *(108)*. A similar treatment of an extract obtained from the edible snail *(Helix pomatia)* yielded a filtrate which contained about half of the initial chitinase and β-phenyl-N-acetyl-D-glucosaminase, while chitobiase, α-phenyl-N-acetyl-D-glucosaminase as well as salicinase remained quantitatively on the column *(111)*.

A considerable relative accumulation of β-phenyl-N-acetyl-D-glucosaminase versus the β-glucosidase content of the snail hepatopancreas was also achieved by NEUBERGER and PITT RIVERS *(55)*.

Tannase and Esterase (ex Aspergillus niger) [TÓTH and BÁRSONY *(84)*]. When the fungus extract was passed through Brockmann's alumina at pH 7,6, the filtrate showed, for example: 0% of the initial esterase activity (substrate, phenyl acetate) but 89% of the initial tannase activity (substrate, methyl gallate).

Tannase and β-Glucosidase (ex emulsin) [TÓTH and BÁRSONY *(84)*]. An artificial mixture of emulsin and *Aspergillus* extract was partially resolved on bauxite. Under favorable conditions the following fractions of the initial enzyme activities were found to be present in the chromatographic filtrate: Action on methyl gallate, 90%; on glucogallin, 49%; and on salicin, 6 to 9%.

Phosphorylase, Phosphatase and Phosphoglucomutase. A separation of liver phosphorylase from some other enzymes was carried out by CORI, CORI and SCHMIDT *(12)* who used "C$_\gamma$" aluminum hydroxide [cf. WILLSTÄTTER, KRAUT and ERBACHER *(98, 99)*]. The centrifuged adsorbate was washed with water, eluted with 0,3 *M*-disodium phosphate and, after adjustment to pH 7, precipitated with ammonium sulfate (till 0,3-saturation). The solution of the precipitate in 0,5% potassium chloride or ammonium sulfate was free of phosphatase and phosphoglucomutase.

Dehydrogenases. A tenfold increase in the ratio, succinodehydrogenase: diaphorase (*ex* heart muscle) was achieved by adsorption on calcium carbonate and elution with disodium phosphate [VON EULER, HELLSTRÖM and GÜNTHER (*22*), VON EULER and HELLSTRÖM (*21*)].

When a solution containing both dehydrogenase and flavine enzyme (*ex* heart muscle) was filtered repeatedly through activated alumina, then a relative accumulation of the dehydrogenase took place in the lower section of the column [ADLER and MICHAELIS (*1*)].

A slight fumarate dehydrogenase content of the "old" yellow enzyme was eliminated by a chromatographic treatment on franconite [WEYGAND and BIRKOFER (*92*)].

Mono- and Diphenolase. A partial separation of monophenolase and *o*-diphenolase (*ex* potato peels or *Psalliota campestris*) on aluminum hydroxide was achieved by ENSELME, CREYSSEL and RAPATEL (*17*); disodium phosphate at pH 7,8 was used as the elutor. Some column sections showed increased diphenolase but much decreased monophenolase activity.

Thiaminases. A partial separation of two thiaminases (termed, on the basis of their pH optima, "thiaminase 3,6" and "thiaminase 6,5") which occur in the fresh water mussel *Lamellidens marginalis*, was carried out by REDDI and GIRI (*57*). Both enzymes were retained quantitatively by aluminum hydroxide "C$_\gamma$," at pH 5; however, at pH 3,5 only 22,8% of the "thiaminase 6,5" but 82,2% of the "thiaminase 3,6" were adsorbed.

Dipeptidase and Aminopolypeptidase. TURBA (*85*) succeeded in resolving some proteolytic enzymes on alumina, hematite or zinc carbonate. For example, when a crude glycerol extract of intestinal mucosa was filtered through a zinc carbonate column, a dipeptidase preparation (substrate, leucylglycine) was obtained which was free of aminopolypeptidase (substrate, leucyldiglycine). First, the dipeptidase was eluted with water and then an aminopolypeptidase-rich fraction could be obtained by washing the adsorbent with disodium phosphate.

Trypsin, Lipase, and Amylase. YOUNG and HARTMAN (*102*) found that by selecting the appropriate adsorbent any two of the enzymes trypsin, lipase and amylase could be removed from the solution, leaving anyone in the filtrate. The following enzymes would pass through the column: trypsin (soapstone), lipase (magnesium silicate), and amylase (bauxite).

V. The Use of Enzymes in the Chromatographic Treatment of Other Substances.

The paper chromatographic study of enzymatic cleavage products can be preceded on the same paper sheet by the enzymatic action. Thus, CHARGAFF and KREAM (6) mixed known volumes of enzyme and substrate solutions on filter paper, and after an incubation period in a moist atmosphere, stopped the reaction by heating to 100°. Paper chromatography was then applied. Some bacterial suspensions can be treated making use of a similar technique.

An interesting application of an enzyme in the resolution of amino acid mixtures and the stereochemical identification of some of the individual amino acids was made by JONES (35) who started from a hydrolysate of the antibiotic, aerosporin. The spots obtained by partition paper chromatography were first sprayed with a 0,06 M-pyrophosphate buffer solution of D-amino acid oxidase (*ex* sheep kidney) and, after an incubation period of $2^1/_2$ hours at 37° in humid oxygen, the paper was dried and sprayed with ninhydrin. Since the spot ordinarily occupying the leucine position was completely removed by enzyme action, D-configuration was indicated in this instance; in contrast, the threonine component remained unchanged, indicating (at least, preponderantly) the presence of the L-configuration. According to LEDERER (43) a similar (unpublished) method was also devised by R. L. M. SYNGE.

References.

1. ADLER, E. u. M. MICHAELIS: Über die Komponenten der Dehydrasesysteme. X. Zur Kenntnis der Milchsäuredehydrase und der Äpfelsäuredehydrase aus Herzmuskel. Z. physiol. Chem. **238**, 261 (1936).
2. AGNER, K.: The Preparation and Properties of a Highly Active Catalase from Horse Liver. Biochemic. J. **32**, 1702 (1938).
3. ALTSCHUL, A. M., A. E. SIDWELL, Jr. and T. R. HOGNESS: Note on the Preparation and Properties of Hemoglobin. J. biol. Chemistry **127**, 123 (1939).
4. BRÜCKE, E.: Beiträge zur Lehre von der Verdauung. 2. Abt. I. Das Pepsin. S.-B. Akad. Wiss. Wien, math.-naturw. Kl. **43**, 601 (1861).
5. CARROLL, B.: Use of Dyestuffs for Determining the Activity of Proteolytic Enzymes. Science **111**, 387 (1950).
6. CHARGAFF, E. and J. KREAM: Procedure for the Study of Certain Enzymes in Minute Amounts and its Application to the Investigation of Cytosine Deaminase. J. biol. Chemistry **175**, 993 (1948).
7. Chromatography. Ann. New York Acad. Sci. **49**, 141 (1948).
8. Chromatographic Analysis. Discuss. Faraday Soc. **7** (1949). London: Gurney and Jackson.
9. CLAESSON, I. and S. CLAESSON: The Adsorption of Some High Molecular Substances on Active Carbon. Ark. Kem., Mineral. Geol. **19 A**, No. 5 (1945).
10. CLAESSON, S.: High Molecular Polymers Separation. Discuss. Faraday Soc. **7**, 321 (1949).

11. Consden, H., A. H. Gordon and A. J. P. Martin: Qualitative Analysis of Proteins: a Partition Chromatographic Method Using Paper. Biochemic. J. **38**, 224 (1944).

12. Cori, G. T., C. F. Cori and G. Schmidt: The Rôle of Glucose-1-phosphate in the Formation of Blood Sugar and Synthesis of Glycogen in the Liver. J. biol. Chemistry **129**, 629 (1939).

13. Cremer, H.-D. u. A. Tiselius: Elektrophorese von Eiweiß auf Filtrierpapier. Biochem. Z. **320**, 273 (1950).

14. Das, N.: Über die Komponenten der Dehydrasesysteme. XI. Zur Kenntnis der Glucosedehydrase aus Leber. Z. physiol. Chem. **238**, 269 (1936).

15. Datta, S. P., C. E. Dent and H. Harris: Apparatus for Mass-Production Two-Way Paper Chromatography. Biochemic. J. **46**, XLII (1950).

16. — — — An Apparatus for the Simultaneous Production of Many Two-dimensional Paper Chromatograms. Science **112**, 621 (1950).

17. Enselme, J., R. Creyssel et A. Rapatel: Isolement d'orthodiphénolases végétales par chromatographie. Bull. Soc. Chim. biol. (Paris) **29**, 939 (1947).

18. Euler, H. von u. E. Adler: Über die Komponenten der Dehydrasesysteme. IX. Die Co-Dehydrasen: Co-Zymase und „Co-Dehydrase II". Co-Zymase als Wasserstoffüberträger. Z. physiol. Chem. **238**, 233 (1936).

19. Euler, H. von u. A. Fonó: Adsorptive reversible Inaktivierung einer tierischen Glycerophosphatase. Ark. Kem., Mineral. Geol. **25** A, No. 15 (1947).

20. Euler, H. von and L. Hahn: Influence of Roentgen Rays on Isolated Cell Nuclei. Acta Radiol. **27**, 269 (1946).

21. Euler, H. von u. H. Hellström: Reinigung tierischer Succinodehydrase. Svensk kem. Tidskr. **51**, 68 (1939) [Chem. Zbl. **1940**, I, 2323].

22. Euler, H. von, H. Hellström and G. Günther: Enzymatische Sarcom-Studien. Ark. Kem., Mineral. Geol. **13** B, No. 8 (1939).

23. Euler, H. von u. F. Schlenk: Co-Zymase. Z. physiol. Chem. **246**, 64 (1937).

24. Feigl, F.: Qualitative Analysis by Spot Tests. Inorganic and Organic Applications, pp. 421—424. New York and Amsterdam: Elsevier. 1946.

25. Fischer, Ed. H., F. Duckert et P. Bernfeld: Isolement et cristallisation de l'α-amylase de pancréas humain. Sur les enzymes amylolytiques. XIV. Helv. chim. Acta **33**, 1060 (1950).

26. Franklin, A. E. and J. H. Quastel: Paper Chromatography of Proteins and Enzymes. Science **110**, 447 (1949).

27. French, D. and D. W. Knapp: The Maltase of *Clostridium acetobutylicum*. Its Specificity Range and Mode of Action. J. biol. Chemistry **187**, 463 (1950).

28. Grassmann, W., L. Zechmeister, G. Tóth u. R. Stadler: Über den enzymatischen Abbau der Zellulose und ihre Spaltprodukte. 2. Mitt. über enzymatische Spaltung von Polysacchariden. Liebigs Ann. Chem. **503**, 167 (1933).

29. Grundmann, Ch.: Chromatographie und verwandte Methoden in der Enzymchemie. In: E. Bamann und K. Myrbäck, Die Methoden der Enzymforschung, Bd. II, S. 1452. Leipzig: G. Thieme. 1941.

30. Grüss, J.: Capillaranalyse einiger Enzyme. Ber. dtsch. bot. Ges. **26**, 8 (1908).

31. — Capillarisation der Fermente. In: C. Oppenheimer und L. Pincussen, Die Fermente und ihre Wirkungen. Bd. III. Die Methodik der Fermente. S. 1442. Leipzig: G. Thieme. 1929.

31 a. Hedén, C.-G.: A Method for Large-scale Separation of Amino Acids on Filter Paper. Nature (London) **166**, 999 (1950).

32. Helferich, B.: Über das Süßmandel-Emulsin und einige verwandte Fermente. Erg. Enzymforsch. **9**, 70 (1943).

32 a. HIRS, C. H. W., W. H. STEIN and S. MOORE: Chromatography of Proteins. Ribonuclease. J. Amer. chem. Soc. **73**, 1893 (1951).

33. INGELMAN, B. and M. S. HALLING: Some Physico-chemical Experiments on Fractions of Dextran. Ark. Kemi **1**, 61 (1949/1950).

34. JONES, J. I. M. and S. E. MICHAEL: Chromatography of Proteins. Nature (London) **165**, 685 (1950).

35. JONES, T. S. G.: The Chemical Nature of "Aerosporin". III. The Optical Configuration of the Leucine and Threonine Components. Biochemic. J. **42**, LIX (1948).

36. KARRER, P. u. H. LIER: Über einen neuen Zucker aus Lichenin: Lichotriose. Helv. chim. Acta **8**, 248 (1925).

37. KLOTZ, I. M., H. TRIWUSH and M. WALKER: The Binding of Organic Ions by Proteins, Competition Phenomena and Denaturation Effects. J. Amer. chem. Soc. **70**, 2935 (1948).

38. KRAUT, H. u. M. ROHDEWALD: Carbohydrasen. In: G.-M. SCHWAB, Handbuch der Katalyse, Bd. III, S. 52. Wien: Springer-Verlag. 1941.

39. KRITSKIĬ, G. A.: Prosthetic Group of Phosphorylase. Doklady Akad. Nauk, S. S. S. R. **61**, 1061 (1948) [Chem. Abstr. **43**, 3049 (1949)].

40. KUHN, R.: Über Spezifität der Enzyme. II. Saccharase- und Raffinasewirkung des Invertins. Z. physiol. Chem. **125**, 28 (1923).

41. — Physikalische Chemie und Kinetik der Fermentreaktionen. In: C. OPPENHEIMER, Die Fermente und ihre Wirkungen. Bd. I. IV. Hauptteil, S. 193. Leipzig: G. Thieme. 1924.

42. LASKOWSKI, M. and J. B. SUMNER: Crystalline Catalase from Beef Erythrocytes. Science **94**, 615 (1941).

43. LEDERER, E.: Progrès récents de la chromatographie. I. Chimie organique et biologique. Paris: Hermann & Cie. 1949.

44. LEYON, H.: Salting out Adsorption of Virus. Ark. Kemi **1**, 313 (1949/1950).

45. LIÉBECQ, C.: Studies of Pseudohemoglobins. III. Absorption Spectra of Pseudohemoglobin and of Different Pseudohemochromogens. Bull. Acad. roy. Méd. Belgique **8**, 325 (1943) [Chem. Abstr. **41**, 1005 (1947)].

46. MacDONNELL, L. R., R. JANG, E. F. JANSEN and H. LINEWEAVER: The Specificity of Pectinesterases from Several Sources with Some Notes on Purification of Orange Pectinesterase. Arch. Biochemistry **28**, 260 (1950).

47. McQUARRIE, E. B., A. J. LIEBMANN, R. G. KLUENER and A. T. VENOSA: Studies on Penicillinase. Arch. Biochemistry **5**, 307 (1944).

48. MEYER, K. H., ED. H. FISCHER, P. BERNFELD et A. STAUB: Purification et cristallisation de l'α-amylase de salive. Experientia **3**, 455 (1947).

49. MEYER, K. H., ED. H. FISCHER, A. STAUB et P. BERNFELD: Sur les enzymes amylolytiques. X. Isolement et cristallisation de l'α-amylase de salive humaine. Helv. chim. Acta **31**, 2158 (1948).

50. MIRSKI, A. and E. WERTHEIMER: Muscle Action and Glycogen Phosphorylysis. Biochemic. J. **36**, 221 (1942).

51. MITCHELL, H. K., M. GORDON and F. A. HASKINS: Separation of Enzymes on the Filter Paper Chromatopile. J. biol. Chemistry **180**, 1071 (1949).

52. MITCHELL, H. K. and F. A. HASKINS: A Filter Paper "Chromatopile". Science **110**, 278 (1949).

53. MORING-CLAESSON, I.: Frontal Analysis Studies on the Adsorption of Crystalline Egg Albumin and its Cleavage Products. Biochim. biophys. Acta **2**, 389 (1949).

54. MUNRO, F. L. and M. P. MUNRO: The Preparation of Prothrombin by Adsorption on, and Elution from, Aluminum Hydroxide. Arch. Biochemistry **15**, 295 (1947).

55. Neuberger, A. and R. V. Pitt Rivers: The Hydrolysis of Glucosaminides by an Enzyme in *Helix pomatia*. Biochemic. J. **33**, 1580 (1939).

56. Nosaka, K.: Some Microchemical Tests for the Detection of the Constituents of Blood and Urine. Microchim. Acta **1**, 78 (1937) [Chem. Abstr. **31**, 6271 (1937)].

56a. Oppenheimer, C.: Die Fermente und ihre Wirkungen. Suppl.-Bd. I, S. 180. Den Haag: Dr. Junk. 1936.

57. Reddi, K. K. and K. V. Giri: Purification and Separation of the Two Thiaminases in Fresh Water Mussel *(Lamellidens Marginalis)*. Enzymologia **13**, 281 (1948/1949).

58. Riley, V. T.: Application of Chromatography to Segregation Studies of the Agent of Chicken Tumor. I. *(Rous Sarcoma Virus)*. Science **107**, 573 (1948).

59. — Chromatographic Studies on the Separation of the Virus from Chicken Tumor. J. Nat. Cancer Inst. (in press).

60. Riley, V. T., M. L. Hesselbach, S. Fiala, M. W. Woods and D. Burk: Application of Chromatography to the Separation of Subcellular, Enzymatically Active Granules. Science **109**, 361 (1949).

61. Rivière, C., G. Gautron et M. Thély: Chromatographie et ioduration artificielle de la thyroglobuline. Bull. Soc. Chim. biol. (Paris) **29**, 600 (1947).

62. Rohdewald, M. and L. Zechmeister: The Detection of Enzymes by the Chromatographic Brush Method. II. Enzymologia (in press).

63. Sastri, B. N. u. M. Sreenivasaya: Entdeckung von Enzymen durch „Fleck"-Proben. Mikrochem. **14**, 159 (1933/1934).

64. Schöberl, A. u. P. Rambacher: Über die Reinigung des Labfermentes durch chromatographische Adsorption. Biochem. Z. **305**, 223 (1940).

65. Schormüller, J.: Adsorption of Crystalline Trypsin by Various Adsorbents. Z. Lebensm.-Untersuch. u. -Forsch. **88**, 576 (1948) [Chem. Abstr. **43**, 3864 (1949)].

66. Schwerdt, C. E.: Chromatography of Hemoglobin and Serum Proteins. Thesis, Stanford Univ. 1940.

67. Schwimmer, S. and A. K. Balls: Isolation and Properties of Crystalline α-Amylase from Germinated Barley. J. biol. Chem. **179**, 1063 (1949).

68. — — Starches and their Derivatives as Adsorbents for Malt α-Amylase. J. biol. Chem. **180**, 883 (1949).

69. Shapiro, B. and E. Wertheimer: Phosphorolysis and Synthesis of Glycogen in Animal Tissues. Biochemic. J. **37**, 397 (1943).

70. Shepard, Ch. C. and A. Tiselius: The Chromatography of Proteins. The Effect of Salt Concentration and pH on the Adsorption of Proteins to Silica Gel. Discuss. Faraday Soc. **7**, 275 (1949).

71. Simonart, P. et K.-Y. Chow: Chromatographie sur papier appliquée à des enzymes. Bull. Soc. chim. Belgique **59**, 417 (1950).

72. Singer, Th. P. and E. B. Kearneyl: The *L*-Aminooxidases of Snake Venom. II. Isolation and Characterisation of Homogeneous *L*-Aminoacid Oxidase. Arch. Biochemistry **29**, 190 (1950).

73. Sober, H. A., G. Kegeles and F. J. Gutter: Chromatographic Analysis of a Mixture of Proteins from Egg White. Science **110**, 564 (1949).

74. Spies, J. R., E. J. Coulson, H. S. Bernton and H. Stevens: The Chemistry of Allergens. II. Isolation and Properties of an Active Protein Component of Cottonseed. J. Amer. chem. Soc. **62**, 1420 (1940).

75. Strain, H. H.: Chromatographic Adsorption Analysis. 2nd printing. New York: Interscience Publ. 1945.

76. SUMNER, J. B., A. L. DOUNCE and V. L. FRAMPTON: Catalase. III. J. biol. Chemistry 136, 343 (1940).

77. SUMNER, J. B., P. S. KRISHNAN and E. B. SISLER: An Improved Method for the Preparation of Coenzyme I. Arch. Biochemistry 12, 19 (1947).

78. SUMNER, J. B. and G. F. SOMERS: Chemistry and Methods of Enzymes. 2nd ed. New York: Academic Press. 1947.

79. TALBOYS, P. W.: Use of "Amberlite IRC-50" for the Partial Purification of a Bacterial Pectinase. Nature (London) 166, 1077 (1950).

80. TISELIUS, A.: Adsorption Analysis of Amino Acids and Peptides. Ark. Kem., Mineral. Geol. 15 B, No. 6 (1941).

81. — Adsorption Separation by Salting Out. Ark. Kem., Mineral. Geol. 26 B, No. 1 (1948).

82. — Some Recent Trends in Biochemistry. Chem. Engng. News 27, 1041 (1949).

83. — Elektrophorese und Adsorptionsanalyse als Hilfsmittel zur Untersuchung hochmolekularer Stoffe und ihrer Zerfallsprodukte. Naturwiss. 37, 25 (1950).

84. TÓTH, G. and J. BÁRSONY: Über die Chromatographie der Tannase. I. Enzymologia 11, 19 (1943).

85. TURBA, F.: Chromatographische Methoden in der Protein-Chemie. Z. Vit.-, Horm.- u. Fermentforsch. 2, 49 (1948/1949).

86. TURBA, F. u. H. J. ENENKEL: Elektrophorese von Proteinen auf Filtrierpapier. Naturwiss. 37, 93 (1950).

87. VEIBEL, S., J. WANGEL and G. ÖSTRUP: On the Difference Between β-Glucosidase and β-Galactosidase. Biochim. biophys. Acta 2, 126 (1947).

88. WALDSCHMIDT-LEITZ, E. u. F. TURBA: Verfahren der Trennung von Peptidgemischen. J. prakt. Chem. 156, 55 (1940).

89. WALLENFELS, K. u. E. VON PECHMANN: Über die Trennung von Enzymgemischen durch Elektrophorese in Filtrierpapier. Angew. Chem. 63, 44 (1951).

90. WEIDENHAGEN, R.: Spezifität und Wirkungsmechanismus der Carbohydrasen. Erg. Enzymforsch. 1, 168 (1932).

91. WESTPHAL, U., P. GEDIGK u. F. MEYER: Über eine chromatographische Methode zur Characterisierung von Serumeiweiß. Z. physiol. Chem. 285, 36 (1950).

92. WEYGAND, F. u. L. BIRKOFER· Über die Reindarstellung von „altem" gelbem Ferment aus Hefe und eine neue Methode zur reversiblen Spaltung. Z. physiol. Chem. 261, 172 (1939).

93. WIELAND, TH. u. L. WIRTH: Retentionsanalyse von Papierelektropherogrammen natürlicher Proteingemische. Angew. Chem. 62, 473 (1950).

94. WILLIAMS, R. J. and H. KIRBY: Paper Chromatography Using Capillary Ascent. Science 107, 481 (1948).

95. WILLIAMS, T. I.: An Introduction to Chromatography. London: Blackie & Son. 1946. Brooklyn: Chem. Publ. Co. 1947.

96. WILLSTÄTTER, R.: Über Isolierung von Enzymen. Ber. dtsch. chem. Ges. 55, 3601 (1922).

97. — Bemerkungen zur Geschichte der Enzymadsorption. In: Untersuchungen über Enzyme, Bd. I, S. 66. Berlin: Springer. 1928.

98. WILLSTÄTTER, R., H. KRAUT u. O. ERBACHER: Über isomere Hydrogele der Tonerde. VII. Mitt. über Hydrate und Hydrogele. Ber. dtsch. chem. Ges. 58, 2448 (1925).

99. — — — Über ein Tonerde-Gel von der Formel AlO · OH. VIII. Mitt. über Hydrate und Hydrogele. Ber. dtsch. chem. Ges. 58, 2458 (1925).

100. WILLSTÄTTER, R., R. KUHN u. H. SOBOTKA: Über die relative Spezifität der Hefemaltase. Z. physiol. Chem. 134, 224 (1924).

101. YANOFSKY, CH., E. WASSERMAN and D. M. BONNER: Large Scale Paper Chromatography. Science 111, 61 (1950).

102. YOUNG, J. H. and R. J. HARTMAN: Adsorption of Pancreatic Enzymes. Proc. Indiana Acad. Sci. 48, 79 (1939).

103. ZECHMEISTER, L.: Progress in Chromatography 1938—1947. London: Chapman & Hall. 1950. New York: J. Wiley. 1951.

104. — Paper Disk Columns in Glass Chromatographic Tubes. Science 113, 35 (1951).

105. ZECHMEISTER, L. and L. CHOLNOKY: Principles and Practice of Chromatography. London: Chapman & Hall. New York: J. Wiley. 3rd impr. 1950.

106. ZECHMEISTER, L., L. CHOLNOKY et E. UJHELYI: Contribution à la chromatographie des substances incolores. Bull. Soc. Chim. biol. (Paris) 18, 1885 (1936).

107. ZECHMEISTER, L. and M. ROHDEWALD: The Detection of Enzymes by the Chromatographic Brush Method. I. Enzymologia 13, 388 (1949).

108. ZECHMEISTER, L. u. G. TÓTH: Chromatographie der in der Chitinreihe wirksamen Enzyme des Emulsins. Enzymologia 7, 165 (1939).

109. ZECHMEISTER, L., G. TÓTH u. M. BÁLINT: Über die chromatographische Trennung einiger Enzyme des Emulsins. Enzymologia 5, 302 (1938).

110. ZECHMEISTER, L., G. TÓTH, P. FÜRTH u. J. BÁRSONY: Über die chromatographische Trennbarkeit einiger β-Glucosidasen. Enzymologia 9, 155 (1940).

111. ZECHMEISTER, L., G. TÓTH u. É. VAJDA: Chromatographie der in der Chitinreihe wirksamen Enzyme der Weinbergschnecke *(Helix pomatia)*. Enzymologia 7, 170 (1939).

(Received, December 30, 1950.)

Namenverzeichnis. Index of Names. Index des Auteurs.

Sachverzeichnis. Index of Subjects. Index des Matières.